Antitumor Potential and Other Emerging Medicinal Properties of Natural Compounds

Evandro Fei Fang · Tzi Bun Ng
Editors

Antitumor Potential and Other Emerging Medicinal Properties of Natural Compounds

Springer

Editors
Evandro Fei Fang
Laboratory of Molecular Gerontology
National Institute on Aging, NIH
Baltimore, MD
USA

Tzi Bun Ng
School of Biomedical Sciences
The Chinese University of Hong Kong
Hong Kong
Hong Kong SAR

ISBN 978-94-007-6213-8 ISBN 978-94-007-6214-5 (eBook)
DOI 10.1007/978-94-007-6214-5
Springer Dordrecht Heidelberg New York London

Library of Congress Control Number: 2013932335

Printed on acid-free paper

Springer is part of Springer Science+Business Media (www.springer.com)

Foreword by Joseph J. Y. Sung

The new millennium is characterized by high-tech gadgets such as Facebook, iPad, TV games, but also as an era of genomic medicine. However, residents in our global village are still harried by different incurable and new diseases that threaten human life. Are there any solutions? One way to circumvent the problem is to return to Mother Nature since it is still a main source of drug leads. According to a recent comprehensive review, from around the 1940s to 2010, a total of 175 small compounds with antitumor activity were approved by FDA and similar organizations. Among them, 74.8 % were other than synthetic compounds and 48.6 % were natural products and their derivatives [2]. However, good reference books in this field are scarce. Therefore, the book 'Health Cornucopia: The Emerging Medicinal Properties of Natural Compounds' arrives at the most opportune moment.

This repertoire of promising medicinal components encompasses a variety of molecules ranging from large proteins to small compounds. Molecules on this list appear sequentially as lectins, ribonucleases, ribosome inactivating proteins, protease inhibitors, and a vast diversity of small compounds (such as triterpenoids). In order to enlighten readers on the medicinal efficacies of the components at the cellular and molecular levels, the chapters in each part of this book not only place a focus on the stirring history of discovery of the components, but also present a detailed description of the molecular mechanisms and clinical research studies wherever possible. Examples include MAP30, BBIC, acetaldehyde, rapamycin, and others. The book also contains information on some new hot research areas which are related but not limited to cancer research, such as the strategy of exploiting autophagy for neurodegenerative diseases, the crosstalk between apoptosis and autophagy, and the prolongation of life span by employing natural compounds. Moreover, the book does not restrict the origin of natural compounds to plants, but extends it to bacteria, fungi, and animals. For instance, there are chapters on animal lectins, lectins of marine organisms, and plant lectins, which render the library of drug candidates more vivid.

Ethnomedicine is emerging as a new trend globally. Though a dominant contribution to human health in the old times, it was displaced by modern medicine and has faded away for many years. However, it is now enjoying a revival as many diseases can be alleviated by 'complementary medicine'. Traditional Chinese medicine (TCM), which constitutes an important part of ethnomedicine, has been credited with some efficacy during its long history of practice. Eminent examples comprise arsenic trioxide (a therapy for human acute promyelocytic leukaemia), artemisinin from *Artemisia annua* (exhibiting anti-malaria activity), bitter melon (*Momordica charantia*), and *Cordyceps sinensis* which manifests many exploitable pharmacological activities. Two chapters in this book are focused on the recent progress of research on medicinal compounds in the bitter melon. There is also a chapter on high-throughput screening for discovery of TCM-based drugs which may provide a guideline for future drug evaluation. The research accomplishments alluded to in these chapters as well as elsewhere furnish a scientific basis for the application of TCM and other ethnomedicine.

The editors of this book are Prof. T. B. Ng who is a professor at the Chinese University of Hong Kong and Dr. Evandro Fei Fang who is a brilliant young scientist in the field. I am confident that this book will provide comprehensive and useful information to researchers in the fields of cancer research, pharmacognosy, drug discovery, preventive medicine, plant nutrition, and others. For undergraduate and graduate students with a curiosity about natural products with medicinal activity, I recommend you this book as a valuable guide to the fascinating, rapidly evolving, and rewarding scientific genre.

Joseph J. Y. Sung MD, Ph.D.
Mok Hing Yiu Professor of Medicine
President and Vice Chancellor
The Chinese University of Hong Kong
Shatin, Hong Kong

Foreword by Rob Verpoorte

The most important human heritage: our knowledge of the use of plants as a resource for drug development

The most important human heritage is the knowledge of nature, allowing us to survive in a hostile environment. The use of plants is probably the most important of this heritage. The knowledge of plants that serve us as food, fuel, shelter, medicines, dyes, fibers, etc., is the basis of our life! At present we use some 30 species for our staple food, about 100 for their fruits and also about 100 as vegetables. It is estimated that some 30,000–70,000 species are used for medicinal purposes [1], of which at present some 100 are used for the production of pure compounds used in western medicine. Much of the latter group originate from the Mediterranean/Middle East area and were already described in old Egyptian, Persian, Greek, Roman, and Arab manuscripts. In the nineteenth and twentieth centuries these medicines were further developed leading to the use of pure compounds and later even to derivatives or synthetic analogues with improved pharmacological properties.

Despite the high score of active compounds from studies of a limited number of medicinal plants, so far most medicinal plants have hardly been studied in-depth for potential novel medicines. This untapped reservoir of traditional knowledge is a very promising resource for drug development and our major challenge is to develop efficient ways to tap this human heritage for the benefit of humanity.

The statistics on the novel drugs developed in the past three decades show clearly the importance of nature as a source for drug development, almost half of all novel small-molecule drugs are natural products or natural products derived compounds [2].

In the last three decades drug development has followed two major strategies. One is rational drug design and is based on the knowledge of targets. This means receptors or enzymes involved in diseases. In this approach, computational methods were developed to design medicines based on the knowledge of stereochemistry of targets. The limitation of this approach is that no novel modes of action are found, only when novel targets are discovered novel drugs can be developed. The other approach is high throughput screening (HTS), in which large numbers of compounds are measured in fast micro/nano scale bioassays, allowing thousands of compounds to be tested per day. Again this approach is based on the knowledge of targets, and thus has the same limitation of the other approach, one may find novel active structures, but still for the same target.

Both approaches are based on the "single target-single compound" paradigm. However, drug development is stalling using these approaches, only novel targets may lead to novel first-in-class medicines. Moreover, except for infectious and parasitic diseases, where an exogenous organism has to be eliminated, most diseases are multifactorial and concern endogenous processes, e.g., cancer and chronic inflammatory diseases. Chances that a single target drug will cure such diseases are small. The success in treating HIV-infected patients is due to the use of multidrug treatments. But if developing a single molecule drug is difficult, two- or even multi-component medicines are exponentially more difficult. At least it seems time for a paradigm shift to develop the next generation of medicines.

Such a paradigm shift should go away from the known targets, and instead focus on observing diseases in a systemic way to be able to see multiple responses to medicines, including synergy between compounds. Such a different approach is in a focus on detailed systemic studies of the various traditional medicines discovered by our ancestors. Many of the presently known targets have their origin in medicinal plants (e.g., opium, belladonna, cannabis, tobacco, curare, and strychnine), so traditional medicines may be the way to novel targets. Moreover, traditional medicines may affect several targets at the same time, e.g., synergy. For example, the pharmacological effects of opium and cannabis cannot only be explained by the single major known active compound, morphine, and THC, respectively. Consequently, novel ways to study traditional medicines may be a gateway to both target and to lead compound discovery, including multidrug treatments based on synergy.

The stalling of drug discovery efforts of the big pharmaceutical companies and the changing global economic reality are in my view important factors to achieve such a paradigm shift in drug development [3, 4]. By assuming that traditional medicines must have some sort of activity and approaching this by a holistic systemic approach instead of the current reductionist approach, at least one may build up the evidence that the traditional treatments are beneficial, and ultimately result in novel first-in-class medicines, including multidrug treatments active on several targets. This means applying all the present day possibilities of the "omics" in studying the effects of traditional medicines in in vivo systems to find correlations between compounds in the traditional medicines and the effect on the in vivo system or whole organism, including clinical trials [5–7]. Recently the standards needed for such studies have been addressed in a series of papers [8].

In a systems biology approach, any type of molecule or combinations of molecules may turn out to be correlated to activity. So far natural products lead finding has mostly been focused on small molecules, but also macromolecules such as oligo- and polysaccharides, peptides, glycosylated peptides, or lipopolysaccharides, should not be overlooked. For example, the plant defense responses against pests and diseases are induced by elicitors made out of such molecules, besides small molecules like jasmonate, salicylate, and ethylene [9]. Here one may see a clear parallel with human responses in which lipopolysaccharides, prostaglandins (close chemical analogues of jasmonate), and salicylate play a role.

The story of salicylic acid from willow bark nicely illustrates how a major drug has its roots in a prodrug from a plant, a drug that would never have been found in the present day reductionist approach of drug development, also because it acts on many different targets with a mild activity. How many more jewels are hidden in the heritage of our ancestors? The present book, edited by Dr. Evandro Fang and Prof. T. B. Ng, shows in a series of excellent contributions by well-known experts the potential of natural products, confirming that nature is a major resource for drug development. Besides the "classical" screening methods of the past three decades, novel systemic observation-based approaches, i.e., systems biology, will play an important role in the future drug development. Moreover, with the improved chemistry able to deal with complex structures also macromolecules will receive more interest.

Many of our future new drugs will come from nature, it is up to us to improve the tools to more rapidly identify the active molecules in complex extracts and understand their mode of action.

Rob Verpoorte
Natural Products Laboratory
IBL, Leiden University, Leiden
The Netherlands
verpoort@chem.leidenuniv.nl

References

1. Verpoorte R, Kim HK, Choi YH (2006) Plants as source of medicines: new perspectives. In: Bogers RJ, Craker LE, Lange D (eds) Medicinal and aromatic plants: agricultural, commercial, ecological, legal, pharmacological and social aspects. Springer, Dordrecht, The Netherlands, pp 261–274
2. Newman DJ, Cragg GM (2012) Natural products as sources of new drugs over the 30 years from 1981 to 2010. J Nat Prod 75:311–335
3. Grayson M (2011) Traditional asian medicine. Nature 480:S81
4. Robinson MM, Zhang X (2011) The world medicines situation 2011. Traditional medicines: global situation, issues and challenges, World Health Organization, Geneva
5. Wang M, Lamers RJ, Korthout HA, van Nesselrooij JH, Witkamp RF, van der Heijden R (2005) Metabolomics in the context of systems biology: bridging traditional Chinese medicine and molecular pharmacology. Phythe´r Res: PTR 19:173–182
6. Verpoorte R, Choi YH, Kim HK (2005) Ethnopharmacology and systems biology: a perfect holistic match. J Ethnopharmacol 100:53-56
7. Verpoorte R (2012) Good practices: the basis for evidence-based medicines. J Ethnopharmacol 140:455–457
8. Xu QH, Bauer R (2012) (Guest editors) Good practice in Traditional Chinese Medicine Research in the post-genomic Era. J Ethnopharmacol 455–642
9. Zhao J, Davis LC, Verpoorte R (2005) Elicitor signal transduction leading to production of plant secondary metabolites. Biotechnol Adv 23:283–333

Preface

Can common people enjoy a healthy life in this world which is permeated with unhealthy food? Can patients afflicted with Parkinson's disease, diabetes, and cancer be given a better and less stressful quality of life? The chapters of this book support an affirmative answer: natural products should be promoted among the top tips in addition to a beneficial diet accompanied by ample physical exercise. The past few decades have witnessed a phenomenal growth in the interest, knowledge, and application of herbal/specialty diet supplements and alternative medicine. The various health benefits of natural components have been corroborated by modern science as evidenced by the voluminous research publications and a long list of drugs with anticancer and other health benefits approved by the US FDA and similar agencies. It is high time to gather together reviews of selected promising bioactive natural compounds.

This book is divided into eight parts, each with a different emphasis. Part I describes medicinal lectins of mammalian, marine, and plant origins. Professor Nathan Sharon, one of the pioneers on lectin research, passed away in 2011, and Prof. Jeremiah Silbert at Harvard University has kindly written a biography in memory of the renowned lectinologist. Some precious materials translated from Prof. Sharon's memoirs written by himself will not only unveil his early days on lectin research but also inspire young scientists. Additionally, we are honored to include two chapters on the history and characteristics of mammalian lectins from Prof. Hans-Joachim Gabius's book 'The Sugar Code: Fundamentals of Glycosciences'. Furthermore, the chapter on marine lectins written by Prof. Koji Muramoto and colleagues expands the scope of bioactive lectins to the ocean. A collection of promising lectins from plants is also included.

Part II contains two chapters which focus on the biochemical characteristics and antitumor activity of plant nucleases and the bovine seminal ribonuclease. They are contributed by Dr. Tomas Podzimek, and Prof. Delia Picone and coworkers, respectively. Part III concentrates on the antitumor and anti-viral activities of ribosome inactivating proteins (RIPs). As the first reporter of the anti-HIV and antitumor protein MAP30 from bitter melon (*Momordica charantia*), Prof. Sylvia Lee-Huang has written a chapter providing readers a complete story of this multifunctional protein. Additionally in Part III, the chapters by Prof. Jose M. Ferreras, Prof. Maria S. Fabbrini, and their colleagues, succinctly review the work on other promising RIPs. Part IV is dedicated to protease inhibitors. Prof. Ann Kennedy shares her expert opinion on the health benefits of the Bowman-Birk inhibitor concentrate and its commercial production. Prof. Krzysztof Rolka and his colleagues give a comprehensive account of the classification, biochemical characterization, and medicinal applications of plant serine protease inhibitors. Furthermore, as described by Prof. Kenei Furukawa and coworkers, synthetic serine protease inhibitors also exhibit promising clinical applications. Some of the other medicinal protease inhibitors are covered in other chapters of Part IV.

Small compounds are in the limelight in Part V which contains the largest number of chapters. Autophagy is at the forefront of our knowledge. The chapter by Prof. David Rubinsztein unveils the potential treatment of neurodegenerative diseases by employing different autophagy-inducing compounds. Among them, rapamycin (the mTOR inhibitor) has received growing attention due to its promising clinical applications, and in the next chapter please enjoy a rapamycin panorama by Dr. Morten Scheibye-Knudsen. Prof. Betty Schwartz and colleagues provide an interesting story about the medicinal applications of mushroom polysaccharides. Furthermore, the antitumor activity of essential oil constituents is underlined by Prof. Daniel Bezerra and coworkers. We thank Prof. Victor Preedy at King's College London and his colleagues Dr. Roshanna Rajendram and Prof. Rajkumar Rajendram for their highly informative chapter on acetaldehyde to alert the general public to the potential adverse effects of this aldehyde.

In view of the fact that most of the above-mentioned chapters are focused on single compounds, with our long-term collaborators, we tried to focus on multifarious medicinal components in one target in Part VI and give a bird's-eye view of a mushroom, a fruit, an insect, and other drug candidates. For instance, there is a spectacular array of medicinal compounds in both *Cordyceps sinensis* and bitter melon, and the diverse medicinal activities have been enumerated in their respective chapters. To highlight the importance of the latter, Ms. Michelle Sanzi Kermes, a famed Baltimore artist, was invited to depict the bitter melon in her creative and impressive drawing which is located ahead of the Table of Contents.

Ethnomedicine often travelled from reality to mythology in the ancient days. With the long history of its practice and the accumulation of experimental results from laboratories from all over the world, some aspects of ethnomedicine have been corroborated and a scientific basis for ethnomedicine has emerged. In this book, traditional Chinese medicine (TCM) is used as a window to facilitate understanding of the modernization of ethnomedicine. Prof. Rudolf Bauer and other GP-TCM board members share with us their experience in the function of '*Good Practice in Traditional Chinese Medicine (GP-TCM) Research in the Post-genomic Era*' associated with clinical applications in the European Union. In addition, the chapter on high-throughput screening of TCM by Prof. Y. H. Wong and his colleagues offers fresh perspectives on ways to elevate the efficiency of drug screening from different TCM materials. In Part VIII which constitutes the finale, some remaining questions are raised and the future perspectives of natural compounds are highlighted.

In addition to our wholehearted gratitude to the contributors of the various chapters, we are especially indebted to Prof. Joseph Sung, Vice-chancellor of

The Chinese University of Hong Kong, and Prof. Rob Verpoorte, Editor-in-Chief of the *Journal of Ethnopharmacology*, for kindly sparing time out of their hectic schedules to write the forewords. We acknowledge with many thanks the generous assistance of the following individuals, who helped us to edit the chapters, and provided inspiring discussions: Vilhelm A. Bohr, Heiko Jahn, Season Jing Fang, Leslie Ferrarelli, Peter Sykora, Clara Bah, Martin Jensen, Chuan-hao Li, Chris Z. Y. Zhang, Kent Horvath, Jennifer Illuzzi, and Rachel Abbotts. Finally, we are much obliged to Melania Ruiz, senior publishing editor of Springer Publishing, for inviting us to be the editors of this meaningful treatise which, as everybody involved in the various aspects of its production sincerely hopes, will serve its purpose as a useful reference to current and future researchers.

30 October 2012

Evandro Fei Fang
Tzi Bun Ng

Contents

Biosketch

Dr. Evandro Fei Fang received his Ph.D. degree from The Chinese University of Hong Kong and currently works with Prof. Vilhelm A. Bohr on DNA repair and genome instability in human aging diseases. His research encompasses: (a) bioassay-guided isolation of antitumor constituents from medicinal plants and the molecular mechanisms involving apoptosis and autophagy pathways, (b) the functions of mitochondria and mitophagy in sustaining human health and the molecular basis of their defects in the etiology of aging diseases. He has published over 30 papers in international journals and 9 book chapters, and serves on the Editorial Board of Medicinal and Aromatic Plants. He was a finalist in the Young Investigator Award 2011 conducted by the Hong Kong Institute of Science.

Tzi Bun Ng is Chair Professor of Biochemistry at the Faculty of Medicine, The Chinese University of Hong Kong, HKSAR. After completing his B.Sc. studies at The University of Hong Kong, he obtained his Ph.D. degree from the Memorial University of Newfoundland in Canada. He pursued his postdoctoral training at University of California, San Francisco, USA, and sabbatical research at the Imperial College London, UK. His major research interests are the health-benefits of proteins and peptides including polysaccharopeptides, lectins, antimicrobial peptides, enzymes and protease inhibitors. In the past 30 years, his team made many contributions to the field of bioactive components, represented in more than 550 papers in international peer-reviewed journals, over 60 invited book chapters, 2 edited books, and numerous other media. He currently sits on the Editorial Board of 9 international journals and contributes actively in many scientific fields.

Part I
Lectins: From Biological Recognition Molecules to Variable Medicinal Applications

Chapter 1
Lectins: Personal Comments of Nathan Sharon Taken from his Memoirs (Translation from Hebrew)

Jeremiah Silbert

1.1 Introduction

1.1.1 Comments by Jeremiah Silbert

I met Nathan Sharon in 1962 when he was on sabbatical for a year in the laboratory of Roger Jeanloz at the Massachusetts General Hospital in Boston. I was a post-doctoral research and clinical fellow in the Arthritis Unit in a laboratory next to the laboratory where Nathan worked. My work concerned the biosynthesis of heparin and chondroitin sulfate and had many overlaps with some of his interests in glycoproteins. We became close friends, and he invited me to spend time in his laboratory at the Weizmann Institute in Rehovoth, Israel. Three years later I took advantage of his invitation and went to Israel with my family for six months (1965–1966) to collaborate with him on the synthesis of an interesting sugar nucleotide. Subsequently my wife and I visited Israel multiple times for research meetings and travel, often staying at the Weizmann or at Nathan's home for part of the time. He and his wife had even more occasions to be in the USA for sabbaticals and other research-related reasons. They often stayed at our home during these visits.

In 2009, we were in Israel for two weeks and stayed with the Sharons for our last three days. He was his usual extraordinarily active, happy and apparently healthy self. A few days after we left, he had a devastating heart attack and was

J. Silbert (✉)
Division of Rheumatology/Immunology/Allergy, Brigham and Women's Hospital, Harvard University, Boston, MA, USA
e-mail: jesilbert@aol.com

J. Silbert
106 Spooner Road, Chestnut Hill, MA 02467, USA

E. F. Fang and T. B. Ng (eds.), *Antitumor Potential and Other Emerging Medicinal Properties of Natural Compounds*, DOI: 10.1007/978-94-007-6214-5_1,

hospitalized for several months. He then returned home, compromised physically, but still as intellectually active as always. The last thing he wrote was a detailed Memoir (in Hebrew) covering his personal and scientific history, taken from detailed notes he wrote throughout his life. He planned to write an English version, but upon the completion of the Memoir, his condition deteriorated. He was hospitalized again for a few months where this grand scientist, and just as grand human being, died.

Several years ago my wife and I attended a scientific symposium held at the Weizmann Institute in honor of Nathan's 80th birthday. A former student, now a professor at the Weizmann, introduced Nathan as Mr. Google because he seemed to know everything not only about science, but also just about everything else. Nathan met and collaborated with numerous scientists from all over the world; many of these became his friends. His Memoirs describe his research, listings of special lectures and courses, and many collaborations and/or sabbaticals in the USA (Boston, NIH, Berkeley, Santa Barbara with Lipmann, Robbins, Isselbacher, Jeanloz, Glick and Koshland), Britain (Phillips, Neuberger), France (Montreuil, Monsigny), Germany (Schauer), and Japan (Osawa). He was invited to visit, lecture, and spend time in many countries and territories including Belgium, Norway, Soviet Union, Switzerland, Mexico, Colombia, Venezuela, India, Japan, and Taiwan. In addition to science and personal subjects, his Memoirs provide a veritable travelogue of all the places he visited.

Together with another colleague of Nathan, Mary Catherine (Susy) Glick, I arranged with Nathan's family to have his Memoirs translated into English. I selected the sections on lectins from these Memoirs, edited them, and organized them for publication with the assistance of Susy Glick, Harry Schachter, and my physician wife, Cynthia.

1.1.2 Personal History, Taken From the Sharon Memoirs

Nathan Sharon was born in 1925 as Nathan Shtrikman in Brest Litovsk, Poland (now part of Belarus) a town of 30,000, about half of whom were Jews. He came with his parents and younger brother to Israel in 1934, leaving behind relatives who were all killed by the Germans early in World War II along with almost all the Jews in the city. In 1934 there were approximately 250,000 thousand Jews and about 500,000 Arabs in Israel (Palestine). He lived and went to school in Tel Aviv, entering high school (gymnasium) on September 1, 1939 the first day of World War II. In 1943, after graduating from high school, he joined the Palmach, which consisted at that time of about 700 members who were the secret military arm of the Jewish population. After 2 years in the Palmach, Sharon entered studies in chemistry at Hebrew University in Jerusalem, the only university in the land. While still at University, he rejoined the Palmach reserves. Shortly afterwards he joined a Palmach underground research unit involved in developing weapons in anticipation of action for independence.

On November 29, 1947 the United Nations voted to partition Palestine into separate Jewish and Arab nations. Organized Arab attacks on Jews followed immediately, initiating the Jewish War of Independence. Several weeks later, University studies were suspended and Sharon was called to service in the Palmach as part of a special chemistry unit for making ammunition, explosives, etc. He remained a Palmach officer of the Israeli army until the war ended in 1949. In the midst of all this he married Rachel after a long courtship and began a marriage that lasted until his death more than 60 years later. In 1951 Nathan changed his name from Shtrikman to Sharon.

In 1950 Nathan began his work for a PhD in Chemistry at the Weizmann Institute with Aharon Katzir. (In May 1972, Katzir was killed by terrorists in Ben Gurion airport along with 15 or so Christians who had come to visit the Holy Land.) After receiving his PhD in 1954, Sharon joined the Department of Biophysics where Ephraim (Katchalski) Katzir, the brother of Aharon Katzir, was Chairman. Sharon stayed in this department at the Weizmann for the rest of his career. In 1973 Ephraim Katzir was named President of Israel, and Sharon took his place as Chairman of Biophysics, a position he held until 1985.

Nathan and Rachel have two daughters who live in Israel. The older daughter, Esti, retired after a career in the Ministry of Finance, and the younger daughter, Osnat, is a research and clinical physician specializing in Hematology. She and her husband Daniel Bairey have three sons and a granddaughter. Esti and Osnat spent time at schools in the USA while Sharon was on sabbaticals.

1.1.3 Lectins, Verbatim as Translated From the Sharon Memoirs

During 1959, Aaron Altschul, an expert on plant proteins, arrived on a visit from the United States. He was a great believer in the use of soy beans to feed humanity, and came by invitation of the Ministry of Commerce and Industry to teach Israelis how, after extracting the oil, to use remains of soy for baking bread and feeding cows and fowl. With his help, we made a plan based on the expertise of Ephraim Katzir, to combine his ideas about soy proteins with ours. We submitted this research plan to the American Agricultural Office and in 1960 received a grant of 30,000 dollar per year, for five years, and later for another five years.

At that time I was joined by Dr. Halina Lis, a Holocaust survivor who arrived at the Weizmann Institute with a PhD from the laboratory of Arne Tiselius (Nobel Laureate) from Uppsala University, Sweden. Halina and I continued to collaborate scientifically until 2005, when she retired from the Institute. During that time together, we published about one hundred scientific papers and two books about lectins.

At the beginning of our collaboration, we focused on the soy flour protein that causes agglutination of red blood cells and other cells by reacting with cell surface sugars. This is the soy agglutinin that later was shown to be in the family of lectins. We found that this protein had some interesting qualities. For example, it was a glycoprotein, containing sugar as part of the molecule. This was the first discovery that there were glycoproteins in plants. We also found that we could use the soy lectin to identify what sugars were on the membranes of cells and detect some changes that occurred there during the growth and division of the cells. Together with Ben-Ami Sela and Leo Sacks of the Weizmann Institute, we discovered that this protein sometimes could differentiate between normal and cancerous cells. Several years later Dr. Rivka Golan, of the faculty of agriculture at the Hebrew University spent a sabbatical in my laboratory. That year she discovered that the soy lectin attached to fungi mildew in a way that was different from other lectins and inhibited fungi growth. Continuing my research on cell membranes, I tried to learn how cells recognized each other and in what way they attached. I asked why they sometime digressed from their position—a behavior that was typical of cancerous growth, thus showing that sugars had an important role in these basic processes. The truth is that many scientists at that time thought I was wasting my time.

By 1971 it was becoming apparent that lectins were of interest. At the annual meeting of the Federation of American Societies for Experimental Biology, one of the main lecturers was Professor Leo Sacks of the Weizmann Institute, whose lecture concerned the function of saccharides on cells in the process of cancerous transformation. He reported about the use of lectins in investigating this phenomenon, based on experiments conducted by Halina and me.

That same year I went to London to work with Albert Neuberger. My plan for work in his laboratory was to prepare saccharides from cell walls and test the effect of lysozyme on them. This work digressed from my original lectin project and involved the study of another protein that was also known to be attached to saccharides and was therefore categorized as a type of lectin. This protein was extracted from wheat germ and caused drying up of various body cells. On my arrival at Neuberger's laboratory, there was great interest in this lectin because other studies showed that it killed cancerous body cells, yet did not interfere with normal cells. These studies showed for the first time that saccharides on cell surfaces might be involved in the cancerous process. Therefore, toward the end of the sixties, many scientists began to use this lectin in basic cancer research.

I decided to investigate the extent of the attachments of various cell wall saccharides to the wheat germ lectin and together with Tony Allen, another scientist in Neuberger's laboratory, we found that this lectin reacted with cell wall saccharides in a way similar to the way that lysozyme reacted. We also found that there were other similarities between the two proteins, but that unlike lysozyme, the wheat germ lectin was not able to dissociate those saccharides. Our findings about this lectin became the basis for the expansion of its use in cancer research and also for further studies on cell surfaces.

In 1973 I was invited to deliver the opening address at the Second Symposium on Glycoconjugates, and chose to lecture on general ideas concerning lectins. The structures of only three of them were described by that time, and each had a different and unique three-dimensional structure. I predicted that all the lectins that would be discovered in the future would have unique three-dimensional structures, although they would share the biological trait of an ability to attach to saccharides. This prediction was fully confirmed in subsequent years.

A great part of our further studies on lectins in the eighties began following a letter I received from Jose Iglesias, a young student at the School of Medicine in Montevideo, Uruguay. In his letter, Jose said that he had read the articles by Halina and me with much interest and decided to try to isolate the lectin from the seeds of the Erythrina Crista-galli trees that were abundant in Uruguay. This tree is of the legume family and bears fiery red flowers in the beginning of the winter and is used widely as a decorative tree. Moreover, the flower is the national flower of Uruguay. Jose wrote that he believed that the seeds of the Erythrina Crista-galli tree contained a lectin, since the essence extracted from the seeds caused agglutination of human red blood cells. He also found that one could prevent such agglutination by the presence of galactose. However, he was not successful in purifying the lectin from the essence of the seeds, because he was unable to obtain the needed equipment and chemicals. He ended his letter with the hope that one day he would be able to come to my laboratory, even for a short time to realize his dream.

Fortunately, the friends of the Weizmann Institute in Uruguay, headed by Dr. Nelson Philosophe, managed to raise the sum needed for Jose to visit Israel and work with me at the Weizmann for several months. He arrived in my laboratory in the middle of 1981. The importation of seeds into Israel was strictly forbidden in order to prevent transmitting agricultural diseases. For this reason Jose brought six kilograms of flour from the Erythrina Crista-galli seeds, and in a short time of hard work at the Institute he succeeded in producing a considerable amount of pure lectin from this flour and also defined some of its traits.

After Jose left, I began to collect seeds of other types of Erythrina (there are about a hundred and ten such sub-species) in order to compare them to the traits of the Erythrina Crista-galli. Fortunately, I met François Jubert at an international carbohydrate conference in Pretoria in South Africa and learned about his interest in proteins from various types of Erythrina. He sent me some of these types. At about the same time, during a visit to Tucson, Arizona, with my friend John Rapley, I learned about another type of Erythrina that grew there, and collected its seeds. We finally had seeds of seven types of Erythrina from which Halina and I extracted the lectins. Comparison of the traits of these proteins demonstrated great similarities, but also some fine but very important differences. We noticed that in Israel most of the Erythrina trees were of the Erythrina Conalrodendrum type that adorn many private and public gardens. Four of those trees were planted by my wife, Rachel, next to our house, and their fiery red blossoms adorn our house during the winter. It was easy to collect the small red seeds and extract pure lectin from them.

I devoted about two dozen years of research to the lectin of the Erythrina Crista-galli, particularly in collaboration with Halina, but also with other scientists from Belgium, Sweden, Germany, Canada, France, England, the United States, Japan and India. Together with Raphael Arrengo, a student from Medellin,

Colombia, and Boaz Shaanan, who was working at that time in the Weizmann Institute, we utilized X-ray crystallography to successfully decipher how the lectin attached to certain saccharides and not to others. This work directed us to methods of genetic engineering in bacteria.

In the beginning of the fifties it was already known that saccharides had crucial importance in the determination of different blood types, e.g., the difference between type A and B red blood cells is dependent on the particular saccharide on the cell surface. Later it was found that saccharides play a major role in the transport of white blood cells and lymphocytes throughout the body. It was shown that the human body had lymphocytes of different types—those that produced immune materials like antibodies, those that attacked foreign bodies, and those that helped other lymphocytes or interfered with their functioning. There was a need to find a way to separate the white cells into their different groups. This would provide a way to understand how these cells protected the body from various illnesses and why they were not acting normally in blood illnesses such as leukemia. Along with many other investigators, I raised the idea that the white cells belonged to different groups because their cell surfaces were covered by different saccharides. I hypothesized that it should be possible to distinguish between different white cells using lectins and that it may be possible to use lectins to isolate different types of white cells.

This objective was achieved in the seventies when Yair Reisner came to my laboratory as a student. We used soy lectin and peanut lectin that were extracted at a high level of purity in my laboratory by Reuven Lotan, another research student. Together with Yair we found that with the help of these lectins it was possible to separate easily and efficiently various types of white cells from both mice and humans. These methods of separation are now used in immunology laboratories throughout the world. Later we made a discovery that was highly important. In cooperation with Professor Asher Meshorer, the head of the Department of Animal Experimentation at the Weizmann Institute, we found that it was possible to use lectins to obtain mouse bone marrow stem cells, the source of all blood cells. We could then use these cells to save mice whose bone marrow had been destroyed by radiation. In a scientific article that we published in 1974, we wrote that the success in mice implied hope that it would be possible to use lectins in the future to obtain bone marrow cells for transplantation into humans. We postulated that the soy lectin and similar proteins could be used to cleanse bone marrow for transplantation into humans, thus helping to prevent the rejection that occurred when the bone marrow of a contributor was not identical to that of the receiver. We tried to write a patent on this but failed.

In 1978, Yair Reisner was invited to the Sloan Kettering Cancer Center in New York to work together with the famous immunology expert, Dr. Robert A. Good, and the bone-marrow transplant specialist, Dr. Richard O'Reily, in order to test our method of separation of stem cells in patients who needed such transplants. After three years of arduous research in New York, the researchers proved that the soy lectin was indeed efficient in removing mature cells from stem cells in the bone marrow and thereby reducing the risk to patients receiving bone marrow transplants. By use of cells that were separated by the soy lectin, it was possible to transplant this bone marrow to patients who needed it. First they showed that such treatment was efficient in transplanting bone marrow in monkeys that were treated by radiation and then they showed that this method was also suitable for transplants in humans.

In April 1986, after the accident at the nuclear reactor in Chernobyl in the Soviet Union, Yair Reisner and bone marrow transplant experts from the United States were invited to come to Hospital No. 6 in Moscow where employees of that reactor were being treated for radiation injuries. According to the little that was known at that time in this field, it was obvious that only bone marrow transplants could save the lives of these people. Although there was no diplomatic connection between Israel and Russia, the Russians, through negotiations with the Jewish American millionaire, Armand Hammer, who was in friendly contact with them, agreed to give Yair a visa to visit their country. This permitted him to work in Moscow together with other transplant experts to perform transplants for the reactor employees with radiation injuries, using the method developed by him in cooperation with me at the Weizmann Institute. The other transplanters used methods that they had developed. Two of the four patients treated by Yair survived.

Starting in 1980, hospitals that had "bubble babies" (severe combined immune deficiency, SCID) used soy lectin developed by us in order to purify bone marrow for transplantation. Until the beginning of the twenty-first century there had been about 200 such cases. These were children that had to be kept in a sterile plastic tent because they were born with a defective immune system and could not otherwise be protected from infection. They could be kept alive for a while but eventually died because of accidental bacterial contamination, usually as young children. Transplantation with bone marrow treated with soy lectin was performed in various hospitals in the United States, England, Germany, and Japan. Seventy-five percent of these babies were successfully treated and went on to lead normal lives. I was particularly excited about this in the beginning of the nineties when I visited the Children's Hospital in Los Angeles. I met the head of the unit for transplants in children, and he told me 50 "bubble children" had been transplanted by the soy lectin method, most of them successfully. He also showed me a container that he took from the freezer that contained commercial soy lectin for these transplants. In recent years, newer and more efficient methods were developed, and the soy lectin method is no longer used. During the nineties, the Sloan Kettering Cancer Center in New York examined the possibility of using the soy lectin method for transplanting bone marrow in leukemia patients and patients with other difficult blood illnesses.

Together with Yitzhak Ofek of the Department of Human Microbiology in Tel Aviv University and David Mirleman of my department in the Weizmann Institute, we showed in 1975 that bacteria attached themselves to membranes and cells of animals. We discovered that it was possible to neutralize this attachment if we incubated these bacteria in suitable solutions of saccharides. From this we concluded that there were cell surface proteins that bonded with saccharides on the surface of the bacteria, namely lectins, so that the bacteria could bind to those cells. On the basis of these findings, in 1979 we conducted the first experiment of its kind with *Escherichia coli*, the bacteria that commonly cause infection in the urinary tract in human beings and animals. We found that these bacteria attach themselves only to a mannose saccharide and to no other chemically similar saccharides. When we injected *E. coli* bacteria that had mannose added to their surface into the bladder of mice, no infection developed in their urinary tract. In contrast, incubation of the *E. coli* in glucose solution did not reduce the rate of infection in mice. This experiment laid the basis for a new and revolutionary method of preventing illness caused by bacteria.

Following our research, many laboratories in various parts of the world have conducted experiments with other bacteria, various saccharides, and various animals such as rats, rabbits and monkeys. The results have suggested the possibility of infectious illness prevention by use of suitable saccharides. Some experiments were also carried out with humans, but not with much success. At present the effort continues at universities and various high-tech companies.

We wanted to register a patent on the idea of prevention of infectious illness by use of saccharides, but we were told that this had already been mentioned in the professional literature. About twelve years later, a company in Philadelphia tried to develop a method using saccharides for treating stomach ulcers caused by bacterial infection. In light of this, an Israeli investment company contacted me because the company in Philadelphia wanted to raise funds, and the company in Tel Aviv wanted to hear my opinion about them. When the founder of the Philadelphia company heard that the advisor was me, he said: "He invented it!"

There is no doubt that expansion of knowledge about lectins and saccharides has caused better understanding of their function in nature, in illness, and in normal situations. This has led to development of new ways of illness prevention and healing.

Author Biography

Jeremiah E Silbert was born in 1931, living in Cleveland Heights, Ohio. He attended Harvard College, graduating in 1953, and then Harvard Medical School, graduating in 1957. After two years of residency in Internal Medicine at Barnes Hospital, Washington University, St. Louis, he spent two years as a Post Doctoral Fellow in the Biochemistry Department at Washington University, and then three years as a Research and Clinical Fellow in Rheumatology at the Massachusetts General Hospital in Boston. During those years he published the first descriptions of the basic assembly of heparin and chondroitin sulfate glycosaminoglycan polymers. In 1964 he went to the Jamaica Plain, Boston VA (Veterans Affairs) hospital, in a program providing funding and research time for physicians. He continued his research on proteoglycan biosynthesis and other aspects of glycobiology at VA hospitals in the Boston area until semi-retirement in 2006. Appointments in the VA included Clinical Investigator, Medical Investigator, and then from 1989 to 1999, Senior Medical Investigator. During the last few years his research has concerned the widespread claims that preparations of glucosamine and chondroitin sulfate are useful for treatment of osteoarthritis. His research has shown that these materials reach joints in amounts too miniscule to have any effect, indicating that their use has no merit. While working and salaried at the VA, he had an academic appointment in Medicine at Tufts

University School of Medicine until 1981, and from then until now has had an appointment as Professor of Medicine at Harvard Medical School, Brigham and Women's Hospital, Division of Rheumatology/Immunology/Allergy. In 2010, he and Harry Schachter shared the 3rd annual Rosalind Kornfeld Award of the Society for Glycobiology for "Lifetime Achievements in Glycobiology." Nathan Sharon had shared the 1st annual award in 2008, and Mary C Glick had shared the 2nd in 2009. Both Glick and Schachter were colleagues and close friends of Sharon and were consulted by Silbert in providing this chapter.

Chapter 2
The History of Lectinology

Hans-Joachim Gabius

A salient precondition for progress and uncomplicated communication in research is the general agreement on definitions and terms. As the headline attests, the term 'lectin' is commonly accepted. This chapter will explain how this term is rooted in the experimental work of the field's pioneers. The impressive demonstration of capacity of agglutinins for selecting distinct blood-group epitopes like an antibody was a cornerstone for lectins to broadly enter laboratories as tools. Further discoveries, e.g. lectin-dependent cell stimulation, paved the way to realizing the enormous scope of lectin-carbohydrate interactions documented in this book. Our historical survey, summarized in Table 2.1, focuses on early developments up to coining the term. Selected more recent advances, which spurred the publication rate as a measure of the field's momentum, are also included in Table 2.1. It presents a graphic overview on the chronology how lectinology developed.

2.1 How Lectinology Started

A seminal report of lasting impact was published in 1860. Therein, the physician S. W. Mitchell gave a detailed account on what happened when performing the following experiment: "One drop of venom was put on a slide and a drop of blood from a pigeon's wounded wing allowed to fall upon it. They were instantly mixed. Within 3 mins the mass had coagulated firmly, and within ten it was of arterial redness" [1]. In the catalogue of his works printed in 1894 he described the situation encountered during the preparation of the cited treatise as follows: "this quarto with its many drawings was the result of 4 years of such small leisure as I could spare amidst

Reprinted chap. 15 from H.-J. Gabius (Ed.)*The Sugar Code: Fundamentals of Glycosciences'*, Wiley-VCH, Weinheim, Germany, 2009. Chapter nos. given in text refer to this source; with permission from Wiley-VCH.

H.-J. Gabius (✉)
Institute of Physiological Chemistry, Faculty of Veterinary Medicine,
Ludwig-Maximilians-University Munich, Veterinarstr. 13, 80539 Munchen, Germany
e-mail: gabius@tiph.vetmed.uni-muenchen.de

E. F. Fang and T. B. Ng (eds.), *Antitumor Potential and Other Emerging Medicinal Properties of Natural Compounds*, DOI: 10.1007/978-94-007-6214-5_2,

Table 2.1 Brief historical account of lectinology

1860	Observation of blood "coagulation" by rattlesnake (*Crotalus durissus*) venom (S. W. Mitchell)
1888	Detection of erythrocyte agglutination by a toxic protein fraction from castor beans (termed ricin) and seeds of related plants (H. Stillmark)
1890	Detection of a toxic lectin in the bark of black locust (*Robinia pseudoacacia*, O. Power, O. Cambier)
1891	Toxic plant agglutinins applied as model antigens (P. Ehrlich)
1898	Introduction of the term "haemagglutinin" or "phytohaemagglutinin" for plant proteins that agglutinate red blood cells (M. Elfstrand)
1902	Detection of bacterial agglutinins (R. Kraus, S. Ludwig) and demonstration that blood "coagulation" by snake venom (seven to nine decades later shown to depend on presence of a C-type lectin) observed in 1860 was due to cell agglutination but not to blood clotting (S. Flexner, H. Noguchi)
1906	Detection of an agglutinin in bovine serum (later characterized as the C-type lectin conglutinin) by use of activated complement-coated erythrocytes (J. Bordet, F. P. Gay); detection of a haemagglutinin in mushrooms (*Amanita* sp., W. W. Ford)
1907/1909	Detection of non-toxic agglutinins in plants, of their nature as proteins and of "deagglutination" of erythrocytes by hog gastric mucin (K. Landsteiner, H. Raubitschek)
1913	Use of intact cells for the purification of the agglutinin ricin (R. Kobert)
1919	Crystallization of a globulin from jack bean, concanavalin A, which was later defined as lectin and used in pioneering studies (please see below) (J. B. Sumner)
1935/1936	Concanavalin A identified as jack bean haemagglutinin; precipitation of starch, glycogen and mucins by concanavalin A defines carbohydrate as ligand and points to "a carbohydrate group in a protein" as binding partner on erythrocytes (J. B. Sumner, S. F. Howell)
1941	Detection of viral agglutinins (G. K. Hirst; L. McClelland, R. Hare)
1944	Description of anti-0(H) haemagglutinating activity in serum of *Anguilla anguilla* (B. Jonsson), following earlier work on a similar activity in *Anguilla japonica* (S. Sugishita; 1935)
1947/1948	Detection of plant agglutinins specific for the human histo-blood group A (W. C. Boyd; K. O. Renkonen), "good keeping qualities" and low "cost of producing them" are emphasized as advantageous properties
1952	Carbohydrate nature of histo-blood group H(0) determinant proven by eel-serum-mediated agglutination of respective erythrocytes and its α-L-fucose-dependent inhibition (W. M. Watkins, W. T. J. Morgan)
1954	Introduction of the term "lectin" for plant (antibody-like) agglutinins, primarily for those which are specific for a distinct histo-blood group (W. C. Boyd)
1956	Detection of an agglutinin specific for the human blood group B in the seeds of the African shrub *Griffonia* (*Bandeiraea*) *simplicifolia* (O. and P. Mäkelä)
1960	Detection of the mitogenic potency of lectins toward lymphocytes (P. C. Nowell)
1963	Introduction of affinity chromatography for the isolation of lectins, published in 1965 (I. J. Goldstein, B. B. L. Agrawal)
1968–1974	Detection of rapid serum clearance of asialoceruloplasmin in rabbits and isolation of a Gal/GalNAc-specific lectin (asialoglycoprotein receptor) from liver, the first mammalian lectin (G. Ashwell, A. G. Morell and colleagues)
1972	Determination of the amino acid sequence and the three-dimensional structure of a lectin, concanavalin A (G. M. Edelman, K. O. Hardman, C. F. Ainsworth et al.)

(continued)

Table 2.1 (contniued)

1972–1977	Detection of impaired synthesis of a marker for glycoprotein (lysosomal enzymes) routing as cause for a human disease (mucolipidosis II) and its identification as Man-6-phosphate, the ligand for P-type lectins (E. F. Neufeld and colleagues; W. S. Sly and colleagues)
1978	First conference focusing on lectins and glycoconjugates, termed Interlec (T. C. Bøg-Hansen)
1979	Detection of endogenous ligands for plant lectins (H. Rüdiger)
1981–1988	Further refinements of the definition of the term "lectin" as carbohydrate-binding protein, separated from antibodies and carbohydrate-processing enzymes/sensor or transport proteins for free sugars (S. H. Barondes, I. J. Goldstein, J. Kocourek et al.)
1982	Introduction of serial lectin affinity chromatography as analytical tool for structural analysis of glycans from glycoproteins (R. D. Cummings, S. Kornfeld)
1983	Detection of the insecticidal action of a plant lectin (L. L. Murdock)
1984	Isolation of lectins from tumors (H.-J. Gabius; R. Lotan, A. Raz)
1985	Immobilized glycoproteins as pan-affinity adsorbents for lectins (H. Rüdiger)
1989	Detection of the fungicidal action of a plant lectin (W. J. Peumans)
1992/1993	Detection of impaired synthesis of lectin (selectin) ligands by defective fucosylation as cause for leukocyte adhesion deficiency type II, a congenital disorder of glycosylation (CDG IIc) (A. Etzioni and colleagues)
1995	Structural analysis of a lectin-ligand complex in solution by NMR spectroscopy (J. Jiménez-Barbero and colleagues)
1996–2003	Detection of differential conformer selection by plant, bacterial and animal lectins (H.-J. Gabius and colleagues; L. Poppe and colleagues)
2001–2005	Development of glycan/lectin microarrays for specificity analysis of lectins/structural analysis of glycans and glycoproteomics (various laboratories worldwide)
2001–2007	Advances in lectinology and glycosciences honored by devoting special issues in *Advanced Drug Delivery Reviews, Biochimica et Biophysica Acta, Biochimie, Biological Chemistry, Cells Tissues Organs, Chemical Reviews, Current Opinion in Structural Biology, Glycoconjugate Journal, Journal of Agricultural and Food Chemistry* ("Liener symposium"), *Nature* and *Science* to these topics

From [11], extended, updated and modified

the cares of constantly increasing practice. The story of the perils and anxieties of this research, embarrassed by want of help and by its great cost, is untold in its pages."

His later experiments using washed erythrocytes and then studies by S. Flexner and H. Noguchi, which S. W. Mitchell himself inspired, revealed that the noted 'coagulation' did not result from procoagulants (clotting factors) in blood. As stated by these authors, "the value of the use of washed corpuscles comes especially from the fact that the suspension of lytic phenomena is eliminated. Agglutination, therefore, may be studied purely" [2]. The "venom-agglutination" was especially strong with rabbit erythrocytes—swine and ox cells being less susceptible—and akin to the reaction with "intermediary bodies" (today known as immunoglobulins) [2]. Its biochemical nature was defined in 1984 by purifying the agglutinin of *Crotalus durissus* venom, a C-type lectin [3]. This feat underscored the pioneering character of S. W. Mitchell's research for lectinology. Groundbreaking as it was, the spirit and attitude, with which he carried out his work, as captured on p. 1 of his report [1], also continue to set commendable example: "for the researches which form the novel part of the following essay, I claim only exactness of detail and honesty of statement. Where the results have appeared to me inconclusive, and where further experimental questioning has not resolved the doubt, I have fairly confessed my inability to settle the matter. This course I have adhered to in every such instance, thinking it better to state the known uncertainty thus created than to run the risk of strewing my path with errors in the garb of seeming truths." Its range was extended to plants as sources for agglutinins by a medical thesis in 1888 (Table 2.1).

Using the same technique of haemagglutination, extracts of plant seeds, initially of toxic castor beans (*Ricinus communis*), were also shown to be active. H. Stillmark described a toxin with agglutinating activity, termed ricin, in his MD thesis in 1888. It was prepared under the guidance of R. Kobert at the University of Dorpat (now Tartu) in Estonia, then belonging to the Russian empire [4]. He defined "ricin" as a protein ("Eiweisskörper, sog. Phytalbumose"), conglomerating (or agglutinating) the red blood corpuscles in defibrinated-serum-containing blood ("Zusammenballung der rothen Blutkörperchen") [4]. Such an activity was also uncovered in *Abrus precatorius* seeds ('abrin'), the respective protein fraction produced and tested in R. Kobert's pharmacological institute, and then preparations of ricin and abrin, made commercially available by the Merck Company in Darmstadt on Kobert's initiative, found immediate usefulness beyond lectinology. They substituted bacterial toxins in P. Ehrlich's fundamental studies on the immune response [5]. Once that activity was measured and degree of purity increased, the need for a name became obvious.

2.2 Early Definitions

On the initiative of R. Kobert, who left Dorpat for Germany in January 1897 due to the russification of the Baltics that had been unleashed by an attempted assassination on czar Alexander III, this issue was addressed in 1898. M. Elfstrand introduced the term "Haemagglutinin (Blutkörperchenagglutinin)" into the literature [6]. He also noted the "striking similarity" between agglutinating proteins from plants and from human/animal sera [6]. Indeed, exactly this period was characterized by an equally dynamic development in serology. It led to the recognition of the ABO blood group system based on detecting and monitoring isoagglutination (for historical reviews, please see [7, 8]). The discovery of the agglutination and lysis of erythrocytes by serum compounds is especially linked with three investigators, i.e. A. Creite (a medial student in Göttingen in 1869), L. Landois (director of the physiological institute at the University of Greifswald in 1875) and K. Landsteiner, a Nobel Laureate in 1930, in 1900 working at the institute for pathological anatomy in Vienna [7, 8]. Their studies and the work on plant agglutinins (K. Landsteiner referred to them as "Normalantikörper (normal antibodies)") revealed that the two protein classes also shared activity as precipitin, selectivity for erythrocytes from different species and inhibition of agglutination by haptens (the info box 2 in Chap. 1 tells the story how an agglutinin from eel serum was instrumental to define α-L-fucose as a structural determinant of the H epitope). On these grounds it becomes rather obvious to look at plant (and other) agglutinins as antibody-like proteins. That they are capable to select epitopes figures as central factor in coining a handy generic term.

2.3 The Current Definition of the Term 'Lectin'

The work on blood-group-specific proteins laid the foundation for the new definition (Table 2.1). It was given by W. C. Boyd in 1954 as follows: "it would appear to be a matter of semantics as to whether a substance not produced in response to an antigen should be called an antibody even though it is a protein and combines specifically with a certain antigen only. It might be better to have a different word for the substances and the present writer would like to propose the word *lectin* from Latin *lectus*, the past principle of *legere* meaning to pick, choose or select" [9]. The author thus intended—in his own words—"to call attention to their specificity without begging the question as to their nature" [10]. Specificity for lectins today means binding activity for sugars. Because not only lectins exhibit this property, they are at present strictly delimited from carbohydrate-specific immunoglobulins, enzymes using carbohydrates as substrates (glycosyltransferases, glycosidases and any enzymes introducing or removing substitutions such as sulfotransferases/sulfatases) and sensor/carrier proteins for free mono- or oligosaccharides [11].

Historically, selectivity of lectins for glycans was unambiguously delineated first in the case of concanavalin A by J. B. Sumner and S. F. Howell in 1936 (for the protocol how to prepare the crystalline lectin, please see info box in Chap. 17) [12]. These authors surmise that dissolving concanavalin A in solutions with sucrose may have masked this activity in any previous studies. Of note in this context, K. Landsteiner and H. Raubitschek had observed "deagglutination" of erythrocyte aggregates, formed by ricin, abrin or bean extracts, by hog gastric mucin in 1909 [13]. This observation, viewed in retrospect, can now be interpreted as early evidence for carbohydrate-binding activity of the tested lectins. In fact, mucins, actually the first protein types detected to be conjugates with sugars in 1865 ("… das Mucin einen gepaarten Stoff darstelle" (that mucin may represent a conjugated compound)) [14], are potent lectin binders exploitable in one-step purification of lectins (please see Chap. 19). Initially, affinity chromatography for lectins used cross-linked dextran (Sephadex®) as a matrix for the isolation of concanavalin A [15]. The enormous potential of this strategy had not immediately been realized. A reviewer judged the report after its initial submission to represent "a modest advance in an obscure area" [16]. In effect, this technical advance was instrumental to markedly increase the publication activity in this field, yielding a surge in the number of papers on new lectins and applications [17]. The weak reactivity to cross-linked agarose of the Charcot-Leyden crystal protein is of interest in this respect, because it indicates reactivity for sugars of this protein, whose typical hexagonal bipyramidal crystals were first described in post-mortem spleen of a leukemia patient in 1853 [18] and sputum of asthmatics in 1872 [19]. Auto-crystallization in situ of this protein constituting 7–10 % of a mature blood eosinophil in protein content (about 8.5 pg/cell) may thus be considered as physiologic purification step. The technical breakthrough of establishing affinity chromatography in lectin research also opened the door for the first experimental purification of a mammalian lectin (Table 2.1) [20]. How its presence and functional significance was traced is recounted in the next paragraph.

2.4 Recent Developments

The mentioned line of research was intended to elucidate the role of ceruloplasmin in maintaining the copper level and details on the metabolism of this transport protein for copper ions. As reagent to address these issues, a radioactive form of the glycoprotein was produced by tritiation. This reaction required desialylation of N-glycan chains (for structural details on these chains, please see Chap. 6) and oxidation of galactose at its C_6 atom catalyzed by galactose oxidase [21]. Amazingly, the performed engineering of the N-glycan chains, which unmasked galactose residues, was not without consequences. It dramatically altered ceruloplasmin's serum clearance: "evidence is presented to show that, in contradistinction to homologous, native ceruloplamin, which survives for days in the plasma of rabbits, intravenously injected asialoceruloplamin disappears from the circulation within minutes and accumulates simultaneously in the parenchymal cells of the liver. The rapidity of this transfer of asialoceruloplamin from plasma to liver has been shown to be dependent upon the integrity of the exposed, terminal galactosyl residues." [22]. Thus, desialylation turned N-glycans into ligands, and—luckily—the enzymatic oxidation did not impair the bioactivity for the endocytic hepatic C-type lectin (for further information on C-type lectins, please see Chaps. 17, 20, 21 and 28). It became a role model for targeted drug delivery by (neo)glycoproteins [23] (for further application of neoglycoproteins and examples of sugars as pharmaceuticals, please see Chaps. 26 and 29).

What presently is covered by the umbrella term 'lectin' is outlined throughout this book. As exemplarily emphasized in Table 2.1, intriguing clinical correlations between status of glycosylation and functional implications via lectins have turned up (please see for example Chaps. 26 and 28). Moreover, the molecular details of glycan recognition are unravelled inspiring drug design and new technologies for measuring lectin specificity are developed (please see Chaps. 14 and 15). Combined, these examples for dynamic research lines in lectinology afford efficient driving forces, to make sure that this field maintains its currently acquired prominent status, honored by a series of special feature issues (please see last lines of Table 2.1), and its momentum.

2.5 Conclusions

Agglutinating cells is common to different types of proteins. Respective assays with erythrocytes led to an early convergence of work on antibodies and on proteins from diverse sources, which are not produced in response to an antigen. In particular, the crucial role of certain haemagglutinins from plants and eel serum in the process to delineate the biochemical basis of the ABO blood-group epitopes attests their selectivity, rivaling that of antibodies. Close inspection of the literature teaches the lesson that this context entailed coining the term 'lectin'. It embodies the aspect of molecular selectivity. At the same time, it separates the agglutinins terminologically from immunoglobulins. As a consequence of the haptenic inhibition of agglutination by sugars, 'lectin' is now the generic name for carbohydrate-specific proteins, different from antibodies, enzymes acting on the ligand and sugar transport/sensor proteins. The current wide scope of structural and functional studies including promising medical perspectives, described throughout this book, ensures lectinology to properly address the challenges of deciphering the sugar code.

References

1. Mitchell SW (1860) Researches upon the venom of the rattlesnake. Smithson Contributions Knowl XII:89–90
2. Flexner S, Noguchi H (1902) Snake venom in relation to haemolysis, bacteriolysis and toxicity. J Exp Med 6:277–301
3. Gartner TK, Ogilvie ML (1984) Isolation and characterisation of three $Ca_2{}^{+}$-dependent beta-galactoside-specific lectins from snake venoms. Biochem J 224:301–307
4. Stillmark H, Ueber R (1888) ein giftiges Ferment aus den Samen von *Ricinus comm.* L. und einigen anderen Euphorbiaceen [Inaugural-Dissertation]. Schnakenburg's Buchdruckerei, Dorpat
5. Ehrlich P (1891) Experimentelle Untersuchungen über Immunität II. Ueber Abrin. Dtsch Med Wschr 17:1218–1219
6. Elfstrand M (1898) Ueber blutkörperchenagglutinierende Eiweisse. In: (ed) Görbersdorfer Veröffentlichungen pp 1–159
7. Hughes-Jones NC, Gardner B (2002) Red cell agglutination: the first description by Creite (1869) and further observations made by Landois (1875) and Landsteiner (1901). Br J Haematol 119:889–893
8. Schwarz HP, Dorner F (2003) Karl Landsteiner and his major contributions to haematology. Br J Haematol 121:556–565
9. Boyd WC (1954) The proteins of immune reactions. In: Neurath H, Bailey K (ed) The Proteins. Academic Press, New York pp 756–844
10. Boyd WC (1963) The lectins: their present status. Vox Sang 8:1–32
11. Gabius H-J et al (2004) Chemical biology of the sugar code. Chem BioChem 5:740–764
12. Sumner JB, Howell SF (1936) The identification of a hemagglutinin of the jack bean with concanavalin A. J Bacteriol 32:227–237
13. Landsteiner K, Raubitschek H (1909) Ueber die Adsorption von Immunstoffen V. Mitteilung. Biochem Z 15:33–51
14. Eichwald E (1865) Beiträge zur Chemie der gewebbildenden Substanzen und ihrer Abkömmlinge. I. Ueber das Mucin, besonders der Weinbergschnecke. Ann Chem Pharm 134:177–211

15. Agrawal BBL, Goldstein IJ (1965) Specific binding of concanavalin A to cross-linked dextran gel. Biochem J 96:23c
16. Sharon N (1998) Lectins: from obscurity into the limelight. Protein Sci 7:2042–2048
17. Rüdiger H et al (2000) Medicinal chemistry based on the sugar code: fundamentals of lectinology and experimental strategies with lectins as targets. Curr Med Chem 7:389–416
18. Charcot JM, Robin C (1853) Observation de leocythemia. C R Mem Soc Biol 5:44–50
19. Leyden E (1872) Zur Kenntnis des Bronchialasthma. Arch Pathol Anat 54:324–344
20. Hudgin RL et al (1974) The isolation and properties of a rabbit liver binding protein specific for asialoglycoproteins. J Biol Chem 249:5536–5543
21. Morell AG et al (1966) Physical and chemical studies on ceruloplasmin. IV. Preparation of radioactive, sialic acid-free ceruloplasmin labeled with tritium on terminal d-galactose residues. J Biol Chem 241:3745–3749
22. Morell AG et al (1968) Physical and chemical studies on ceruloplasmin. V. Metabolic studies on sialic acid-free ceruloplasmin in vivo. J Biol Chem 243:155–159
23. Yamazaki N et al (2000) Endogenous lectins as targets for drug delivery. Adv Drug Deliv Rev 43:225–244

Author Biography

Hans-Joachim Gabius studied biochemistry in Hannover, Germany and received his PhD degree for work on proofreading mechanisms of phenylalanyl-tRNA synthetase in the lab of Prof. F. Cramer (Max-Planck-Institute for Experimental Medicine, Goettingen) in 1982. Work on tRNA splicing in Prof. J. Abelson's lab at UC San Diego was followed by starting research on lectins at the Max-Planck-Institute. Following a postdoc period with Prof. S. H. Barondes (UC San Diego/VA Medical Center) he was appointed as Assistant Professor in 1988 in Goettingen, then in 1991 as Associate Professor at the University of Marburg, followed by the chair position for Physiological Chemistry (Faculty of Veterinary Medicine) at the Ludwig-Maximilians-University Munich. His track record includes over 690 peer-reviewed publications, which have received over 17,000 citations at an h-factor of 71, and edition of five books, among them the textbook "The Sugar Code". Besides research awards he recently received an honorary degree in medicine by the Charles University in Prague (founded 1348) for his achievements in glycosciences.

Chapter 3
Animal and Human Lectins

Hans-Joachim Gabius

Abstract This chapter provides an overview on protein folds with lectin activity and next on lectin functions. Triggering such responses depends on the binding of distinct glycoconjugates, the bioactive lectin ligands, as will be exemplified. Not only their presence but also spatial parameters will be identified to be crucial for bioactivity. As a consequence, six levels, at which the affinity of a glycan for a lectin can be adjusted, will be presented at the end of this chapter. This synopsis can serve as a guideline to interpret structural changes in glycosylation as regulatory events.

Previous (Chaps. 6–11) have illustrated the enormous structural diversity of glycans. An elaborate synthetic machinery including enzymes for introducing distinct substitutions and remodeling underlies the establishment of the exceptionally high coding capacity of glycans. The question immediately arises on how this information is turned into biological responses. An intermolecular contact within the decoding process is especially favored at branch ends of mature chains. They are spatially readily accessible (for a primer on the interaction of glycans with proteins, please see Chap. 14). The limited flexibility around glycosidic linkages in glycans accounts for the presentation of only few key-like conformers suited for contact to a lectin (for conformational aspects of glycans, please see Chaps. 2 and 14; for definition of the term 'lectin', please see Chap. 16). Given all these favorable characteristics of glycans at branch ends to enable a role as biochemical signals (structural variability, spatial accessibility

Reprinted Chap. 19 from H.-J. Gabius (Ed.) *The Sugar Code: Fundamentals of Glycosciences'*, Wiley-VCH, Weinheim, Germany, 2009. Chapter nos. given in text refer to this source; with permission from Wiley-VCH.

H.-J. Gabius (✉)
Institute of Physiological Chemistry, Faculty of Veterinary Medicine,
Ludwig-Maximilians-University Munich, Veterinarstr. 13, 80539 Munchen, Germany
e-mail: gabius@tiph.vetmed.uni-muenchen.de

E. F. Fang and T. B. Ng (eds.), *Antitumor Potential and Other Emerging Medicinal Properties of Natural Compounds*, DOI: 10.1007/978-94-007-6214-5_3,

and limited flexibility), then the level of diversity reached on the sugar side should be matched in numbers by lectins. What held true for bacteria and viruses (please see Chap. 18) and plants (please see Chap. 19) in which numerous lectins had been detected, was questioned for higher animals as late as 1973 [1] (1 year before the first mammalian lectin was isolated; please see Chap. 16 for details). The following decades witnessed a remarkable surge in our knowledge about lectins in higher animals, both in number of families and family members [2]. With lectins now finally being accepted as "a common cell component" [1], lectin-carbohydrate interactions are expected to entail a wide range of functions, a key topic of this chapter. It first provides an overview on protein folds with lectin activity and next on lectin functions. Triggering such responses depends on the binding of distinct glycoconjugates, the bioactive lectin ligands, as will be exemplified. Not only their presence but also spatial parameters will be identified to be crucial for bioactivity. As a consequence, six levels, at which the affinity of a glycan for a lectin can be adjusted, will be presented at the end of this chapter. This synopsis can serve as a guideline to interpret structural changes in glycosylation as regulatory events.

3.1 Protein Folds with Lectin Activity

The concept for a complex network of productive protein-carbohydrate interactions would be strongly supported, if several folding patterns were adapted to binding glycans. Then sugar-binding modules are clearly not a singular invention of restricted relevance. The sheer size of the compilation of proven cases, presented in Table 3.1, is compelling evidence for glycan binding to be a rather common aspect of protein evolution (a view on the secondary-structure patterns of C- and P-type lectins, the antiparallel β-sandwich and β-barrel is given in the figure of Chap. 17). Starting from ancestral modules, gene duplications and sequence variations then spurred the process of intrafamily diversification. Already subtle sequence deviations in otherwise closely related C-type lectins (for example a Val/Ser substitution in dendritic and endothelial cell receptors) have a significant bearing on glycan specificity (for details, please see [3]). As recounted in Chap. 21, the C-type domain is present in widely divergent subgroups of respective lectins and has even developed into a module accommodating tyrosine sulfates in addition to carbohydrates as in P-selectin or peptides in natural killer (NK) cell receptors. Equally intriguing, dynamic binding-site evolution holds true for other lectin families such as the galectins or siglecs, fulfilling the structural prerequisite for non-redundant activity profiles [2–5]. And there are more, not yet structurally well-defined lectin sites, found in the α_M-integrin (please see also Chap. 30), α/Θ-defensins (please see Chap. 27) or the sea urchin 350 kDa sperm-binding protein and a nucleocytoplasmic GlcNAc-binding protein, the two latter proteins belonging either to the hsp (heat shock protein) 110 or hsp 70 groups, respectively.

Table 3.1 Overview of protein folds with lectin activity

Type of fold	Example for lectin	Example for ligand[a]
C-type	Asialoglycoprotein receptor, collectins, selectins	Fuc, Gal, GalNAc, Man, heparin tetrasaccharide
I-type (Ig fold)	N-CAM, TIM-3, siglecs	$Man_6GlcNAc_2$, HNK-1 epitope, α2,3/6-sialylated glycans
P-type	Mannose-6-phosphate receptors (MR) and proteins with MR homology domain (erlectin, OS-9)	Man-6-phosphate, $Man_{5,8}GlcNAc_2$
β-sandwich (jelly-roll)	(a) Galectins	β-galactosides
	(b) Calnexin, calreticulin	$Glc_1Man_9GlcNAc_2$
	(c) ERGIC-53	$Man_xGlcNAc_2$
	(d) CRD[b] of Fbs1 in SCF E3 ubiquitin ligase and peptide-N-glycanase	$Man_3GlcNAc_2$; mannopentaose
	(e) Pentraxins	Glycosaminoglycans, MOβDG, 3-sulfated Gal, GalNAc and GlcA, Man-6-phosphate
	(f) G-domains of the LNS family (laminin, agrin)	Heparin
β-trefoil	(a) Fibroblast growth factors	Heparan sulfate
	(b) Cystein-rich domain of C-type macrophage mannose receptor	GalNAc-4-sulfate in LacdiNAc
	(c) Lectin domain in GalNAc-Ts[c] involved in mucin-type O-glycosylation	GalNAc
	(d) Hemolytic lectin CEL-III of sea cucumber and lectin EW29 of earthworm	Gal
β-propeller	(a) 4-bladed: tachylectin-3	S-type lipopolysaccharide
	(b) 5-bladed: tachylectin-2	GlcNAc/GalNAc
	(c) 6-bladed: tachylectin-1	KDO
β-prism II	Pufferfish (fugu) lectin	Man
β-barrel with jelly-roll topology	Tachylectin-4, eel (*Anguilla anguilla*) agglutinin, *Xenopus*	Fuc
Fibrinogen-like domain	(a) Ficolins	GlcNAc
	(b) Intelectins (mammalian, *Xenopus*)	Gal*f*, pentoses
	(c) Tachylectin-5	N-acetylated sugars
	(d) Slug (*Limax flavus*) lectin	sialic acid

(continued)

Table 3.1 (continued)

Type of fold	Example for lectin	Example for ligand[a]
Link module	CD44, TSG-6, LYVE-1, aggregating proteoglycans	Hyaluronic acid
Hevein-like domain	Tachycytin and spider (*Selenocosmia huwena*) neurotoxin; cobra venom cardiotoxin	GalNAc; heparin-derived disaccharide
$(\beta/\alpha)_8$ barrel (glycoside hydrolase family 18)	YKL-40 (human cartilage glycoprotein-39; chitinase-like lectin)	$(GlcNAc)_n$
Short consensus repeat (complement control protein module)	Factor H (complement regulator)	Glycosaminoglycans, sialic acid

[a] for abbreviations of names for sugars, please see Chap. 1,
[b] carbohydrate recognition domain,
[c] N-acetylgalactosaminyltransferases

When looking at occurrence of folds in plant lectins (Chap. 19), several motifs are shared among plant and animal lectins such as the β-sandwich fold of leguminous agglutinins, the classical example being concanavalin A (for illustration, please see Chap. 17), the β-trefoil fold (first detected in soybean trypsin inhibitor) in the lectin subunit of AB-toxins (ricin and others) and in amaranthin as well as the hevein-like domain in wheat germ agglutinin and also chitinases. Close inspection of positioning of the binding sites and/or the binding mode intimates a convergent rather than divergent course of evolution in the mentioned cases [6]. This aspect should be reckoned with in any future suggestions for a terminology system. With these multiple folds engaged in sugar recognition, the expectation is nourished for multiple functions.

3.2 Functions of Animal and Human Lectins

In concert with the glycans of cellular glycoconjugates, lectins can turn their carbohydrate-binding activity into specific recognition already in the endoplasmic reticulum and then at the cell surface. As already noted in Chap. 6, the nascent and processed N-glycans, originating from the precursor common to all Asn-X-Ser/Thr-defined acceptor sites, are signals for quality control by virtue of their ligand capacity. The information at each stage of glycan processing is translated by lectins into efficient monitoring for correct folding of nascent glycoproteins and into ensuing intracellular transport. This intracellular activity profile fills the first part of Table 3.2. Of note, the glycans added in a cotranslational manner may well guide folding pathways and later influence secretory efficiency and protein activity/stability, these processes in principle also involving intramolecular protein-carbohydrate interactions or a switching-off of intermolecular interactions. When glycoproteins with mature glycans finally reach the cell surface, cell adhesion and diverse cellular responses can be attributed to *cis/trans* interactions (Table 3.2). Examples for lectin-elicited intracellular signaling routes leading to responses such as growth control are further explained with illustrations in Chaps. 26 and 28.

Toward these ends, a factor different from domain folding comes into play. In addition to direct ligand binding by a lectin's carbohydrate recognition domain (CRD) spatial aspects of CRD arrangement contribute markedly to in vivo lectin functionality. They not only increase the affinity by multivalent interactions but also the selectivity for distinct types of ligand display, for example to distinguish self from non-self glycan signatures [3]. The ways Nature has compensated for the weakness of a single protein-carbohydrate contact are illustrated in Fig. 3.1 (please see legend for further details). A firm grip on glycans will be facilitated by the suitably spaced CRDs. This topological arrangement is realized and drawn for galectins, a serum collectin and two other depicted C-type lectins endowed with capacity for endocytic uptake (Fig. 3.1). The shown arrangement, matched by branched glycans, readily explains the rapid removal of asialoglycoproteins with triantennary N-glycans from circulation by the hepatic lectin (please see Chap. 16). And Fig. 3.1 then shows a different and equally effective way to turn sugar binding into a host of cell actions, that is the modular arrangement

Table 3.2 Functions of animal and human lectins

Activity	Example of lectin
Recognition of stem region of N-glycans, a signal for ubiquitin conjugation when accessible in incorrectly folded glycoproteins	F-box proteins Fbs1/2 (Fbx2/FBG1, Fbx6b/FBG2) as a ligand-specific part of SCF ubiquitin ligase complexes
Molecular chaperones with dual specificity for $Glc_1Man_9GlcNAc_2$ and protein part of nascent glycoproteins in endoplasmic reticulum (ER)	Calnexin, calreticulin
Targeting of misfolded glycoproteins with $Man_{8\text{-}5}GlcNAc_2$ as carbohydrate ligand to ER-associated degradation (ERAD)	EDEM1,2/Mnl1 (Htm1) (lectins or glycosidases?), Yos9 protein (MRH domain)
Intracellular routing of glycoproteins and vesicles and apical delivery	ERGIC-53 and VIP-36 (probably also ERGL and VIPL), P-type lectins, comitin, galectins-3 and -4
Intracellular transport and extracellular assembly	Non-integrin 67 kDa elastin/laminin-binding protein
Enamel formation and biomineralization	Amelogenin
Inducer of membrane superimposition and zippering (formation of Birbeck granules)	Langerin (CD207)
Cell type-specific endocytosis	Hepatic and macrophage asialoglycoprotein receptors, dendritic cell and macrophage C-type lectins (mannose receptor family members (tandem-repeat type) and single CRD lectins such as trimeric langerin/CD207 or tetrameric DC-SIGN/CD209), cysteine-rich domain (β-trefoil) of the dimeric form of mannose receptor for GalNAc-4-SO_4-bearing glycoprotein hormones in hepatic endothelial cells, P-type lectins
Recognition of foreign glycans (β1,3-glucans, LPS)	CR3 (CD11b/CD18, Mac-1 antigen), C-type lectins such as DC-SIGN and dectin-1, immulectins, intelectins, *Limulus* coagulation factors C and G, earthworm CCF, tachylectins
Recognition of foreign or aberrant glycosignatures on cells (including endocytosis or initiation of opsonization or complement activation)	Collectins, ficolins, C-type macrophage and dendritic cell lectins, CR3 (CD11b/CD18, Mac-1 antigen), α/Θ-defensins, pentraxins (CRP, limulin), RegIIIγ (HIP/PAP), siglecs, tachylectins
Targeting of enzymatic activity in multimodular proteins	Acrosin, C-terminal β-sandwich lectin domain of mouse peptide-N-glycanase (PNGase), *Limulus* coagulation factor C, laforin, β-trefoil fold ($(QxW)_3$ domain) of GalNAc-Ts involved in mucin-type O-glycosylation (frequent in microbial glycosylhydrolases for plant cell wall polysaccharides)
Bridging of molecules	Galectins, C-type lectins, cerebellar soluble lectin

(continued)

Table 3.2 (continued)

Activity	Example of lectin
Induction or suppression of effector release (H_2O_2, cytokines etc.)	Galectins, selectins and other C-type lectins such as CD23, BDCA-2 and dectin-1, I-type lectins (CD33 (siglec-3), siglecs-7 and -9), Toll-like receptor 4
Modulation of enzymatic activities/receptor endocytosis via glycan recognition	Mannan-binding lectin (acting on neprins) galectins (acting on growth factor receptors)
Cell growth control, induction of apoptosis/anoikis and axonal regeneration	Galectins, C-type lectins, amphoterin-like protein, hyaluronic acid-binding proteins, cerebellar soluble lectin, CD22 (siglec-2), MAG (siglec-4)
Cell migration and routing	Galectins, selectins and other C-type lectins, I-type lectins, hyaluronic acid-binding proteins (RHAMM, CD44, hyalectans/lecticans)
Cell–cell interactions	Selectins and other C-type lectins such as DC-SIGN, galectins, I-type lectins (siglecs, N-CAM, P_0 or L1), gliolectin
Cell–matrix interactions	Galectins, heparin- and hyaluronic acid-binding lectins including hyalectans/lecticans, calreticulin
Matrix network assembly	Proteoglycan core proteins (C-type CRD and G1 domain of hyalectans/lecticans), galectins (e.g. galectin-3/hensin), non-integrin 67 kDa elastin/laminin-binding protein

connecting CRDs with other types of modules. This is encountered in various lectins. As seen for selectins and I-type lectins such as siglecs (Fig. 3.1), the presence of spacers ensures to separate the CRD from the cell surface, a prerequisite for accessibility in cell–cell interactions. In selectins, interdomain contacts and the force-dependent equilibrium between closed-angle and extended positioning of the two distal types of module have a strong bearing on the kinetics of binding. The transition to the extended form at low force level prolongs the lifetime of selectin-ligand association. The counterintuitive effect that applied force stabilizes a cell–cell contact leads to regular rolling of leukocytes on a vascular surface (for further details on selectins, please see Chap. 28). The so-called catch bonds reduce k_{off}-values, and a structural force-dependent change distant to the CRD, yielding a high-affinity state, underlies the unique and counterintuitive feature of selectins to support rolling and tethering against a hydrodynamic force [7, 8], In siglecs, ligand binding accounts for covalent alterations by a posttranslational modification far away, that is in the intracellular section of these lectins (please see below).

Modular design has a wide range of functional implications. It aids routing of collectin-opsonized particles, targeting of enzyme activities (for example in laforin, the product of the EMP2A gene, a dual-specificity protein phosphatase with a lectin domain; defects are linked to an autosomal recessive progressive myoclonus epilepsy, the lafora disease, on the clinical level, to lymphoma development in nude mice) or the just mentioned transmembrane signaling reaching intracellular

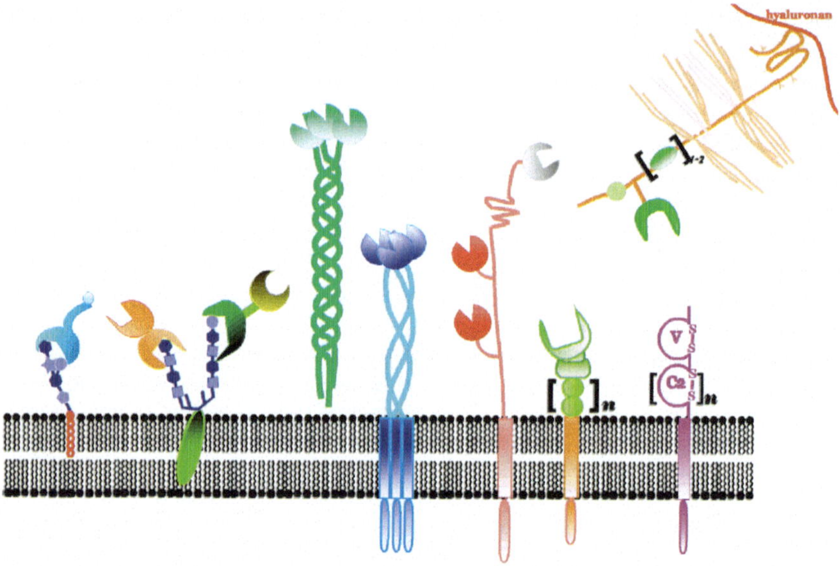

immunoreceptor tyrosine-based activating/inhibitory motifs (ITAM/ITIM) found in several C-type lectins or siglecs (Table 3.2) [3]. Embedded in modular display, the C-type lectin-like domain of aggregating proteoglycans reacts with sugar or peptide motifs and contributes to their role as molecular glue in the matrix (Fig. 3.1, upper right part). In cooperation with the selection of the ligand by a CRD spatial aspects of the lectin inspect glycosignatures and control the lectins' reactivity to glycans. Implicitly, endogenous lectins are expected to show considerable selectivity for binding partners. Compared to plants as source for lectins (please see Chap. 19 for details), animal tissues contain much less lectin so that recombinant expression has become a valued tool to be able to address this issue.

3.3 Lectin Ligands and Affinity Regulation

Purification of tissue lectins, often using affinity chromatography as crucial step (please see Chap. 16 for its introduction to lectin research and Chap. 19 for further practical details), and also their labeling, as similarly done with plant lectins, were thus crucial to trace endogenous ligands. The range of functions presented in Table 3.2 intimates a correlation between certain lectins and their targets, sometimes already reflected in lectin names. A lysosomal enzyme will present the epitope with mannose-6-phosphate as postal code for P-type-lectin-mediated transfer to its final destination, and mannan on the cell surface, a characteristic of the mannose-rich glycophenotype of yeast cells, is a high-affinity matrix for the mannan(or mannose)-binding lectin [9, 10]. Functions in cell adhesion and growth or mediator release call for

◂ **Fig. 3.1** Illustration of the strategic ways how carbohydrate recognition domains (CRDs) in animal lectins are positioned to reach optimal ligand selection (for example to separate self from non-self glycan profiles in innate immunity) and topological complementarity. From left to right, the CRD display in the three subtypes within the galectin family (chimeric, proto-type and tandem-repeat-type arrangements binding to a ganglioside or a branched complex-type N-glycan without or with terminal α2,3-sialylation), the presentation of CRDs (C-type or fibrinogen-like domain) in serum and surfactant collectins or ficolins connected to their collagenous stalks and the non-covalent association of binding sites in transmembrane C-type lectins by α-helical coiled-coil stalks (for example asialoglycoprotein and Kupffer cell receptors, the scavenger receptor C-type lectin, CD23, DC-SIGN or DC-SIGNR) are given. Similar to tandem-repeat-type galectins the C-type family of lectins also has a branch of members with this design, i.e. immulectins-1, -2 and 3. Next, the tandem-repeat display in the mannose-specific macrophage receptor (also found on dendritic cells, hepatic endothelial cells, kidney mesangial cells, retinal pigment epithelial cells and tracheal smooth muscle cells) and the related C-type lectin Endo180 with eight domains as well as in the cation-independent P-type lectin with 15 domains is presented. Capacity for sugar binding is confined to only few domains as depicted. The occurrence of lectin activity for GalNAc-4-SO_4-bearing pituitary glycoprotein hormones in the cysteine-rich domain, a member of the β-trefoil fold family with one $(QxW)_3$ domain in the N-terminal section of the macrophage mannose receptor (amino acids 8–128), which is linked via a fibronectin-type-II-repeat-containing module to the tandem-repeat section, is also included into the schematic drawing for these lectins with more than one type of CRD per protein chain. Moving further to the right side, the association of a distal CRD in selectins (attached to an epidermal-growth-factor (EGF)-like domain and two to nine complement-binding consensus repeats) or in the siglec subfamily of I-type lectins using 1–16 C2-set immunoglobulin-like units as spacer equivalents to let the CRD reach out to contact ligands and to modulate capacity to serve in *cis*- or *trans*-interactions on the cell surface is shown. The force-dependent alterations of the topological arrangement of the two distal domains in selectins accounts for catch bonds of selectins (please see text), a canonical immunoreceptor tyrosine-based inhibitory motif (ITIM) together with a putative tyrosine-based motif is frequently present in the intracellular portion of siglecs. C2-set domains linked to fibronectin-type-III repeats establish the extracellular section of the I-type lectins L1 and neural cell adhesion molecule (NCAM). In the matrix, the modular proteoglycans (hyalectans/lecticans: aggrecan, brevican, neurocan and versican) interact a.) with hyaluronan (and also link protein) via the link-protein-type modules of the N-terminal G1 domain (and an Ig-like module), b.) with receptors binding to the glycosaminoglycan chains in the central region and c.) with carbohydrates or proteins (fibulins-1 and -2 and tenascin-R) via the C-type lectin-like domain flanked by EGF-like and complement-binding consensus repeat modules (kindly provided by H. Kaltner)

particular sets of glycoligands in the respective context. The summary of currently known binding partners for two multifunctional galectins lists glycoconjugates known for adhesion and cell signaling capacities (Table 3.3). From the wide array of cell surface glycoconjugates, a tissue lectin is capable to select few binding partners. For example, interaction with the extracellular matrix glycoproteins fibronectin and laminin can modulate adhesion (here a galectin acts as a non-integrin receptor), binding to CD7 or ganglioside GM1 initiates growth inhibition in activated T cells or in neuroblastoma cells (please see Chaps. 26 and 28 for details) [4]. This principle figures as decisive driving force not only for galectins but also for other lectins such as selectins (Table 3.2). As further example, intercellular contacts via the dendritic cell lectin DC-SIGN (dendritic cell-specific ICAM-3 grabbing non-integrin) similarly hinge on a cell-type-specific selection of the target glycoprotein. The dendritic cell engages the lectin to bind glycans of ICAM (intercellular adhesion molecule)-2 on endothelial cells, of ICAM-3 on T cells and of the α_M-subunit of the Mac-1 antigen (CR3 receptor, $\alpha_M\beta_2$ integrin) on neutrophils and of various pathogens with clustered Man or Lewis epitopes [11]. In addition to glycans, the two galectins, as noted above for the C-type lectin-like NK cell receptors or hyalectans/lecticans (please see legend to Fig. 3.1), also bind peptide motifs. This property facilitates intracellular functions in pre-mRNA splicing, placement of oncogenic ras or apoptosis regulation (Table 3.3) [4]. Overall, the details emerging teach the following salient lesson: despite the abundance of glycan chains on the cell surface animal and human lectins have a preference to particular binding partners in a distinct context.

In this sense, shifts in the assembly line for glycans and their substitution pattern can have an impact on affinity toward lectins. Not only the presence of a certain epitope counts but also its topological aspects, starting with the conformation of the lectin-binding determinant and the complete structure of the glycan chain including branching and core substitutions. These parameters have a bearing on lectin binding, although not physically involved in the molecular rendezvous [12]. It is therefore imperative to systematically profile factors with likely bearing on affinity regulation. If moving in a stepwise manner from the smallest interaction partner of a lectin, a monosaccharide, to the level of microdomains in membranes,

Table 3.3 Cellular glycoconjugates and proteins as ligands for endogenous lectins: case study of galectins-1 and -3

Type of ligand	Galectin-1	Galectin-3
Glycan	Ovarian carcinoma antigen CA125, CD2, CD3, CD4, CD7, CD43, CD45, carcinoembryonic antigen (CEA), fibronectin (tissue), gastrointestinal mucin, hsp90-like glycoprotein, α_1/α_4/α_5/$\alpha_7\beta_1$- and $\alpha_4\beta_7$-integrins, cell adhesion molecule L1, laminin, lamp-1, Mac-2-binding protein, neuropilin-1, thrombospondin, Thy-1, tPA, chondroitin sulfate proteoglycan, distinct neutral glycolipids, ganglioside GM_1	CD7, CD11b of CD11b/CD18 (Mac-1 antigen, CR3), CD13 (aminopeptidase N), CD32, CD43, CD45, CD66a,b, CD71, CD95, CD98, CEA, colon cancer mucin and MUC1-D (N-glycan at Asn36), cubilin, C4.4A (member of Ly6 family), epidermal growth factor receptor, haptoglobin β-subunit (after desialylation), hensin (DMBT-1), glycoform of IgE, β_1-integrin (CD29) and $\alpha_4\beta_1$-integrin, LI-cadherin, laminin, lamp-1/-2, Mac-2-binding protein, Mac-3, MAG, MP20 (tetraspanin), NG2 proteoglycan, TCR complex, tenascin, tPA, transforming growth factor-β receptor, ganglioside GM_1
Protein	Gemin4, oncogenic H-Ras, OCA-B, pre-B cell receptor (human, not murine system)	AGE products, Alix/AIP-1, axin, bcl-2, b-catenin, Cys/His-rich protein, Gemin4, mSufu, nucling, oncogenic K-Ras, OCA-B, pCIP, PIAS1, synexin (annexin VII), TTF-1

Table 3.4 Six levels of regulation of affinity of glycan binding to a lectin

1.	Mono- and disaccharides (including anomeric position)
2.	Oligosaccharides (including branching and substitutions)
3.	Shape of oligosaccharide (differential conformer selection)
4.	Spatial parameters of glycans in natural glycoconjugates
	a. shape of glycan chain (examples: modulation of conformation by substitutions not acting as lectinligand such as core fucosylation or introduction of bisecting GlcNAc in N-glycans, influence of protein part)
	b. cluster effect with bi- to pentaantennary N-glycans or branched O-glycans (including modulation by substitutions, please see a.)
5.	Cluster effect with different but neighboring glycan chains on the same glycoprotein (for example in mucins)
6.	Cluster effect with different glycoconjugates on the cell surface in spatial vicinity forming microdomains

six different layers with regulatory potential are then identified (Table 3.4). They give reason to imply an exquisitely tuned interplay between structural glycan tailoring and lectin expression, for example in fertilization (Chap. 25), malignancy (Chap. 26) or inflammation (Chap. 28).

3.4 Conclusions

The assumption that lectin-carbohydrate recognition is a common mode for intermolecular association and biological information transfer is convincingly backed by the large number of folds with lectin activity. Ensuing intrafamily diversification within more than a dozen folds with identified CRDs led to a toolbox of specialized effectors and a broad range of covered functions. Toward this end, covalent and non-covalent CRD clustering opened the way to optimize ligand selection and to distinguish glycosignatures. With lectins thus being also sensors of topological aspects, the affinity of the binding process can swiftly be regulated by altering the local density on the level of individual glycoconjugates and even cell surfaces. Picking functional ligands from the glycomic complexity thus includes context-dependent parameters, making cell specificity of lectin binding possible.

3.5 Summary Box

More than a dozen folds of animal and human proteins can accommodate a carbohydrate ligand. Binding entails a broad range of functions. Its affinity is regulated by sequence changes and topological aspects on the levels of the CRD and the glycoepitope. Respective changes on both sides can act in concert to optimize target specificity.

References

1. Roth S (1973) A molecular model for cell interactions. Quart Rev Biol 48:541–563
2. Gabius H-J (1997) Animal lectins. Eur J Biochem 243:543–576
3. Gabius H-J (2006) Cell surface glycans: the why and how of their functionality as biochemical signals in lectin-mediated information transfer. Crit Rev Immunol 26:43–80
4. Cooper DNW (2002) Galectinomics: finding themes in complexity. Biochim Biophys Acta 1572:209–231
5. Angata T, van der Brinkman-Linden ECM (2002) I-type lectins. Biochim Biophys Acta 1572:294–316
6. Loris R (2002) Principles of structures of animal and plant lectins. Biochim Biophys Acta 1572:198–208
7. Phan UT, Waldron TT, Springer TA (2006) Remodeling of the lectin-EGF-like domain interface in P- and L-selectin increases adhesiveness and shear resistance under hydrodynamic force. Nat Immunol 7:883–889
8. Yago T, Wu J, Wey CD, Klopocki AG, Zhu C, McEver RP (2004) Catch bonds govern adhesion through L-selectin at threshold shear. J Cell Biol 166:913–923
9. Dahms NM, Hancock MK (2002) P-type lectins. Biochim Biophys Acta 1572:317–340
10. Dommett RM, Klein N, Turner MW (2006) Mannose-binding lectin in innate immunity: past, present and future. Tissue Antigens 68:193–209
11. Cambi A, Koopman M, Figdor CG (2005) How C-type lectins detect pathogens. Cell Microbiol 7:481–488
12. André S, Kozár T, Schuberth R, Unverzagt C, Kojima S, Gabius H-J (2007) Substitutions in the N-glycan core as regulators of biorecognition: the case of core-fucose and bisecting GlcNAc moieties. Biochemistry 46:6984–6995

Author Biography

Hans-Joachim Gabius studied biochemistry in Hannover, Germany and received his PhD degree for work on proofreading mechanisms of phenylalanyl-tRNA synthetase in the lab of Prof. F. Cramer (Max-Planck-Institute for Experimental Medicine, Goettingen) in 1982. Work on tRNA splicing in Prof. J. Abelson's lab at UC San Diego was followed by starting research on lectins at the Max-Planck-Institute. Following a postdoc period with Prof. S. H. Barondes (UC San Diego/VA Medical Center) he was appointed as Assistant Professor in 1988 in Goettingen, then in 1991 as Associate Professor at the University of Marburg, followed by the chair position for Physiological Chemistry (Faculty of Veterinary Medicine) at the Ludwig-Maximilians-University Munich. His track record includes over 690 peer-reviewed publications, which have received over 17,000 citations at an h-factor of 71, and edition of five books, among them the textbook "The Sugar Code". Besides research awards he recently received an honorary degree in medicine by the Charles University in Prague (founded 1348) for his achievements in glycosciences.

Chapter 4
Lectins of Marine Origin and Their Clinical Applications

Yasuharu Watanabe, Takako Naganuma, Tomohisa Ogawa and Koji Muramoto

Abstract During the past several decades, intensive investigations have been conducted to clarify the biochemical and physiological properties of lectins from marine organisms, including cyanobacteria, algae, and invertebrates and fish. These investigations have revealed that lectins are highly diversified in terms of not only structural aspects but also functional aspects, including unique carbohydrate-binding specificities. Lectins are still being intensively investigated to understand their biological roles in cell recognition and biodefense as well as to employ them as valuable tools for studying complex carbohydrates in solution and on cell surfaces. Here, we review the structures and activities of lectins from marine organisms and their applications as carbohydrate recognition molecules and medicinal agents with antitumor and antiviral activities.

4.1 Introduction

Marine organisms have been extensively explored for the last several decades as potential sources of novel biologically active substances. In fact, various specific and potent bioactive molecules such as alkaloids, terpenoids, polyethers, polyketides, and peptides have been isolated from marine microorganisms, algae, and various invertebrates. These molecules have been found to possess various pharmacological functions as potent antitumor, antiviral, antimicrobial, and enzyme-inhibiting drugs. Extensive research has been conducted on relatively small molecules; however, macromolecules such as proteins, glycoproteins, and polysaccharides are also promising candidates for such functional agents because it is evident that they play critical roles in many biological processes. Among these, lectins are the most outstanding group because of their multiple functions.

Lectins are a group of carbohydrate-binding proteins, except for antibodies and enzymes that exhibit specific and reversible carbohydrate-binding activity.

Y. Watanabe · T. Naganuma · T. Ogawa · K. Muramoto (✉)
Graduate School of Life Sciences, Tohoku University, Sendai 980-8577, Japan
e-mail: muramoto@biochem.tohoku.ac.jp

E. F. Fang and T. B. Ng (eds.), *Antitumor Potential and Other Emerging Medicinal Properties of Natural Compounds*, DOI: 10.1007/978-94-007-6214-5_4,

They exist in all types of organisms, ranging from viruses to bacteria to plants and animals, and play important roles in diverse biological processes such as cell–cell adhesions, cell communications, signaling events, nutritional effects, cancer metastasis, microbial infection, and others as counterparts of carbohydrates. Investigations of their role in biological systems and their employment as invaluable tools for studying complex carbohydrates in solution and on cell surfaces are markedly contributing to advances in glycobiology.

During the past several decades, intensive investigations have been conducted to clarify the biochemical and physiological properties of lectins from marine organisms, including cyanobacteria, algae, invertebrates, and fish (Tables 4.1, 4.2 and 4.3) [1]. Based on the structural similarity of carbohydrate-recognition domains (CRDs) and their characteristics, lectins from plants and animals are classified into various classes and families, respectively [2]. Lectins from fish and marine invertebrates are mainly classified into C-type lectins (CTLs), galectins, F-type lectins, rhamnose-binding lectins (RBLs), and intelectin families. Here, we review lectins from marine organisms in terms of their structures, activities, and applications as tools and medicinal agents.

4.2 Lectin Family

4.2.1 CTL

CTLs comprise one of the major animal lectin families. Members of this family bind mono and oligosaccharides, mostly in a Ca^{2+}-dependent manner. They generally adopt multidomain structures and contain one or more highly conserved CRDs consisting of 115–130 amino acid residues. They are involved in a broad range of biological processes such as adhesion, endocytosis, and pathogen neutralization. The CTL domain (CTLD) superfamily is a large group of extracellular proteins with conserved CRD sequences but different functions and is now classified into many subgroups by Drickamer et al. [3]. For example, group III CTLs are known as collectins, which consist of an amino-terminal collagen domain and a carboxyl-terminal CTLD. They participate in host defense mechanisms through complement activation and include the serum mannose-binding protein (MBP). Many marine invertebrate lectins are classified into Group VII CTLs, which are soluble single CRD proteins.

CTLs are diversified and widely distributed in the animal kingdom, including marine invertebrates and fish. They have unique structural motifs and functional domains, regardless of whether or not they possess sugar-binding properties. The ability of CTLs to react with bacterial or viral pathogens has been reported in shellfish and fish species. A CTL (Ec-CTL) messenger RNA (mRNA) was abundant in the liver and skin of the orange-spotted grouper *Epinephelus coioides*,

Table 4.1 A part of diverse lectins from marine invertebrates

Organism	Lectin	Binding specificity	Lectin family	Remarks
[Protozoa]				
Sponge				
(*Aphrocallistes vastus*)	LECCI	Gal	C-type	Aggregation factor
(*Axinella polypoides*)	AP-I/II/III/V	Gal	Ricin-type	Formation of intercellular matrix
	AP-IV	Hexuronic acid		
(*Suberites domuncula*)	Suberites lectin	GlcNAc	6xβ-propeller/Tectonin	Antibacterial activity
(*Geodia cydonium*)	GCLT1	LacNAc	Galectin	Aggregation factor
[Cnidaria]				
Colonical hydroid				
(*Hydractinia symbiolongicarpus*)	Rsp1, 2		RBL (chimera)	Expressed in the mouth
Octocoral				
(*Sinularia lochmodes*)	SLL-2	Gal/GalNAc	Discoidin type	Mediate the symbiosis with microalgae
Jellyfish				
(*Nemopilema nomurai*)	NnL	GalNAc/NANA	No homology	Agglutinate bacteria
[Mollusc]				
Tridacna				
(*T. maxima*)	tridacin	GalNAc	C-type	Symbiotic factor
Abalone				
(*Haliotis laevigata*)	PLC	Gal/Man	C-type	Promote $CaCO_3$ precipitation
(*H. discus discus*)	perlucin		C-type	Promote $CaCO_3$ precipitation and direct calcite morphology
Penguin wing oyster				
(*Pteria penguin*)	PPL	Gal	RBL	Agglutinate *gram*-negative bacteria
Bay scallop				
(*Argopecten irradians*)	AiGal2	Gal/Lac	Galectin (4CRD)	Upregulated by bacteria, agglutinate bacteria
	AiCTL5-7	(Ca^+)	C-type	Upregulated by fungi/bacteria, agglutinate bacteria

(continued)

Table 4.1 (continued)

Organism	Lectin	Binding specificity	Lectin family	Remarks
Zhikong scallop				
(*Chlamys fareri*)	Cflec-2	Gal	C-type	Agglutinate bacteria
	Cflec-3	Man (Ca^{+})	C-type	Upregulated by bacteria
	Cflec-4	Man (Ca-)	C-type (4CRD)	Upregulated by bacteria
Eastern oyster				
(*Crassostrea virginica*) CrML		C-type		Upregulated by starvation or bacteria, capture particles
	CvGal	LacNAc	Galectin (4CRD)	Bind microbial pathogens, promote phagocytosis
Tropical clam				
(*Codakia orbicularis*)	codakine	Man (Ca^{+})	C-type	Symbiotically associated with sulfur oxidizing bacteria
Manila clam				
(*Venerupis philippinarum*)	VpSABL	Sialic acid	I-typ	Upregulated by bacteria
Razor clam				
(*Solen grandis*)	SgSABL-1,2	Sialic acid	I-type	Induced by bacterial glycan (LPS), contain C1q domain
Pearl oyster				
(*Pinctada martensii*)	PmF-lectin		F-type	Upregulated by bacteria
[Arthropod]				
Barnacle				
(*Megabalanus rosa*)	BRA-1-3	Gal	C-type	Inhibit $CaCO_3$ precipitation
Horseshoe crab				
(*Tachypleus tridentatus*)				
	TL-1	LPS (KDO), LTA	6xβ-propeller/Tectonin	Plasma lectin, agglutinate bacteria antibacterial activity
	TL-2	GlcNAc/GalNAc, LTA	5xβ-propeller	Plasma lectin, agglutinate a bacterium
	TL-3	LPS (O-antigen)	F-type	Plasma lectin
	TL-4	LPS(O-antigen), LTA		Plasma lectin, agglutinate human blood type A

(continued)

Table 4.1 (continued)

Organism	Lectin	Binding specificity	Lectin family	Remarks
	TL-5	Acetyl group	Ficolin	Agglutinate bacteria
American horseshoe crab				
(*Limulus polyphemus*)	Limulin	Sialic acid	I-type	Plasma lectin, mediator of hemolysis
Shrimp				
Penaeus monodon)	PmLec	LPS	C-type	Agglutinate bacteria, opsonic effect, enhance phagocytosis
(*Litopenaeus vannamei*)	LvLectin-1,2		C-type	Upregulated by bacteria
	LvCTLD		C-type (multidomain)	Opsonic effect
(*Fenneropenaeus chinensis*)	FcLec3	Muramic acid	C-type	Upregulated by bacteria/virus agglutinate bacteria
[Echinoderm]				
Sea cucumber				
(*Cucumaria echinata*)	CEL-I/CEL-II	GlcNAc (Ca^+)	C-type	Induce the secretion of TNF-α, granulocyte colony stimulating factor from macrophage cells
	CEL-IV	GlcNAc/Gal (Ca^+)	C-type	Plasma lectin as other CELs
	CEL-III	Gal (Ca^+)	Ricin type	Hemolytic and cytotoxic activity
(*Stichopus japonicus*)	SPL-1	Uronic acid (Ca^+)	C-type	Bind to coelomocytes
	SPL-2	GalNAc (Ca^+)	C-type	Bind to coelomocytes
Urchin				
Anthocidaris crassispina)	SUEL	Gal	RBL	Egg lectin
	Echinoidin (ECH)	Gal	C-type	
Starfish				
(*Asterina pectinifera*)		GalNAc (Ca^+)	C-type	

KDO 2-keto-3-deoxyoctonate; *LPS* lipopolysaccharide; *LTA* lipoteichoic acid; *Gal* D-galactose; *Man* D-mannose; *GalNAc* N-acetyl-D-galactosamine; *GlcNAc* N-acetyl-D-glucosamine; *Lac* lactose; *LacNAc* N-acetyl lactosamine; *NANA* N-acetylneuraminic acid; *Ca*$^+$ Ca^{2+} dependent; *Ca-* Ca^{2+} independent; *4CRD* four tandem-repeat CRDs

Table 4.2 A part of diverse lectins from marine chordate and protochordates

Organism	Lectin	Binding specificity	Lectin family	Remarks
[Chordate]				
Amphioxus				
(*Brachiostoma belcheri*)	AmphiCTL1	Peptidoglycan (Ca-)	C-type	Upregulated by bacteria/yeast, agglutinate bacteria, antimicrobial
	AmphiITLN	LPS (Ca^+)	Interlactin	Upregulated by bacteria, agglutinate bacteria
[Protochordate]				
Ascidian				
(*Polyandrocarpa misakiensis*)	TC14	Gal (Ca^+)	C-type	Antibacterial activity
(*Botryllus schlosseri*)	BsRBL-1-5 Rha		RBL (2CRD) Opsonic activity	
	P36	Gal	C-type	
(*Halocynthia roretzi*)		Gal	ficolin	Plasma lectin, enhance phagocytosis
		Gal	Intelectin	Plasma lectin, enhance phagocytosis
(*Styela plicata*)		Gal	C-type (collectin like)	Plasma lectin
(*Ciona intestinalis*)	CiLgals-a, b		Galectin (2CRD)	Upregulated by LPS

LPS lipopolysaccharide; *Gal* D-galactose; *Rha* L-rhamnose; *Ca*$^+$ Ca^{2+} dependent; *Ca-* Ca^{2+} independent; *2CRD* two tandem-repeat CRDs

Table 4.3 A part of diverse lectins from fish

Organism	Lectin	Binding specificity	Lectin family	Remarks
[Fish]				
Japanese eel (*Anguilla japonica*)	AJL-1	β-Galactoside	Galectin	Agglutinate bacteria
	AJL-2	Lac	C-type	
	eCL-1, 2	Lac	C-type	
	eFL	Fuc	F-type	Serum lectinLectin, upregulated by LPS
Conger eel (*Conger myriaster*)	Congerin I, II	Lac	Galectin	Skin mucus lectins, opsonic activity
	Congerin P	Lac/Man		Galectin Peritoneal fluid lectin
	ConCL-s	Man	C-type	Skin mucus lectin
Windowpane flounder (*Lophopsetta maculate*)		Gal	Galectin	
Orange-spotted grouper (*Epinephelus coioides*)	Ec-CTL	Gal	C-type	Upregulated by yeast/bacteria/virus
				agglutinate yeast/bacteria
Shishamo smelt (*Osmerus lanceolatus*)	OLABL	Gal/Rha (Ca-)	C-type (heterodimer)	Egg lectin
	OLL	Rha	RBL	Egg lectin
Steelhead trout (*Oncorhynchus mykiss*)	STL-1-3	Rha	RBL	Egg lectins, agglutinate bacteria
Chum salmon (*Oncorhynchus keta*)				
White-spotted charr (*Salvelinus leucomaenis*)	CSL-1-3	Rha	RBL	Egg lectins, agglutinate bacteria
Atlantic salmon (*Salmo salar*)	WCL-1-3	Rha	RBL	Egg lectins

(continued)

Table 4.3 (continued)

Organism	Lectin	Binding specificity	Lectin family	Remarks
Spanish mackerel (*Scomberomorus niphonius*)	SSL	Man (Ca^{+})	C-type	Serum lectin, opsonic activity
Ponyfish (*Leiognathus nuchalis*)	SML	Rha	RBL (2CRD)	Egg lectin
Striped bass (*Morone saxatilis*)	PFL-1,2	Rha	RBL (2CRD)	Skin mucus lectin
Sea bass (*Dicentrarchus labrax*)		Fuc	F-type	
Japanese sea perch (*Lateolabrax japonicus*)	DlFBL	Fuc	F-type (2CRD)	Opsonic activity
Rock bream (*Oplegnathus fasciatus*)	JspFL	Fuc	F-type	
Pufferfish (*Fugu rubripes*)	RbFTL-1, 2		F-type	Upregulated by bacteria/virus
Scorpionfish (*Scorpaena plumieri*)	Pufflectin	Man	Monocot B-lectin	Bind parasitic trematode
Far-East dace (*Tribolodon brandti*)	Plumieribetin	$\alpha 1\beta 1$ integrin	Monocot B-lectin	Integrin inhibition
[Fresh water fish]	TBL-1-3	Rha	RBL (2,3CRD)	Egg lectins
Rainbow trout (*Oncorhynchus mykiss*)			Galectin (2CRD)	
	RTInt	GlcNAc/Man	Intelectin	Induced by parasitic infection, bind bacteria/chtinous targets
	RTLL	Galactan/Man	C-type (ladder lectin-Lectin)	Serum lectin
Catfish (*Arius thalassinus*)	MBL-H2, 3		C-type (MBL like)	
		Gal	Galectin	

(continued)

Table 4.3 (continued)

Organism	Lectin	Binding specificity	Lectin family	Remarks
(*Silurus asotus*)	SAL	Rha/Gb3	RBL (2CRD)	Egg lectin
	saIntL	Man	Intelectin	
(*Ictalurus punctatus*)		High-mannose type		L-type
Sweet fish (ayu)				
(*Plecoglossus altivelis*)	SFL	Rha/Gb3	RBL	Egg lectinLectin, bind to micro-sporidia
Carp				
(*Cyprinus carpio*)	Ca-CTL	Man	C-type	Upregulated by LPS
	carpFEL	GlcNAc	6xβ-propeller/Tectonin	
(*Ctenopharyngodon idellus*)		Rha	RBL	Immunostimulatory activity
Zebrafish				
(*Danio rerio*)	Drgal1-L1-L3	LacNAc	Galectin	

LPS lipopolysaccharide; *LTA* lipoteichoic acid; *Gal* D-galactose; *Man* D-mannose; *Rha* L-rhamnose; *Fuc* L-fucose; *GlcNAc* N-acetyl-D-glucosamine; *Lac* lactose; *LacNAc* N-acetyl lactosamine; *Gb3* Galα1-4Galβ1-4Galβ1-Cer; *Ca+* Ca^{2+} dependent; *Ca-* Ca^{2+} independent; *2CRD* two tandem-repeat CRDs; *3CRD* three tandem-repeat CRDs and Ec-CTL expression was upregulated by challenge with different microbes, including the DNA virus Singapore grouper iridovirus (SGIV) [4]. Two CTLs, LvLectin-1 and-2, from shrimp (*Litopenaeus vannamei*) were highly expressed in the hepatopancreas and were upregulated by challenge with both *Listonella anguillarum* and white spot syndrome virus (WSSV) [5]. These results suggest that CTLs are related to defense mechanisms and involved in the initial response to invasion by bacterial and viral pathogens.

Typical mammalian CTLs such as MBP, macrophage mannose receptor, and selectins require Ca^{2+} for their activity. However, it has become evident that some CTLs do not require Ca^{2+} to function. For example, OLABL, a CTL from shishamo smelt (*Osmerus lanceolatus*) eggs, does not require Ca^{2+} to function [6]. In some CTLs, Ca^{2+} is required for agglutinating activity but not for sugar-binding activity. AmphiCTL1 from amphioxus requires Ca^{2+} for its hemagglutinating and microbial aggregation activities but not for its microbial-binding and growth suppression activities [7]. A CTL of conger eel (*Conger myriaster*), ConCL-s, binds to sugars in a Ca^{2+}-dependent manner; however, it shows Ca^{2+}-independent activity during yeast binding [8]. CTLs are sometimes classified into two groups according to binding preferences either to D-mannose and related sugars or to D-galactose and related sugars [3]. These two CTL groups have the motifs Asp–Pro–Asn (EPN) and Gln–Pro–Asp (QPD), respectively. However, the presence of several discrepancies in the combination of binding specificity and the motif reveals remarkable plasticity in sugar binding and divergent evolution of carbohydrate specificity.

Ladderlectin was named for its characteristic ladder-like electrophoretic pattern of equally spaced oligomers under nonreducing SDS-PAGE conditions with a 32 kDa disulfide-linked dimer. Rainbow trout ladderlectin (RTLL) was identified by its Ca^{2+}- and mannose-dependent affinity for pathogen-associated molecular patterns (PAMPs) such as lipopolysaccharide (LPS) isolated from *Aeromonas salmonicida* and galactan-based polymers [9]. The sequences of ladderlectins were identified as those of group VII mannose-binding CTLs.

4.2.2 Galectin Family

Galectins are a family of carbohydrate-binding proteins characterized by their Ca^{2+}-independent affinity for β-galactoside, sharing of a conserved sequence motif within their CRDs of approximately 130 amino acid residues, lack of a signal peptide and attached carbohydrate, and predominant cytoplasmic location. They have been found in various eukaryotic organisms, ranging from fungi to mammals [10]. Based on the structural features, galectins can be classified into three types: prototype (monomer or homodimer of single carbohydrate-binding domain), tandem-repeat type (two carbohydrate-binding domains on a single chain), and chimera type (a carbohydrate-binding domain and an extra N-terminal domain on a single chain). Galectins reportedly participate in diverse physiological phenomena such as development, differentiation, morphogenesis, immunity, apoptosis, metastasis of malignant cells, and others.

Two prototype galectins, congerins I (Con I) and II (Con II), were isolated from the conger eel (*Conger myriaster*) skin mucus. As components of biological defense systems, these proteins mainly exist in the frontier organs and tissues that delineate the body from the outer environment, e.g., the epidermal club cells of the skin, wall of the oral cavity, pharynx, esophagus, and gills, and they agglutinate marine pathogenic bacteria such as *Vibrio anguillarum* and exibit opsonic and cytotoxic activities against cells [11]. Con I and Con II form a homodimer that exhibits divalent cross-linking activity. Molecular evolutionary and structural studies have suggested that Con I and Con II have evolved by accelerated amino acid substitutions under positive selection to acquire their diversified properties such as carbohydrate-binding specificity and thermostability [12]. The structure of Con I demonstrated protein-fold evolution by swapping β-strands between subunits, which altered the β-sheet topology at the dimer interface from entirely antiparallel to partially parallel for entangling the two subunits, although Con I and Con II adopt similar subunit structures with the "jellyroll" motif consisting of five-stranded and six-stranded β-sheets [13]. An important feature of strand swapping is an increase in the stability of the quaternary structure, which is essential for divalent cross-linking activity required for agglutination of pathogenic bacteria. In contrast, crystal structure analysis of Con II revealed extension of the carbohydrate-binding site at the nonreducing end of lactose. Furthermore, a new congerin isotype, congerin P (Con-P), was identified from the peritoneal cells of conger eel [14]. Con-P displayed substitution of seven out eight amino acid residues in a CRD, which are conserved in all other known galectins, and allosteric carbohydrate-binding ability by mannoside.

Local inoculation of LPS through the ascidian body wall stimulates a prompt local inflammatory reaction that includes upregulation of collagen-IX-like, C3a, TNF-α-like, and MBL-like genes, together with enhanced phenoloxidase activity. Tandem-repeat type galectins (CiLgals-a and CiLgals-b) were promptly upregulated in *Ciona intestinalis* [15]. Furthermore, a novel galectin-related protein, CvGal, which contains four canonical galectin CRDs, has been discovered in hemocytes of the eastern oyster *Crassostrea virginica* [16]. CvGal recognizes endogenous and exogenous ligands including bacteria, algae, and the protozoan parasite *Perkinsus marinus* as soluble opsonins for pathogens or as hemocyte surface receptors for both microbial pathogens and algal food ingested into digestive ducts and as modulators for upregulation of CvGal itself. Unique gene and protein domain architecture of the galectin CRD nematogalectin was also found in the freshwater hydrozoan *Hydra* and the marine hydrozoan *Clytia* [17]. Nematogalectin, a 28-kDa protein with an N-terminal GlyXY domain that can form a collagen triple helix followed by galectin CRD, is a major component of the nematocyst tubule, and it is transcribed by nematocyte-specific alternative splicing. Thus, the galectin family proteins also diversified by unique evolutionary processes, including tandem duplication and accelerated evolution.

4.2.3 F-Type Lectin Family

F-type lectins (fucolectins), which bind L-fucose and share characteristic sequence motifs, have been described with members in invertebrates and vertebrates such as horseshoe crab (*Tachypleus tridentatus*) and Japanese eel (*Anguilla japonica*) as immunorecognition molecules. The F-type CRD has also been associated with pentraxin, CTL, or "sushi" domains, yielding complex chimeric proteins. The crystal structures of single CRD and tandem CRDs of F-type lectins with a jellyroll β-barrel topology were reported for *A. japonica* agglutinin (AAA) and MsaFBP32 from striped bass (*Morone saxatilis*), respectively. In contrast, F-type lectin CRD motifs are absent in genomes of higher vertebrates such as reptiles, birds, and mammals. Two distinct F-type lectins, RbFTL-1 and RbFTL-2, have been identified in rock bream (*Oplegnathus fasciatus*) [18] and are predominantly expressed in the head kidney and the liver, respectively. Their expression is upregulated by some bacterial and viral challenges. Importantly, sea bass (*Dicentrarchus labrax*) fucolectin (DlFBL) consisting of two tandemly arranged CRDs was found to enhance the phagocytosis of formalin-killed *Escherichia coli* by *D. labrax* peritoneal macrophages [19].

4.2.4 RBL Family

RBLs are a family of animal lectins that show specific binding activities to L-rhamnose or D-galactose and have been predominantly isolated from eggs and ovary cells of fish and invertebrates. Till date, the RBL family has been found in the oocytes, ovaries, and skin mucus of over 20 species of fish (Table 4.3). RBLs have been also found in the mantle of penguin wing oyster and ascidians. Except for the reproductive cells, including oocytes and eggs, RBLs are mainly located in the tissues and organs related to the immune system, e.g., mucous cells of the gill, goblet cells of the intestine, the spleen, thrombocytes, lymphocytes, monocytes, and neutrophils. When the genomic database was retrieved, genes containing the distinctive structural motif of RBL CRDs were found to be broadly distributed in almost all animals, including invertebrates, chordates, and vertebrates, and also in bacteria and plants.

Most RBLs are composed of two or three tandem-repeat CRDs, which consist of approximately 95 amino acid residues, and share the conserved topology of four disulfide bonds. Three RBLs, CSL1, CSL2, and CSL3, composed of four, 18, and two subunits by noncovalent binding, respectively, have been isolated from chum salmon (*Oncorhynchus keta*) eggs [20]. CSLs have high binding specificity for globotriaosyl ceramide (Gb3; Galα1-4Galβ1-4Galβ1-Cer, also known as CD77), which is located in lipid rafts, is upregulated through immune responses, and is also known as the functional receptor for various toxins such as Shiga toxin (Stx). The homodimer of CSL3 has a kinked dumbbell shape, in which two lobes are connected through linkers composed of two five-residue peptides (-QQQET-) [21]. Each lobe seems to be a single globular protein with a pseudo two-fold axis and includes two antiparallel β-sheets with two (β2 and β4) and three (β1, β3, and $\beta5$) strands and three helices (α1-3). It was found that the monosaccharide (rhamnose) or nonreducing end residues (Gal1 of melibiose and Gb3) share the same conserved primary binding sites in CSL3 (Glu7/107, Tyr27/127, Lys86/186, and Gly83/183). In addition, Asn74/174 and Asp79/179 are used in the primary site of CSL3. The complex structures of melibiose and Gb3 reveal the oligosaccharide recognition mechanism of RBL. Arg39/139 and Gln43/143 residues of CSL3, which bind to carbohydrates in the 2nd and 3rd sites, are the key residues in determining specificity for the oligosaccharides Gb3. The total numbers of hydrogen bonds between CSL3 and rhamnose, melibiose, and Gb3 are 7, 8, and 10, respectively, which are consistent with the observed high affinity ($Kd = 2.6 \times 10^{-5}$ M) of CSL3 to Gb3.

RBLs have been proposed to be involved in biodefense systems. CSLs were found to induce proinflammatory cytokines, including IL-1β1, IL-1β2, TNF-α1, TNF-α2, and IL-8, by recognizing Gb3 on the surface of a peritoneal macrophage cell line (RTM5) from rainbow trout and an established fibroblast-like cell line (RTG-2) from gonadal tissues of fish [22]. The receptor of RBL from amago (*O. rhodurus*) was expressed on the peritoneal macrophage after inflammatory stimulation, and RBL from grass carp (*Ctenopharyngodon idellus*) roe induced a dose-dependent increase in phagocytic activity of seabream macrophages [23].

These RBL CRDs are diversified and widely distributed in functional proteins as unique structural motifs. Rsps, composed of an N-terminal serine-rich domain, eight tandem-repeated thrombospondin type 1 repeats (TSRs), and a C-terminal RBL CRD, have been identified as immune recognition molecules in the colonial hydroid *Hydractinia symbiolongicarpus* [24].

4.2.5 Intelectin

Intelectins, or X-lectins, are a recently identified group of animal lectins possessing a fibrinogen-like motif similar to that of members of the ficolin/opsonin/p35 lectin family and displaying Ca^{2+}-dependant carbohydrate binding in the absence of CTLD. Membrane-bound, intracellular and soluble intelectin-like complementary DNA (cDNA) sequences and proteins with high predicted amino acid sequence homology have been reported from various species, including mammals, frogs, fish, and ascidians. In fish, the potential role of intelectin-like proteins as innate immune molecules is fast being realized. In rainbow trout, intelectin-like cDNA sequences and other putative acute phase genes were upregulated in the liver tissue following injection of an emulsified, killed, gram-negative bacterium (*L. anguillarum*) [9].

4.2.6 Novel Lectin Family

L-type lectins, which have a leguminous lectin domain and can bind to high-mannose type oligosaccharides, have been hypothesized to play roles in infectious responses and phagocytosis in mammalian systems. In the secretory pathway, L-type lectins play crucial roles in selective protein trafficking, sorting, and targeting. Three L-type lectins, namely a 53 kDa endoplasmic reticulum (ER)–Golgi intermediate compartment protein (ERGIC-53), a vesicular integral protein of 36 kDa (VIP36), and a VIP36-like protein, were cloned in the channel catfish *Ictalurus punctatus* [25]. These proteins were abundantly expressed in the liver, head kidney, and brain, respectively. Upon infection by the gram-negative intracellular bacterium *Edwardsiella ictaluri*, expression of all the three genes was upregulated in the head kidney with distinct patterns. Catfish L-type lectins have 50–70 % identity with other vertebrate L-type lectin protein sequences such as those of rats, humans, birds and frogs.

Most known animal lectins have been found to internally; however, some are also found in the skin mucus of several animal species, in particular, fish. It is generally believed that the skin mucus serves as a mechanical and biochemical barrier. Four different families of lectin (CTLs, galectins, RBLs, and monocot mannose-binding B-type lectins) have been found in the skin mucus of fish. With respect to skin mucus galectins, congerins and AJL-1 can agglutinate various bacterial strains. Such lectin activity suggests that they agglutinate and trap pathological bacteria and restrict their growth until such organisms are washed away with old mucus into the water of the surrounding environment [26]. A mannose-binding B-type lectin, pufflectin, binds to the surface of the parasitic trematode *Heterobothrium okamotoi*, indicating its role in host defense mechanisms. Recently, a novel mannose-binding lectin, FHL, was purified from the skin mucus of flathead (*Platycephalus indicus*). FHL is structurally similar to the heavy-chain domain of mammalian plasma kallikrein, which has not yet been identified in any teleost species [27].

4.3 Bioactivity of Lectins

4.3.1 Antimicrobial Activity

Many studies have shown that marine lectins bind and agglutinate fungi, bacteria, blue-green algae, diatoms, dinoflagellates, and other unicellular algae, thereby regulating growth and movement. These lectins act in a similar manner to eliminate invading pathogenic bacteria. Red algal lectins from *Eucheuma serra* (ESA) and *Galaxaura marginata* (GMA) strongly inhibited the fish pathogen *Vibrio vulnificus* but were inactive against *V. Pelagius* and *V. neresis* [28]. A CTL from amphioxus directly killed *Staphylococcus aureus* and *Saccharomyces cerevisiae* in a Ca^{2+}-independent manner by binding to their cell wall polysaccharides such as peptidoglycan and glucan [7].

RBLs from fish eggs such as STLs and CSLs also interacted and agglutinated gram-negative and gram-positive bacteria by recognizing cell-surface LPS and lipoteichoic acid, respectively [29]. These RBLs recognize the O-antigen, which is the immunodominant structure exposed to the environment that is highly variable among bacterial strains, by diverse carbohydrate recognition ability. In addition, RBLs bind to glycolipids and glycoproteins of microsporidian fish pathogens, and the RBL receptor was found to be expressed on peritoneal macrophages of fish after inflammatory stimulation [30]. More recently, it was found that CSLs induced the production of radical oxygen species in the rainbow trout macrophage RTM5 cells in a dose-dependent manner.

4.3.2 Antitumor Activity

Lectins function as surface markers for tumor cell recognition, cell adhesion, signal transduction across the membrane, mitogenic cytotoxicity, and apoptosis. In addition, they modulate the growth, proliferation, and apoptosis of premalignant and malignant cells in vitro and in vivo. Therefore, lectins have great potential in the treatment, prevention, and diagnosis of chronic diseases such as cancer [31]. As early as 1983, acorn barnacle (*Megabalanus rosa*) lectins (BRAs) were shown to have antitumor activity in vivo [32]. Mouse mammary tumor MM46 cells were inoculated into C3H/He mice after in vitro incubation with or without BRAs. The survival rate of the lectin-treated group increased significantly, and these mice became resistant to subsequent challenge with fresh tumor cells. This activity was attributed to lectin-dependent macrophage-mediated cytolysis of tumor cells.

Apoptotic cells are characterized on the basis of biological and morphological features such as externalization of phosphatidylserine (PS), activation of caspase, fragmentation of chromosomal DNA, cell shrinkage, plasma membrane blebbing, and nuclear condensation. RBL from catfish (*Silurus asotus*) eggs (SAL) bound Gb3 in the glycosphingolipid-enriched microdomain of Burkitt's lymphoma Raji cells and induced externalization of PS, followed by cell shrinkage [33]. CSLs showed cytotoxicity against Gb3-displaying human colon cancer cell lines, Caco-2 and Lovo cells, by an apoptotic pathway through the recognition of Gb3 on cell surfaces in a dose-dependent manner. However, this activity was not observed in DLD-1 and HCT-15 human colonic tumor cell lines lacking Gb3 [21].

Delivery of targeted therapeutics by drug delivery systems (DDSs) to specific sites provides numerous advantages over traditional nontargeted therapeutics [34]. Targeted drug delivery increases the efficacy of treatment by enhancing drug exposure to targeted sites while limiting side effects of drugs on normal and healthy tissues. Because human and experimental tumors display increased levels of *N*-linked-1,6-GlcNAc oligosaccharides, *N*-glycan may be a possible moiety for lectin targeting anticancer DDS. Cytotoxicity of lectin-based DDS may be exploited by two mechanisms. One mechanism would involve a nontoxic lectin conjugated to a drug, which would become toxic upon activation within the target cell. The second mechanism would involve a toxic lectin that would function as a homing moiety and toxic agent by apoptosis induction.

4.3.3 Antiviral Activity

Galactose-specific lectins isolated from the marine sponge *Chondrilla nucula* and from the sea worm *Chaetopterus variopedatus* showed antiviral activity in vitro by stimulating (22–52) oligoadenylate metabolism and by inhibiting cytopathic effects induced by human immunodeficiency virus (HIV)-1 and the production of viral p24 antigen, respectively. This anti-HIV action may be associated with the blockade of viral entry into cells. A GlcNAc-specific lectin isolated from the sea worm *Serpula vermicularis* (Annelida) inhibited the production of viral p24 antigen and cytopathic effect induced by HIV-1 at the half-maximal effective concentrations (EC_{50}) of 0.23 and 0.15 μg/mL, respectively [35].

Although the anti-HIV activity of mannose-specific plant lectins, including concanavalin A and *Galanthus nivalis* agglutinin (GNA), which target the "glycan shield" of surface envelope proteins of HIV to inhibit membrane fusion and infection, have been reported, these lectins generally show a low micromolar range of EC_{50} values in whole-cell anti-HIV assays. In contrast, high-mannose-binding cyanobacterial (blue-green algal) or eukaryotic algal lectins are very promising as antiviral agents because they inhibit HIV replication with EC_{50} values in the low nanomolar to picomolar range. The red algal lectin griffithsin (GRFT) from *Griffithsia* sp. is the most potent inhibitor of HIV multiplication, and its activity is displayed at picomolar concentrations [36]. A lectin from the green alga *Boodlea coacta* (BCA) consists of three internal tandem-repeated domains with sequence motifs similar to the carbohydrate-binding site of GNA-related lectins [37]. BCA inhibited HIV-1 entry into host cells at EC_{50} of 8.2 nM and showed potent anti-influenza activity against various strains, including a clinical isolate of pandemic H1N1-2009 virus, by directly binding to viral envelope hemagglutinin.

Many high-mannose-binding lectins that exhibit anti-HIV activity have been isolated from cyanobacteria by Boyd and other researchers, e.g., CV-N from *Nostoc ellipsosporum*, scytovirin from *Scytonema varium*, MVL from *Microcystis viridis*, and OAA from *Oscillatoria agardhii*. These prokaryotic lectins share common structural features and internal multiplication of amino acid sequences, but with no sequence homology. Three-dimensional structures of these lectins commonly exhibit characteristic domain swapping.

Cyanovirin-N (CV-N) inhibits HIV infection by binding to high-mannose oligosaccharides on the surface of the viral envelope glycoprotein gp120 [38]. CV-N is also active against a number of other enveloped viruses, including influenza, Ebola, hepatitis C, and herpes virus. Each CV-N monomer contains two symmetrically related structural domains (A and B), with each domain containing a carbohydrate-binding site that specifically interacts with α(1–2)-linked oligomannose moieties of Man-8 and Man-9 glycans. The two binding sites in monomeric CV-N exhibit distinct affinities for carbohydrates in solution: the binding site in domain B is located distant from the N and C termini with a K_D of approximately 140 nM for Manα1–2Man disaccharide, whereas the binding site in domain A is located near the termini with a K_D of approximately 1.5 μM. Both sites are required for viral neutralization [39].

The interest in developing CV-N as a medicinal agent has led to investigations of its large-scale production using bacterial, yeast, and plant expression systems. Stabilization of CV-N has implications for its recombinant production and application as a medicinal agent. Compared with wild-type CV-N, a tandem repeat of two CV-N molecules (CVN_2), linked together in a head-to-tail fashion, increased HIV-1 neutralization activity by up to 18-fold [39]. The stabilized domain-swapped dimer showed extensive cross-clade reactivity and was more potent than broadly neutralizing anti-HIV antibodies. CV-N variants were rationally designed to be stabilized against thermal and chemical denaturation by substituting the buried polar side chains with aliphatic groups, while maintaining high affinity and selectivity for glycan targets [40]. The removal of the buried polar groups improved packing within the protein core to yield increased stability.

4.3.4 Biomineralization

Calcified hard tissues and skeletons such as shells and pearls provide structural support and protection for many marine invertebrate phyla, including Mollusca, Crustacea, and Echinodermata. A calcified shell layer is composed of two polymorphs of calcium carbonate, aragonite or calcite, and an organic matrix. This organic matrix associated with shells is thought to function in shell formation (biomineralization) by guiding the nucleation and growth of calcium carbonate crystals under ambient conditions and to associate the aragonite tablet with organic matrix in its peculiar structure. CTL-like proteins have been isolated from calcium carbonate biominerals as hard tissue-associated proteins of Mollusca, Crustacea, and Echinodermata as well as vertebrates. These proteins are postulated to interact with specific faces of the calcite crystal when occluded within the mineral and control spicule growth. A 17 kDa protein, perlucin, was isolated from the shell of the mollusc, *Haliotis laevigata* (abalone) by demineralization [41]. Perlucin, consisting of a single CTL motif, accelerates the precipitation of calcium carbonate from a saturated solution, indicating that perlucin may promote the nucleation and/or growth of calcium carbonate crystals and may combine the chitin and aragonite layers. In contrast to molluscan shells, few matrix proteins have been isolated and characterized from crustacean shells. Among these, the CTLs BRA-1–3, isolated from the acorn barnacle *Megabalanus rosa*, inhibited the nucleation and growth of calcium carbonate crystals at concentrations of 4–26 μg/mL [42]. Although aragonite and calcite crystals were formed in the presence of lower concentrations of BRA, the shape and size of the crystals were changed. These results further indicate the participation of lectins in biomineralization. Although it is generally accepted that invertebrate lectins play roles in biodefense mechanisms by recognizing the carbohydrate structures on pathogens, the functions of BRA in biomineralization as well as biodefense have been proposed [43]. CTLs may be applicable to artificial crystallization in the field of nanotechnology.

Furthermore, fish-specific CTLD proteins have been identified as type II antifreeze proteins, which inhibit freezing by binding to ice in plasma of coldwater-living fish species such as herring (*Clupea harengus*), rainbow smelt (*Osmerus mordax*), Japanese smelt (*Hypomesus nipponensis*), sea raven (*Hemitripterus americanus*), and longsnout poacher (*Brachyopsis rostratus*).

4.4 Conclusion and Perspectives

Numerous lectins have been isolated from various marine organisms and have been revealed to be highly diversified in terms of not only structure but also function, including specific and unique carbohydrate-binding properties. The interactions between lectins and monosaccharides are relatively weak, with poor specificity and dissociation constants, usually in the millimolar range. However, lectins frequently exhibit both high affinity and exquisite specificity for oligosaccharide structures of glycoproteins and glycolipids. The high affinity and specificity of protein-carbohydrate interactions may attribute to multivalent binding that produce the cluster effect. Lectins are now intensively investigated to understand their biological roles in cell recognition or biodefense as well as to employ them as invaluable tools for the study of complex carbohydrates in solution and on cell surfaces.

Soluble lectins are regarded as pattern recognition receptors (PRRs) capable of binding spatially organized microbial surface polysaccharides and pathogen-associated molecular patterns (PAMPs) such as LPS and lipoteichoic acid on gram-negative and gram-positive bacteria, respectively. Mannose-binding lectins and ficolins are two well-characterized lectins, which bind PAMPs on various microbial targets and mediate resistance to pathogens through neutralization, agglutination, and opsonization. Other lectin families, including the RBL family, are also likely to act as PRRs. The observed structural and functional diversity of lectins is consistent with their function as PRRs in innate immunity. The emergence of multiple isoforms of lectins by various means, resulting in diversification of ligand affinities, may provide an arsenal of proteins to a wide range of organisms and may compensate for the lack of more highly developed and efficient adaptive immune responses.

The enormous developments that could be made possible by inhibiting, activating, and exploiting protein–carbohydrate interactions, in particular, in medicinal chemistry, push the study of lectins and carbohydrates to the forefront of research. In fact, the cellular protein glycosylation pattern is influenced by several physiological changes such as disease occurrence. Thus, the altered glycoform population of a given glycoprotein may be diagnostic of the disease responsible for the alteration itself. Abnormal glycosylation has been detected in cancer development, e.g., changes in prostate-specific antigen (PSA) glycosylation during malignant transformation. In addition, RBL family lectins are useful for diagnosis of pathologies associated with Gb3 ceramide, e.g., highly malignant Burkitt's lymphoma. Further research, including research on clinical trials, molecular mechanisms of action, and structure–function relationships, should help researchers in examining and elucidating the therapeutic effects, nutritional benefits, and toxic consequences of lectins. The isolation and characterization of novel lectins from unexplored marine organisms will greatly contribute to this research.

Acknowledgments This work was supported by Grants-in-Aid for Scientific Research (23380121) from the Ministry of Education, Science, Sports, and Culture of Japan.

References

1. Ogawa T, Watanabe M, Naganuma T, Muramoto K (2011) Diversified carbohydrate-binding lectins from marine resources. J Amino Acids. doi:10.4061/2011/838914
2. Sharon N, Lis H (2003) Lectins. Kluwer Academic Publishers, Dordrecht
3. Zelensky AN, Gready JE (2005) The C-type lectin-like domain superfamily. FEBS J 272:6179–6217
4. Wei J, Xu D, Zhou J, Cui H, Yan Y, Ouyang Z, Gong J, Huang Y, Huang X, Qin Q (2010) Molecular cloning, characterization and expression analysis of a C-type lectin (Ec-CTL) in orange-spotted grouper Epinephelus coioides. Fish Shellfish Immunol 28:178–186
5. Wei X, Liu X, Yang J, Fang J, Qiao H, Zhang Y, Yang J (2012) Two C-type lectins from shrimp *Litopeneaeus vannamei* that might be involved in immune response against bacteria and virus. Fish Shellfish Immunol 32:132–140
6. Hosono M, Sugawara S, Ogawa Y, Kohno T, Takayanagi M, Nitta K (2005) Purification, characterization, cDNA cloning, and expression of asialofetuin-binding C-type lectin from eggs of shishamo smelt (Osmerus lanceolatus). Biochim Biophys Acta 1725:160–173
7. Yu Y, Yu Y, Huang H, Feng K, Pan M, Yuan S, Huang S, Wu T, Guo L, Dong M, Chen S, Xu A (2007) A short-form C-type lectin from amphioxus acts as a direct microbial killing protein via interaction with peptidoglycan and glucan. J Immunol 179:8425–8434
8. Tsutsui S, Iwamoto K, Nakamura O, Watanabe T (2007) Yeast-binding C-type lectin with opsonic activity from conger eel (*Conger myriaster*) skin mucus. Mol Immunol 44:691–702
9. Russell S, Hayes MA, Lumsden JS (2009) Immunohistochemical localization of rainbow trout ladderlectin and intelectin in healthy and infected rainbow trout (*Oncorhynchus mykiss*). Fish Shellfish Immunol 26:154–163
10. Vasta GR (2009) Roles of galectins in infection. Nat Rev Microbiol 7:424–438
11. Nakamura O, Matsuoka H, Ogawa O, Muramoto K, Kamiya H, Watanabe T (2006) Opsonic effect of congerin, a mucosal galectin of the Japanese conger, *Conger myriaster* (Brevoort). Fish Shellfish Immunol 20:433–435
12. Konno A, Yonemaru S, Kitagawa A, Muramoto K, Shirai T, Ogawa T (2010) Protein engineering of conger eel galectins by tracing of molecular evolution using probable ancestral mutants. BMC Evol Biol 10. article 43. doi:10.1186/1471-2148-10-43
13. Shirai T, Shionyu-Mitsuyama C, Ogawa T, Muramoto K (2006) Structure based studies of the adaptive diversification process of congerins. Mol Div 10:567–573
14. Watanabe M, Nakamura O, Muramoto K, Ogawa T (2012) Allosteric regulation of the carbohydrate-binding ability of a novel conger eel galectin by D-mannoside. J Biol Chem 287:31061–31072
15. Vizzini A, Parrinello D, Sanfratello MA, Salerno G, Cammarata M, Parrinello N (2012) Inducible galectins are expressed in the inflamed pharynx of the ascidian *Ciona intestinalis*. Fish Shellfish Immunol 32:101–109
16. Tasumi S, Vasta GR (2007) A galectin of unique domain organization from hemocytes of the eastern oyster (*Crassostrea virginica*) is a receptor for the protistan parasite *Perkinsus marinus*. J Immunol 179:3086–3098
17. Hwang JS, Takaku Y, Momose T, Adamcyk P, Ozbek S, Ikeo K, Khaturin K, Hemmrich G, Bosch TCG, Holstein TW, David CN, Gojobori T (2010) Nematogalectin, a nematocyst protein with

GlyXY and galectin domains, demonstrates nematocyte-specific alternative splicing in Hydra. Proc Nat Aca Sci U S A 107:18539–18544

18. Park HJ, Kim JW, Kim EG, Kim HN, Chae YS, Jeong JM, Kim DH, Park CI (2012) Molecular cloning and expression analysis of two distinct F-type lectins from the rock bream *Oplegnathus fasciatus*. Dev Comp immunol 36:230–235
19. Salerno G, Parisi MG, Parrinello D, Benenati G, Vizzini A, Vazzana M, Vasta GR, Cammarata M (2009) F-type lectin from the sea bass (*Dicentrarchus labrax*): purification, cDNA cloning, tissue expression and localization, and opsonic activity. Fish Shellfish Immunol 27:143–153
20. Shiina N, Tateno H, Ogawa T, Muramoto K, Saneyoshi M, Kamiya H (2002) Isolation and characterization of L-rhamnose-binding lectins from chum salmon (*Oncorhynchus keta*) eggs. Fish Sci 68:1352–1366
21. Shirai T, Watanabe Y, Lee MS, Ogawa T, Muramoto K (2009) Structure of rhamnose-binding lectin CSL3: unique pseudo-tetrameric architecture of a pattern recognition protein. J Mol Biol 391:390–403
22. Watanabe Y, Tateno H, Nakamura-Tsuruta S, Kominami J, Hirabayashi J, Nakamura O, Watanabe T, Kamiya H, Naganuma T, Ogawa T, Naudé RJ, Muramoto K (2009) The function of rhamnose-binding lectin in innate immunity by restricted binding to Gb3. Dev Comp Immunol 33:187–197
23. Ng TB, Lam YW, Woo NYS (2003) The immunostimulatory activity and stability of grass carp (*Ctenopharyngodon idellus*) roe lectin. Vet Immunol Immunopathol 94:105–112
24. Schwarz RS, Hodes-Villamar L, Fitzpatrick KA, Fain MG, Hughes AL, Cadavid LF (2007) A gene family of putative immune recognition molecules in the hydroid Hydractinia. Immunogenet 59:233–246
25. Zhang H, Peatman E, Liu H, Feng H, Chen L, Liu Z (2012) Molecular characterization of three L-type lectin genes from channel catfish, *Ictalurus punctatus* and their responses to Edwardsiella ictaluri challenge. Fish Shellfish Immunol 32:598–608
26. Suzuki Y, Tasumi S, Tsutsui S, Okamoto M, Suetake H (2003) Molecular diversity of skin mucus lectins in fish. Comp Biochem Physiol Pt B 136:723–730
27. Tsutsui S, Okamoto M, Ono M, Suetake H, Kikuchi K, Nakamura O, Suzuki Y, Watanabe T (2011) A new type of lectin discovered in a fish, flathead (*Platycephalus indicus*), suggests an alternative functional role for mammalian plasma kallilrein. Glycobiol 21:1580–1587
28. Hori K, Sato Y, Ito K, Fujiwara Y, Iwamoto Y, Makino H, Kawakubo A (2007) Strict specificity for high-mannose type *N*-glycan and primary structure of a red alga *Eucheuma serra* lectin. Glycobiol 17:479–491
29. Tateno H, Ogawa T, Muramoto K, Kamiya H, Saneyoshi M (2002) Rhamnose-binding lectins from steelhead trout (*Oncorhynchus mykiss*) eggs recognize bacterial lipopolysaccharides and lipoteichoic acid. Biosci Biotechnol Biochem 66:604–612
30. Watanabe Y, Shiina N, Shinozaki F, Yokoyama H, Kominami J, Nakamura-Tsuruta S, Hirabayashi J, Sugahara K, Kamiya H, Matsubara H, Ogawa T, Muramoto K (2008) Isolation and characterization of L-rhamnose-binding lectin, which binds to microsporidian *Glugea plecoglossi*, from ayu (*Plecoglossus altivelis*) eggs. Dev Comp Immunol 32:487–499
31. Ghazarian H, Idoni B, Oppenheimer SB (2011) A glycobiology review: carbohydrates, lectins and implications in cancer therapeutics. Acta Histochem 113:236–247
32. Yamazaki M, Esumi-Kurisu M, Mizuno D, Ogata K, Kamiya H (1983) Marine animal lectin-dependent tumor recognition by macrophages. Gann 74:405–411
33. Kawano T, Sugawara S, Hosono M, Tatsuta T, Ogawa Y, Fujimura T, Taka H, Murayama K, Nitta K (2009) Globotriaosylceramido-expressing Burkitt's lymphoma cells are committed to early apoptotic status by rhamnose-binding lectin from catfish eggs. Biol Pharm Bull 32:345–353
34. Bies C, Lehr CM, Woodley JF (2004) Lectin-mediated drug targeting: history and applications. Adv Drug Deliv Rev 56:425–435
35. Molchanova V, Chikalovets I, Chernikov O, Belogortseva N, Li W, Wang JH, Yang DYO, Zheng YT, Lukyanov P (2007) A new lectin from the sea worm *Serpula vermicularis*:isolation, characterization and anti-HIV activity. Comp Biochem Physiol Pt C 145:184–193

36. Mori T, O'Keefe BR, Sowder BC, Bringans S, Gardella R, Berg S, Cochran P, Turpin JA, Buckheit RW, McMahon JB, Boyd MR (2005) Isolation and characterization of griffithsin, a novel HIV-inactivating protein, from the red alga *Griffithsia* sp. J Biol Chem 280:9345–9353
37. Sato Y, Hirayama M, Morimoto K, Yamamoto N, Okuyama S, Hori K (2011) High mannose-binding lectin with preference for the cluster of α1-2-mannose from the green alga *Boodlea coacta* is a potent entry inhibitor of HIV-1 and influenza viruses. J Biol Chem 285:19446–19458
38. Boyd MR, Gustafson KR, McMahon JB, Shoemaker RH, O'Keefe BR, Mori T et al (1997) Discovery of cyanovirin-N, a novel human immunodeficiency virus-inactivating protein that binds viral surface envelope glycoprotein gp120: potential applications to microbicide development. Antimicrob Agents Chemother 41:1521–1530
39. Keeffe JR, Gnanapragasam PNP, Gillespie SK, Yong J, Bjorkman PJ, Mayo SL (2011) Designed oligomers of cyanovirin-N show enhanced HIV neutralization. Proc Natl Acad Sci U S A 108:14079–14084
40. Patsalo V, Raleigh DP, Green DF (2011) Rational and computational design of stabilized variants of cyanovirin-N that retain affinity and specificity for glycan ligands. Biochemistry 50:10698–10712
41. Wang N, Lee YH, Lee J (2008) Recombinant perlucin nucleates the growth of calcium carbonate crystals: molecular cloning and characterization of perlucin from disk abalone, Haliotis discus discus. Comp Biochem Physiol Pt B 149:354–361
42. Matsubara H, Hayashi T, Ogawa T, Muramoto K, Jimbo M, Kamiya H (2008) Modulating effect of acorn barnacle C-type lectins on the crystallization of calcium carbonate. Fish Sci 74:418–424
43. Kamiya H, Jimbo M, Yako H, Muramoto K, Nakamura O, Kado R, Watanabe T (2002) Participation of the C-type hemolymph lectin in mineralization of the acorn barnacle *Megabalanus rosa*. Mar Biol 140:1235–1240

Author Biography

Yasuharu Watanabe studied Biology at Tohoku University, Japan and completed his Ph.D. at the same University in 2008. There his thesis work focused on the role of rhamnose-binding lectins in innate immunity. From 2008 to 2010 he spent a postdoctoral researcher at Nelson Mandela Metropolitan University, South Africa, and University of Alberta, Canada. He currently works as postdoctoral researcher at University of Toyama, Japan. His research focus is on chronic inflammation in the metabolic syndrome.

Takako Naganuma studied Biochemistry at Tohoku University, Japan. From 1993 to 1996 she worked at Tohoku University as a technical worker, and appointed as an Assistant Professor in 1997 at the same University. She completed her Ph.D. in Biomolecular Science at Tohoku University in 2007. Her research interests have focused on the structure and biological functions of lectins from plant and animal, in particular, marine animals, and on the utilization of lectins as detection reagents of sugar chains that exist in biological objects.

Tomohisa Ogawa studied Chemistry at Kyushu University, Japan, and completed his Ph.D. in Chemistry (Protein chemistry and Molecular biology) at Kyushu University, Japan, in 1991. After he did postdoctoral research as JSPS Fellowships, he joined the Department of Chemistry, Faculty of Science, Kyushu University as Assistant Professor (1991–1996). He is now an Associate Professor of Laboratory of Biofunctional Molecules, Graduate School of Life Sciences, Tohoku University (1997–Present). He was also visiting associate professor of Institute for Protein Research, Osaka University (2002–2004), and Program leader at Center for Interdisciplinary Research, Tohoku University (2004–2009). His research interests have focused on the molecular evolution and structure–function relationships of proteins in bio-defense and offense (toxin) systems, and also focused on the biomineralization mechanisms and their application for nano-materials.

Koji Muramoto graduated from Tohoku University, Faculty of Agriculture in 1974 and received his Ph.D. from the same university in 1979. From 1979 to 1982 he worked at University of California, San Francisco as a postdoctral fellow. He was Lecturer from 1982 to 1985 and Associate Professor from 1985 to 1992 at Kitasato University, School of Fisheries Sciences. He returned to Tohoku University, Faculty of Agriculture in 1992. He was promoted to Professor in 1995 and moved to Graduate School of Life Sciences when it was founded in 2001. His research interests focus on the structures and functions of lectins, and also on the utilization of protein resources.

Chapter 5
Medicinal Applications of Plant Lectins

Clara Shui Fern Bah, Evandro Fei Fang and Tzi Bun Ng

Abstract Plant lectins are a unique group of proteins and glycoproteins with potent biological activity and have received widespread attention for many years. They can be found in wheat, corn, tomatoes, peanuts, kidney beans, bananas, peas, lentils, soybeans, mushrooms, tubers, seeds, mistletoe and potatoes among many others. Due to their ability to bind reversibly with specific carbohydrate structures and their abundant availability, plant lectins have commonly been used as a molecular tool in various disciplines of biology and medicine. Whilst once thought of being a dietary toxin, the focus on plant lectins has since shifted to understanding the useful properties of these lectins and utilizing them in medicinal applications to advance human health. This chapter reviews the current and potential applications of plant lectins in various areas of medically related research.

5.1 Introduction

The word "lectin" originates from the Latin word "*lego*" which means, "to choose" or "pick out" [1]. Lectins are defined as sugar-binding proteins that are neither antibodies nor enzymes [2]. According to Rudiger and Gabius (2001), a glycoprotein must meet three distinct requirements to qualify as a lectin. Firstly, a lectin is a protein/glycoprotein that binds carbohydrate(s). Secondly, lectins are not immunoglobulins (antibodies). Thirdly, lectins do not biochemically modify the carbohydrates which they bind [3].

C. S. F. Bah
Department of Food Science, Division of Sciences, University of Otago, North Dunedin, New Zealand

E. F. Fang (✉)
Laboratory of Molecular Gerontology, National Institute on Aging, National Institutes of Health, Baltimore, MD 21224, USA
e-mail: fange@mail.nih.gov

E. F. Fang · T. B. Ng
School of Biomedical Sciences, The Chinese University of Hong Kong, Hong Kong
e-mail: b021770@mailserv.cuhk.edu.hk

E. F. Fang and T. B. Ng (eds.), *Antitumor Potential and Other Emerging Medicinal Properties of Natural Compounds*, DOI: 10.1007/978-94-007-6214-5_5,

Lectins were first discovered more than 100 years ago in plants, when in 1888, Stillmark found that extracts of castor bean (*Ricinus communis*) seeds contained a protein that could agglutinate animal red blood cells [4]. Following the discovery of Stillmark, a number of other plant seed extracts demonstrated the same ability to agglutinate red blood cells, but escalating interest in lectins was sparked during the Second World War as a result of the awareness of using lectins for blood typing. Because some lectins were found to be specific for various blood types (A, B or O), and others were found to have specificities for different glycans, lectins were used for blood typing before blood transfusions were performed [5, 6]. Lectins have since been found in almost every plant species studied and are particularly abundant in the seeds of leguminous plants. They have also been found in various tissues and organs of many vertebrates and invertebrates [7]. Due to their biochemical properties, lectins have become a beneficial tool in several fields of biological research including immunology [8], study of membrane structure, cell recognition [9], cancer research [10], and clinical microbiology [11]. Chapter 2 in this book provides more interesting stories regarding the history of lectinology.

Despite a lack of complete understanding of their biological roles, lectins have been exploited for several years in many applications. The use of lectins has facilitated advancements in many areas of medical research. Lectins are promising candidates as useful therapeutic agents because they can recognize specific carbohydrate structures such as proteoglycans, glycoproteins, and glycolipids, resulting in the regulation of various cells via glycoconjugates and their physiological and pathological phenomena through host-pathogen interactions and cell–cell communications [12].

5.2 Types of Plant Lectins

Plant lectins can be categorized based on their overall mature structure into merolectins, hololectins, chimerolectins and superlectins (Fig. 5.1a). Firstly, merolectins are small monomeric lectins consisting exclusively of a single carbohydrate-binding domain. Due to their monovalent nature, they do not possess agglutinating activity, such as hevein isolated from the latex of *Hevea brasiliensis* [13]. Figure 5.1b shows the crystal structure of hevein with (4s)-2-methyl-2,4-pentanediol. Next are hololectins, also exclusively built up of carbohydrate-binding domains but containing at least two such domains that are either identical or very homologous and bind either the same or structurally similar sugar(s), and most plant lectins are included in this subgroup. One example is Concanavalin A which is a tetrameric protein and binds specifically α-D-mannosyl and α-D-glucosyl residues (two hexoses differing only by the alcohol on carbon 2). A refined structure of Concanavalin A with mannose at 2.0 angstroms resolution is shown in Fig. 5.1b [14]. Another example is peanut (*Arachis hypogaea*) agglutinin [15]. It is a 110-kDa, homotetrameric non-glycosylated protein (without RIP activity) and shows a specificity for the tumor-associated T-antigenic disaccharide Galβ1,3GalNAc. The third

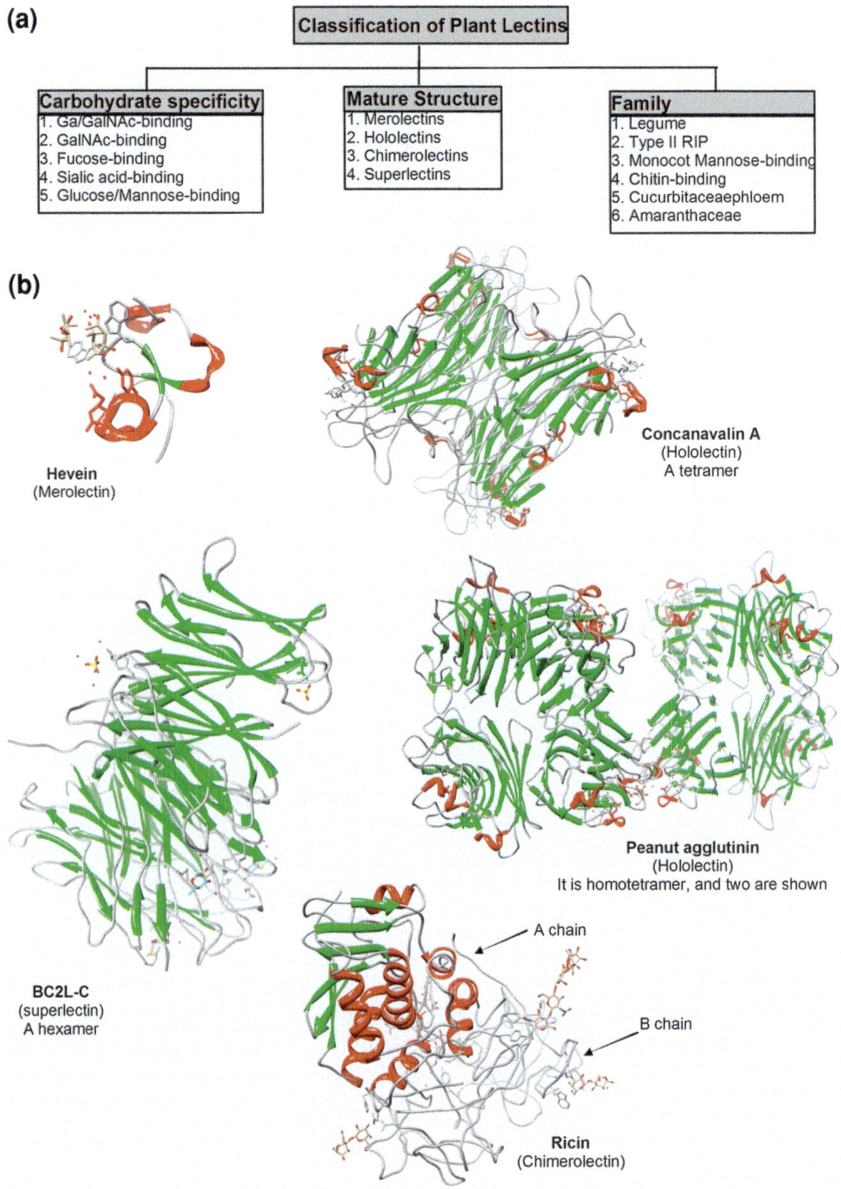

Fig. 5.1 Classifications of plant lectins and the crystal structures of representative plant lectins under the classification standard of mature structure. Crystal data were from RCSB protein data bank and visualized by the UCSF chimera software as used previously [23]. Protein crystal structures were from PROTEIN DATA BANK: Hevein (PDB ID: 1Q9B); Concanavalin A (PDB ID: 5CNA) [14]; Ricin (PDB ID: 2AAI) [16]; Peanut agglutinin (PDB ID: 1CR7); BC2L-C (N-terminal domain: 2WQ4; C-terminal domain: 2XR4) [18]. Please note that BC2L-C is a bacterial lectin since currently no crystal structure of plant superlectin has been released. The coil, helix and strand are shown in *dark grey, red and green*, respectively

group is chimerolectins, which are fusion proteins consisting of one or more carbohydrate-binding domain(s) tandemly arrayed to an unrelated domain. An example of this is ricin, which can be classified as both lectin and type II ribosome inactivating protein (RIP) [16]. As shown in Fig. 5.1b, ricin consists of two parts; an A chain (with N-glycosidase activity/RIP activity) and a B chain (hemagglutinating/lectin activity) with the B chain capable of binding different carbohydrates, such as β-D-glucose and β-D-galactose as shown in the figure. Lastly, superlectins are a type of chimerolectins which consist of at least two structurally different domains which recognize structurally unrelated carbohydrates [17]. For instance, TxLC-I is a superlectin isolated from tulip bulbs, and it consists of a mannose-binding domain and an unrelated GalNAc-binding domain. Since there is no crystal data on plant superlectin, the crystal structure of a superlectin purified from *Burkholderia cepacia* (a Gram-negative bacterium), named BC2L-A, is used for demonstration (Fig. 5.1b) [18]. The N-terminal domain of BC2L-A is a novel TNF-α-like fucose-binding lectin while the C-terminal is a calcium-dependent bacterial lectin.

Plant lectins can also be grouped into different families according to some common features (legume lectins, type II ribosome-inactivating proteins, monocot mannose-binding lectins, and other lectins) [19]. Legume lectins are the best known lectin family. Lectin content in seeds is higher compared to the content in the bark, leaves, roots and stems of leguminous plants. Other smaller families of plants whose lectins have been characterized are Gramineae (cereals, such as wheat germ) and Solanaceae (potatoes and tomatoes). Monocot-binding lectins exhibit an exclusive specificity towards mannose and are built up of 1, 2, 3, or 4 subunits of about 12 kDa while chitin-binding lectins are composed of hevein domains [20]. For more information, readers are referred to an excellent review by Van Damme et al. (1998) describing in detail how plant lectins are structurally and evolutionarily related [21].

Because of the tremendous diversity of carbohydrate-binding specificities among the plant lectins, some researchers classify them according to the small carbohydrate haptens they recognize, e.g., galactose-binding lectins or GlcNAc-binding lectins (Fig. 5.1a). The specificity of a lectin is usually defined in terms of the monosaccharide(s) or glycosaccharides that inhibit lectin-induced agglutination [22]. This specificity is usually determined by comparing sugars on the basis of the minimum concentration needed to inhibit hemagglutination. If the lectin-binding carbohydrate is present freely dissolved and at a sufficiently high concentration, it will compete with the red cell glycoconjugates for the lectin and agglutination will not take place.

5.3 Medicinal Applications of Plant Lectins

Due to their versatility, lectins are frequently used in biological and medical research. The overwhelming success of plant lectins is based firstly on their highly specific carbohydrate-binding activity, and the biological effects they provoke in various organisms. Being able to procure reasonable amounts of pure lectin preparations also contributes to their success [20]. Consequently, screening of plant species in the search for lectins with new and useful biological properties has been taking place for several decades.

5.3.1 Plant Lectins in Serology

Because of their ability to distinguish carbohydrate determinants in human blood cells, lectins have historically been used for blood typing. Serology was the first discipline of medicine that relies on the specific biological activity of lectins [24]. Lectins agglutinate erythrocytes by binding to a carbohydrate-containing moiety on the surface of the cells and forming cross-bridges between them [4]. This useful property of lectins endows them the ability to discriminate between red blood cells that have different terminal, non-reducing sugars in the major glycoprotein that carries the blood group antigens [25]. For example, blood group A is determined by the presence of N-acetyl-galactosamine in the terminal, non-reducing end of the oligosaccharide portion of the receptor protein while group B is determined by galactose in this position. If both of these sugars are lacking, fucose is the terminal sugar and this determines blood group O. In this way, lectins with different specificities can be used in blood typing to differentiate between blood types. For readers seeking a comprehensive compilation of lectins that have been studied for blood group serology, the review by Judd (1980) is recommended [26].

5.3.2 Plant Lectins as Mitogens

A limited number of lectins from plants possess the unique ability to induce quiescent lymphocytes to grow and divide, a phenomenon known as mitogenic stimulation [27]. Mitogenic lectins mimic the action of antigens on lymphocytes, except that they activate a large proportion (as much as 70–80 %) of the cells, whereas antigens stimulate only specific clones, each of which compromise a tiny proportion, 0.1 % or less, of the total number of lymphocytes. Because of their ability to stimulate multiple lymphocyte clones, lectins are classified as polyclonal mitogens. One of the most valuable outcomes of this proliferative ability of lectins has been an increased understanding of the relationship between chromosomal abnormality and human diseases, which has tremendously helped in diagnosis [28]. Besides other cells, lymphocytes have been the usual target cells for mitogenic assays, and the study of lectin-lymphocyte interaction has made a significant contribution to elucidating the mechanism of lymphocyte activation and its control, further contributing to the understanding of cell growth and development.

The discovery of the first mitogenic lectin was made by Nowell in 1960, who found that the lectin of red kidney bean (*Phaseolus vulgaris*), known as phytohemagglutinin (PHA), possessed the ability to stimulate lymphocytes to undergo mitosis [29]. This discovery had a groundbreaking impact on immunology as it shattered the then prevailing opinion that lymphocytes were dead-end cells incapable of dividing or differentiating further [29]. Within a short time thereafter, several other lectins were proven to be mitogenic. Table 5.1 provides a few examples of plant lectins with mitogenic activity as such a lectin from the tuber of wild cobra lily [30], red kidney bean lectin [31], mushroom lectins [32, 33], Brazilian camaratu bean lectin [34], and jackfruit seed lectin [35] .

Concanavalin A, a lectin from the Jack bean, is an important mitogen because, in contrast to PHA, its activity can be inhibited by low concentrations of monosaccharides, for example, mannose. This finding provided proof that mitogenic stimulation is the result of binding of lectins to sugars on the surface of the lymphocytes and was among the earliest demonstrations for a biological role of cell surface sugars [36]. It has been suggested that mitogenic lectins interact with unique membrane components that may act as 'stimulating receptors', and that non-mitogenic lectins may not bind to these membrane components [28]. Mitogenic lectins are now tools for the study of signal transmission into cells and for the analysis of the biochemical events that occur during lymphocyte stimulation in vitro [36]. Such properties make them useful tools for the isolation and characterization of polysaccharides and glycoconjugates, in cancer research, as diagnostic tools for the investigation of early cell-membrane alterations and carbohydrate changes that accompany neoplastic processes, and in immunological studies.

Table 5.1 Mitogenic activity of selected plant lectins

Lectin	MW (kDa)	Carbohydrate specificity	Cells	Activity	References
AFL	13.5	Asialofetuin	BALB/c splenocytes	increase in IL-2	[30]
DRKBL	67	–	BALB/c splenocytes	Lower than Con A	[31]
GCL	18	Galactose	BALB/c splenocytes	More potent than Con A	[32]
VVL	12.6	Thyroglobulin	Mouse T cells	10 fold more effective than Con A	[33]
CML	–	Mannose/Glucose	Human lymphocytes	Similar activity to Con A at 0.78 – 25 μg/ml dosage	[34]
Jacalin	50	Galactose Mannose	Human lymphocytes	$33.7 \pm 15 \times 10^3$ ct/min	[35]

AFL a lectin from tubers of *Arisaema flavum* (Schott.); *DRKBL* a lectin from *Phaseolus vulgaris* cv. Dark Red Kidney Bean; *GCL Ganoderma capense* lectin; *Con A* Concanavalin A; *VVL Volvariella volvacea* lectin; *CML Cratylia mollis* seed lectin; *Jacalin* Jackfruit (*Artocarpus heterophyllus*) seed lectin

5.3.3 Plant Lectins in Cancer Therapy

Cancer is one of the leading causes of death worldwide. Cancer is a deadly disease, where the abnormal behavior of a single cell type is difficult to treat by chemotherapy. It is important in cancer therapy that the treatment targets only the affected cells, leaving the normal cells undisturbed, which is quite difficult, especially in chemotherapy. Anti-cancer drugs available in the current market are not target-specific and elicit several side-effects and complications encountered in the clinical management of various forms of cancer, which highlights the urgent need for novel effective and less-toxic therapeutic approaches [37]. Recently, focus has shifted from using lectins to detect cancer to actually using lectins to combat cancer.

Evidence is now emerging that lectins are dynamic contributors to tumor cell recognition (surface markers), cell adhesion and localization, signal transduction across membranes, mitogenic stimulation, augmentation of host immune defense, cytotoxicity, and apoptosis [38]. A review by De Mejia and Prisecaru (2005) provides a comprehensive appraisal of the inhibitory effects of plant lectins on malignant cells in vitro and vivo [39]. Table 5.2 below adds some more recent selected plant lectins and lists their inhibitory effects on malignant cells.

Small glossy black soybean (*Glycine max*) lectin inhibited the proliferation of breast cancer and hepatoma cells [48]. An anti-tumor action mechanism of soybean lectins has been proposed involving the action of the lectins on tumor cell membranes, the reduction of tumor cell proliferation, the induction of tumor-specific cytotoxicity of macrophages, and apoptosis. Thus, tumor cells are more susceptible to attack by macrophages after treatment with lectins. Furthermore, lectins exert an immunomodulatory effect by altering interleukins production [54].

Mistletoe lectin (ML) is one of the most studied lectins in clinical trials and has demonstrated beneficial effects against cancer development. Additionally their mechanisms of action towards cancer treatment have been extensively studied. For example, in a study using European mistletoe lectin which demonstrated antiproliferative activity towards human melanoma cells, a significant number of melanoma cells started rounding up and exhibited cell shrinkage, chromatin condensation and nuclear fragmentation typical for apoptotic body formation indicating apoptotic cell death [45]. Three mistletoe lectins, I, II and III (ML-I, II, III) have been isolated [55]. Mistletoe lectin-I, which belongs to the type II ribosome-inactivating protein (RIP II) family and is composed of a catalytically active A-chain with rRNA N-glycosidase activity and a B-chain with carbohydrate binding properties, exerted potent cytotoxic effects on tumor cells. It also induced apoptosis through both caspase-8/FLICE independent of a death receptor pathway and via a p53-independent pathway following ionizing radiation. Meanwhile, mistletoe lectin-II induced apoptotic death in cancer cells involving the generation of intracellular hydrogen peroxide (H2O2) and activation of a caspase- 9-caspase-3 cascade [46].

The oral consumption of mistletoe lectins as an alternative therapy towards cancer therapy has been advocated by some parties as these lectins are resistant towards low pH in the stomach and are not affected by proteolytic enzymes in

Table 5.2 Inhibitory effects of selected plant lectins on malignant cells

Lectin	MW (kDa)	Carbohydrate specificity	Antiproliferative activity (IC_{50})	Mechanism of action	References
BTKL	60	Polygalacturonic acid	HepG2 (7.9 μM)	Apoptosis induction Mitochondria damage	[40]
FVML	12	–	L1210 (13 μM)	–	[41]
GCML	18	Galactose/galactosamine	L1210 (8 μM) Hep G2 (16.5 μM) M1 (12.5 μM)	–	[32]
PAL	32	Inulin	Hep G2 (2.1 μM) MCF7 (3.2 μM)	–	[42]
CNL	15.9	Asialofetuin and lactose	leukemic T (100 μg/ml)	binding to carbohydrate receptors on leukemic T cells	[43]
HCSL	29	D-galactose and *N*-acetyl-D galactosamine	HeLa (9 μg/ml) FemX (11 μg/ml)	Displayed a cytotoxic effect	[44]
ML-1	63	Galactose	UISO-Mel6 (73.6 ng/ml) MeWo (5084.0 ng/ml) FemX-1 (1.9 ng/ml) Lox (10.5 ng/ml) G361 (0.6 ng/ml)	Induction of apoptosis	[45]
Con A	104	Mannose/Glucose	A375 (25 μg/ml)	Inducing apoptosis in both casp-8 and casp-9-dependent ways	[46] [47]
DRKBL	67	–	L1210 (1.6 μM)	–	[31]
GML	50	Melibiose	MCF7 (2.6 μM) HepG2 (4.1 μM)	–	[48]
DBL	30	Fructose	L1210 (3.6 μM) HepG2 (25 μM)	Cytokine-inducing No production	[49]
EAPL	60	Galactose	Hep G2 (34.8 μM)	Apoptosis-inducing	[50]
AFL	13.5	Asialofetuin	J774 P388D1		[30]

(continued)

Table 5.2 (continued)

Lectin	MW (kDa)	Carbohydrate specificity	Antiproliferative activity (IC_{50})	Mechanism of action	References
MCL	130	D-galactose and α-lactose	CNE 1 (6.9 μM) CNE 2 (7.4 μM)	Inducing apoptosis in both casp-8 and casp-9-dependent pathways; Damage of mitochondria	[51]
TDL	48	Mannose	Pro-01 (56.7 μM) Bre-04 (41.5 μM) Lu-04 (11.4 μM)		[52]
SAL	40	Galactose	HeLa (6.25 μM)	–	[53]

BTKL Phaseolus vulgaris cv. Blue tiger king lectin; *FVML Flammulina velutipes* mushroom lectin; *GCML Ganoderma capense* mushroom lectin; *PAL Pholiota adipose* mushroom lectin; *CNL* Ricin B-like lectin from mushroom *Clitocybe nebularis*; *HCSL Haliclona crater* sponge lectin; *ML-1* European mistletoe lectin; *Con A* Concanavalin A; *DRKBL* Dark red kidney bean (*Phaseolus vulgaris* cv.) lectin; *GML* small glossy black soybean (*Glycine max*) lectin; DBL, *Del Monte* banana lectin; *EAPL Phaseolus vulgaris* cv. Extralong autumn purple bean lectin; *AFL* lectin from tubers of Wild Cobra Lily *Arisaema flavum*; *MCL Momordica Charantia* lectin; *TDL* lectin from fresh tubers of a medicinal herb *Typhonium Divaricatum*; *SAL* lectin from seeds of *Sophora alopecuroide* the stomach [56]. Pryme et al. (2007) provide a meticulous case report with more information on this topic [56]. However, the usefulness of mistletoe extracts in the treatment of malignant melanoma is still controversial for some. This may be in part due to the fact that the full molecular mechanisms underlying mistletoe treatment and how it works *in vivo* are still not completely clarified.

Lectins from several types of mushroom have demonstrated anti-proliferative activity including those from *Flammulina velutipes* [41], *Ganoderma capense* [32], *Pholiota adipose* [42] and *Clitocybe nebularis* [43]. A homodimeric 32.4-kDa lectin was isolated from fresh fruiting bodies of the mushroom *Pleurotus citrinopileatus* [57]. The lectin exerted strong anti-tumor activity in mice bearing sarcoma 180, and caused approximately 80 % inhibition of tumor growth when administered intraperitoneally at 5 mg/kg daily for a period of 20 days.

As many plant lectins have demonstrated anticancer properties in vitro, and *in vivo*, there clearly is a huge potential for their use as therapeutic agents in cancer treatment. Mechanisms of plant lectin action elucidated thus far include preferential binding to cancer cell membranes or their receptors, causing cytotoxicity, apoptosis, and inhibition of tumor growth. Plant lectins can be internalized into cells, causing cancer cell agglutination and/or aggregation [54]. Ingested lectins can also sequester the available body pool of polyamines, in this manner they prevent cancer cell growth. The immune system is also affected by alterations in the production of various interleukins, or by activation of certain protein kinases. Additionally, lectins can bind to ribosomes and inhibit protein synthesis. They also alter the cell cycle by inducing non-apoptotic G1-phase accumulation mechanisms, G2/M phase cell cycle arrest and apoptosis, and can trigger the caspase cascade. Lectins can also down-regulate telomerase activity and inhibit angiogenesis [54].

Future advances in cancer prevention, detection, and treatment could potentially be achieved by using plant lectins. These substances possess antitumor activity and anti-carcinogenic activity that could be beneficial in cancer treatment.

5.3.4 Plant Lectins as Antiviral Agents

Compounds with antiviral activity are generally of great medical interest and different modes of pharmaceutical actions have been described. The antiviral activity of plant lectins can be based on several mechanisms. The surfaces of retroviruses such as human immunodeficiency virus (HIV) and many other enveloped viruses are covered by virally-encoded glycoproteins. Glycoproteins gp120 and gp41 present on the HIV envelope are heavily glycosylated, with glycans estimated to contribute almost 50 % of the molecular weight of gp120. Agents that specifically and strongly interact with the glycans may disturb interactions between the proteins of the viral envelope and the cells of the host [58]. Sugar-binding proteins can crosslink glycans on the viral surface and prevent further interactions with the co-receptors.

HIV RT is a key enzyme of the HIV life cycle. Screening of HIV RT inhibitors is currently a strategy to search for anti-HIV drugs. Strikingly, the vast majority of plant lectins that are active against HIV possess carbohydrate specificity directed against mannose oligomers. Most HIV-inhibitory plant lectins are derived from the monocot families *Amaryllidaceae*, *Orchidaceae* and *Alliaceae* or the dicot families *Fabaceae*, *Moraceae*, *Urticaceae* and *Cecropiaceae* [58]. Table 5.3 lists several plant lectins such as *Phaseolus vulgaris* cv. Extralong autumn purple bean lectin [50], Del Monte banana lectin [49], black soybean lectin [48] and *Pholiota adipose* mushroom lectin [42] which all possessed anti-HIV RT activity.

Studies on plant lectins such as those from leek (Table 5.3) have shown that they can be potent inhibitors of coronaviruses by interfering with two targets in the viral replication cycle [59]. The first target was located early in the replication cycle, most probably viral attachment, and the second target was located at the end of the infectious virus cycle. The antiviral activity spectrum of plant lectins varies considerably, depending on the nature of their sugar specificity. In general, the mannose-specific plant lectins were found to be highly effective against coronaviruses [59].

A lectin from blue green algae (CV-N) exhibited a broad range of antiviral activities. It has been shown that CV-N binds with high affinity to HIV envelope protein gp120 [61] and also interacts with another surface glycoprotein, gp41 [63]. The specificity of CV-N, however, is not limited only to different strains of HIV and related retroviruses. CV-N inhibited the development of viral cytopathic effects of Ebola virus, binding to its surface envelope glycoprotein [62] and also blocked influenza A and B strains by binding to the hemagglutinin surface glycoprotein [63]. The results of viral pretreatment studies indicated that CV-N directly neutralized both influenza A and B viruses, including both H1N1 and H3N2 strains. However, the CV-N resistant influenza virus strain A/PR/8/34 was completely resistant to any direct neutralizing activity. Results suggested that CV-N bound directly to and inactivated the viral particle, preventing subsequent infection, and that both the likely molecular target for CV-N and the basis of CV-N resistance resided in the viral particle [63].

Unlike the majority of current antiviral therapeutics that act through inhibition of the viral life cycle, lectins can prevent penetration of the host cells by the viruses [60]. Antiviral lectins are best suited to topical applications and can exhibit lower toxicity than many currently used antiviral therapeutics. Additionally, these proteins are often resistant to high temperatures and low pH, as well as being odorless, which are favorable properties for potential microbicide drugs [60]. The mechanistic details on how lectins operate at a molecular level to inhibit virus growth need to be further explored to know the basis of their biological activity.

5.3.5 Plant Lectins as Antibacterial and Antifungal Agents

The growing resistance of microorganisms to convectional antimicrobial agents is a source of concern to clinical microbiologists all over the world. As a result, efforts are being made to develop antimicrobial agents from local sources for better chemotherapeutic effects. Lectins from plants could satisfy the demand for more natural antimicrobials as several studies have demonstrated the effectiveness of lectins as

Table 5.3 Antiviral activity of selected plant lectins

Name	MW (kDa)	Carbohydrate specificity	Anti-Viral activity (shown in MIC or as indicated)	Mechanism of action	References
EAPL	60	Galactose	HIV-1 (1.8 μM)	–	[50]
DBL	30	Fructose	HIV-1 (3 mM)		[49]
APA	–	Mannose	Sars CoV corona virus (0.45 μg/ml)	Disruption of virus replication cycle	[59]
SGBSL	50	Melibiose	HIV-1 (2.82 μM)		[48]
PAL	32	Inulin	HIV-1 (1.9 μM)	–	[42]
TDL	48	Mannose	HSV-II (3.1 μg/ml)	Bind on high-mannose glycans on the surface of the virus particles	[52]
CVN	11	–	HIV-1 laboratory strain RF (0.1 nM)	Binding on HIV surface envelope glycoprotein gp120	[60, 61]
CVN	11	–	EboZV GP (50 nM) Marburg GP (6-25 nM)	–	[62]
CVN	11	–	Influenza A Virus Strains: Sydney/05/07 (H3N3, 0.04 μg/ml) Victoria/3/75 (H3N3, 0.005 μg/ml) Mem/8/99 (H3N3, 0.03 μg/ml) Mem/2/99 (H3N3, 0.15 μg/ml) Beijing/262/95 (H1N1, 0.11 μg/ml) Influenza B Virus Strains: Hong Kong/5/72 (0.5 μg/ml) Yamanashi/166/98 (0.04 μg/ml) Mem/3/99 (0.02 μg/ml) Lee/40 (0.026 μg/ml)	CV-N bound directly to and inactivated the viral particle	[63]
CLL	12	Mannose	Poxvirus (around 30 μg/ml)	–	[64]

EAPL Phaseolus vulgaris cv. Extralong autumn purple bean lectin; *DBL Del Monte* banana lectin; *APA* plant lectin isolated from leek; *SGBSL* small glossy black soybean (*Glycine max*) lectin; *PAL Pholiota adiposa* mushroom lectin; *TDL* lectin from fresh tubers of a medicinal herb *Typhonium divaricatum* (L.) Decne; *CVN* blue green algae *Nostoc ellipsosporum* lectin; *CLL* lectin from *Crinum latifolium* bulb; *HIV-1* human immunodeficiency virus type 1; HSV-II herpes simplex virus type 2 inhibitory compounds towards bacterial and fungal growth such as the examples listed in Table 5.4. Plant lectins have demonstrated antibacterial activity against many pathogens including *Escherichia coli*, *Shigella dysenteriae*, *Staphylococcus aureus*, *Pseudomonas aeruginosa*, *Bacillus subtilis*, and *Klebsiella sp*. Lectins have also demonstrated antifungal activity towards *Aspergillus flavus, Trichoderma viride, Fusarium oxysporum, Fusarium moniliforme, Coprinus comatus, Rhizoctonia olani, Penicillium digitatum, Alternaria alternata*, and *Valsa mali.* The use of lectins in clinical microbiology in a review by Slifkin and Doyle (1990) [11] as well as the antimicrobial activity from plant lectins article by Paiva et al. (2010) [65] is recommended for readers who seek further detailed information.

The cell wall of bacteria not only precludes any interaction between the glycoconjugates on their membrane and carbohydrate-binding proteins but also prevents these proteins from penetrating the cytoplasm. Therefore, plant lectins cannot alter the structure and/or permeability of the membrane or disturb the normal intracellular processes of invading microbes. Therefore, if lectins play a role in the plant's defense against bacteria, it must be through an indirect mechanism that is based on interactions with cell wall carbohydrates or extracellular glycans [73].

Similarly, plant lectins are not capable of binding to glycoconjugates on the fungal membranes or penetrating the cytoplasm due to the barrier formed by the cell wall. Thus, it is not likely that lectins directly inhibit fungal growth by changing the structure and/or permeability of the fungal membrane. However, there may be indirect effects produced by the binding of lectins to carbohydrates on the surface of the fungal cell wall [74].

This activity was concluded to be related to the lectin carbohydrate binding property, that might endow lectin molecules with binding activity towards certain carbohydrate components in the fungal cell wall affecting its activity and viability as most lectins recognize either *N*-acetylneuraminic acid, *N*-acetylglucosamine, *N*-acetylgalactosamine, galactose, mannose, or fucose [75]. Alternatively, it was stated that antifungal activity of some proteins or peptides is associated with chitin binding property and the active proteins should have a specific amino acid sequence and a cysteine/glycine rich chitin binding domain. This chitin binding property might simulate the carbohydrate binding property as chitin is composed of modified glucose subunits (*N*-acetyl glucose amine) which can be equally recognized by lectin as glucose [69]. Chitin binding might lead to the disruption of the fungal cell wall that increases toxicity, since chitin, which is a major component of the fungal wall, is a polymer of *N*-acetylglucosamine.

5.3.6 Plant Lectins in Drug Targeting

The idea to use lectins for drug delivery began when the use of tomato lectin (TL) to target the luminal surface of the small intestine was proposed [76]. The underlying principle behind lectin-mediated drug targeting is that most cell surface proteins and many lipids in cell membranes are glycosylated and these glycans can

Table 5.4 Antimicrobial activity of selected plant lectins

Name	Antimicrobial activity	Lectin Specificity	Anti-microorganism activity (MIC or as indicated)	References
MOL	Antibacterial	–	*E. coli*	[66]
			Shigella dysenteriae	
			Staphylococcus aureus	
EUL	Antibacterial	–	*S. aureus* (1.5 μg/ml)	[67]
			P. aeruginosa (1.5 μg/ml)	
			Klebsiella sp. (1.5 μg/ml)	
			Bacillus subtilis (16.5 μg/ml)	
			Streptococcus sp. (16.5 μg/ml)	
			Escherichia coli (16.5 μg/ml)	
AJL	Antibacterial		*Bacillus subtilis* (0.227 mg/ml)	[68]
	Antifungal		*S. aureus* (0.0567 mg/ml)	
			C. albicans (0.0567 mg/ml)	
PSL	Antifungal	Mannose	*Aspergillus flavus* (0.1 mg/ml)	[69]
			Trichoderma viride(0.01 mg/ml)	
			Fusarium oxysporum. (0.01 mg/ml)	
CFL	Antifungal	Mannose/glucose	*Fusarium. moniliforme* (-)	[70]
			Aspergillus. flavus. (-)	
RKBL	Antifungal		*Fusarium oxysporum* (-)	[71]
			Coprinus comatus (-)	
			Rhizoctonia solani (-)	
SAL	Antifungal	Galactose	*Penicillium digitatum* ($IC_{50} = 3.125$ μM)	[53]
			Alternaria alternate ($IC_{50} = 3.338$ μM)	
CSL	Antifungal	Galactose, lactose, arabinose, etc.	*Valsa mali* ($IC_{50} = 18$ μM)	[72]

MOL lectin purified from Drumstick (*Moringa oleifera* Lam.) leaves; *EUL Eugenia uniflora* seed lectin; *AJL* lectin from seeds of *Archidendron jiringa*; *PSL* lectin from Egyptian *Pisum sativum* seeds; *CFL* lectin from red cluster pepper (*Capsicum frutescens*); *RKBL* lectins from red kidney bean; *SAL* Lectin from seeds of *Sophora alopecuroides*; *CSL* lectin from caper (*Capparis spinosa*) seeds

Table 5.5 Potential drug targeting use of selected plant lectins

Name	Source	Carbohydrate specificity	MW (kDa)	Potential use	References
AEL	Tomato	$(GlcNAc)_4$	71	Intestinal wall Bioadhesive in the GI tract	[76] [81]
STL	Potato	N-acetyl-D-glucosamine	–	Ocular (Corneal and conjunctival)	[78]
WGA	Wheat germ	N-acetyl-D-glucosamine Sialic acid	38	Blood brain barrier Intestinal mucosal surface	[79] [82]
PNA	Peanut	D-galactose	110	Intestinal mucosal surface	[80]

AEL Lycopersicon esculenttun lectin; *STL Solanum tuberosum* lectin. *WGA* Wheat germ agglutinin; *PNA Arachis hypogaea* lectin

be the binding sites for lectins. The combination of a small number of sugars can produce a vast range of different chemical structures. Different cell types express different glycan arrays and diseased cells, such as transformed or cancerous cells, often express different glycans compared with their normal counterparts. Therefore, lectins could be used as carrier molecules to target drugs specifically to different cells and tissues [77]. By targeting cell types exclusively, the side effects of drugs could be minimized. Besides targeting specific cells, the lectin–sugar interaction can also been used to trigger vesicular transport into or across epithelial cells. The concept of bio-adhesion via lectins may be applied not only for the gastrointestinal tract [76] but also for other biological barriers like the nasal mucosa, the lungs, the buccal cavity, the eye [78] and the blood–brain barrier [77, 79]. Table 5.5 lists some examples of plant lectins from tomato [76], potato [78], wheat germ [79] and peanut [80] which have been studied to investigate their potential of use as drug targeting agents.

Attempts to evaluate the binding of lectin candidates to non-histologically processed cell surfaces, must be done in order to systematically identify the appropriate receptors and lectin types for further studies. Further work will more precisely locate the lectin binding site on the tissue surface and will quantify the binding. Lectin toxicity and *in vivo* binding will then be considered prior to selecting the most promising candidates for formulation studies [78]. Lectins such as WGA could be useful as specific bioadhesive ligands for lipid nanoparticles intended for oral administration [82] once the bioadhesive properties and oral bioavailability efficiency are determined.

Results from preliminary studies performed so far can be taken as an indication that it may indeed be possible to exploit lectins of certain carbohydrate specificities for oral drug delivery and intestinal targeting. However, a great deal of research remains to be done before lectins can be used in practice. For a more thorough understanding, readers are directed to reviews by Lehr (2000) and Bies et al. (2004) which provides more information on lectin-mediated drug delivery and targeting, from their history to applications [77, 83].

5.4 Outlooks and Perspectives

Plant lectins have been and still are a subject of intense investigation. They have come a long way, since their first detection in plants as hemagglutinins, to their present status as recognition molecules with innumerable exciting functions and applications. As more plant lectins are isolated and further studies are conducted on the biological activities and mechanisms of action of lectins, the production of lectins can be improved and new applications of these lectins will be found or conceived. Lectins could be used as the next generation of medicines once research has contributed to their full understanding.

More research is still needed and a genomic and proteomic approach to elucidate and support the potential shown by lectins as anticancer, antimicrobial and drug delivery agents is warranted. The mechanistic details of how lectins operate at a molecular level need to be further explored to know the basis of their biological activity. Although there is still much to be learned about the effects of lectins, the area of research concerning plant lectins is constantly evolving and being updated. Thus the medicinal applications of plant lectins hold considerable potential and exciting discoveries lie ahead.

References

1. Boyd WC, Shapleigh E (1954) Antigenic relations of blood group antigens as suggested by tests with lectins. J Immunol 73:226–231
2. Barondes SH (1988) Bifunctional properties of lectins: lectins redefined. Trends Biochem Sci 13:480–482
3. Rüdiger H, Gabius HJ (2001) Plant lectins: occurrence, biochemistry, functions and applications. Glycoconj J 18:589–613
4. Liener IE (1964) Seed hemagglutinins. Econ Bot 18:27–33
5. Renkonen KO (1948) Studies on hemagglutinins present in seeds of some representatives of the family of Leguminoseae. Annales Medicinae Experimentalis et Biologiae Fenniae 26:66
6. Boyd WC, Reguera RM (1949) Hemagglutinating substances for human cells in various plants. J Immunol 62:333–339
7. Goldstein IJ, Hughes RC, Monsigny M, Osawa T, Sharon N (1980) What should be called a lectin. Nature 285:66
8. Fujita T (2002) Evolution of the lectin-complement pathway and its role in innate immunity. Nat Rev Immunol 2:346–353
9. Sharon N, Lis H (1989) Lectins as cell recognition molecules. Science 246:227–234
10. Mody R, Joshi SH, Chaney W (1995) Use of lectins as diagnostic and therapeutic tools for cancer. J Pharmacol Toxicol Methods 33:1–10
11. Slifkin M, Doyle RJ (1990) Lectins and their application to clinical microbiology. Clin Microbiol Rev 3:197–218
12. Ogawa T., Watanabe M., Naganuma T., Muramoto K. (2011) Diversified carbohydrate-binding lectins from marine resources. J Amino Acids 2011
13. Reyes-Lopez CA, Hernandez-Santoyo A, Pedraza-Escalona M, Mendoza G, Hernandez-Arana A, Rodriguez-Romero A (2004) Insights into a conformational epitope of Hev b 6.02 (hevein). Biochem Biophys Res Commun 314:123–130

14. Naismith J.H., Emmerich C., Habash J., Harrop S.J., Helliwell J.R., Hunter W.N., et al. (1994) Refined structure of concanavalin A complexed with methyl alpha-D-mannopyranoside at 2.0 A resolution and comparison with the saccharide-free structure. Acta crystallographica. Section D, Biol crystallogr 50 847–858
15. Ravishankar R, Thomas CJ, Suguna K, Surolia A, Vijayan M (2001) Crystal structures of the peanut lectin-lactose complex at acidic pH: retention of unusual quaternary structure, empty and carbohydrate bound combining sites, molecular mimicry and crystal packing directed by interactions at the combining site. Proteins 43:260–270
16. Rutenber E, Katzin BJ, Ernst S, Collins EJ, Mlsna D, Ready MP et al (1991) Crystallographic refinement of ricin to 2.5 A. Proteins 10:240–250
17. Van Damme E.J.M., Peumans W.J., Pusztai A., Bardocz S. (1998) Handbook of plant lectins: prop biomed appl, Wiley
18. Sulak O, Cioci G, Lameignere E, Balloy V, Round A, Gutsche I et al (2011) Burkholderia cenocepacia BC2L-C is a super lectin with dual specificity and proinflammatory activity. PLoS Pathog 7 e1002238
19. Lam SK, Ng TB (2011) Lectins: production and practical applications. Appl Microbiol Biotechnol 89:45–55
20. Peumans W, Van Damme EJ (1988) Plant lectins: versatile proteins with important perspectives in biotechnology. Biotechnol Genet Eng Rev 15:199–228
21. Van Damme EJM, Peumans WJ, Barre A, Rougé P (1998) Plant lectins: A composite of several distinct families of structurally and evolutionary related proteins with diverse biological roles. Crit Rev Plant Sci 17:575–692
22. Goldstein I.J., Hughes R.C., Monsigny M., Osawa T., Sharon N. (1980) What should be called a lectin?, Published online: 08 May 1980; | doi:10.1038/285066b0 285: 66
23. Fang EF, Ng TB (1815) Ribonucleases of different origins with a wide spectrum of medicinal applications. Biochim Biophys Acta 2011:65–74
24. Końska G, Wójtowicz U, Pituch-Noworolska A (2008) Possible application of lectins in diagnostics and therapy. Part I. Diagnostic application. Przeglad lekarski 65:189–194
25. Sequeira L (1978) Lectins and their role in host-pathogen specificity. Annu Rev Phytopathol 16:453–481
26. Judd WJ, Issitt PD (1980) The role of lectins in blood group serology. Crit Rev Clin Lab Sci 12:171–214
27. Kilpatrick DC (1999) Mechanisms and assessment of lectin-mediated mitogenesis. Mol Biotechnol 11:55–65
28. Ashraf M.T., Khan R.H. (2003) Mitogenic lectins. Med Sci Monit 9 RA265-269
29. Sharon N (2007) Lectins: Carbohydrate-specific reagents and biological recognition molecules. J Biol Chem 282:2753–2764
30. Singh J, Singh J, Kamboj SS (2004) A novel mitogenic and antiproliferative lectin from a wild cobra lily. Arisaema flavum. Biochem Biophys Res Commun 318:1057–1065
31. Xia L, Ng TB (2006) A hemagglutinin with mitogenic activity from dark red kidney beans. J Chromatogr B 844:213–216
32. Ngai PHK, Ng TB (2004) A mushroom (Ganoderma capense) lectin with spectacular thermostability, potent mitogenic activity on splenocytes, and antiproliferative activity toward tumor cells. Biochem Biophys Res Commun 314:988–993
33. Ho J.C.K., Sze S.C.W., Shen W.Z. , Liu W.K. (2004) Mitogenic activity of edible mushroom lectins. Biochimica et Biophysica Acta (BBA) - General Subjects 1671: 9–17
34. Maciel EVM, AraÃºjo-Filho VS, Nakazawa M, Gomes YM, Coelho LCBB, Correia MTS (2004) Mitogenic activity of Cratylia mollis lectin on human lymphocytes. Biologicals 32:57–60
35. Pineau N, Aucouturier P, Brugier JC, Preud'homme JL (1990) Jacalin: a lectin mitogenic for human CD4 T lymphocytes. Clin Exp Immunol 80:420–425
36. Sharon N, Lis H (2004) History of lectins: from hemagglutinins to biological recognition molecules. Glycobiology 14:53R–62R

37. Patel S, Goyal A (2012) Recent developments in mushrooms as anti-cancer therapeutics: a review. 3. Biotech 2:1–15
38. Mody R, Joshi S, Chaney W (1995) Use of lectins as diagnostic and therapeutic tools for cancer. J Pharmacol Toxicol Methods 33:1–10
39. De Mejía EG, Prisecaru VI (2005) Lectins as bioactive plant proteins: a potential in cancer treatment. Crit Rev Food Sci Nutr 45:425–445
40. Fang EF, Pan WL, Wong JH, Chan YS, Ye XJ, Ng TB (2011) A new Phaseolus vulgaris lectin induces selective toxicity on human liver carcinoma Hep G2 cells. Arch Toxicol 85:1551–1563
41. Ng TB, Ngai PHK, Xia L (2006) An agglutinin with mitogenic and antiproliferative activities from the mushroom Flammulina velutipes. Mycologia 98:167–171
42. Zhang GQ, Sun J, Wang HX, Ng TB (2009) A novel lectin with antiproliferative activity from the medicinal mushroom Pholiota adiposa. Acta Biochim Pol 56:415–421
43. Pohleven J, Obermajer N, Sabotic J, Anzlovar S, Sepcić K, Kos J et al (2009) Purification, characterization and cloning of a ricin B-like lectin from mushroom Clitocybe nebularis with antiproliferative activity against human leukemic T cells. Biochim Biophys Acta 1790:173–181
44. Pajic I, Kljajic Z, Dogovic N, Sladic D, Juranic Z, Gasic MJ (2002) A novel lectin from the sponge Haliclona cratera: isolation, characterization and biological activity. Comp Biochem Physiol C: Toxicol Pharmacol 132:213–221
45. Thies A, Nugel D, Pfüller U, Moll I, Schumacher U (2005) Influence of mistletoe lectins and cytokines induced by them on cell proliferation of human melanoma cells in vitro. Toxicology 207:105–116
46. Liu B, Min M-W, Bao J-K (2009) Induction of apoptosis by Concanavalin A and its molecular mechanisms in cancer cells. Autophagy 5:432–433
47. Liu B (2009) Li C.-y., Bian H.-j., Min M.-w., Chen L.-f., Bao J.-k., Antiproliferative activity and apoptosis-inducing mechanism of Concanavalin A on human melanoma A375 cells. Arch Biochem Biophys 482:1–6
48. Lin P, Ye X, Ng T (2008) Purification of melibiose-binding lectins from two cultivars of Chinese black soybeans. Acta Biochim Biophys Sin 40:1029–1038
49. Cheung AHK, Wong JH, Ng TB (2009) Musa acuminata (Del Monte banana) lectin is a fructose-binding lectin with cytokine-inducing activity. Phytomedicine 16:594–600
50. Fang EF, Wong JH, Lin P, Ng TB (2010) Biochemical and functional properties of a lectin purified from korean large black soybeans–a cultivar of glycine max. Protein Pept Lett 17:690–698
51. Fang EF, Zhang CZY, Ng TB, Wong JH, Pan WL, Ye XJ et al (2012) Momordica Charantia lectin, a type II ribosome inactivating protein, exhibits antitumor activity toward human nasopharyngeal carcinoma cells in vitro and *in vivo*. Cancer Prev Res (Phila) 5:109–121
52. Luo Y, Xu X, Liu J, Li J, Sun Y, Liu Z et al (2007) A novel mannose-binding tuber lectin from Typhonium divaricatum (L.) Decne (family Araceae) with antiviral activity against HSV-II and anti-proliferative effect on human cancer cell lines. J Biochem Mol Biol 40:358–367
53. Li T, Yin X, Liu D, Ma X, Lv H, Sun S (2012) Isolation and characterization of a novel lectin with antifungal and antiproliferative activities from Sophora alopecuroides seeds. Acta Biochim Biophys Sin 44:606–613
54. de Mejia EG, Bradford T, Hasler C (2003) The anticarcinogenic potential of soybean lectin and lunasin. Nutr Rev 61:239–246
55. Schumacher U, Stamouli A, Adam E, Peddie M, Pfuller U (1995) Biochemical, histochemical and cell biological investigations on the actions of mistletoe lectins I, II and III with human breast cancer cell lines. Glycoconj J 12:250–257
56. Pryme IF, Tilrem P (2007) Oral mistletoe lectins: A case for their use in cancer therapy. Cancer Ther 5:287–300
57. Li YR, Liu QH, Wang HX, Ng TB (2008) A novel lectin with potent antitumor, mitogenic and HIV-1 reverse transcriptase inhibitory activities from the edible mushroom Pleurotus citrinopileatus. Biochim Biophys Acta 1780:51–57

58. Balzarini J (2006) Inhibition of HIV entry by carbohydrate-binding proteins. Antiviral Res 71:237–247
59. Keyaerts E, Vijgen L, Pannecouque C, Van Damme E, Peumans W, Egberink H et al (2007) Plant lectins are potent inhibitors of coronaviruses by interfering with two targets in the viral replication cycle. Antiviral Res 75:179–187
60. Ziółkowska NE, Wlodawer A (2006) others, Structural studies of algal lectins with anti-HIV activity. Acta Biochim Pol 53:617–626
61. Boyd MR, Gustafson KR, McMahon JB, Shoemaker RH, O'Keefe BR, Mori T et al (1997) Discovery of cyanovirin-N, a novel human immunodeficiency virus-inactivating protein that binds viral surface envelope glycoprotein gp120: potential applications to microbicide development. Antimicrob Agents Chemother 41:1521–1530
62. Barrientos LG, Lasala F, Otero JR, Sanchez A, Delgado R (2004) In vitro evaluation of cyanovirin-N antiviral activity, by use of lentiviral vectors pseudotyped with filovirus envelope glycoproteins. J Infect Dis 189:1440–1443
63. O'Keefe BR, Smee DF, Turpin JA, Saucedo CJ, Gustafson KR, Mori T et al (2003) Potent anti-influenza activity of cyanovirin-N and interactions with viral hemagglutinin. Antimicrob Agents Chemother 47:2518–2525
64. Kaur A, Singh R (2008) Singh Kamboj S., Singh J., J. Kotwal G., In vitro antiviral activity of *Crinum latifolium* lectin against poxvirus replication. Journal of Biological Sciences 8:1236–1240
65. Paiva P.M.G., Gomes F.S., Napole ~ ao T.H., S\'a R.A., Correia M.T.S., Coelho L. (2010) Antimicrobial activity of secondary metabolites and lectins from plants. Curr Res Technol Educ Top Appl Microbiol Microb Biotechnol Badajoz: Formatex 396–406
66. Khatun S., Khan M.M.H., Ashraduzzaman M., Pervin F., Bari L., Absar N. (2009) Antibacterial Activity and Cytotoxicity of Three Lectins Purified from Drumstick (Moringa oleifera Lam.) Leaves. J Bio-Sci 17 89–94
67. Oliveira M.d.l., Andrade C.a.s., Santos-Magalhães N.s., Coelho L.c.b.b., Teixeira J.a., Carneiro-da-Cunha M.g., et al. (2008) Purification of a lectin from Eugenia uniflora L. seeds and its potential antibacterial activity. Lett Appl Microbiol 46 371–376
68. Charungchitrak S, Petsom A, Sangvanich P, Karnchanatat A (2011) Antifungal and antibacterial activities of lectin from the seeds of Archidendron jiringa Nielsen. Food Chem 126:1025–1032
69. Sitohy M, Doheim M, Badr H (2007) Isolation and characterization of a lectin with antifungal activity from Egyptian Pisum sativum seeds. Food Chem 104:971–979
70. Ngai PHK, Ng TB (2006) A lectin with antifungal and mitogenic activities from red cluster pepper (Capsicum frutescens) seeds. Appl Microbiol Biotechnol 74:366–371
71. Ye XY, Ng TB, Tsang PW, Wang J (2001) Isolation of a homodimeric lectin with antifungal and antiviral activities from red kidney bean (Phaseolus vulgaris) seeds. J Protein Chem 20:367–375
72. Lam Sze K., Han Qi F., Ng Tzi B. (2009) Isolation and characterization of a lectin with potentially exploitable activities from caper (Capparis spinosa) seeds. Biosci Rep 29: 293–299
73. Peumans W.J., Van Damme E.J. (1995) Lectins as plant defense proteins. Plant physiology 109
74. Wong JH, Ng TB, Cheung RCF, Ye XJ, Wang HX, Lam SK et al (2010) Proteins with antifungal properties and other medicinal applications from plants and mushrooms. Appl Microbiol Biotechnol 87:1221–1235
75. Lis H, Sharon N (1998) Lectins: Carbohydrate-Specific Proteins That Mediate Cellular Recognition. Chem Rev 98:637–674
76. Naisbett B, Woodley J (1994) The potential use of tomato lectin for oral drug delivery. 1. Lectin binding to rat small intestine in vitro. Int J Pharm 107:223–230
77. Bies C, Lehr C-M, Woodley JF (2004) Lectin-mediated drug targeting: history and applications. Adv Drug Deliv Rev 56:425–435
78. Nicholls TJ, Green KL, Rogers DJ, Cook JD, Wolowacz S, Smart JD (1996) Lectins in ocular drug delivery: An investigation of lectin binding sites on the corneal and conjunctival surfaces. Int J Pharm 138:175–183

79. Fischer D, Kissel T (2001) Histochemical characterization of primary capillary endothelial cells from porcine brains using monoclonal antibodies and fluorescein isothiocyanate-labelled lectins: implications for drug delivery. Eur J Pharm Biopharm 52:1–11
80. Cai Q, Zhang Z-R (2005) Lectin-mediated cytotoxicity and specificity of 5-fluorouracil conjugated with peanut agglutinin (5-Fu-PNA) in vitro. J Drug Target 13:251–257
81. Lehr CM, Bouwstra JA, Kok W, Noach AB, de Boer AG, Junginger HE (1992) Bioadhesion by means of specific binding of tomato lectin. Pharm Res 9:547–553
82. Liu Y, Wang P, Sun C, Zhao J, Du Y, Shi F et al (2011) Bioadhesion and enhanced bioavailability by wheat germ agglutinin-grafted lipid nanoparticles for oral delivery of poorly water-soluble drug bufalin. Int J Pharm 419:260–265
83. Lehr CM (2000) Lectin-mediated drug delivery: the second generation of bioadhesives. Journal of controlled release: official journal of the Controlled Release Society 65:19–29

Author Biography

Clara Bah completed her Bachelor's Degree in Chemical-Bioprocess Engineering at the University of Technology, Malaysia (UTM). After working as a research and development engineer, she completed a Postgraduate Diploma and Master's of Science (Food Science) at the University of Otago, New Zealand. Her research topic explored the availability of bioactive compounds from fish eggs, with a particular focus on lectins. This work led to collaboration with Dr Evandro Fang and Prof. T.B. Ng of the School of Biomedical Sciences in the Chinese University of Hong Kong. Currently Clara is working towards completing her PhD at the University of Otago.

Dr. Evandro Fei Fang Dr. Evandro Fei Fang received his Ph.D. degree from The Chinese University of Hong Kong and currently works with Prof. Vilhelm A. Bohr on DNA repair and genome stability in human aging diseases. His research encompasses: (a) bioassay-guided isolation of antitumor constituents from medicinal plants and the molecular mechanisms involving apoptosis and autophagy pathways, and (b) the functions of mitochondria and mitophagy in sustaining human health and the molecular basis of their defects in the etiology of aging diseases. He has published over 30 papers in International Journals and 9 Book Chapters, and serves on the editorial board of Medicinal and Aromatic Plants. He was a finalist in the Young Investigator Award 2011 conducted by the Hong Kong Institute of Science.

Tzi Bun Ng Tzi Bun Ng is Professor of Biochemistry at the Faculty of Medicine, The Chinese University of Hong Kong, HKSAR. After completing his B.Sc. studies at The University of Hong Kong, he obtained his Ph.D. degree from the Memorial University of Newfoundland in Canada. He pursued his postdoctoral training at University of California, San Francisco, USA, and sabbatical research at the Imperial College London, UK. His major research interests are the health-benefits of proteins and peptides including polysaccharopeptides, lectins, antimicrobial peptides, enzymes and protease inhibitors. In the past 30 years, his team made many contributions to the field of bioactive components, represented in more than 550 papers in International Peer-Reviewed Journals, over 60 invited Book Chapters, 2 Edited Books, and numerous other media. He currently sits on the editorial board of 9 International Journals and contributes actively in many scientific fields.

Part II
Structural and Functional Insights into Medicinal Ribonucleases

Chapter 6
Plant Nucleases from *Nuclease* I Family: Their Properties and Antitumor Potential

Tomas Podzimek

Abstract Plant nucleases belonging to the *nuclease I* family (E.C. 3.1.30.x) are very potent catalysts, involved in many processes, such as programmed cell death, stress response, recycling of nucleic acids, etc. This review is focused on three recombinant zinc-dependent nucleases: R-TBN1, R-HBN1, R-ABN1 belonging to *nuclease I* family, additionally exerting antitumorigenic effects on some types of tumors in vivo. These nucleases are sugar non-specific nucleases degrading both DNA and RNA and both single- and double-stranded substrates. Moreover, they cleave viroid RNA and some artificial substrates, such as poly(A), poly(U) and poly(C). However, poly(G) is resistant to cleavage and completely inhibits their ssDNase, dsDNase and RNase activities. The nucleases are *N*-glycoproteins with a molecular mass of about 37 kDa. The 3-D structure of R-TBN1 has been solved being the first known crystal structure of a plant nuclease. This structure can serve as a template for other related nucleases. The previous experiments showed that the free or PEG-conjugated nucleases caused an inhibition of growth of melanoma C-32 and human prostate carcinoma. The evaluation of negative side effects supports the use of these substances as therapeutical agents. Finally, PN nuclease is described in this review as an example of native antitumor nuclease.

6.1 Introduction

Ribonucleases comprise a vast number of enzymes that act on ribonucleic acids. They appear in all living organisms and a part of this superfamily is composed of plant enzymes possessing ribonuclease activity. These enzymes are usually divided into four major families: *RNase I, RNase II, nuclease I* and *exonuclease I* [1]. The nucleases outlined here belong to the *nuclease I* family. They are sugar non-specific

T. Podzimek (✉)
Department of Biochemistry and Microbiology, Institute of Chemical Technology, Prague, Czech Republic
e-mail: podzimet@vscht.cz

E. F. Fang and T. B. Ng (eds.), *Antitumor Potential and Other Emerging Medicinal Properties of Natural Compounds*, DOI: 10.1007/978-94-007-6214-5_6,

nucleases, cleaving both DNA and RNA, and are defined to prefer single-stranded substrates. Their molecular masses lie in the range 31–39 kDa and pH optimum 5.0–6.5. They are sensitive to EDTA and possess 3′-nucleotidase activity. To date, the family includes many nucleases, for example bifunctional nuclease BFN1 from *Arabidopsis thaliana* [2], ZEN1 nuclease from *Zinnia elegans* [3], BEN1 [3] and Bnuc1 from *Hordeum vulgare* [4], CEL I from celery [5], DSA6 from daylily [6] and nuclease A from *Agropyron elongatum* [7]. The classification of the nucleases from *nuclease I* family is based on their abovementioned biochemical properties and on their amino acid sequence homology. In addition, these nucleases can be divided into groups according to their metal requirement, where Zn^{2+}, Mg^{2+} and Ca^{2+} cations usually serve as cofactors [8].

The biological functions of some of them have been proposed, however, more information is needed to describe the mechanism of their regulation, whether they interact with other proteins, and whether they are involved in multiple biological processes. Here will be presented the properties of three recombinant nucleases—R-TBN1, R-HBN1 and R-ABN1 and one native plant nuclease—PN nuclease.

6.2 Plant Nucleases

6.2.1 Sequence Similarity of the Nucleases

As mentioned above, one of the common features of the nucleases from the *nuclease I* family is a similarity of their amino acid (aa) sequences. Figure 6.1 shows some aspects emerging from amino acid sequence alignment of several plant nucleases and P1 nuclease, which is fungal single-strand specific nuclease. Plant nucleases mentioned in this paper are members of one enzyme family, but they originate from various families of plants. They show more than 44 % homology among their aa sequences, moreover, we can see higher similarities of aa sequences of nucleases within each plant family. For example, tomato TBN1 and potato StEN1, belonging to *Solanaceae* family, share 97 % aa sequence homology. Similarly, a homology of ABN1 (*Arabis brassica*) and BFN1 (*Arabidopsis thaliana*) from *Brassicaceae* family is 94 %. Interestingly, BEN1 and Bnuc1 nucleases originating even from the same species share only 84 % aa sequence homology. Specially, these two nucleases show homology lower than 50 % with other mentioned nucleases. However, available information indicates the presence of a family of related nucleases possibly across all higher plant species.

6.2.2 Functions of Plant Nucleases

The action of nucleases from the *nuclease I* family is linked mostly with programmed cell death (PCD). They are associated with developmental PCD during

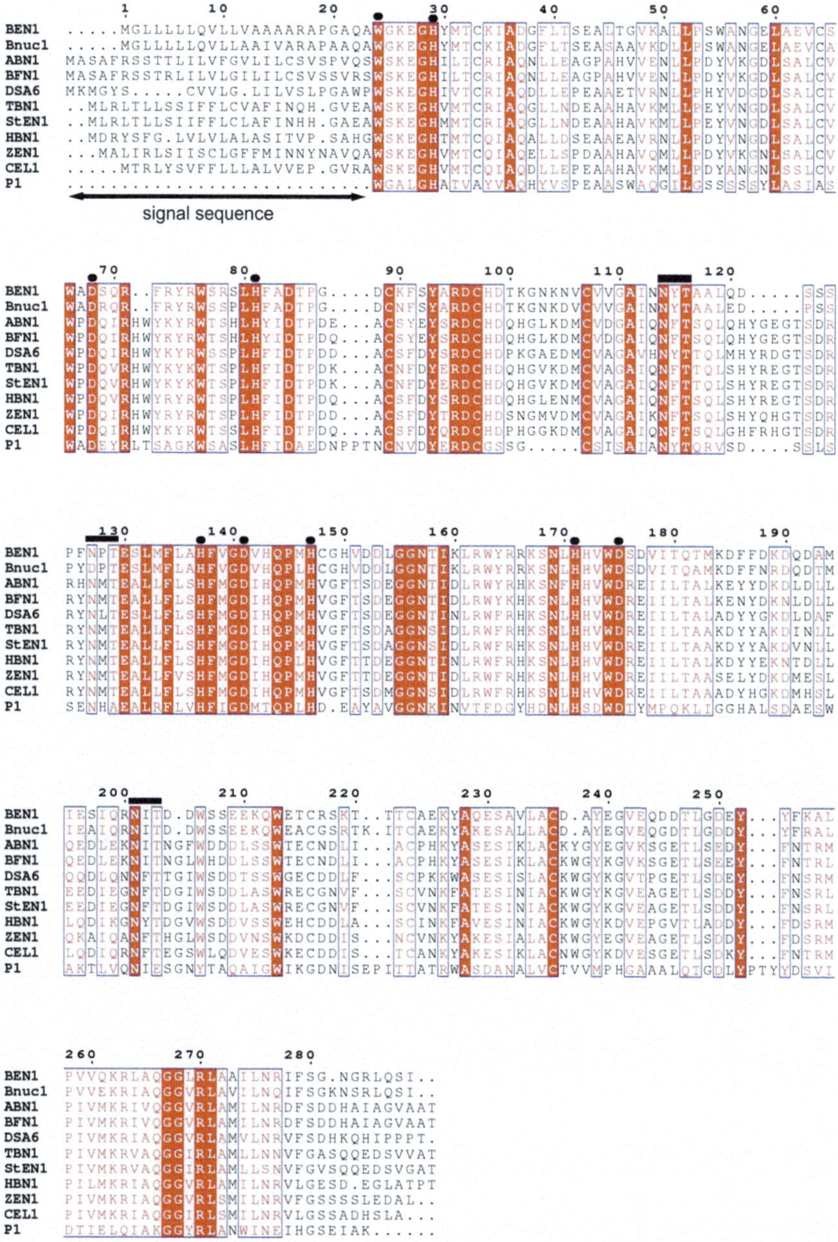

Fig. 6.1 Amino acid sequence alignment of plant nucleases BEN1 (BAA28942), Bnuc1 (BAA82696), ABN1 (FN690739), BFN1 (AAD00693), DSA6 (AAC34856), TBN1 (AM238701), StEN1 (AAT79582), HBN1 (AM909687), ZEN1 (BAA28948) and CEL I (AAF42954) and fungal P1 nuclease (AAB19975). Conserved residues in all sequences are highlighted by *red color*, ● the residues interacting with three zinc cations in the molecules of P1 nuclease and TBN1,— three *N*-glycosylation sites (sequence motif NXT/S) in the molecules of TBN1, HBN1 and ABN1

flowering, reproduction and other tissue transformations, and with PCD connected with a hypersensitive response (during pathogenesis), senescence or stress response (e.g. salt stress) [9]. The nucleases participate in degradation of both nuclear DNA and RNA, which leads to cell death. An example of such a nuclease acting during developmental PCD is the ZEN1 nuclease, which was described to participate in nuclear DNA degradation of tracheary elements [10]. As a result of its action, the content of the cell is totally degraded, but the cell walls are thickened and they remain in place to form tracheids. The tracheids then form xylem, which is responsible for transport of water and nutrients within the plant.

BEN1 nuclease was found to be secreted by the aleurone layer of barley seeds into the endosperm, where it degrades DNA. This degradation takes place in overall utilization of polysaccharides, proteins and other nutrients during seed germination [11]. Bnuc1, the closest known homolog of BEN1, is probably involved in two processes: its presence was observed in barley plants during salt stress induced by NaCl and also during senescence induced by abscisic acid [4]. The action of nucleases (and RNases) during salt stress and senescence has a common feature, which lies in degradation of nucleic acids and salvage of nucleotides as nutrients from damaged or old leaves [12]. Another senescence-associated nuclease is BFN1. This nuclease is expressed mainly in senescent leaves of *A. thaliana*. The senescence of *A. thaliana* leaves is linked with the time of flowering and fruit development. No wonder that mRNA coding BFN1 was observed also in developing flowers. BFN1, similar to Bnuc1, could be involved in a system providing the nucleotides from degraded DNA and RNA to other parts of a plant [2]. More information is needed to explain a precise role of BFN1 during flower development.

These examples describe proposed natural roles of nucleases in plants. However, they could also exhibit other biological effects, when added to different systems, such as animal or human tumors. Except mung bean nuclease [13] and PN nuclease (below) [14], R-TBN1, R-HBN1 and R-ABN1 belong to the first plant recombinant nucleases tested for antitumorigenic properties.

6.3 Discovery of TBN1, HBN1 and ABN1

The action of nucleases TBN1 and HBN1 is probably associated with plant pathogenesis. TBN1 was abnormally expressed in the leaves of viroid-infected (PSTVd-potato spindle tuber viroid) tomato plants. The PSTVd strain AS1 caused a stop effect on the growth of experimental plants, which was confirmed by observation of insufficient development of xylem. The authors proposed that TBN1 can play a role in development of xylem, similar to ZEN1, however, the action of viroid caused deregulation of TBN1 and stopped tissue differentiation [15]. The second nuclease, HBN1, was revealed during a similar experiment with viroid-caused pathogenesis. HLVd (hop latent viroid) was used for infection of hop (*Humulus lupulus*) plants. This viroid did not affect plant phenotype as PSTVd, but it was able to enter the anther and pollen. However, there was no observation of viroid in the next generation of hop plants. It was found that viroid was able to enter the immature pollen, but during pollen maturation its level decreased to zero. On the other hand, the level of RNase together with ssDNase activity increased during the experiment, suggesting the presence of a *nuclease I* enzyme. RNA transcript for this nuclease was isolated and designated as HBN1 [16]. Transcript for ABN1 was isolated from senescent leaves of *A. brassica* (unpublished). Its role has not yet been suggested but one can imagine some apoptotic properties of this nuclease.

6.3.1 Purification of Plant Nucleases

A number of purification strategies are known from the literature and only a few examples are mentioned in this paper. Plant nucleases were usually purified by ammonium sulfate precipitation, ion-exchange, affinity and gel chromatography. However, some exceptions exist. Ammonium sulfate precipitation precedes the use of chromatography techniques to remove a part of contaminant proteins, and to decrease the initial volume of an extract.

The purification of nuclease A from *Agropyron elongatum* consisted of ammonium sulfate precipitation, cation-exchange and affinity (heparin) chromatography with the yield of about 34 % of initial activity [7]. Extracellular nuclease from pollen of *N. tabacum* was purified by ammonium sulfate precipitation and cation-exchange chromatography [17]. In contrast, corn nuclease I was precipitated by mercaptoethanol and then applied onto a CM-cellulose column and finally onto a gel filtration column yielding 20 % of initial activity [18]. In the example of nuclease from rice, the proteins were precipitated by acetone. Further purification included affinity chromatography on heparin-Sepharose [19]. Finally, CEL I was extracted from 105 kg of fresh celery and purified by ammonium sulfate precipitation, affinity chromatography (concanavalin A-Sepharose), cation-exchange chromatography (DEAE and phosphocellulose), hydrophobic interaction chromatography (phenyl-Sepharose), anion-exchange chromatography (Mono Q) and gel chromatography. This strategy included seven steps and the yield was 5 μg of CEL I [5].

6.3.2 Production, Isolation and Purification of R-TBN1, R-HBN1 and R-ABN1

The cDNA of each nuclease was cloned into plant expression vector pLV07 [20]. The cells of *Agrobacterium tumefaciens* were transformed with this expression vector and also with a vector containing gene for protein p19, the inhibitor of gene silencing. The mixed suspension of transformed *A. tumefaciens* was infiltrated into *Nicotiana benthamiana* leaves using a syringe without needle. Infiltrated leaves were harvested 5 days post infiltration and lyophilized for long-term storage.

The number of plants used repeatedly for infiltration experiments usually exceeded 100. Nearly the whole area of 4–6 leaves/per plant was utilized for production of nucleases and the fresh weight of harvested leaves reached as much as 1 kg. Nucleases were extracted from lyophilized leaves and purified by ammonium sulfate precipitation (35–80 % saturation), ion-exchange (strong anex) chromatography and affinity chromatography (heparin). Proteins were eluted from the column by a linear gradient of NaCl (0–1 mol/l). The fractions collected after each chromatography process were screened for ssDNase activity. Concentrated samples were stored at −20 °C [21].

R-TBN1, R-HBN1 and R-ABN1 showed similar properties during the purification steps and no radical modification of the procedure was necessary. Even though they are probably zinc-dependent nucleases (confirmed only in TBN1) zinc was not added to the used buffers and the activity was still detectable during the whole purification process. Efforts have been made to produce nucleases in larger quantities, because the yields from 100 g fresh weight were usually in milligrams. Contemporary overall yields of the nucleases from the leaves are in milligrams to tens of milligram. However, production of TBN1 in bacteria (*E. coli*) failed due to absence of *N*-glycosylation. Although, an expression of TBN1 occurred, the product was soluble only in 8 M urea (used for extraction from inclusion bodies) and without any activity (unpublished). The alternative possibility is the expression of the nucleases in yeasts, which perform protein glycosylation. However, production of these nucleases in plants is now the only way to study them. Although they are produced recombinantly in *N. benthamiana*, the only difference between native and recombinant form will consist of the glycosylation pattern of each studied nuclease.

6.3.3 Primary Structure and N-glycosylation of R-TBN1, R-HBN1 and R-ABN1

R-TBN1, R-HBN1 and R-ABN1 are monomeric proteins, with a molecular weight of 37.2 kDa for both R-TBN1 and R-HBN1 and 37.1 kDa for R-ABN1. All three nucleases contain a signal sequence comprising 25, 24 and 28 residues for R-TBN1, R-HBN1 and R-ABN1, respectively. The N-terminal sequence WSKEG is common for all three nucleases and also for some of the related nucleases from *nuclease I* family (Fig. 6.1). The molecular weights of the mature proteins calculated from their aa sequence are 31.6 kDa for both R-TBN1 and R-HBN1 and 32 kDa for R-ABN1. This result indicated some kind of posttranslational modification. All three aa sequences contain three glycosylation sites—N94, N112 and N 186 (numbering does not include the signal sequence) with the characteristic sequence motif NXT (Fig. 6.1). *N*-glycosylation of all three nucleases has been proved by removing the oligosaccharides from the surface of the nucleases using PNGase F. Subsequently, MALDI-TOF analysis confirmed the glycosylation of all three sites for R-TBN1 and R-HBN1 and possibly two sites for R-ABN1. It suggests that each oligosaccharide chain has a molecular mass of approximately

1.8 kDa in R-TBN1 and R-HBN1 and approximately 2.5 kDa in R-ABN1. No activity of deglycosylated nucleases was detected in polyacrylamide gel containing ssDNA as a substrate. However, we cannot exclude any negligible activity of deglycosylated nuclease in solution. It is important to note that revealing of the glycoprotein nature of R-TBN1 explained a failure of its production in bacterial system and directed the attempts to another production system. As mentioned above, the recombinant and native forms of studied nucleases may differ only in glycosylation pattern between tobacco plant and the original source plant. Interestingly, the two forms of R-TBN1 with molecular weights of 37.2 and 35.4 kDa were detected on a polyacrylamide gel, both with ssDNase activity. We assumed that two forms of R-TBN1 are produced in tobacco, one fully glycosylated and another modified only at two sites [21]. A similar situation was observed after preparation of two mutant forms using site-directed mutagenesis—TBN1 D34S and TBN1 D34S/E104 N, with one and two additional *N*-glycosylation sites, respectively (numbering does not include a signal sequence). A sample of TBN1 D34S/E104 N contained also two forms with activity toward ssDNA, indicating the presence of fully and partially glycosylated mutant forms, which both arose during biosynthesis in a tobacco plant (unpublished). The mutant forms are being characterized now.

6.3.4 3-D Models of R-TBN1, R-HBN1 and R-ABN1

A considerable number of plant nucleases have been described, however, information about their structure and catalytic mechanism is still missing. The most related nuclease with a known 3-D structure is single-strand specific P1 nuclease (PDB ID: 1AK0) [22]. Although, P1 nuclease originates from fungi *Penicillium citrinum*; it shares 28, 29 and 26 % of sequence identity with R-TBN1, R-HBN1 and R-ABN1, respectively. This is sufficient to build reliable models of all three nucleases using homology modeling (Modeller 9v3 software). Figure 6.2 shows the models of all three nucleases and a 3-D structure of P1 nuclease. All four nucleases are composed mainly of α-helices and their general fold is very similar. The models further revealed the importance of N-terminal tryptophan as part of the nuclease active site, where it coordinates one of the three zinc ions. For this reason, no tag can be attached to the N-terminus of these nucleases and it must remain free for correct folding. Zinc ions are necessary for the catalytic activity of these nucleases. P1 nuclease contains nine residues coordinating three zinc ions and these residues are conserved in all three nucleases and also in other related nucleases (Fig. 6.1). P1 nuclease is also a glycoprotein with four sites for *N*-glycosylation, where three sites are occupied by oligosaccharide chains and only one (N92) of these sites is homologous to that in our studied nucleases (N94). In relation to other plant nucleases, sequence motifs for *N*-glycosylation of these enzymes are highly conserved (Fig. 6.1). The structures of P1 nuclease and all three nucleases are stabilized by several disulfide bridges. P1 nuclease contains two disulfide bridges and four disulfide bridges have been proposed for R-TBN1, R-HBN1 and R-ABN1, respectively (Fig. 6.2) [21].

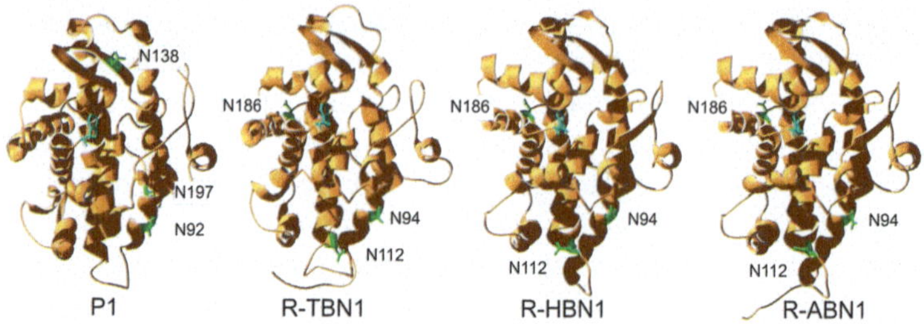

Fig. 6.2 3-D models of R-TBN1, R-HBN1 and R-ABN1 built on the basis of the crystal structure of P1 nuclease (PDB ID: 1AK0). Tryptophan residue (W1, *light blue*) is an important part of the active site of all four nucleases. Asparagines (N) of *N*-glycosylation sites (*green*) are numbered [21]

6.3.5 Three-Dimensional Structure of R-TBN1

The preliminary crystallization studies have shown the ability of R-TBN1 to crystallize, even if oligosaccharides attached to the protein surface may complicate formation of crystals. The crystals of R-TBN1 were obtained using the hanging-drop vapour-diffusion method in orthorhombic, rhombohedral and trigonal forms with a resolution of 5.2, 3.2 and 2.16 Å, respectively. The native presence of zinc ions was confirmed. Phase problem was solved by a combination of MAD (multiple anomalous dispersion) and molecular replacement techniques [23, 24] (Koval' et al. submitted).

R-TBN1 has Phospholipase C/P1 nuclease fold with three zinc ions (Zn1, Zn2 and Zn3) in the active site. All nine amino acids (W1, H6, D45, H60, H122, D126, H132, H156 and D160) coordinating with all three zinc ions in the active site of R-TBN1 are in analogous positions, as in P1 nuclease (W1, H6, D45, H60, H116, D120, H126, H149 and D153) (Fig. 6.1). As follows from the position numbers of these amino acid residues, Zn1, Zn2 and Zn3 connect distant parts of the structure, and thus stabilize it. The N-terminal tryptophan is a part of the active site, as mentioned above and participates in the coordination of Zn1. All three *N*-glycosylation sites were determined in a crystal to be glycosylated. The catalytic mechanism of R-TBN1 is now under investigation.

6.3.6 Substrate Specificity of R-TBN1, R-HBN1 and R-ABN1

The specific activity is usually determined with calf thymus dsDNA, with its heat-denatured form (ssDNA) and yeast ribosomal RNA. R-TBN1 preferred both ssDNA and dsDNA over RNA. R-HBN1 cleaved ssDNA three times and dsDNA two times better than RNA, and the preference of R-ABN1 was in the order ssDNA > dsDNA ≈ RNA. Unlike single-strand specific nucleases, these nucleases were able to cleave double-stranded substrates as well. Further, they were able to cleave dsRNA and viroid RNA (PSTVd and HLVd), which can form highly structured and stable forms [21]. Viroid RNA is a single-stranded circular molecule, composed of double-stranded regions, stem-loops and a certain number of mismatched base pairs. These abilities observed *in vitro* point to possible participation in PCD during pathogenesis. It is clear that these nucleases are not only sugar non-specific, but also "secondary structure non-specific" nucleases [25]. Degradation of polymeric artificial substrates such as poly(A), poly(U), poly(C) and poly(G) was tested and compared with RNase activity. All three nucleases cleaved poly(A) and poly(U) rather than RNA. R-TBN1 cleaved poly(A) and poly(U) 4.4-fold and 4.5-fold better than RNA, respectively, R-HBN1 3.9-fold and 2.2-fold, respectively and R-ABN1 3-fold and 3.6-fold, respectively. Poly(C) was cleaved with all three nucleases only with 30 % efficiency, relative to RNase activity and poly(G) was completely resistant to attack. We found >90 % inhibition effect of poly(G) on dsDNase and RNase activity and >80 % on ssDNase activity of all three nucleases. Poly(G) possibly strongly binds to active sites of the nucleases and prevents the access of a substrate. This inspired us to use poly(G) as a ligand for R-TBN1 in co-crystallization experiments. However, the length of poly(G) (hundreds base pairs) was not suitable for this purpose. Therefore, shorter oligodeoxynucleotides—oligo(dG)2, oligo(dG)10, oligo(dG)15 and oligo(dG)25 were tested for their inhibition effect. Unfortunately, the inhibitory effect decreased with the decreasing in length of an oligodeoxynucleotide, indicating their lower affinity for the active or binding site [21].

It is worth mentioning that activity of these nucleases was detected also in a polyacrylamide gel containing substrate. After SDS-PAGE and SDS wash out, the nucleases were able to renaturate to their active conformation [21].

6.3.7 *Factors Influenced the Activity of R-TBN1, R-HBN1 and R-ABN1*

The factors affecting an enzyme activity include pH, temperature, ionic strength and presence of activators. The influences of these factors on dsDNase, ssDNase and RNase activity have been determined. All three enzymes were active in the range of pH 4.4–7.7 with a maximal activity at pH 6. Only the dsDNase activity of R-ABN1 was shifted to pH 5. Even though, their temperature optimum around 60 °C was quite high, incubation of the enzyme itself at this temperature strongly affected the activity. The loss of activity of R-TBN1, R-HBN1 and R-ABN1 was about 80, 60 and 55 %, respectively, after incubation for 5 min. The most stable nuclease was R-ABN1 with 40 % of remaining activity after 30 min at 60 °C. R-TBN1 was practically inactive after 30 min at the same temperature. The nucleases were quite sensitive to increasing concentrations of NaCl. The activity of R-HBN1 and R-ABN1 toward dsDNA decreased faster than the activity toward the other two substrates and was completely abolished by adding 0.95 M NaCl.

The effect of various metal ions on nuclease activity has been tested. Zn^{2+}, Ca^{2+}, Mg^{2+} and Mn^{2+} ions added to the reaction mixture (1 mmol/l) caused inhibition or had no effect on nuclease activity. Only the dsDNase activity of R-ABN1 was stimulated nearly two-fold by manganese ions. Ni^{2+} and Co^{2+} ions (1 mmol/l) inhibited the reaction as expected. EDTA (1 mmol/l) caused an inhibition, however, it did not completely abolish the activity of these metal-dependent nucleases. The effect of EDTA was increased, when the nuclease was incubated with EDTA for longer period and subsequently removed by dialysis [26]. The activity of nucleases in such treated samples was significantly restored by adding zinc in the range of tens μmol/l. This result proves the detection of zinc atoms in the structure of R-TBN1 by X-ray fluorescence during crystallographic studies. It is likely that R-HBN1 and R-ABN1 are also zinc-dependent nucleases.

6.4 Biological Effects of R-TBN1, R-HBN1 and R-ABN1

MBN nuclease from mung bean sprouts [13] and PN nuclease from black pine (described below) [14] were the only members of *nuclease I* family tested for potential antitumor properties to date. Recently, R-TBN1, R-HBN1 and R-ABN1 were

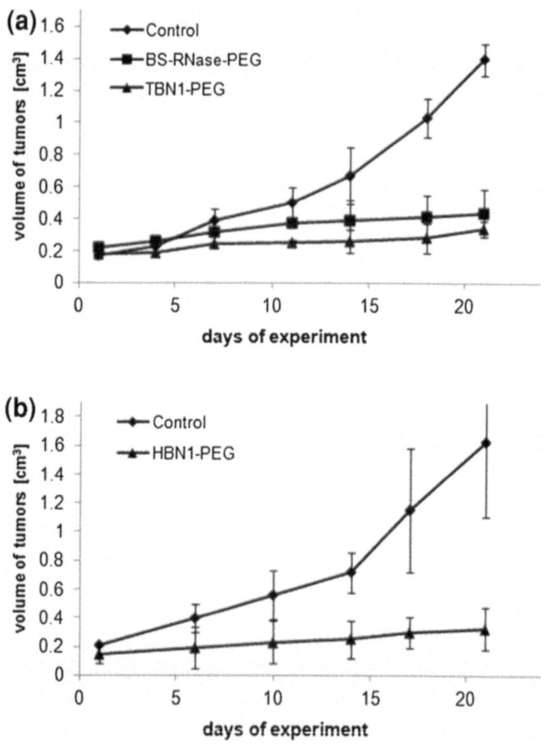

Fig. 6.3 Plant recombinant nucleases produced inhibitory effect on the growth of human melanoma. Human melanoma tumors were grown in athymic nu/nu mice. 15 μg of conjugated R-TBN1-PEG, R-HBN1-PEG and 100 μg of BS-RNase-PEG was intravenously administered seven times for a period of 21 days. Phosphate buffered saline was used as a control. **a** R-TBN1-PEG and BS-RNase-PEG, **b** R-HBN1-PEG [27]. Results represent mean ± SD, n = 3–5 mice in both control and treated groups

tested and their biological effects published [21, 27, 28]. The experiments were performed *in vitro* on myeloid leukemia cell line ML-2 and in vivo on athymic (nu/nu) mice CD-1 bearing human melanoma CD-32 and prostate carcinoma tumor. The nucleases were used in both free form and conjugated form (activated by NHS-PEG, 5 kDa). It was found that the free form of all three nucleases were ineffective *in vitro*. On the other hand, both free (intratumorally applicated) and conjugated (intravenous applicated) form significantly inhibited the tumor growth in mice. Figure 6.3 shows the effect of both R-TBN1-PEG and R-HBN1-PEG on human melanoma C-32 growing in athymic mice CD-1, compared with BS-RNase-PEG. In this experiment, 15 μg of both R-TBN1-PEG and R-HBN1-PEG was applied seven times over a period of 21 days. The overall inhibitory effect of both R-TBN1-PEG and R-HBN1-PEG was about 80 % [27]. In comparison, BS-RNase-PEG produced the same effect, but at a dose seven times higher. The effect of all three nucleases on prostate carcinoma was tested using a similar procedure and it resulted in more than 60 % inhibition of tumor growth (Fig. 6.4) [21]. Equally important are the potential negative side effects of antitumor drugs. R-TBN1, R-HBN1 and R-ABN1 exhibited lower negative side effects (e.g. aspermatogenicity, embryotoxicity or immunogenicity) in comparison with BS-RNase, mung bean nuclease and PN nuclease [27, 28]. These properties make them promising anticancer therapeutics. It is worth mentioning that R-HBN1 has been patented as "a recombinant plant nuclease as antitumor therapeutic with low side effects" (No. 302164, Industrial Property Office, Czech Republic).

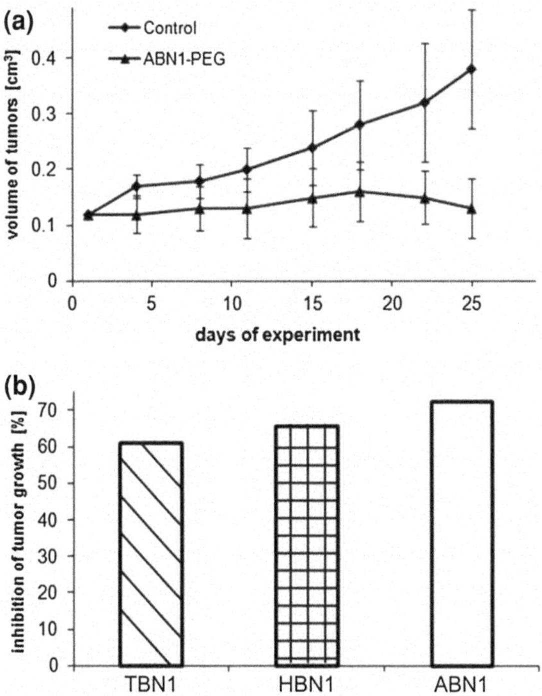

Fig. 6.4 Plant recombinant nucleases produced inhibitory effect on the growth of human prostate carcinoma. Prostate carcinoma tumors were grown in athymic nu/nu mice. 10 μg of conjugated R-TBN1-PEG, R-HBN1-PEG and R-ABN1-PEG was intravenously administered two times per week for a period of 25 days. Phosphate buffered saline was used as a control. **a** R-ABN1, **b** R-TBN1, R-HBN1 and R-ABN1 [21]. Results represent mean ± SD, n = 4–6 mice in both control and treated groups

6.5 PN Nuclease

PN nuclease isolated from the pollen of black pine (*Pinus nigra*) is another example of plant nuclease from *nuclease I* family possessing antitumor properties. This nuclease was isolated in its native form, because efforts to read its nucleotide sequence have failed. PN nuclease was purified by ion exchange (cation exchanger), affinity (heparin) and gel chromatography (unpublished).

6.5.1 Biochemical Properties of PN Nuclease

The molecular weight of PN nuclease was estimated by SDS-PAGE to be 35 kDa. On the other hand, the molecular weight as determined by gel filtration was about 70 kDa. The results indicated the presence of dimeric PN nuclease under non-reducing and non-denaturing conditions. The activity gel assay revealed that the form of PN nuclease migrating at about 35 kDa was active after renaturation.

PN nuclease was active toward both single- and double-stranded substrates and both DNA and RNA. Its activity was maximal at 70 °C and pH 4. Contrary to R-TBN1, R-HBN1 and R-ABN1, PN nuclease was stable at 60 °C and its activity was completely abolished after 30 min at 80 °C. EDTA (10 mmol/l) added to the sample of PN nuclease and subsequently removed by dialysis, completely suppressed its activity. Magnesium ions (10 mmol/l) then restored the activity of PN nuclease up to 87 %, while manganese, zinc and calcium ions had no or minimal stimulation effect. This indicates that PN nuclease is a Mg^{2+}-dependent nuclease (unpublished).

6.5.2 Biological Effects of PN Nuclease

Antitumor properties of PN nuclease were tested in vitro on ML-2 cell line (myeloid leukemia) and *in vivo* using athymic (nu/nu) mice CD-1 bearing human melanoma C-32 [14]. PN nuclease showed no significant antitumor effect in vitro, similar to R-TBN1 and R-HBN1 [27]. On the other side, pegylated PN nuclease applied seven times into melanoma tumors growing in athymic mice caused the inhibition of tumors growth. As in the case of R-TBN1, the result was compared with the effect of BS-RNase. BS-RNase-PEG caused 79 % inhibition of melanoma growth with a dose of 100 μg. 10 μg of pegylated PN nuclease inhibited the melanoma growth by 46 %. The antigenicity of PN nuclease was lower than the antigenicity of BS-RNase. Moreover, PN nuclease did not lose its antitumor properties when bound to its antibody. On the other hand, PN nuclease exhibited aspermatogenic and embryotoxic effects. It significantly decreased the width of the spermatogenic layer of injected testes and the number of splitted embryos, in comparison with a control (only 31 vs. 70 % of splitted control embryos). Even if these negative side effects were milder in comparison with BS-RNase (which is strongly embryotoxic), the lack of knowledge of nucleotide sequence (necessary for recombinant production) led to a termination of further anticancer experiments with this nuclease [14].

6.6 Conclusion

A variety of experiments have demonstrated the antitumor properties and medicinal potential of plant nucleases. They are able to inhibit the growth of various types of tumors in vivo. Some of them exert mild side effects, which is an advantage, when compared with the overall effect of chemotherapeutics. The biochemical properties of R-TBN1, R-HBN1 and R-ABN1 comply with the definition of enzymes from *nuclease* I family with one exception—the efficient cleavage of double-stranded substrates. In the future, a new family named *nuclease II* could be defined, as mentioned in the work of Siwecka [29]. Because plant nucleases seem to be mostly glycoproteins, their production is limited to cells with a glycosylation machinery. R-TBN1, R-HBN1 and R-ABN1 are currently produced in a recombinant form in tobacco plants. Efforts are being made to produce these substances in yeasts to obtain higher yields in a shorter period. Solving of the R-TBN1 structure can elucidate the catalytic mechanism of this nuclease. This knowledge can be used for crystallographic studies of other related plant nucleases, and also for proposing their catalytic mechanism.

Further efforts should be made to elucidate a mechanism of action against tumors, which comprise penetration of the nuclease into the tumor cell, its intracellular routing, degradation of target molecules and/or potential interactions with other cellular components. Finally, it is clear that the plant nucleases described here comprise only a small sample of all existing (including those not yet discovered) plant nucleases with antitumor effect. Tremendous potential lies in the *nuclease I* family.

Acknowledgments Thanks are due to the following colleagues who participated in different experiments. I thank Dr. Poučková and Dr. Zadinová for carrying out the biological tests in vivo, Dr. Matoušek and Dr. Kalous for their evaluation of the negative side-effects. Thanks go to Dr. Dohnálek and Dr. Koval for crystallographic studies on R-TBN1 and to Dr. Lipovová, Dr. Matoušek and Assoc. Prof. Spiwok for critical reading and English corrections.

This work was supported by Ministry of Education, Youth and Sports grant MSM6046137305 and LC 06034, by specific university research (MSMT no. 21/2010) and by Czech Science Foundation grant 521/09/1214 and AV0Z50510513. The research leading to these results has also received funding from the European Community's Seventh Framework Programme (FP7/2007-2013) under/grant agreement/no. 229518.

References

1. Green PJ (1994) Ribonucleases of higher plants. Annu Rev Plant Physiol Plant Mol Biol 45:421–425
2. Pérez-Amador MA, Abler ML, Rocher EJD, Thompson DM, Hoof AV, LeBrasseur ND (2000) Identification of BFN1, a bifunctional nuclease induced during leaf and stem senescence in *Arabidopsis*. Plant Physiol 122:169–179

3. Aoyagi S, Sugiyama M, Fukuda H (1998) BEN1 and ZEN1 cDNAs encoding S1-type DNases that are associated with programmed cell death in plants. FEBS Lett 429:134–138
4. Muramoto Y, Watanabe A, Nakamura T, Takabe T (1999) Enhanced expression of a nuclease gene in leaves of barley plants under salt stress. Gene 234:315–321
5. Yang B, Wen X, Kodali NS, Oleykowski CA, Miller CG, Kulinski J et al (2000) Purification, cloning, and characterization of the CEL I nuclease. Biochemistry 39:3533–3541
6. Panavas T, Pikula A, Reid PD, Rubinstein B, Walker EL (1999) Identification of senescence-associated genes from daylily petals. Plant Mol Biol 40:237–248
7. Yupsanis T, Symeonidis L, Kalemi T, Moustaka H, Yupsani A (2004) Purification, properties and specificity of an endonuclease from *Agropyron elongatum* seedlings. Plant Physiol Biochem 42:795–802
8. Sugiyama M, Ito J, Aoyagi S, Fukuda H (2000) Endonucleases. Plant Mol Biol 44:387–397
9. Pennell RI, Lamb C (1997) Programmed cell death in plants. Plant Cell 9:1157–1168
10. Ito J, Fukuda H (2002) ZEN1 is a key enzyme in the degradation of Nuclear DNA during programmed cell death of tracheary elements. Plant Cell 14:3201–3211
11. Brown PH, Ho T-HD (1986) Barley aleurone layers secrete a nuclease in response to gibberellic acid. Plant Physiol 82:801–806
12. Gan S, Amasino RM (1997) Making sense of senescence. Plant Physiol 113:313–319
13. Souček J, Škvor J, Poučková P, Matoušek J, Slavík T, Matoušek J (2006) Mung bean sprout (*Phaseolus aureus*) nuclease and its biological and antitumor effects. Neoplasma 53:402–409
14. Lipovová P, Podzimek T, Orctová L, Matoušek J, Poučková P, Souček J et al (2008) Antitumor and biological effects of black pine (*Pinus nigra*) pollen nuclease. Neoplasma 55:158–164
15. Matoušek J, Kozlová P, Orctová L, Schmitz A, Pešina K, Bannach O et al (2007) Accumulation of viroid-specific small RNAs and increase in nucleolytic activities linked to viroid-caused pathogenesis. Biol Chem 388:1–13
16. Matoušek J, Orctová L, Škopek J, Pešina K, Steger G (2008) Elimination of hop latent viroid upon developmental activation of pollen nucleases. Biol Chem 389:905–918
17. Matoušek J, Tupý J (1984) Purification and properties of extracellular nuclease from tobacco pollen. Biol Plant 26:62–73
18. Wilson CM (1968) Plant Nucleases. I. separation and purification of two ribonucleases and one nuclease from corn. Plant Physiol 43:1332–1338
19. Hosseini R, Mulligan BJ (2002) Application of rice (*Oryza sativa* L.) suspension culture in studying senescence in vitro (I). Single strand preferring nuclease activity Electron. J Biotechnol 5:42–54
20. Matoušek J, Vrba L, Škopek J, Orctová L, Pešina K, Heyerick A et al (2006) Sequence analysis of a "True" Chalcone Synthase (chs_H1) Oligofamily from hop (*Humulus lupulus* L.) and PAP1 activation of chs_H1 in Heterologous systems. J Agric Food Chem 54:7606–7615
21. Podzimek T, Matoušek J, Lipovová P, Poučková P, Spiwok V, Šantrůček J (2011) Biochemical properties of three plant nucleases with anticancer potential. Plant Sci 180:343–351
22. Romier C, Dominguez R, Lahm A, Dahl O, Suck D (1998) Recognition of single-stranded DNA by nuclease P1: high resolution crystal structures of complexes with substrate analogs. Protein Struct Funct Genet 32:414–424
23. Dohnálek J, Koval T, Lipovová P, Podzimek T, Matoušek J (2010) Structure analysis of group I plant nucleases. J Sync Rad 18:29–30
24. Koval T, Lipovová P, Podzimek T, Matoušek J, Dušková J, Skálová T et al (2010) Crystallization of recombinant bifunctional nuclease TBN1 from tomato. Acta Crystallogr F67:124–128
25. Rangarajan ES, Shankar V (2001) Sugar non-specific endonucleases. FEMS Microbiol Rev 25:583–613
26. Fujimoto M, Kuninaka A, Yoshino H (1974) Identity of phosphodiesterase and phosphomonoesterase activities with P1 nuclease (a nuclease from *Penicillium citrinum*). Agr Biol Chem 38:785–790

27. Matoušek J, Podzimek T, Poučková P, Stehlík J, Škvor J, Souček J et al (2009) Antitumor effects and cytotoxicity of recombinant plant nucleases. Oncol Res 18:163–171
28. Matoušek J, Podzimek T, Poučková P, Stehlík J, Škvor J, Lipovová P et al (2010) Antitumor activity of apoptotic nuclease TBN1 from *L. esculentum*. Neoplasma 57:339–348
29. Siwecka MA (1997) Purification and some properties of a novel dsRNA degrading nuclease bound to rye germ ribosomes. Acta Biochim Pol 44:61–68

Author Biography

Tomas Podzimek was born in the town Ceske Budejovice, Czech Republic. After finishing the Masaryk Secondary School of Chemistry, he was accepted to study biochemistry at the Institute of Chemical Technology Prague. His diploma thesis, made in the Laboratory of Protein Biochemistry and Biotechnology, was concerned with isolation, purification and characterization of mutated forms of β-galactosidase from psychrotrophic bacteria *Arthrobacter sp.* C2-2-1. After the defense of diploma thesis, he entered postdoctoral studies in the same laboratory. He has joined the project concerned with characterization of three plant nucleases with anticancerogenic effects.

Chapter 7
Bovine Seminal Ribonuclease and Its Special Features: When Two is Better Than One

Delia Picone, Antonello Merlino and Roberta Spadaccini

Abstract Bovine Seminal Ribonuclease (BS-RNase) is a structurally well-characterized protein which has been isolated in Naples in the sixties. It is a homodimer whose subunits are linked by two disulfide bridges and exists as an equilibrium mixture of two isomers, with and without swapping of the N-termini. The protein has an enzymatic activity very similar to that of the well-known Ribonuclease A, but it also displays a potent antitumor activity. We describe our current understanding of how BS-RNase exerts its cytotoxicity against malignant cells, which is strictly related to the RNA hydrolysis occurring in the host cytosol. Structural and biochemical data collected on the BS-RNase isoforms and on some mutants suggest that the swapped form is responsible for the citotoxicity, mainly because its compact 3D structure allows the protein to evade the RNase inhibitor, a protein acting as a cell sentry against exogenous ribonucleases. Structural comparisons among dimeric ribonucleases and site-directed mutagenesis studies suggest that only a few residues are critical to stabilize this compact structure even in the cytosol, where the reducing environment promotes the selective cleavage of the interchain disulfides. However, further engineering studies are needed to develop new potential anticancer drugs based on BS-RNase and its derivatives.

7.1 Introduction

Among mammalian ribonucleases the bovine seminal enzyme, henceforth indicated as BS-RNase, shows two main distinctive features. It is indeed the only native dimeric protein and has a natural cytotoxic activity selective for tumor cells.

D. Picone (✉) · A. Merlino
Department of Chemical Sciences, University of Naples Federico II,
via Cintia, 80126 Naples, Italy
e-mail: delia.picone@unina.it

A. Merlino
Institute of Biostructure e Bioimages, CNR, Via Mezzocannone 16, 80100 Naples, Italy

R. Spadaccini
Department of Biological, Geological and Environmental Sciences, University of Sannio,
via Port'Arsa 11, 82100 Benevento, Italy

E. F. Fang and T. B. Ng (eds.), *Antitumor Potential and Other Emerging Medicinal Properties of Natural Compounds*, DOI: 10.1007/978-94-007-6214-5_7,

These two properties are inter-twined, because the dimeric structure is an essential requisite for the antitumor activity, and is critical also for the other "special" functions i.e. additional with respect to the basal ribonucleolytic activity, such as immunosuppressive activity and embryotoxicity. In the monomeric state, indeed, BS-RNase is lacking of all the special functions, including antitumor activity, although it displays an enzymatic activity even higher than that of the native, dimeric enzyme. Another singular feature of BS-RNase is the presence of multiple forms, showing small differences in all the structural levels: primary, secondary, tertiary and even quaternary structure. Some of the structural and functional properties of BS-RNase have been recently reviewed by Giuseppe D'Alessio [1], who first characterized this protein, and also provided a historical retrospective of the related research results. While D'Alessio's review has mainly covered aspects of biochemical features of the native protein, herein we discuss the structural and/or biological progresses carried out on BS-RNase in the last few years, with particular emphasis on the mechanism of antitumor action of the protein. We describe our current understanding of how BS-RNase exerts its cytotoxicity against malignant cells, as a detailed understanding of this mechanism provides new avenues for the development of powerful anticancer molecules. Finally, we close the review with a perspective on the future challenges for this field.

7.2 Structure and Function of the Native Protein

7.2.1 General Features

Native BS-RNase, as isolated either from seminal plasma or vesicles, is a mixture of three isoforms, which can be easily separated by cation-exchange chromatography [2]. The three isoforms originate by a deamidation process occurring spontaneously at level of Asn67, a well characterized reaction [3] which gives rise to two products, containing either a "normal" or an "iso" peptide bond, which are formed in a 3:2 ratio. The deamidated, more acidic, subunit is indicated as α, independently on the nature of the peptide bond involving the CO group of Asp 67, whereas the unreacted, Asn-containing, subunit is indicated as β. As a consequence, three dimers differing for the net charge (αα, αβ and ββ) are formed, the molar ratio between them being about 1:3:6 in the natural enzyme. The ratio αα:αβ:ββ is constant in all preparations of seminal ribonuclease, independent of the purification procedure as well as the biological source of the enzyme (seminal plasma or seminal vesicles) [2]. Thus, deamidation occurs in vivo before the protein is secreted from the seminal glands. No functional differences have been reported for the three isoforms, so that the biological function of this reaction has not yet been clarified. From a structural point of view, all the X-ray structures solved so far were obtained either from forms containing "normal" peptide bonds, due to their higher population in the natural enzyme, or containing Asp in position

67, in the case of recombinant proteins. In this latter case the choice to express the protein using a cDNA coding for the Asp67 variant is motivated by the opportunity to work with homogeneous samples, i.e. to avoid side reactions during the experiments, even in consideration that the deamidation of Asn67 occurs also in vitro [2], converting the most cationic ββ isoform into the αα isoform via the intermediate αβ isoform. Moreover, in all the structural and functional characterizations of BS-RNase performed so far, no differences were detected among the Asn and Asp containing variants.

Besides the differences originated by the deamidation reaction, native BS-RNase samples are enriched of additional isomers. BS-RNase is in fact isolated as an equilibrium mixture of two dimeric forms, indicated as M = M and M × M, which have the same quaternary structure, since the two protein subunits are held together by the additional constraints of two inter-chain disulfide bridges formed between Cys31 and Cys32. The M × M/M = M equilibrium mixture contains a slight molar excess of the M × M form [4]. In the M = M form, the two subunits are linked only by the two interchain disulfide bonds, whereas in the M × M form, the two subunits share additional non-covalent interactions established by the swapping of their N-terminal α-helices. Under mild reducing conditions, the two interchain disulfides are selectively cleaved and the M = M isoforms readily is converted into monomers, whilst M × M retains a dimeric structure because the two subunits are still intertwined through the interchanged N-terminal tails. This interesting feature is the basis of the special biological activities of BS-RNase, and notably of the antitumor activity associated to its M × M isoform, because the steric hindrance of the dimers prevents their inactivation by the cytosolic ribonuclease inhibitor (RI) [5], as we will discuss in detail in the following sections. The multiple forms of BS-RNase and their interconversions and interactions are depicted in Fig. 7.1.

7.2.2 Enzymatic Activity and Other Biological Functions

The enzymatic mechanism of BS-RNase is identical to that of the well-known prototype bovine pancreatic RNase (RNase A), consisting in the cleavage of RNA at pyrimidine sites in a transphosphorolytic-hydrolytic two-step process. The two enzymes share most of their residues, showing more than 80 % of sequence identity, and have identical catalytic and sub-catalytic sites. However significant differences are found in the kinetic aspects: the second, rate-limiting hydrolytic reaction step follows a non-hyperbolic behaviour in the case of M × M, whereas it has a Michaelian kinetics in M = M. The presence of a non-hyperbolic behaviour is a distinctive feature of the M × M isomer, because monomeric BS-RNase also displays a Michaelian kinetics [6].

A deeper investigation of the reaction kinetics revealed a mixed cooperativity for M × M, i.e. negative cooperativity at low substrate concentration (<1 mM) and positive cooperativity at high substrate concentration (>30 mM).

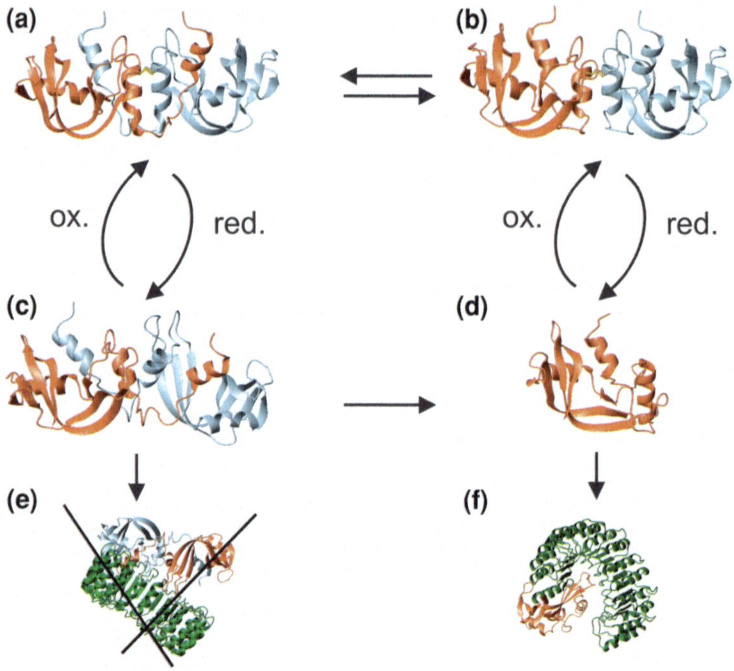

Fig. 7.1 The multiple forms of BS-RNase and their properties. The native enzyme is isolated as an equilibrium mixture of swapped (**a**), and unswapped (**b**), isoforms. The selective reduction of the interchain disulfides converts the swapped form into a non-covalent swapped dimer (**c**), and the unswapped one into two monomers (**d**), the non-covalent dimer is able to evade the binding by the Ribonuclease Inhibitor (**e**), while the monomeric derivative is inactivated (**f**)

This result has been explained by postulating the presence of an allosteric site in addition to the two catalytic sites which, as we will describe in detail below, are mostly contributed by residues belonging to both subunits. To explain the mixed cooperativity between the sites, it has been suggested that the binding of the substrate at one site reduces the affinity for the second site and exposes others, weaker sites, to the binding. The subsequent saturation of the allosteric subsite(s), at a high substrate concentration, makes again available the second catalytic site and elicits a positive cooperativity.

Remarkably, since the first studies it was demonstrated that BS-RNase was very active also on double-stranded RNA and on DNA-RNA hybrids [1]. The catalytic function is a requisite for the special biological activities of BS-RNase, which behaves like antitumor, aspermatogenic and immunosuppressive agents [7]. Since their discovery, all these additional properties, and specifically the cytotoxicity for tumor cells, have been confirmed over and over not only in BS-RNase, but also in other natural ribonucleases isolated both from vertebrates and microbes [8]. In the case of mammalian proteins, it seems that bovine and human pancreatic enzymes acquire the antitumor activity upon dimerization, which can be induced by protein

Table 7.1 Summary of BS-RNase mutants produced and characterized

Mutant	M/D[a]	%[b]	Catalytic activity[d]	Antitumor activity[d]	Other activities	Refs.
N67D, P19A	D	70	Comparable	n.a.[7]	n.a.	[38]
N67D, L28Q	D	50	Comparable	n.a.	n.a.	[38]
N67D, P19A, L28Q	D	50	Comparable	n.a	n.a.	[38]
N67D,G16S/N17T/P19A/S20A	D	70	Comparable	n.a.	n.a.	[39]
G16S/N17T/P19A/S20A	D	70	Comparable	Comparable (MLC, K562, ML-1 cells)	Same ims[e]	[48]
M35L/Q37 K/G38D/K39R	D	n.a.	Comparable	Comparable (MLC, K562, ML-1 cells)	Same ims[e]	[48]
K62 N/T64A	D	n.a.	Comparable	Comparable (MLC, K562, ML-1 cells)	Same ims[e]	[48]
R80S/V102A/E103 N	D	n.a.	Comparable	Comparable (MLC, K562, ML-1 cells)	Same ims[e]	[48]
G111E/K113 N/S115Y	D	n.a.	Comparable	Reduced cytotoxic activity	Same ims[e]	[48]
G16S/N17T/P19A	D	n.a.	Comparable	Comparable (MLC, K562, ML-1 cells)	Same ims[e]	[48]
N67D, R80S	D	60	Comparable	Less cytotoxic (SVT2 cell)	n.a	[41, 42]
N67D, R80S, G16S/N17T/P19A/S20A	D	30	Comparable	Less cytotoxic (SVT2 cell)	n.a	[42]
C31S	D	21–23	Identical	Reduced cytotoxic activity	Modest asp[f]	[18, 48]
C32S	D	23–29	Identical	Reduced cytotoxic activity	Modest asp[f]	[18][48]
H119D	D	60	Big decrease	Inactive	Little ims[e], asp[f]	[18, 48]
H12D	D	n.a.		Inactive	Little ims[e], asp[f]	[18]
A4C, V118C	D	0	Less active	3-times less cytotoxic	n.a.	[49]

(continued)

Table 7.1 (continued)

Mutant	M/D[a]	%[b]	Catalytic activity[d]	Antitumor activity[d]	Other activities	Refs.
G88R	D	n.a.	20 %	More active (K562 cells)	n.a	[50]
G38 K	D	70	20 %	More active (SVT2 cells)	n.a	[15]
G38 K,K39G,G88R	D	n.a.	20 %	More active (K562 cells)	n.a	[50]
G38R,K39D,G88R	D	n.a.	20 %	More active (K562 cells)	n.a	[50]
G88R, S89E	D		Comparable	Comparable	n.a	[51]
H119D	D	60	Big decrease	No activity	No activity	[52]
BS-RNase-KDEL[c]	n.a.	n.a.	10 %	No effects on SVT2 cells	n.a.	[11]
Monomeric derivatives						
Substitution of cysteine residues						
C31 K, C32S, G38D	M	0	1.7-fold increase	n.a.	n.a.	[53]
C31S						
C31A, C32A	M	0	20 %	No cytotoxic activity	n.a.	[50]
C31A,C32A,G88R	M	0	20 %	More active on K562 cells	n.a.	[50]
C31A,C32A,G38 K,K39G,G88R	M	0	20 %	More active on K562 cells	n.a.	[50]
C31A,C32A,G38R,K39D,G88R	M	0	20 %	More active on K562 cells	n.a.	[50]
C31A,C32A,G38 K,K39G,N67R,G88R	M	0	20 %	More active on K562 cells	n.a.	[50]
C31 K, C32S	M	0	1.2-fold increase	Less cytotoxic	No ims[e]	[48, 51, 54]
C31 K, C32S,T87 W, G88R, S89E, S90 W	M	0	Comparable	More active on SVT2 cells	n.a.	[51]
C31 K, C32S,G88R, S89E	M	0	Comparable	More active on SVT2 cells	n.a	[51]
Variants with modificated cysteines						
AcetylatedCys31	M	0	Comparable	No activity	1/10 of ims[e]	[55, 56]

(continued)

Table 7.1 (continued)

Mutant	M/D[a]	%[b]	Catalytic activity[d]	Antitumor activity[d]	Other activities	Refs.
Acetylated Cys32	M	0	Comparable	No activity	1/10 of ims[e]	[56]
S-carboxymethylated bis-Cys31,Cys32	M	0	Comparable	Inactive	n.a	[57]
S-carboxyamidomethylated bis-Cys31,Cys32	M	0	Comparable	Inactive	n.a	[57]
S-aminoethylated bisCys31,Cys32	M	0	Comparable	Inactive	n.a	[57]
S-ethylamine bis-Cys31,Cys32	M	0	Comparable	Cytotoxic on malignant SVT2-3T3	n.a	[57]
Bis-Cys31,Cys32-S-glutathione	M	0	Comparable	Cytotoxic on malignant SVT2-3T3	n.a	[57]
N67D, bis-Cys31,Cys32-S glutathione	M	0	Comparable	Inactive	n.a.	[42]
N67D, P19A, bis-Cys31,Cys32-S-glutathione	M	0	Comparable	n.a. (possibly inactive)	n.a.	[38]
N67D, L28Q, bis-Cys31,Cys32-S-glutathione	M	0	Comparable	n.a. (possibly inactive).	n.a.	[38]
N67D, P19A, L28Q, bis-Cys31,Cys32-S-glutathione	M	0	Comparable	n.a. (possibly inactive)	n.a.	[38]
N67D,G16S/N17T/P19A/S20A, bis-Cys31,Cys32-S-glutathione	M	0	Comparable	n.a. (possibly inactive)	n.a.	[39]
G38 K, bis-Cys31,Cys32-S-glutathione	M	0	20 %	More active on SVT2 cells	n.a.	[15]

[a] *M* monomer; *D* dimer;
[b] percentage of domain swapped dimer at equilibrium;
[c] BS-RNase in which a Lys-Asp-Glu-Leu peptide segment is inserted at the C-terminal ends of the subunit chains;
[d] activity in comparison with wide type;
[e] *ims* immunosuppressive activity;
[f] asp, aspermatogenic activity;
[g] n.a., not available

engineering [9] or by physico-chemical conditions [10]. Table 7.1 reports most of the BS-RNase derivatives and their main structural and functional features.

7.2.3 Antitumor Activity

The knowledge of the mechanism underlying the cytotoxic activity of BS-RNase, which is selective for tumor cells, can be helpful to engender this property in other, pancreatic-like proteins, and to exploit their potential as anti-tumor agents. Although the studies on BS-RNase cytotoxic mechanism are limited, some hypotheses have emerged and some features of BS-RNase, besides the catalytic activity, have been proved to be necessary for its capability to kill malignant tumor cells. These features, sketched in Fig. 7.2, can be summarized as follows: (a) selective binding to the surface of tumor cells; (b) internalization into the cytosol and selective cleavage of the interchain disulfides; (c) RNA degradation by the form resistant to the RNase inhibitor (RI) binding, whilst (d) the monomeric derivative is inactivated by RI binding.

It seems that the enzyme concentrates on the surface of malignant cells, then it crosses the cell membrane and is packaged in endosomes. From the endosomes the protein reaches the Golgi, and then the cytosol, where it exerts its ribonuclease activity [11]. It is not clear how BS-RNase is transported from the Golgi

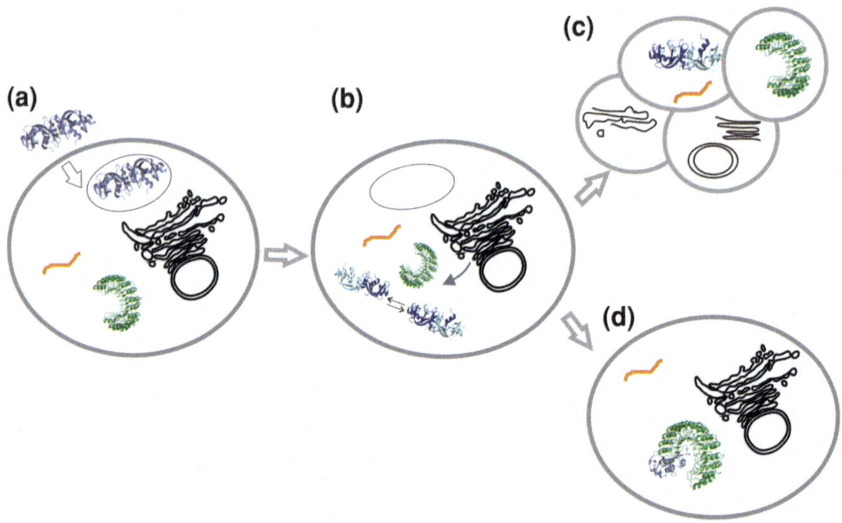

Fig. 7.2 Schematic representation of the biological steps required for the antitumor activity of BS-RNase; **a** binding to the tumor cell membrane; **b** internalization into the cytosol and selective cleavage of the interchain disulfides; **c** RNA degradation by the swapped isoform, which is resistant to the RI binding; **d** inactivation by RI binding of the monomeric derivative, formed upon reduction of the unswapped isoform

to the cytosol and why it is able to kill only the malignant cells. The former hypothesis that antitumor action could be based on the ability of BS-RNase to interact at specific receptor sites on the tumor cell membrane [12] has been promptly confuted, because the enzyme binds cell surface and is internalized also by non-malignant cells, although it is not able to kill them [13]. This should be due to a different intracellular BS-RNase management in normal cells, where the cytotoxic protein displays no effects on RNA stability and on protein biosynthesis. Nevertheless, the differences in membrane composition between malignant and non-malignant cells could also play a crucial role. In fact, already in 1994, it has been shown that BS-RNase is able to destabilize negatively charged membranes [14] and recent publications have provided proof-of-concept for the role of charges on the protein surface in the mechanism of antitumor activity of BS-RNase [15]. Given the higher negatively charge lipid content of the tumor cells with respect to non-malignant ones, and the high isoelectric point of BS-RNase, the interaction with the membrane tumor cells is driven essentially by electrostatic interactions. Several studies have suggested that BS-RNase approaches the negative charge cell membrane by its N-terminal face, which is the most cationic [15, 16]. The enhancement of the protein positive charge by chemical modification [17] or protein engineering [15] results in an increased cytotoxic activity. As mentioned before, the biochemical basis of the antitumor activity relies on the RNA degrading function, because chemically inactivated BS-RNase molecules are devoid of any antitumor activity. Moreover, the cytosol of mammalian cells is very rich in a horseshoe-shaped protein which binds very tightly to monomeric ribonucleases [5], notably monomeric pancreatic type proteins, to preserve the integrity of the different RNA molecules necessary for the protein biosynthesis machinery. An outstanding property of BS-RNase is the capability to evade the binding of this molecule, hereafter referred as RI (RNase inhibitor), due to its structural features. The swapping of N-terminal helices allows indeed the $M \times M$ isomer to keep a dimeric structure and to evade RI binding even in the reducing cytosol environment, thus preserving the enzymatic activity and the capability to kill the tumor cells. This characteristic is directly related to its structure, as confirmed by mutagenesis studies showing that the cytotoxic activity of BS-RNase is correlated to the amount of swapped isoform [18].

7.2.4 Structural Features of the Different Forms

The structures of both $M = M$ and $M \times M$ isoforms of BS-RNase have been extensively studied in the last 30 years to find the structural basis of the special biological activities of the enzyme, in particular its strong cytotoxic activity towards malignant tumor cells, and potentially to develop drugs that can overcome tumor resistance and minimize toxic effects to normal cells. A complete list of X-ray structures of BS-RNase deposited in the Protein Data Bank is reported in Table 7.2.

Table 7.2 List of BS-RNase structures deposited in the Protein Databank

Structure description	PDB code	Method	Resolution (Å)	Refs
Monomeric forms				
N67D mutant	1QWQ	NMR[1]		[33]
N67D mutant	2LFJ	NMR[1]		[34]
S-carboxyamidomethyl-Cys31, S-carboxyamidomethyl-Cys32	1N1X	X-ray[2]	1.45	[31]
S-carboxyamidomethyl-Cys31, S-carboxyamidomethyl-Cys32 in complex with 3′-uridine monophosphate and adenosine	1N3Z	X-ray[2]	1.65	[31]
Swapped forms				
Sulphate bound form	1BSR	X-ray[2]	1.90	[21]
Ligand free form	1R5D	X-ray[2]	2.50	[26]
Complexed with uridylyl-2′,5′-adenosine	11BA	X-ray[2]	2.06	[22]
Complexed with uridylyl-2′,5′-guanosine	11BG	X-ray[2]	1.90	[23]
Complexed with 2′-deoxycytidine-2′-deoxyadenosine-3′,5′-monophosphate	1R5C	X-ray[2]	2.10	[26]
P19A/N67D mutant	1Y92	X-ray[2]	2.20	[39]
G16S/N17T/P19A/S20A/N67D mutant	1Y94	X-ray[2]	2.20	[39]
P19A/L28Q/N67D mutant	3BCO	X-ray[2]	2.25	[40]
Complexed with Uridine 2′ phosphate	3DJO	X-ray[2]	1.60	[28]
Complexed with Uridine 3′ phosphate	3DJP	X-ray[2]	1.60	[28]
Complexed with Uridine 5′ diphosphate	3DJQ	X-ray[2]	1.53	[28]
Complexed with Cytidine 3′ phosphate	3DJV	X-ray[2]	1.60	[28]
Complexed with Cytidine 5′ phosphate	3DJX	X-ray[2]	1.69	[28]
Non-covalent Swapped forms				
Non covalent swapped dimer in complex with 2′-deoxycytidine-2′-deoxyadenosine-3′,5′-monophosphate	1TQ9	X-ray[2]	2.00	[37]
P19A/L28Q/N67D mutant	3BCP	X-ray[2]	2.57	[40]
Unswapped form				
Phosphate bound form	1R3 M	X-ray[2]	2.20	[29]
P19A/L28Q/N67D mutant	3BCM	X-ray[2]	2.25	[40]

[1] Nuclear magnetic resonance;
[2] X-ray crystallography

7.2.4.1 Swapped Isoform

The first structure of M × M isomer of BS-RNase was refined by Mazzarella and co-workers in 1983 [19]. This structure was the first one leading to the hypothesis of a mutual exchange of N-terminal hands between the two subunits of a dimeric protein. Unfortunately the low resolution of the X-ray data precluded a detailed description of this structural feature, that later on was observed also in another protein, the difteria toxin, and was indicated as "3D domain swapping" [20]. To date, about 300 3D domain swapped protein structures have been reported on PDB (for a Database: http://caps.ncbs.res.in/3dswap). With respect to all these proteins, however, BS-RNase retains still a singular feature: it is indeed the only protein whose un-swapped counterpart, *i.e.* M = M, is a dimer too, whereas in all other cases the unswapped form is a monomer.

A more detailed structure of the swapped dimer was published in 1993 [21]. This model confirmed the swapping of the N-terminal α-helices (residues 1–15 of the two subunits) that are hinged on the main body of the enzyme through the peptide 16–22 (hinge peptide), the region that changes conformation in the two isomers. Moreover it provided more details on the geometry of the additional constraints constituted of two disulfide bridges formed by Cys31 and Cys32 linked to Cys32 and Cys31 of the other subunit, respectively. The two BS-RNase chains showed the typical bean-shaped fold of pancreatic-like ribonucleases. Because of the swapping of the N-terminal ends, each of the two active sites results formed from residues coming from both subunits, but the geometry stayed the same as in RNase A. Also the position of sulphate ions in the crystallized structure confirmed the great similarity with RNase A in terms of RNA binding.

In order to add information to the structural basis of interactions of pancreatic-like ribonucleases and to get information on the peculiar structure–activity relationship in BS-RNase, several structures of M × M in complex with substrate analogues have been published. In particular, in 1998 the structure of the complex of M × M with uridylyl (2′,5′) adenosine (2′,5′-UpA) was solved [22]. The overall structure of this complex is very similar to that of native swapped dimer. However, some differences at level of tertiary and quaternary structure are observed. Indeed, when compared to the native enzyme, the complex shows a slight variation of the intersubunit rearrangement. The position of the dinucleotide in the complex is very similar to that expected on the basis of the comparison with similar complexes of RNase A. In a following paper published by Mazzarella's group [23] new insights on the allosteric regulation of BS-RNase came out. Indeed, the crystal structure of M × M in complex with uridylyl (2′,5′)guanosine (2′,5′-UpG) reveals the presence of 4 dinucleotide molecules: two are bound to the active site, as in RNase A [24], while the other two molecules are located outside the active site, the former on the surface of the protein, in a pocket generated by crystal packing, the latter at the interface between the 2 subunits of the swapped dimer. The binding of 2′,5′-UpG does not induce those modifications of the quaternary structure of the enzyme observed for the productive binding of 2′,5′-UpA [22]. The authors hypothesized that the interface site plays a role in the allosteric regulation exhibited by BS-RNase. As previously mentioned, biochemical studies had shown that the M × M isomer, differently from M = M, is endowed with a complicated regulatory mechanism [25]. The structural data suggest that the binding of the substrate at the first catalytic site produces small but significant changes in the second active site of the enzyme (negative cooperativity). At higher substrate concentrations, binding at the interface site also occurs. Since the interface is linked through Asn24 to the hinge peptide and interacts through Arg33 with the N-terminal α-helix (which contains the catalytically important His12), it can alter the binding properties of the active sites and induce a positive cooperativity. This finding suggests that domain swapping may not merely be a mechanism that proteins adopt for the transition from a monomeric to oligomeric state but it can be used to achieve modulations in catalytic function. Another interesting aspect emerged from the comparison of the crystal structures of the ligand free M × M structure and its complex with 2-deoxycitidylyl(3′,5′)-2′-deoxyadenosine [26]. Comparing these structures with the ones formerly published, with and without substrate analogues [21–23], the intrinsic flexibility of the protein was studied. Because the different orientations of the subunits and the subsequent modulation of quaternary structure are not only a result of ligand binding but are also observed comparing different ligand-free structures, the authors suggest that BS-RNase is an example of population shift mechanism, which describes the ligand-binding process in terms of an equilibrium shift of pre-existing conformations to one that best fits incoming ligand [27].

A detailed structural information on the substrate binding modes can be helpful to design inhibitors of the enzyme able to switch on/off the activity of BS-RNase since, as already mentioned, the enzyme has not only a powerful antitumor activity but also several other special properties, like aspermatogeicity and immunosuppression. Recently, Leonidas and coworkers [28] have solved the structure of M × M isomer of BS-RNase in complex with 5 different oligonucleotides. The phosphonucleotide ligands used are competitive inhibitors of the enzyme. All of them bind the active site of BS-RNase and their structural mode of binding suggests ways for further optimization through structure-assisted rational design.

7.2.4.2 Unswapped Isoform

The structure of the unswapped form (M = M), resolved at 2.2 Å resolution, was published in 2003 (pdb code 1R3 M) [29]. The overall structure of this isomer is superimposable to that of the swapped form. The geometry of the active sites in two homodimers is also comparable; for this reason, it is possible that the different catalytic activities of the two isomers could be connected to cooperativity effects. With respect to M×M, in the M = M isoform the hinge peptide is more flexible, and a more symmetric organization of the two subunits is observed. This finding has been attributed to an increased flexibility of the unswapped protein due to the lack of constraints between the two subunits. This increased flexibility could explain the lack of cooperativity [4, 30] in the unswapped protein (M = M presents Michaelis–Menten kinetic), since small conformational variations in one subunit induced upon substrate binding can be easily dissipated before reaching the other subunit.

7.2.4.3 Monomeric Derivative

To complete the definition of the unswapped forms of BS-RNase, it is necessary to describe the structure of the monomer. This structure has been solved both by X-ray crystallography and NMR. First, the crystal structure of the monomeric derivative carbossimethylated at cysteine 31 and 32 and that of its complex with 2′,5′-UpG were reported (pdb code 1N1X and 1N1Y) [31]. In both structures, the hinge peptide is fully disordered, as already suggested by preliminary solution NMR studies [32]. The tertiary structure is different from that found in the MxM dimer mainly in loop regions. A detailed investigation of the structural features of monomeric BS-RNase in solution [33] was carried out in order to find the structural details which induce swapping of the N-terminal arms of BS-RNase. The solution structure determined by 3D heteronuclear NMR (pdb code 1QWQ), shows close similarity with that of RNase A in all regions of regular secondary structure. The results indicate a large difference in the backbone flexibility of the hinge peptide segment 16–22 of the two proteins, which could provide the molecular basis to explain the ability of BS-RNase subunits to swap their N-terminal arms. Other differences are located in the loop 65–72 that is more flexible than the corresponding region of RNase A, maybe because of the mutation N67D that was introduced to avoid sample heterogeneity arising from the spontaneous deamidation of Asn67 [2]. Recently, a detailed study on the stability and dynamics of the monomeric form of BS-RNase [34] was published. The main aim of this work is to understand if the BS-RNase subunit has an intrinsic propensity to undergo a local unfolding and to be converted into an open monomer, according to the Eisenberg hypothesis [20].

Structural and relaxation data indicate that the monomer behaves like a compact, globular protein, and the calculated correlation time allowed to exclude not only the presence of dimeric forms, but also of a partial opening of the protein. The high compactness of the BS-RNase monomeric structure is consistently suggested by the analysis of the H/D exchange data and TEMPOL surface accessibility profiles, as secondary structure elements are all protected from the solvent [34]. On the whole, these data definitely confirm a close similarity between BS-RNase monomer and RNase A, suggesting also that there is no evidence of a pre-opening of the monomeric structure in solution. Moreover relaxation data confirm the presence of two main flexible regions, corresponding to the 16–22 and 65–72 loops; in addition, extra-mobility is found at the C-terminal region (112–115 loop and 119–121 strand), i.e. in the region involved in the swapping of the C-terminal strand in RNase A [35]. A *cis-trans* isomerisation of proline at position 114 is involved in this process [36] but, despite the presence of a *cis* Pro in the same position, the swapping of the C-terminal ends has never been observed in BS-RNase.

Actually, we hypothesize that in different and eventually more severe conditions, the swapping of the C-terminal region could be observed also in BS-RNase.

7.2.4.4 Non-covalent Swapped Dimer

A very crucial paper, aiming at elucidating the structure-cytotoxic activity relationship of BS-RNase, reported the crystal structure of the so-called swapped non covalent dimer (NCD, pdb code 1TQ9) [37]. In the reducing environment of the cytosol, the two disulfide bridges keeping together the two subunits in BS-RNase are expected to be reduced and, while in these conditions the M = M dimer dissociates to the monomers, M $\times$ M is transformed in NCD. A detailed analysis of the structure of NCD reveals that its quaternary assembly prevents the interaction with the ribonuclease inhibitor and allows the enzyme to exert its cytotoxic activity. A fundamental role in determining this structural feature is played by Pro19 and Leu28. Indeed, Pro19 side chain is buried in a cavity formed from Tyr25 and Gln101 of the other subunit, eventually this interaction reduces the flexibility of the hinge peptide and keeps the quaternary structure similar to that of M $\times$ M. On the other hand, the side chains of Leu28 of both subunits, located at the helix interface, have hydrophobic interactions that stabilize the intersubunit interface.

7.3 Design of Mutants by Homology Mutation Approach

As already mentioned, BS-RNase shares 80 % of residues with RNase A, including the catalytic residues (Fig. 7.3a). RNase A and BS-RNase represent a suitable model system to identify the structural determinants which elicit the special biological functions of mammalian ribonucleases, with the long term aim of designing proteins with improved antitumor activity.

7.3.1 From BS-RNase to RNase A

In this direction, several mutants of BS-RNase, in which residues of the enzyme where replaced with corresponding residues of RNase A, have been characterized, as reported in Fig 7.3b and Table 7.1. Considering that the two protein sequences differ for 23 residues, only a limited number of the possible substitutions have been evaluated. Among these, the effect of the mutations Pro19 → Ala (P19A) and Leu28 → Gln (L28Q) on the swapping propensity of the protein have been extensively studied [38]. Combining several spectroscopic techniques, the structure and stability of monomeric forms of native and mutant proteins have been compared. Also, the swapping kinetics of the dimeric forms of variants and parent enzyme have been investigated. The similarity of kinetic behaviour between BS-RNase and P19A seems to rule out the involvement of a *cis–trans* Pro19 isomerization as a crucial step in the swapping mechanism of BS-RNase. On the other hand, the mutation slightly improves the thermal stability of the mutant. The mutation of L28Q instead affects the swapping extent and kinetics. The two side chains of Leu28 stabilize with hydrophobic interactions the interface between the two subunits and the swapping process is faster then in the mutant. These results seem to prove that the swapping propensity of the enzyme is not connected to a particular sequence of the hinge peptide but depends basically on residues located in other regions of the protein. A further proof that the hinge sequence does not interfere with the swapping propensity of BS-RNase came from the work by Picone et al. [39] where the crystal structures of the P19A mutant and that of the mutant in which the full hinge sequence of BS-RNase is replaced with the corresponding one of RNase A are reported. Both mutants again show equilibrium and kinetic parameters of the swapping similar to those of the parent protein. The X-ray structures of the M × M dimer of both mutants indicate a greater mobility of the hinge residues than in the parent protein, whereas the quaternary structure stays the same in all the proteins. The authors suggest that the results indirectly give credit to the hypothesis that the major role of P19 resides in directing the assembly of NCD, i.e. the species produced by selective reduction of the interchain disulfides, and considered responsible for the antitumor activity of BS-RNase.

(a)

```
                   10        20        30        40        50        60
                    |         |         |         |         |         |
BS-RNase  KESAAAKFERQHMDSGNSPSSSSNYCNLMMCCRKMTQGKCKPVNTFVHESLADVKAVCSQKKVTCKNGQ
RNaseA    KETAAAKFERQHMDSSTSAASSSNYCNQMMKSRNLTKDRCKPVNTFVHESLADVQAVCSQKNVACKNGQ

            70        80        90       100       110       120
             |         |         |         |         |         |
BS-RNase  TNCYQSKSTMRITDCRETGSSKYPNCAYKTTQVEKHIIVACGGKPSVPVHFDASV
RNaseA    TNCYQSYSTMSITDCRETGSSKYPNCAYKTTQANKHIIVACEGNPYVPVHFDASV
```

(b)

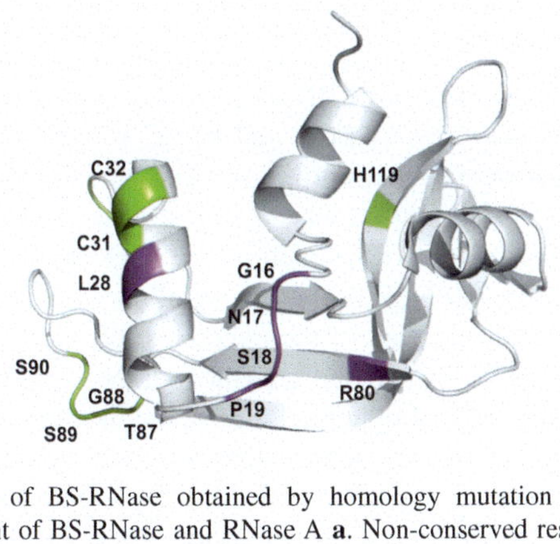

Fig. 7.3 Mutants of BS-RNase obtained by homology mutation approach, guided by the sequence alignment of BS-RNase and RNase A **a**. Non-conserved residues are coloured in *red*. The mutated positions are reported in colour on the structure of the monomeric derivative shown in **b**, using the following colour-code: in purple the residues corresponding to mutants that have been also structurally characterized, in *green* those for which only biochemical data have been collected

The importance of Leu28 and Pro19 for the stability of the NCD structure, suggested on the basis of the model refined by Sica et al. in [37], was confirmed by the structure of the mutant in which both residues were replaced by the corresponding ones of RNase A. Indeed, the crystal structures of the different forms of P19A/L28Q mutant (PALQ) proved that M $\times$ M and M $=$ M dimers have the same structure of the native protein, but the quaternary structure of the non covalent swapped dimer of PALQ (NCD-PALQ) is different from that of the parent protein [40]. This difference has a dramatic effect on the interactions of these forms with the protein inhibitor. Indeed, in the modelled complex with RI, NCD-PALQ is nicely trapped from the inhibitor. This structural result is also in agreement with the functional data, since the PALQ mutant has a cytotoxic activity lower than BS-RNase, thus confirming the importance of Pro19 and Leu28 for the special biological activity of the enzyme.

Also Arg80 has a stabilizing effect on the swapped form, as suggested by the X-ray structure of the native protein [21]. Indeed, it has been shown that in the BS-RNase mutant obtained by replacing Arg80 with Ser, and containing already the 16–22 hinge sequence of RNase A, the M $\times$ M:M $=$ M equilibrium ratio is inverted to 1:2 [41]. The main effect of the substitution of Arg by Ser has been attributed to a decreased stability of the NCD form since both the kinetic and thermodynamic stabilities are reduced in the Ser80 variant with respect to the parent BS-RNase [42].

7.3.2 From RNase A to BS-RNase

Engineering the sequences of RNase A with corresponding residues of BS-RNase has been another way to identify the structural determinants which elicit the special biological functions of mammalian ribonucleases. For example, cytotoxic mutants of RNase A were obtained by introducing Leu28, Cys31, Cys32 and Pro19 in the sequence of the pancreatic enzyme [43–45]. However, as cytotoxins these proteins are not as powerful as BS-RNase [43]. The crystal structure analysis of the non-covalent swapped form of LCC and PLCC mutants, complexed with the substrate analogue 2′-deoxycytidylyl(3′,5′)-2′-deoxyguanosine, has revealed that the dimers adopt an opened quaternary structure, with the two Leu residues fully exposed to the solvent [45]. This structure does not prevent the binding of RI, thus providing the molecular basis of the low cytotoxic activity. Similar results have been obtained for a third mutant of the pancreatic enzyme, engineered with the hinge peptide sequence of the seminal enzyme and the two cysteines in position 31 and 32, but lacking the hydrophobic Leu residue in position 28. According to the authors, the comparison between these three structures and those previously reported for other ribonuclease swapped dimers strongly suggests that, in addition to Pro19 and Leu28, the presence of a glycine at the N-terminal end of the hinge peptide is also important to force the swapped form of RNase A dimer into the compact quaternary organization observed for NCD. These data were further and indirectly confirmed by another, related paper [44], which considered the effect on the swapping propensity and cytotoxicity of RNase A mutants in which the full N-terminal hinge sequence was replaced with the one of BS-RNase together with the mutation S80R. All the proteins have a very poor cytotoxic activity, independently of the swapping propensity, which can even reach the same extent as that in native BS-RNase. In addition, the BS-RNase and RNase A variants with Arg80 have a slightly increased percentage of swapped form in comparison with the corresponding Ser80 variants, and also the dissociation rate of the corresponding NCD forms is decreased [42, 44]. These results suggest that Arg80 has a stabilizing effect on the swapped form, both under reducing and non-reducing conditions. Overall the data suggest that the swapping represents still an essential requisite for the cytotoxic activity, because it keeps the dimeric structure stable even in the reducing cytosolic environment, but other features are essential to design dimeric antitumor ribonucleases, including a strong positive potential at the N-terminal face and a quaternary structure able to evade the cytosolic ribonuclease inhibitor, with or without the interchain disulfide bridges.

7.4 Concluding Remarks, Challenges, Perspective and Future Directions

The results of about 30 years of studies on the structure–function relationship of BS-RNase have been briefly summarized. We have described our understanding of the special biological functions BS-RNase gained through structural biology analyses. Collectively, the reviewed studies have shown that ribonucleolytic activity is required for BS-RNase cytotoxicity and that the swapped dimer is the form responsible for the BS-RNase antitumor activity. It has also been suggested that the structure of the non-covalent dimeric form, obtained by reduction of intersubunit disulfide bridges, plays a crucial role in evading RI, thus providing the structural features needed for the antitumor activity of BS-RNase M $\times$ M form. Only a few residues are critical for stabilizing NCD quaternary structure that allows the acquisition of cytotoxic activity. In particular, crucial roles are played by Pro19 and Leu28. These suppositions have been confirmed by systematic site-directed mutagenesis studies in which individual or combinations of BS-RNase residues have been replaced with the corresponding residues of RNase A [40]. Along this line, site-directed mutagenesis studies should be continued to further the understanding of antitumor properties of the protein and to develop ways to improve the stability of NCD. Recent data have also evidenced some other residues which contribute to the shape and stability of the reduced swapped dimer, such as Gly16 and Arg80, however further studies are necessary in view of designing proteins with better cytotoxicity for tumor cells.

The engineering of BS-RNase should also be continued to gain more insight into the molecular mechanism of domain swapping. In this respect, it should be recalled that BS-RNase represents a unique example since it is the only known domain swapped protein existing in an equilibrium between unswapped dimeric and swapped dimeric form.

Work on the key stations encountered by BS-RNase in its journey from outside the cell to the cytosol should be continued as well. The finding that basic charge on the protein surface is a determinant for the entrance of BS-RNase to malignant cells and hence could be responsible, at least in part, for the biological action of BS-RNase, suggests a possible use of higher BS-RNase oligomers as therapeutic agents. In this respect, it should be underlined that experimental evidence for the antitumor activity of BS-RNase tetramers and higher oligomers in vitro and in vivo has been reported very recently [46]. The potential biological activity of BS-RNase oligomers deserves further investigation, together with their structure and stability. Furthermore, recent data suggested the possibility that cross-linked RNases also may acquire the antitumor activity through the formation of high molecular weight oligomeric species that are not inhibited by RI [47]. Although many other factors are also likely to be involved in controlling the cytotoxicity of RNases, these evidences suggest that a cross-linked high-molecular weight BS-RNase could also exhibit enhanced cytotoxic activity. Another aspect for the future is to perform antitumor tests on different cancer cell lines and to develop clinical trials. In this respect, it should be underlined that the discovery of the antitumor activity of BS-RNase, has been, in our opinion, largely underappreciated in the past. Although antitumor activity exerted by BS-RNase is comparable to that of Onconase, the mammalian protein has never been evaluated in human clinical trials, whereas the frog enzyme, which shows renal toxicity and a possible immunogenicity, has been and is currently tested in diverse Phases of human clinical trials for the treatment of numerous solid tumors.

In conclusion, although studies on BS-RNase have started about 30 years ago, there is definitively plenty of room for new experiments to be developed.

References

1. D'Alessio G (2009) It takes two to flirt with a dimeric RNase. Biopolymers 91:989–994
2. Di Donato A, D'Alessio G (1981) Heterogeneity of bovine seminal ribonuclease. Biochemistry 20:7232–7237
3. Di Donato A, Galletti P, D'Alessio G (1986) Selective deamidation and enzymatic methylation of seminal ribonuclease. Biochemistry 25:8361–8368
4. Piccoli R, Tamburrini M, Piccialli G, Di Donato A, Parente A, D'Alessio G (1992) The dual-mode quaternary structure of seminal RNase. Proc Natl Acad Sci U S A 89:1870–1874
5. Haigis MC, Kurten EL, Raines RT (2003) Ribonuclease inhibitor as an intracellular sentry. Nucleic Acids Res 31:1024–1032
6. Piccoli R, Di Donato A, D'Alessio G (1988) Co-operativity in seminal ribonuclease function kinetic studies. Biochem J 253:329–336
7. Matousek J (1973) The effect of bovine seminal ribonuclease (AS RNase) on cells of crocker tumour in mice. Experientia 29:858–859
8. Fang EF, Ng TB (1815) Ribonucleases of different origins with a wide spectrum of medicinal applications. Biochim Biophys Acta 2011:65–74
9. Di Donato A, Cafaro V, Romeo I, D'Alessio G (1995) Hints on the evolutionary design of a dimeric RNase with special bioactions. Protein Sci 4:1470–1477
10. Ercole C, Lopez-Alonso JP, Font J, Ribo M, Vilanova M, Picone D et al (2011) Crowding agents and osmolytes provide insight into the formation and dissociation of RNase A oligomers. Arch Biochem Biophys 506:123–129

11. Bracale A, Spalletti-Cernia D, Mastronicola M, Castaldi F, Mannucci R, Nitsch L et al (2002) Essential stations in the intracellular pathway of cytotoxic bovine seminal ribonuclease. Biochem J 362:553–560
12. Piccoli R, Vescia S, Bridges SH, D'Alessio G (1990) The antitumor action of seminal ribonuclease tested with the plasmacytoma spleen colonization assay. Ital J Biochem 39:242–249
13. Mastronicola MR, Piccoli R, D'Alessio G (1995) Key extracellular and intracellular steps in the antitumor action of seminal ribonuclease. Eur J Biochem 230:242–249
14. Mancheno JM, Gasset M, Onaderra M, Gavilanes JG, D'Alessio G (1994) Bovine seminal ribonuclease destabilizes negatively charged membranes. Biochem Biophys Res Commun 199:119–124
15. D'Errico G, Ercole C, Lista M, Pizzo E, Falanga A, Galdiero S et al (1808) Enforcing the positive charge of N-termini enhances membrane interaction and antitumor activity of bovine seminal ribonuclease. Biochim Biophys Acta 2011:3007–3015
16. Notomista E, Mancheno JM, Crescenzi O, Di Donato A, Gavilanes J, D'Alessio G (2006) The role of electrostatic interactions in the antitumor activity of dimeric RNases. FEBS J 273:3687–3697
17. Futami J, Maeda T, Kitazoe M, Nukui E, Tada H, Seno M et al (2001) Preparation of potent cytotoxic ribonucleases by cationization: enhanced cellular uptake and decreased interaction with ribonuclease inhibitor by chemical modification of carboxyl groups. Biochemistry 40:7518–7524
18. Kim JS, Soucek J, Matousek J, Raines RT (1995) Structural basis for the biological activities of bovine seminal ribonuclease. J Biol Chem 270:10525–10530
19. Capasso S, Giordano F, Mattia CA, Mazzarella L, Zagari A (1983) Refinement of the structure of bovine seminal ribonuclease. Biopolymers 22:327–332
20. Schlunegger MP, Bennett MJ, Eisenberg D (1997) Oligomer formation by 3D domain swapping: a model for protein assembly and misassembly. Adv Protein Chem 50:61–122
21. Mazzarella L, Capasso S, Demasi D, Di Lorenzo G, Mattia CA, Zagari A (1993) Bovine seminal ribonuclease: structure at 1.9 A resolution. Acta Crystallogr D Biol Crystallogr 49:389–402
22. Vitagliano L, Adinolfi S, Riccio A, Sica F, Zagari A, Mazzarella L (1998) Binding of a substrate analog to a domain swapping protein: X-ray structure of the complex of bovine seminal ribonuclease with uridylyl(2',5')adenosine. Protein Sci 7:1691–1699
23. Vitagliano L, Adinolfi S, Sica F, Merlino A, Zagari A, Mazzarella L (1999) A potential allosteric subsite generated by domain swapping in bovine seminal ribonuclease. J Mol Biol 293:569–577
24. Vitagliano L, Merlino A, Zagari A, Mazzarella L (2000) Productive and nonproductive binding to ribonuclease A: X-ray structure of two complexes with uridylyl(2',5')guanosine. Protein Sci 9:1217–1225
25. Di Donato A, Piccoli R, D'Alessio G (1987) Co-operativity in seminal ribonuclease function: binding studies. Biochem J 241:435–440
26. Merlino A, Vitagliano L, Sica F, Zagari A, Mazzarella L (2004) Population shift vs induced fit: the case of bovine seminal ribonuclease swapping dimer. Biopolymers 73:689–695
27. Ma B, Nussinov R (2010) Enzyme dynamics point to stepwise conformational selection in catalysis. Curr Opin Chem Biol 14:652–659
28. Dossi K, Tsirkone VG, Hayes JM, Matousek J, Pouckova P, Soucek J et al (2009) Mapping the ribonucleolytic active site of bovine seminal ribonuclease. The binding of pyrimidinyl phosphonucleotide inhibitors. Eur J Med Chem 44:4496–4508
29. Berisio R, Sica F, De Lorenzo C, Di Fiore A, Piccoli R, Zagari A et al (2003) Crystal structure of the dimeric unswapped form of bovine seminal ribonuclease. FEBS Lett 554:105–110
30. D'Alessio G, Di Donato A, Piccoli R, Russo N (2001) Seminal ribonuclease: preparation of natural and recombinant enzyme, quaternary isoforms, isoenzymes, monomeric forms; assay for selective cytotoxicity of the enzyme. Methods Enzymol 341:248–263
31. Sica F, Di Fiore A, Zagari A, Mazzarella L (2003) The unswapped chain of bovine seminal ribonuclease: crystal structure of the free and liganded monomeric derivative. Proteins 52:263–271

32. D'Ursi A, Oschkinat H, Cieslar C, Picone D, D'Alessio G, Amodeo P et al (1995) Assignment and secondary-structure determination of monomeric bovine seminal ribonuclease employing computer-assisted evaluation of homonuclear three-dimensional 1H-NMR spectra. Eur J Biochem 229:494–502
33. Avitabile F, Alfano C, Spadaccini R, Crescenzi O, D'Ursi AM, D'Alessio G et al (2003) The swapping of terminal arms in ribonucleases: comparison of the solution structure of monomeric bovine seminal and pancreatic ribonucleases. Biochemistry 42:8704–8711
34. Spadaccini R, Ercole C, Gentile MA, Sanfelice D, Boelens R, Wechselberger R et al (2012) NMR studies on structure and dynamics of the monomeric derivative of BS-RNase: new insights for 3D domain swapping. PLoS ONE 7:e29076
35. Liu Y, Gotte G, Libonati M, Eisenberg D (2001) A domain-swapped RNase A dimer with implications for amyloid formation. Nat Struct Biol 8:211–214
36. Merlino A, Picone D, Ercole C, Balsamo A, Sica F (2012) Chain termini cross-talk in the swapping process of bovine pancreatic ribonuclease. Biochimie 94:1108–1118
37. Sica F, Di Fiore A, Merlino A, Mazzarella L (2004) Structure and stability of the non-covalent swapped dimer of bovine seminal ribonuclease: an enzyme tailored to evade ribonuclease protein inhibitor. J Biol Chem 279:36753–36760
38. Ercole C, Avitabile F, Del Vecchio P, Crescenzi O, Tancredi T, Picone D (2003) Role of the hinge peptide and the intersubunit interface in the swapping of N-termini in dimeric bovine seminal RNase. Eur J Biochem 270:4729–4735
39. Picone D, Di Fiore A, Ercole C, Franzese M, Sica F, Tomaselli S et al (2005) The role of the hinge loop in domain swapping. The special case of bovine seminal ribonuclease. J Biol Chem 280:13771–13778
40. Merlino A, Ercole C, Picone D, Pizzo E, Mazzarella L, Sica F (2008) The buried diversity of bovine seminal ribonuclease: shape and cytotoxicity of the swapped non-covalent form of the enzyme. J Mol Biol 376:427–437
41. Ercole C, Spadaccini R, Alfano C, Tancredi T, Picone D (2007) A new mutant of bovine seminal ribonuclease with a reversed swapping propensity. Biochemistry 46:2227–2232
42. Giancola C, Ercole C, Fotticchia I, Spadaccini R, Pizzo E, D'Alessio G et al (2011) Structure-cytotoxicity relationships in bovine seminal ribonuclease: new insights from heat and chemical denaturation studies on variants. FEBS J 278:111–122
43. Cafaro V, Bracale A, Di Maro A, Sorrentino S, D'Alessio G, Di Donato A (1998) New muteins of RNase A with enhanced antitumor action. FEBS Lett 437:149–152
44. Ercole C, Colamarino RA, Pizzo E, Fogolari F, Spadaccini R, Picone D (2009) Comparison of the structural and functional properties of RNase A and BS-RNase: a stepwise mutagenesis approach. Biopolymers 91:1009–1017
45. Merlino A, Russo Krauss I, Perillo M, Mattia CA, Ercole C, Picone D et al (2009) Toward an antitumor form of bovine pancreatic ribonuclease: the crystal structure of three noncovalent dimeric mutants. Biopolymers 91:1029–1037
46. Gotte G, Helmy AM, Ercole C et al (2012) Double domain swapping in bovine seminal RNase: formation of distinct N- and C-swapped tetramers and multimers with increasing biological activities PLoS One 7:e46804
47. Ellis GA, Hornung ML, Raines RT (2011) Potentiation of ribonuclease cytotoxicity by a poly(amidoamine) dendrimer. Bioorg Med Chem Lett 21:2756–2758
48. Soucek J, Raines RT, Haugg M, Raillard-Yoon SA, Benner SA (1999) Structural changes to ribonuclease A and their effects on biological activity. Comp Biochem Physiol C Pharmacol Toxicol Endocrinol 123:103–111
49. Russo A, Antignani A, Giancola C, D'Alessio G (2002) Engineering the refolding pathway and the quaternary structure of seminal ribonuclease by newly introduced disulfide bridges. J Biol Chem 277:48643–48649
50. Lee JE, Raines RT (2005) Cytotoxicity of bovine seminal ribonuclease: monomer versus dimer. Biochemistry 44:15760–15767
51. Antignani A, Naddeo M, Cubellis MV, Russo A, D'Alessio G (2001) Antitumor action of seminal ribonuclease, its dimeric structure, and its resistance to the cytosolic ribonuclease inhibitor. Biochemistry 40:3492–3496

52. Kim JS, Soucek J, Matousek J, Raines RT (1995) Catalytic activity of bovine seminal ribonuclease is essential for its immunosuppressive and other biological activities. Biochem J 308(Pt 2):547–550
53. Opitz JG, Ciglic MI, Haugg M, Trautwein-Fritz K, Raillard SA, Jermann TM et al (1998) Origin of the catalytic activity of bovine seminal ribonuclease against double-stranded RNA. Biochemistry 37:4023–4033
54. Ciglic MI, Jackson PJ, Raillard SA, Haugg M, Jermann TM, Opitz JG et al (1998) Origin of dimeric structure in the ribonuclease superfamily. Biochemistry 37:4008–4022
55. Soucek J, Marinov I, Benes J, Hilgert I, Matousek J, Raines RT (1996) Immunosuppressive activity of bovine seminal ribonuclease and its mode of action. Immunobiology 195:271–285
56. Slavik T, Matousek J, Fulka J, Raines RT (2000) Effect of bovine seminal ribonuclease and bovine pancreatic ribonuclease A on bovine oocyte maturation. J Exp Zool 287:394–399
57. Bracale A, Castaldi F, Nitsch L, D'Alessio G (2003) A role for the intersubunit disulfides of seminal RNase in the mechanism of its antitumor action. Eur J Biochem 270:1980–1987

Author Biography

Delia Picone graduated in Chemistry and completed her PhD in Chemical Sciences at the University of Naples "Federico II", Italy, in 1987. She was appointed as Research Assistant in 1990 and as Associated Professor of Chemistry in 1992 at the same university. From 1981 to 1982 she worked at the Ecole Polytechniqhe of Palaiseau, France, as a research fellow. She was visiting researcher at the Bijvoet Center for Biomolecular Research of Utrecht, The Netherland (1991) and at the Molecular Division of NIMR of London, United Kindom (2009,and 2011). Her research interests have always focused on structure and function relationships of proteins and peptides using different biochemical and biophysical approaches, notably biomolecular NMR.

Antonello Merlino is Assistant Professor of Physical Chemistry in the Department of Chemical Sciences at the University of Naples Federico II, where he received his degree in Chemistry (1998) and his PhD in Chemical Sciences (2002) for X-ray crystallography and Molecular Dynamics studies of monomeric and dimeric RNases, under the guidance of Professor Lelio Mazzarella. He was visiting PhD student at Mount Sinai School of Medicine, New York, and at University of Rome La Sapienza (2002), Post-doc of the University of Salerno (2003–2004) and of Bioteknet SCpA (2005–2006), visiting post-doc of the University of Buenos Aires (2009 and 2011). His current research topics are related to characterization of structural and dynamic features of proteins and protein-nucleic acid complexes. He belongs to the Editorial Board of Scientifica (Hindawi Publisher) and is Editor in chief of World Research Journal of Peptide and Protein (Bioinfo Publication).

Roberta Spadaccini studied Chemistry at the University of Napoli "Federico II" where she completed also her PhD in "Chemical Sciences" in 2002. During her graduate studies she was visiting the Molecular Division of NIMR in London, United Kindom (2001). From 2002 to 2005 she worked as a post-doc (EMBO longterm fellowship) at EMBL-Heidelberg in the group of Prof. Sattler. In 2005 she was appointed as Assistant Professor in the University of Sannio. In 2009 she was visiting scientist at the department of Biochemistry of the University of Marburg. Her research interests have always focused on structure and function relationships of biomolecules, mainly proteins and peptides, using different biochemical, biophysical an bioinformatics approaches, mainly biomolecular NMR.

Part III
The Perspectives for Ribosome Inactivating Proteins: Unveiling the Working Mechanisms and Potential Medicinal Benefits

Chapter 8
The Discovery of MAP30 and Elucidation of its Medicinal Activities

Sylvia Lee-Huang, Philip Lin Huang and Paul Lee Huang

Abstract In this chapter, we review the discovery of MAP30, a novel multi-functional antiviral and anti-tumor plant protein, and elucidation of its medicinal properties. MAP30, ***M****omordica* ***A****ntiviral* ***P****rotein* 30 kD, is isolated from the medicinal plant *Momordica charantia*. Biochemical and biophysical studies have yielded important insights into the structure, function and mechanism of MAP30. Column chromatography by ionic exchange, affinity binding and size fractionation were applied for the isolation and purification of MAP30. Protein and nucleotide sequencing were used to elucidate the structure of this protein and its gene. The gene encoding MAP30 was cloned and expressed by recombinant technology. Structure activity mapping was performed using 3D-structure determination, bioinformatics and molecular modeling, together with bioactivity confirmation. Results from these studies demonstrate how cutting-edge modern approaches can be applied to discover medicinal activities from natural products and to define their molecular structure, biological function, and mechanisms of action.

This Chapter is dedicated to the memory of Professor Severo Ochoa, our adored mentor, and to Dr. An Fu Lee, our beloved mother and grandmother, for their inspiration and encouragement.

S. Lee-Huang (✉)
Department of Biochemistry and Molecular Pharmacology,
New York University School of Medicine, New York, NY 10016, USA
e-mail: sylvia.lee-huang@med.nyu.edu

P. Lin Huang
American BioScience, Boston, MA, USA

P. Lee Huang
Department of Medicine, Harvard Medical School and Massachusetts General Hospital,
Boston, MA 02114, USA

E. F. Fang and T. B. Ng (eds.), *Antitumor Potential and Other Emerging Medicinal Properties of Natural Compounds*, DOI: 10.1007/978-94-007-6214-5_8,

8.1 Introduction

Over 75 % of the world population relies mainly on medicinal plants for health care [1]. The global market for traditional medicines was estimated at US $83 billion annually in 2008 [1]. In the United States, botanical extracts and dietary supplements are regarded as part of complementary and alternative medicine (CAM) because they are not presently considered an integral part of conventional medicine [2–6]. CAM is neither routinely available in U.S. hospitals nor taught at U.S. medical schools. However, data from medical surveys reveal that the use of natural products and CAM in therapy has increased substantially in the U.S. [4]. Currently, more than 50 % of the US general population uses CAM for healthcare, including herbs, vitamins and other dietary supplements [4–6]. Americans spend about $34 billion annually on CAM, according to the first national estimate of such out-of-pocket spending in more than a decade [7]. Many CAM therapies are centuries old, and ages of practical experience have led to today's applications. Despite the widespread use of CAM, little is known about their structure, mechanism, efficacy and safety or toxicity. Thus, many CAM users are at risk for potential adverse effects, or unintended interactions with other CAM or with prescription medications. To enhance the role of CAM in modern health care, it is essential to use the latest technologies to identify biologically active components, to characterize their structure/function and to define their molecular mechanisms. Our laboratories have been engaged in natural product research in recent years. We use a broad spectrum of modern approaches to identify CAM products and to determine their structure/function, mechanism, efficacy and safety. We identified several lead CAM components against a variety of diseases, from HIV/AIDS to other viral infections, cancers, cardiovascular diseases, diabetes and obesity [8–40]. We determined the structure of these compounds, conducted structural-activity mapping and defined their molecular mechanisms [8–40]. We believe that results from these studies are useful in expanding the options of CAM products to modern medicine. In this chapter, we review the discovery of MAP30 and its medicinal activities.

8.2 CAM of MAP30 Plant: *Momordica charantia*

MAP30, or ***Momordica Antiviral Protein***, 30 kD, is a protein we first described and named [15]. It is isolated from the mature fruit and seed of *Momordica charantia*, also known as bitter melon, or bitter gourd. This medicinal plant is indigenous to China. It has been used in China for centuries as medicinal and health food against cancer and viral infections as well as for immunopotentiating actions. Bitter melon is also widely grown in Africa, Southeast Asia, South America and the Southwestern regions of the United States. It has been used as CAM for various purposes in different regions including digestive disorders, dyspepsia and constipation in South America, anti-malaria, antiviral and anti-HIV in Africa, and anti-diabetes and anti-obesity by American Indians. Our studies on the medicinal components of bitter melon started from bioactivity discovery and verification, and progressed to purification, characterization and standardization, to cloning and expression, 3D structure determination, structure activity mapping, bioinformatics and molecular modeling. Our results demonstrate how cutting-edge approaches can be applied to the discovery of medicinal activities of natural products, to elucidate their structures and functions, and to translate CAM into twenty-first century medicine.

8.3 Isolation, Purification and Characterization

We designed and developed an efficient procedure for the isolation and purification of MAP30 [15, 18]. Briefly, this consists of four main steps which yield homogeneous protein with constant anti-viral and anti-tumor specific activity and no toxicity. Step 1 is extraction and fractionation. Step 2 is anionic exchange chromatography on DEAE-Sepharose. Step 3 is size fractionation on Sephadex G75 superfine. Step 4 is cationic exchange chromatography on CM 52 cellulose. The size, homogeneity and subunit structure of MAP30 were determined by SDS-PAGE. A single band, with molecular weight corresponding to 30 kD, was obtained under both reducing and non-reducing conditions, indicating that the purified MAP30 exists as a single chain polypeptide (Fig. 8.1). Recently, preparations of pure MAP30 have been reported by others [41, 42].

8.4 Medicinal Activities of MAP30

8.4.1 Anti-Viral Activity

Anti-HIV-1. MAP30 inhibits not only *de novo* infection by HIV-1 and cell-to-cell transmission of the virus but also blocks viral replication in already infected cells [15]. Anti-HIV activity was examined in terms of viral infection and replication by syncytial formation and HIV-p24 expression, respectively. MAP30 is potent against HIV-1 with an EC_{50} on the order of 0.2–0.3 nM [15]. However, it is not toxic to uninfected target cells, or to intact animals or humans [15, 18–21, 25]. Furthermore, MAP30 is not toxic to human spermatozoa [24], suggesting that MAP30 may be developed into useful

Fig. 8.1 Purification of MAP30. *Momordica charantia*, the source of MAP30 (*Left*). SDS-PAGE shows purified MAP30 (*Right*). Lanes: *S* molecular weight standards, *M* purified MAP30

non-spermicidal protective agent against the transmission of HIV-1 and other sexually transmitted diseases. It can also be used in sperm washes for *in vitro* fertilization.

Anti-HSV and Anti-HHV8. MAP30 is active against Herpes Simplex Viruses (HSV) [26, 27] and Human Herpes Virus 8 (HHV8) [28]. MAP30 inhibits the infection and replication of HSV-1, HSV-2, and ACV-resistant strains. It also inhibits the proliferation of AIDS-related primary effusion lymphoma (PEL) BC-2 cells [28]. These cells are latently infected with Kaposi's sarcoma-associated herpes virus (KSHV), also known as human herpes virus 8 (HHV8). MAP30 is equally effective in BC-2 cells with or without induction by 12-O-tetradecanoylphorbol-13-acetate (TPA), indicating that it is effective against the HHV8-infected KS tumor cells in both the latent and lytic states. Northern blot and RT-PCR analyses indicate that MAP30 inhibits the expression of viral genes vCD, vIL-6, and v-FLIP [28].

8.4.2 Anti-Tumor Activity

Results from NCI. MAP30 was tested by the Cancer Drug Discovery and Development Program of the National Cancer Institute (NCI) against a panel of 58 human tumor cell lines representing nine cancer types [21]. Potent inhibition was detected with brain (SF539, EC_{50} 0.01 μg/ml), breast (MDA-MB-231, EC_{50} 0.01 μg/ml; HS578T, 0.05–0.07 μg/ml), melanoma (MALME-3M, EC_{50} 0.01–0.02 μg/ml) and renal (ACHN, EC_{50} 0.05–0.07 μg/ml) tumor cell lines.

Preclinical studies-in vivo animal efficacy. The anti-cancer activity of MAP30 against estrogen-independent and highly metastatic human breast tumor MDA-MB-231 was studied *in vitro* and *in vivo* [21, 22]. MAP30 showed dose-dependent inhibition on the proliferation of the tumor cells with an EC_{50} of 0.035 nM (21). *In vivo* anti-tumor activity was studied using MDA-MB-231 xenografts in SCID mice [22]. The mice developed tumors and extensive metastases, began to die after day 35 and all mice succumbed to tumors by day 45. Treatment with MAP30 at 10 μg per injection every other day for a total 10 injections increased the survival of tumor bearing mice to 64 days. In addition, over 25 % of the SCID mice remained disease-free for over 100 days. No toxicity was observed after MAP30 treatment. These results indicate that MAP30 may be therapeutically useful against breast carcinomas. MAP30 has also been reported to promote apoptosis of liver cancer cells both *in vitro* and *in vivo* [41].

8.5 Cloning and Recombinant Expression

We cloned the MAP30 gene and expressed the recombinant protein in bacteria [19], as well as in edible cucurbit plants [20]. Recombinant MAP30 is as active as naturally occurring MAP30 in medicinal activities. The MAP30 gene consists of 1044 base pairs, and encodes 286 amino acids with a leader sequence of 23 amino acids. The mature protein consists of 263 amino acids with one N-glycosylation site (Asn-Val-Thr) at position 250, and no cysteine residues [19].

8.6 Structural–Functional Mapping, Bioinformatics and Modeling

We determined the solution and crystal structures of MAP30 [29, 30]. To define the structural basis of MAP30 activities, we conducted bioinformatic analysis by searching for MAP30 homologues in the non-redundant protein sequence database. A total of 150 reasonable homologies were generated. Each sequence was aligned with MAP30. In so doing, it became evident that Tyr91 (Y91) and Asp124 (D124) are uniquely conserved only in MAP30 and GAP31, ***Gelonium Anti-HIV Protein*** 31 kD, an antiviral protein from *Gelonium multiflorum*, whereas they are mutated in all other ribosome-inactivating proteins (RIPs). Structural-activity mapping and modeling of HIV-LTR DNA binding are shown in Fig. 8.2.

8.7 Molecular Mechanism of Actions

MAP30 exhibits multiple actions, including DNA topological inactivation [27, 29], DNA Glycoslyase/Apurinic lyase (DGAL) [31] and RIP [21] activities. DNA topological inactivation and DGAL are critical for anti-viral and anti-tumor actions [29, 31]. MAP 30 binds to the HIV-LTR, and converts supercoiled HIV-LTR to relaxed and linear forms, rendering the DNA topologically inactive as substrate for HIV integrase [27, 29, 31]. The DGAL activity is distinct from RIP activity.

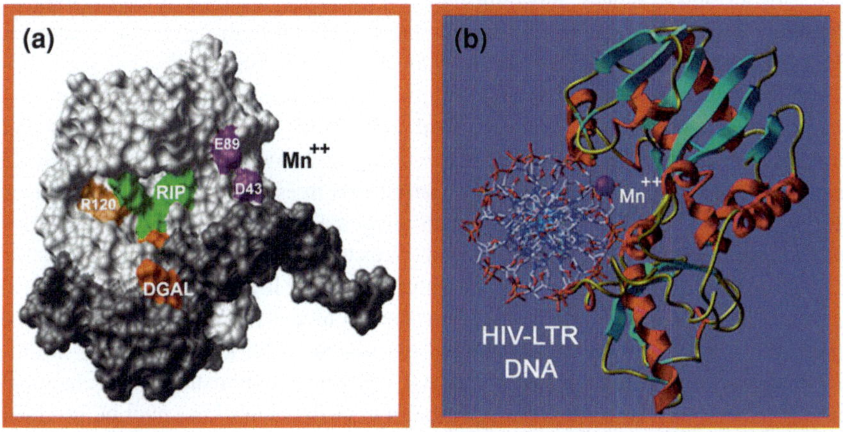

Fig. 8.2 Structural-activity mapping and modeling of MAP30. a represents the 3D surface structure of MAP30 showing active cleft and multiple action sites. RNA N-glycosidase site (RIP, *Green*), DNA Glycoslyase/Apurinic lyase site (DGAL, *Red*) and Mn^{2+} coordination site (E89 and D43, *Magenta*). **b** Modeling the binding of a HIV-LTR DNA fragment into the binding groove of MAP30. *Green* represents residues whose chemical shifts were most sensitive to the addition of the LTR DNA. Magenta represents Mn^{2+} coordinated by D43 and E89

It depurinates viral and tumor DNA, removes 5′adenine and renders the DNA unsuitable for integration into host genome. RIP acts on 28S rRNA of eukaryotic ribosome, cleaves the N-glycosidic bond between a ribose and adenine or guanine at positions A4324 or G4323 in the GAGA loop, inactivates 60S ribosome and inhibits the elongation step in protein biosynthesis. RIP activity is distinct from anti-HIV/tumor activity, as MAP30 and GAP31 inhibit HIV-1, other viruses and tumors at concentrations that show little effect on ribosome function [15, 17, 21, 26–28]. These observations indicate that mechanisms unrelated to ribosome inactivation, contribute to MAP30 anti-HIV/anti-tumor activities.

To define molecular events in viral pathogenesis and to identify host cell targets, we conducted cDNA microarrays to profile gene expression modulated by HIV-1 infection and MAP30 treatment. HIV-1 infection up-regulates the expression of heat shock proteins hsp27, hsp90 in the stress pathway; p53 binding protein MDM2, while it down-regulates apoptosis regulators Bax, GADD45, CD95 and caspases 3, 6, 10 in the apoptotic pathway, Cyclin B1, D2, egr-1 in cell cycle pathway and cytokines IL-2, Il-6 and IL-1α. Treatment with MAP30 is capable of reversing HIV-1 infection-associated changes. Treatment of HIV-1 infected cells with MAP30 also up-regulates the expression of activating transcription factor 2 (ATF2), IκBa and NFκB. No changes were noted in the expression of housekeeping genes GAPDH, PPIA, RPL13A or β-actin.

8.8 Conclusion and Perspectives

MAP30, an antiviral and anti-tumor protein isolated from the medicinal plant *Momordica charantia*, is a multi-functional agent with high therapeutic index. It differentiates itself from other single-chain RIPs, including α-, β- and γ-MMCs from the same plant, because of its unique conserved amino acid sequence, distinct molecular mechanisms, specific viral and cellular targets, lack of toxicity and high bioavailability [15, 18–36]. Our discovery of MAP30 and elucidation of its molecular, biochemical, and structural-activity features demonstrate how modern technology can be applied to investigate medicinal components of natural products. Future challenges include pharmacokinetic and dynamic evaluations for clinical applications. In view of its high efficacy and low toxicity, the structure-activity relationship of MAP30 would be a useful model for other RIPs to improve their therapeutic potential by increasing plasma half life, and decreasing immunogenicity. To achieve these goals, rational structural modifications, and chemical cross-linkings may be applied. In addition, nanomaterial packing may be used for targeted delivery and controlled release [42].

It is estimated that about one-third of the prescription drugs in the U.S. contain plant components, and more than 120 important prescription drugs are derived from plants. Most of these were developed because of their use in traditional medicine. According to recent World Health Organization studies, over 30 % of the world's plant species have, at one time or another, been used for medical proposes. Thus, of the 250,000 recorded higher plant species on Earth, more than 80,000 species are medicinal. Of the 80,000 medicinal plants, only 120 have been developed into useful drugs. It is thus clear that there is a wealth of potential medicines in our plants that are waiting to be discovered.

Acknowledgments Work done in this review was supported in part by NIH grants R01 AI31343, R01 AT01383, and by Blackout Funding. The authors are grateful to Helen C. Lin for her support and encouragement.

References

1. Robinson MM, Zhang X (2011) Traditional medicines, issues and challenges. WHO, Geneva
2. Eisenberg DM, Davis RB, Ettner SL, Appel S, Wilkey SA, Van Rompay M, Kessler RC (1998) Trends in alternative medicine use in the United States, 1990–1997: results of a follow-up national survey. JAMA 280:1569–1575
3. Eisenberg DM, Kessler RC, Foster C, Norlock FE, Calkins DR, Delbanco TL (1993) Unconventional medicine in the United States. Prevalence, costs, and patterns of use. N Engl J Med 328(4):246–252
4. Wolsko PM, Eisenberg DM, Davis RB, Ettner SL, Phillips RS (2002) Insurance coverage, medical conditions, and visits to alternative medicine providers: results of a national survey. Arch Intern Med 162(3):281–287
5. Eisenberg DM, Kessler RC, Van Rompay MI, Kaptchuk TJ, Wilkey SA, Appel S, Davis RB (2001) Perceptions about complementary therapies relative to conventional therapies among adults who use both: results from a national survey. Ann Intern Med 135(5):344–351
6. Kessler RC, Davis RB, Foster DF, Van Rompay MI, Walters EE, Wilkey SA, Kaptchuk TJ, Eisenberg DM (2001) Long-term trends in the use of complementary and alternative medical therapies in the United States. Ann Intern Med 135(4):262–268
7. Nahin RL, Barnes PM, Stussman BJ, Bloom B (2009) Cost of CAM and frequency of visits to CAM prectitioners: United States, 2007 National health statistics reports; no 18. National Center for Health Statistics, Hyattsville
8. Lee-Huang S, Huang PL, Zhang D, Lee JW, Bao J, Sun YT, Chang Y-T, Zhang J, Huang PL (2007) Discovery of small-molecule HIV-1 fusion and integrase inhibitors oleuropein and hydroxytyrosol: Part I. Fusion inhibition. Biochem Biophys Res Commun 354:872–878
9. Lee-Huang S, Huang PL, Zhang D, Lee JW, Bao J, Sun YT, Chang Y-T, Zhang J, Huang PL (2007) Discovery of small-molecule HIV-1 fusion and integrase inhibitors oleuropein and hydroxytyrosol: Part II. Integrase inhibition. Biochem Biophys Res Commun 354:879–884
10. Bao J, Zhang DW, Zhang JZH, Huang PL, Huang PL, Lee-Huang S (2007) Computational study of bindings of olive leaf extract (OLE) to HIV-1 fusion protein gp41. FEBS Lett 581:2737–2742
11. Lee-Huang S, Zhang L, Huang PL, Chang YT, Huang PL (2003) Anti-HIV activity of olive leaf extract (OLE) and modulation of host cell gene expression by HIV-1 infection and OLE treatment. Biochem Biophys Res Commun 307:1029–1037
12. Lee-Huang S, Maiorov V, Huang PL, Lee HC, Ng A, Chang Y-T, Kallenbach N, Huang PL, Chen HC (2005) Structural and functional modeling of human lysozyme reveals a unique nonapeptide, HL9, with anti-HIV activity. Biochemistry 44:4648–4655
13. Lee-Huang S, Huang PL, Huang PL, Sun YT, Kung HF, Blithe DL, Chen HC (1999) Lysozyme and ribonucleases as anti-HIV components in beta-core preparations of hCG. Proc Natl Acad Sci U S A 96:2678–2681
14. Bourinbaiar AS, Lee-Huang S (1995) Anti-HIV effect of beta subunit of human chorionic gonadotropin (βhCG) in vitro. Immunol Let 44:13–18
15. Lee-Huang S, Huang PL, Nara PL, Chen HC, Kung HF, Huang P, Huang HI, Huang PL (1990) MAP 30: a new inhibitor of HIV-1 infection and replication. FEBS Letts 272:12–18

16. Lee-Huang S, Huang PL, Kung HF, Li BQ, Huang PL, Huang P, Huang HI, Huang PL, Chen HC (1991) TAP 29: an anti-human immunodeficiency virus protein from *Trichosanthes kirilowii* that is nontoxic to intact cells. Proc Natl Acad Sci U S A 88:6570–6574
17. Lee-Huang S, Kung HF, Huang PL, Huang PL, Li B-Q, Huang P, Huang HI, Chen HC, Huang PL (1991) A new class of anti-HIV agents: GAP 31, DAP 30 and DAP 32. FEBS Letts 291:139–144
18. Huang PL, Sun YT, Chen HC, Huang PL, Lee-Huang S (1999) Proteolytic fragments of anti-HIV and anti-tumor proteins MAP30 and GAP31 are biologically active. Biochem Biophys Res Commun 262:615–623
19. Lee-Huang S, Huang PL, Chen HC, Huang PL, Bourinbaiar AS, Huang HI, Kung HF (1995) Anti-HIV and anti-tumor activities of recombinant MAP 30 of bitter melon. Gene 161:151–156
20. Arazi T, Huang PL, Huang PL, Zhang L, Shiboleth YM, Gal-On A, Lee-Huang S (2002) Production of antiviral and antitumor proteins MAP30 and GAP31 in cucurbits using the plant virus vector ZYMV–AGII. Biochem Biophys Res Commun 292:441–448
21. Rybak SM, Lin JJ, Newton DL, Kung HF, Monks A, Chen HC, Huang PL, Huang PL, Lee-Huang S (1994) *In vitro* anti-tumor activity of the plant ribosome inactivating proteins MAP 30 and GAP 31. Int J Onc 5:1171–1176
22. Lee-Huang S, Huang PL, Sun YT, Kung HF, Huang PL, Murphy WJ (2000) Inhibition of human breast tumor MDA-MB-231 xenografts and HER2 expression in SCID mice by anti-tumor agents GAP31 and MAP30. Anticancer Res 20:653–659
23. Lee-Huang S, Huang PL, Chen HC, Kung HF, Nara PL, Li BQ, Huang PL, Huang HI, Huang PL (1992) Plant proteins with antiviral activity against HIV. In: Chu CK, Cutler HG (eds) Natural products as antiviral agents. Plenum, New York, pp 153–170
24. Schreiber CA, Sun YT, Wan L, Krey LC, Lee-Huang S (1999) Anti-HIV agents, MAP30 and GAP31 are not toxic to human spermatocytes: potential use of these agents against sexual transmission of HIV-1. Steril Ferti 72:686–690
25. Lee-Huang S, Huang PL, Kong XP, Schapira M, Huang PL (2003) Novel anti-HIV and anti-tumor proteins MAP30 and GAP31: molecular, biochemical and structural insights into biological activities and mechanism. Recent Res Devel Biophys Biochem 81:235–257
26. Bourinbaiar AS, Lee-Huang S (1995) Potentiation of anti-HIV activity of anti-inflammatory drugs dexamethasone and indomethacin, by MAP 30, the antiviral agent from bitter melon. Biochem Biophys Res Commun 208:779–785
27. Bourinbaiar AS, Lee-Huang S (1996) The activity of plant-derived antiretroviral proteins MAP30 and GAP31 against herpes simplex virus infection in vitro. Biochem Biophys Res Commun 219:923–929
28. Sun YT, Huang PL, Li JJ, Huang YQ, Zhang L, Huang PL, Lee-Huang S (2001) Anti-HIV agent MAP30 modulates the expression profile of viral and cellular genes for proliferation and apoptosis in AIDS-related lymphoma cells infected with kaposi's sarcoma-associated virus. Biochem Biophys Res Commun 287:983–994
29. Huang PL, Chen HC, Kung HF, Huang PL, Huang P, Huang HI, Lee–Huang S (1992) Anti-HIV plant proteins catalyze topological changes of DNA into inactive forms. BioFactors 4:37–41
30. Lee-Huang S, Kung HF, Huang PL, Bourinbaiar AS, Morell JL, Brown JH, Huang PL, Tsai WP, Chen AY, Huang HI, Chen HC (1994) HIV-1 inhibition, DNA binding, RNA binding and ribosome inactivation activities in the N-terminal segments of the plant anti-HIV protein GAP 31. Proc Natl Acad Sci U S A 91:12208–12212
31. Lee-Huang S, Huang PL, Huang PL, Bourinbaiar AS, Chen HC, Kung HF (1995) Inhibition of HIV-1 integrase by plant antiviral proteins MAP 30 and GAP 31. Proc Natl Acad Sci U S A 92:8818–8822
32. Lee-Huang S, Kung HF, Chen HC, Huang PL, Rybak SM, Huang PL, Bourinbaiar AS, Musayev F, Liaw YC (1994) Crystallization and preliminary X-ray analysis of GAP 31: an anti-HIV protein. J Mol Biol 240:92–94

33. Wang YX, Nouri N, Jacob J, Palmer I, Stahl SJ, Kaufman JD, Huang PL, Huang PL, Heather E, Pommier Y, Wingfield PT, Lee-Huang S, Bax A, Torchia DA (1999) Solution structure of anti-HIV and anti-tumor protein MAP30: structural insights into its multiple functions. Cell 99:433–442
34. Wang YX, Jacob J, Wingfield PT, Palmer I, Stahl SJ, Kaufman JD, Huang PL, Huang PL, Lee-Huang S, Torchia DA (2000) Anti-HIV and anti-tumor protein MAP30, a 30 kDa single strand type-1 RIP, shares similar secondary structure and β-sheet topology with the A chain of ricin, a type-II RIP. Prot Sci 9:138–144
35. Wang XY, Marquardt J, Wingfield P, Stahl SJ, Lee-Huang S, Torchia D, Bax A (1999) Simultaneous measurement of 1H–15N, 1H–13C′ and 15N–13C′ dipole coupling in a per-deuterated 30 kDa protein dissolved in a dilute liquid crystalline phase. J Am Chem Soc 120:7385–7386
36. Wang XY, Jacob J, Cordier F, Wingfield P, Stahl SJ, Lee-Huang S, Torchia D, Grzesiek S, Bax A (1999) Measurement of (3 h)J(NC′) connectivities across hydrogen bonds in a 30 kDa protein. J Biomol NMR 14:181–184
37. Huang PL, Huang PL, Lee-Huang S (2010) Oleuropein and related compounds reduce atherosclerosis. Open Conf Proc J 1:81–86
38. Lee-Huang S, Huang PL, Zhang D, Zhang JZH, Huang PL (2012) Oleuropein and related compounds from olive plants modulate adipogenesis. Open Conf Proc J (in press)
39. Hartono YD, Lee AN, Lee-Huang S, Zhang D (2011) Computational study of bindings of HL9, a nonapeptide fragment of human lysozyme, to HIV-1 fusion protein gp41. Bioorg Med Chem Lett 21:1607–1611
40. Lee-Huang S, Huang PL, Huang PL (2011) Live-cell real-time imaging reveals role of mitochondria in cell-to-cell transmission of HIV-1. Biochem Biophys Res Commun 415(2):384–389
41. Fang EF, Zhang CZ, Wong JH, Shen JY, Li CH, Ng TB (2012) The MAP30 protein from bitter gourd (Momordica charantia) seeds promotes apoptosis in liver cancer cells in vitro and in vivo. Cancer Lett 324(1):66–74 (Epub 9 May 2012)
42. Meng Y, Liu S, Li J, Meng Y, Zhao X (2012) Preparation of an antitumor and antivirus agent: chemical modification of α-MMC and MAP30 from *Momordica charantia* L. with covalent conjugation of polyethyelene glycol. Int J Nanomed 7:3133–3142 (Epub 27 June 2012)

Author Biography

Professor Sylvia Lee-Huang received her PhD in Biophysics from the University of Pittsburgh. For many years, she worked closely with Nobel Laureate Severo Ochoa in the Biochemistry Department at New York University School of Medicine. She is currently Professor of Biochemistry and Molecular Pharmacology with permanent tenure at New York University School of Medicine. Professor Lee-Huang is renowned for her contributions to biomedical science including first proposing the concept and providing experimental evidence for reverse transcription; discovery of messenger specific initiation factors; discovery of small RNAs; purification, cloning, and expression of human erythropoietin; discovery and mechanism of action of novel anti-tumor, antiviral and anti-obesity agents; and discovery of the role of mitochondria in cell-to-cell transmission of HIV-1.

Professor Paul Lee Huang received his MD and PhD degrees from Harvard Medical School. He completed an Internal Medicine residency and Cardiology fellowship at Massachusetts General Hospital (MGH), and was Chief Medical Resident. He is currently Professor of Medicine at Harvard Medical School, and Director of the MGH Cardiac Metabolic Syndrome Program. Professor Huang is known internationally as a scientific leader in the area of nitric oxide (NO) biology.

Dr. Philip Lin Huang received his MD degree from Harvard Medical School and Harvard-MIT Division of Health Sciences and Technology. He received his MBA degree from the Wharton Business School, University of Pennsylvania. He completed an Internal Medicine residency at Massachusetts General Hospital and Harvard Medical School, and served as an Attending Physician at Brigham and Women's Hospital. He is currently Vice President, American BioScience.

Chapter 9
Ribosome Inactivating Proteins from Plants: Biological Properties and their Use in Experimental Therapy

Lucía Citores, Rosario Iglesias and José Miguel Ferreras

Abstract Ribosome-inactivating proteins (RIPs) are 28 S rRNA N-glycosidases isolated mainly from plants that irreversibly inactivate ribosomes, thereby impairing protein synthesis. In recent years, polynucleotide:adenosine N-glycosidase activity and induction of apoptosis have been reported and may have a particular significance. There are two classes of RIPs: type 1 RIPs, consisting of single-chain proteins, and type 2 RIPs, consisting of an A chain with RIP properties covalently linked to a B chain with lectin properties. Type 2 RIPs may be very toxic or non toxic, whereas type 1 RIPs are always non-toxic. Due to the diverse activities of RIPs, research has been conducted to investigate their use as antiviral and antitumor agents or as the toxic part of conjugates. Conjugates consist of a targeting portion such as an antibody, a lectin or a growth factor linked to a toxic portion. RIPs have been used as the toxic portion in conjugates that have been tested in several experimental therapies against various malignancies. Although some important disadvantages still need to be improved, recent clinical trials encourage the use of these conjugates as efficacious agents in the treatment of cancer and other diseases.

9.1 Introduction

Ribosome-inactivating proteins (RIPs) have been initially studied as proteins widely distributed in plants that inhibit protein synthesis in mammalian cell-free systems [1, 2]. However, work done in recent years revealed that RIPs can also be

L. Citores · R. Iglesias · J. M. Ferreras (✉)
Department of Biochemistry and Molecular Biology and Physiology, Faculty of Sciences, University of Valladolid, E-47005 Valladolid, Spain
e-mail: rosario@bio.uva.es

L. Citores
e-mail: luciac@bio.uva.es

R. Iglesias
e-mail: riglesia@bio.uva.es

E. F. Fang and T. B. Ng (eds.), *Antitumor Potential and Other Emerging Medicinal Properties of Natural Compounds*, DOI: 10.1007/978-94-007-6214-5_9,

found in algae, fungi and bacteria [3, 4]. It has been reported that RIPs can also inhibit protein synthesis in other animal and fungi cell-free systems, and some of them can inhibit protein synthesis in plants and bacteria [3]. They also display other important biological activities. Among them, the polynucleotide:adenosine glycosidase activity [5] and the ability to induce apoptosis in several types of animal cells [6] could play a significant role, perhaps more important than that of protein synthesis inhibition. Despite all the work done in the field of RIPs, evidence is not yet available for a convincing role played by RIPs in plants. They seem to play different roles in different species, so antiviral, antifungal, storage, programmed senescence, antifeedant, stress protection and development regulation roles have been proposed for these proteins [3]. Stirpe and Barbieri [1] have proposed to classify these proteins into two types: type 1 RIPs, consisting of single-chain proteins, and type 2 RIPs, consisting of an A (active) chain with RIP properties covalently linked to a B (binding) chain with lectin properties. The latter RIPs can enter cells more easily because the B chain allows the binding to sugar-containing cell surface receptors, and for this reason they could be potent toxins. However as discussed later, recently a surprising variety of RIP and related lectin structures have been found. For example, the type 2 RIPs from *Sambucus* are less toxic to animals than the type 1 RIPs from *Saponaria* [7]. Alternatively, four-chain type 2 RIPs [7, 8] and type 1 RIPs with an inner removable peptide [9] also have been reported.

Because of their diverse activities, RIPs either alone or as part of conjugates are good candidates for developing selective antiviral and anticancer agents. Conjugates consist of a targeting portion such as an antibody, a lectin or a growth factor linked to a toxic portion. RIPs have been used as the toxic portion in several conjugates that have been tested in experimental therapies against various malignancies. In agriculture, RIPs have been shown to increase resistance against virus and other parasites in transgenic plants [3].

9.2 Distribution

RIPs have been found mainly in flowering plants (angiosperms) and, to a lesser extent, in fungi, algae and bacteria (reviewed in [3]). To date, type 1 RIPs have been isolated from at least 70 species of plants of both monocotyledons and dicotyledons. They are distributed among all flowering plant families without a pattern permitting their location in a particular taxon. However, the highest number of RIPs is present in a small number of families, namely Caryophyllaceae, Cucurbitaceae, Euphorbiaceae, Phytolacaceae and Poaceae. RIPs have been found in all types of plant tissues. Some species contain type 1 RIPs in only one tissue of the plant and others contain RIPs in several parts of the plant. However, the fact that no RIPs have been found in a particular species, family or tissue does not mean that they do not contain them since the amount of RIP could be too low to be isolated or even detected. The detection method usually consists in assaying the inhibition of protein synthesis in a cell-free system such as a rabbit reticulocytes lysate. The presence of inhibitors has been detected in greater than 70 plants but it has not been possible to determine if the protein synthesis inhibition is due to a RIP or to another inhibitor such as an RNase or a protease. Finally, the presence of RIPs depends on different factors like senescence, virus infection, development and stress [3]. Type 2 RIPs are less widespread but they have been isolated from at least 20 species of plants belonging to 10 different families. At least five species (*Sambucus ebulus*, *Sambucus nigra*, *Cinnamomum camphora*, *Momordica charantia* and *Iris hollandica*) contain both type 1 and type 2 RIPs.

RIPs are also present in bacteria [4], fungi [10] and algae [11] but to date the presence of these proteins seems to be limited to a few species. All these findings favour the hypothesis that RIPs are broadly distributed and therefore could play an important biological role in the RIP-producer organism.

9.3 Enzymatic and Biological Activities

9.3.1 rRNA N-glycosidase

Initially RIPs were studied as the most potent inhibitors of protein synthesis in mammalian cell-free systems [1, 2]. They have been shown to inactivate the 60S subunit of mammalian ribosomes in a catalytic and irreversible manner. The inactivated ribosomes are not able to carry out the elongation cycle in protein synthesis. Later it was also reported that RIPs inactivate ribosomes from other animal species and that some of them could also inactivate ribosomes from plants and bacteria [3]. The mechanism of protein synthesis inhibition was elucidated by Endo and Tsurugi [12]. They found that RIPs are 28S rRNA N-glycosidases (EC 3.2.2.22) that cleave the N-glycosidic bond between the adenine No. 4324 from the 28S rRNA and its ribose in the 60S subunit of rat ribosomes (or the equivalent adenine in sensitive ribosomes from other organisms) [3]. This adenine is located in the α-sarcin–ricin loop (SRL) that is crucial for anchoring the elongation factor G or the elongation factor 2 on the ribosome during mRNA–tRNA translocation in prokaryotes and eukaryotes respectively [13].

Most type 1 and type 2 RIPs depurinate ribosomes at one site (the adenine 4324), whereas other RIPs such as saporins, PAP-R and trichokirin depurinate the ribosomal RNA at multiple sites [14]. However, the depurinating activity seems to be greater on the adenine 4324 than at other sites [14].

9.3.2 Polynucleotide: Adenosine Glycosidase

Recently other substrates for RIPs have been reported [5]. Some RIPs release adenines from viral genomic RNAs of MS 2, TMV and AMCV [5]. Many RIPs can also depurinate polyadenylic acid. Most RIPs release adenines from rRNA and all of them extensively depurinate herring sperm DNA [5]. Thus, Barbieri has proposed the name of polynucleotide: adenosine glycosidases for these proteins [5].

Some RIPs remove adenine from the poly(ADP-ribosyl)ated poly(ADP-ribose) polymerase (PARP), which is involved in DNA repair and apoptosis [15]. This damage to activated PARP may have a role in the inhibition of DNA repair by RIPs, which seems to be independent of the inhibition of protein synthesis [16].

9.3.3 Apoptosis

In addition to their N-glycosidase activity on nucleic acids, both type 1 and type 2 RIPs are also capable of inducing cell death by apoptosis [17]. RIP-treated cells exhibit the morphological and biochemical events associated with apoptosis. Emerging evidence suggests that induction of apoptosis does not directly correlate with the protein synthesis inhibition [18]. RIPs trigger apoptosis in different cell types via different mechanisms. Despite a large number of studies on RIP-induced apoptosis, the exact mechanism by which these toxins induce apoptosis is not very clear. Generally, the apoptosis induced by RIPs involves the caspase dependent mitochondrial pathway leading to loss of mitochondrial membrane potential, rapid release of cytochrome c, activation of caspase-9 and an increase in the production of reactive oxygen species (ROS) in cells [17]. However, the death receptor-mediated apoptosis pathway also seems to be involved in the killing of cells by some RIPs [6, 17].

9.3.4 Other Activities

Other activities associated with some RIPs are chitinase activity, topological activity on DNA, HIV integrase inhibitory activity, superoxide dismutase activity, DNase activity, and lipase activity (reviewed in [3]). Some of them may have a significant role as has been proposed for the lipase activity in ricin toxicity or the HIV integrase inhibitory activity in the antiviral properties of some type 1 RIPs.

9.4 Structure

Figure 9.1 summarizes the different structures of RIPs and related lectins. Type 2 RIPs may be dimeric (such as ricin) or tetrameric. In the latter case the two A-B dimers may be linked by their A chains (*Ricinus communis* or *Viscum album* agglutinins) or by their B-chains (*Sambucus* tetrameric type 2 RIPs). In this case the B-chain 1 alpha site does not bind sugar-containing compounds [8]. Type 2 RIP related lectins are proteins which do not show enzymic activity and show only

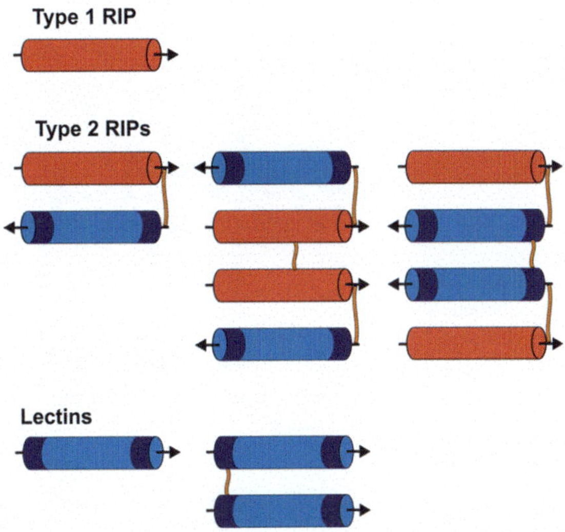

Fig. 9.1 Graphical representation of RIPs and related lectins. Type 1 RIPs consist of single-chain catalytic proteins whereas type 2 RIPs consist of catalytic(s) or A chain(s) (coloured in *red*) linked to binding or B chain(s) (coloured in *cyan*) by disulphide bond(s) (coloured in *orange*). The sugar binding subdomains 1 alpha and 2 gamma are coloured in blue. Pure lectins may have one or two B chains linked by a disulphide bond. The *arrow* indicates the C-terminal end

lectin activity. The structure of these lectins has a striking homology with that of the B chain of type 2 RIPs. In fact they are encoded by a truncated type 2 RIP gene which has lost almost all the sequence coding for the A chain. They can be monomeric (one single B-chain) or dimeric (two B-chains held together by a disulphide bridge). In this case a new cysteine appeared which is responsible for the dimerization of the two polypeptide chains through a disulphide bond. The different B chains vary with respect to sugar specificity. Type 2 RIPs from *Ricinus*, *Abrus* or *Adenia* are specific for galactose containing sugars and are very toxic. Type 2 RIPs and related lectins from *Sambucus* are much less toxic and may be specific for galactose, sialic acid or unable to bind sugars [8].

The structure of ebulin l (a heterodimeric type 2 RIP presents in *S. ebulus* leaves) is shown in Fig. 9.2. In the A chain, ebulin l has roughly the same positioning of key active site residues as type 1 RIPs and other type 2 RIPs such as ricin, abrin or volkensin [19]. Tyr 77 and 116, Arg 127, Glu 163, Arg 166 and Trp 197 are highly conserved in both type 1 and type 2 RIPs. Additionally all type 2 RIPs conserve the Cys 249 of ebulin l A chain that links it to the B chain by a disulphide bond. This Cys is absent in type 1 RIPs such as saporin or PAP. Some tetrameric type 2 RIPs (i.e., *R. communis* agglutinin) present a seven amino acid loop with an additional Cys (located in a position corresponding to Gly148 and 149 in ebulin l) that allows the formation of a disulphide bridge with another A chain [8]. The overall fold of the ebulin B chain is also very similar to that of type 2 RIP B-chains and lectins and is composed of two beta trefoil domains (I and II) with sugar-binding ability [19]. The domain I presents the 1 alpha sugar binding site (Trp 39, Asp 24, Gln 37, Asn 46 and Gln 47) and the domain II the 2 gamma sugar binding site (Asp 235, Phe 249, Asn 256 and Gln 257). These aminoacids are also well conserved in the B chain of other type 2 RIPs and lectins but the sugar-binding ability of the different B chains varies considerably. Tyr29 and

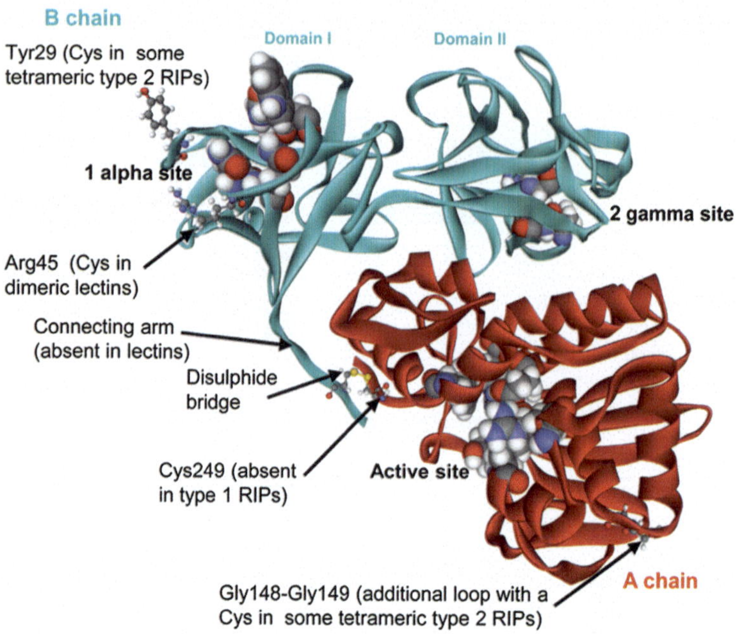

Fig. 9.2 Three-dimensional model of the type 2 RIP ebulin l (PDB ID: 1HWP). The A (*red*) and B (*cyan*) chains and the disulphide bridge linking both chains are indicated. The active site and sugar-binding sites are represented in CPK (*balls*). The residues in ebulin l that change in type 1 RIPs, tetrameric RIPs and dimeric lectins are represented in balls and sticks

Arg 45 are substituted by Cys in some tetrameric type 2 RIPs and dimeric lectins, respectively, thus allowing the formation of an additional disulphide bridge between B chains [8].

9.5 Toxicity and Intracellular Pathway of Type 1 and 2 RIPs

As previously indicated, type 1 RIPs are single-chain toxins (A-chain) with N-glycosidase activity while type 2 RIPs consist of an A-chain with N-glycosidase activity and a galactoside binding lectin (B chain) linked together by a disulphide bond. Type 1 RIPs have a high cell-free translation inhibitory potency but a low cytotoxicity (Table 9.1). Because of the absence of a B chain, type 1 RIPs, such as saporin, poorly enter into cells, consequently causing low toxicity to cells and animals. However, the presence of the B chain is not sufficient to confer a high level of cytotoxicity on all type 2 RIPs. Based on their toxicity to mammals, type 2 RIPs has been divided into two groups: the toxic and nontoxic type 2 RIPs. The former group include ricin, abrin, viscumin and

Table 9.1 Toxicity of type 1 and type 2 RIPs

	Rabbit lysate IC_{50} (nM)[a]	HeLa cells IC_{50} (nM)	Mouse LD_{50} (mg/kg)
Type 1 RIPs			
Saporin	0.037	2,300	4
PAP	0.037	33,000	2.6
Gelonin	0.4	34,000	40
Type 2 RIPs			
Ebulin l	0.15	64.3	2
Nigrin b	0.03	27.6	12
Ricinus communis agglutinin	0.046	n.d	>0.033
Sieboldin b	0.015	11.8	>1.6
Abrin	0.5	0.0037	0.00056
Ricin	0.1	0.00067	0.003
Viscumin	3.5	0.008	0.0024
Volkensin	0.37	0.0003	0.00138

The table shows the effects of type 1 and type 2 (either toxic or non-toxic) RIPs on protein synthesis by a cell-free system derived from rabbit reticulocytes lysates and their toxicity to intact HeLa cells and mice. The data were obtained from Refs. [1, 2, 8]

[a]Reduced toxin (only in type 2 RIPs): the toxin was pre-incubated with a reducing agent (such as 2-mercaptoethanol or dithiothreitol) before testing activity

volkesin, which are among the most potent plant toxins. In contrast, nigrin b, ebulin l, sieboldin b and *R. communis* agglutinin belonging to the latter group show little or no toxicity in intact cells and higher animals (Table 9.1). Their lack of toxicity has been attributed in part to a defective B chain with reduced affinity for the galactosides present at the surface of plasma membrane proteins and to a different intracellular routing and processing of the toxin [8].

The cell surface receptors differ from one RIP to another. However they share the need to enter the cytosol to exert their toxicity. The molecular mechanisms involved in intracellular traffic of RIPs have been studied extensively for toxic type 2 RIPs, especially for ricin, which has been taken as a model of highly toxic RIPs. Ricin enters target mammalian cells by receptor-mediated endocytosis and is transported to endosomes. From here part is recycled back to the cell surface, another part is transported to the lysosomes and degraded and only a small fraction is eventually translocated to the cytosol (Fig. 9.3). Only 5 % of the endocytosed toxin is transported to the trans-Golgi network [20] and then backward through the Golgi apparatus to the endoplasmic reticulum (ER). In ER the catalytic A chain is reductively separated from the cell-binding B chain. The A chain is then retrotranslocated to the cytosol by utilizing the ER-associated degradation pathway (ERAD) usually followed by misfolded proteins, which in the cytosol are polyubiquitinated and degraded by the proteasome. Once in the cytosol ricin A-chain escapes, in part, proteasomal degradation probably due to its low content in lysine residues. Once in the cytosol the enzymatic A chain inactivates the ribosomes [21].

Ricin toxicity is sensitive to brefeldin A and to low temperature. In contrast to ricin, the non toxic type 2 RIPs nigrin b and ebulin l follow a pathway that is insensitive to brefeldin A and to temperatures below 37 °C indicating that transport from endosomes to the Golgi complex is not required for nigrin b and ebulin l A-chain translocation [8]. In fact, nigrin b was found to enter cells like ricin, but was more rapidly and extensively degraded, and when excreted by HeLa cells the nigrin b-derived material was completely inactive [22]. Thus, cell protein synthesis inhibition by nigrin b seems to be a consequence of the spontaneous translocation of nigrin b from the endosome when the extracellular concentration of RIP is high (Fig. 9.3).

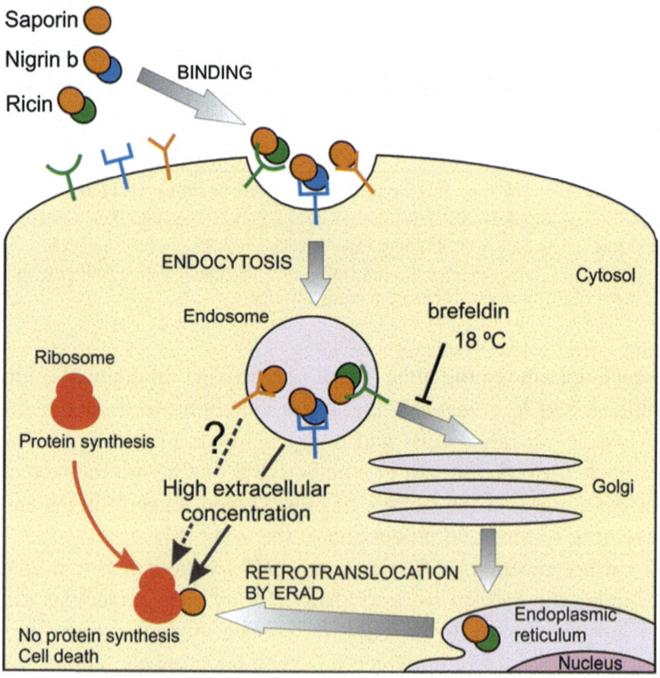

Fig. 9.3 Intracellular trafficking of ricin, nigrin b and saporin. Ricin binds to glycoproteins of the plasma membrane and internalizes into the cell. Only a small number of protein molecules are transported first to the Golgi network and then to the endoplasmic reticulum (ER). In the endoplasmic reticulum, the disulphide bridge is reduced and the A chain translocates to the cytosol by the endoplasmic reticulum-associated degradation (ERAD) pathway. In the cytosol, the A chain inactivates the ribosomes, inhibiting protein synthesis and causing cell death. This pathway is sensitive to low temperature and brefeldin A. Nigrin b can bind to different glycoproteins on the plasma membrane and internalizes into the cell. From endosomes, protein molecules are mostly transported to lysosomes for degradation. However, at much higher extracellular concentration, the saturation of the endosome with nigrin b can lead to a spontaneous release of nigrin b into the cytosol, causing ribosome inactivation. This pathway is not sensitive to low temperature and brefeldin A. Saporin binds in part to low-density lipoprotein receptor-related proteins and is internalized into the cell. Saporin cytotoxicity cannot be blocked by brefeldin A, indicating that the protein reaches the cytosol following a yet unknown intracellular route that does not involve transport to the Golgi. The orange circles represent the A chain of the three RIPs. Ricin B-chain is represented by green circles and nigrin b B-chain by blue circles

Less is known about the pathway followed by type 1 RIPs, which could vary, because of their different interaction with cells due to the lack of the lectin chain that facilitates binding to intact cells. However, studies done with saporin have shown a clear role for cell surface receptors belonging to the low density lipoprotein related receptor family in mediating saporin internalization in different cell lines [23]. Also, saporin cytotoxicity cannot be blocked by brefeldin A indicating that the protein reaches the cytosol following an intracellular route that does not include transport to the Golgi [23] (Fig. 9.3).

9.6 Use of RIPs in Experimental Therapy

Due to the diverse activities of RIPs, extensive research has been conducted to investigate their use as antiviral and antitumor agents. The most promising applications of RIPs in experimental medicine, especially in anticancer therapy, are related to their use as components of immunotoxins, conjugates or recombinant chimeras in which the enzymatic RIPs are linked to tumor targeting ligands or antibodies that mediate their binding and internalization by malignant cells.

On the down side, because of their cytotoxic nature, type 2 RIPs present a significant public health concern due to their potential use as bioterrorism agents. Ricin, for example, is a potent cytotoxin, easy to purify in large amounts, easy to handle and stable. For this reason extensive efforts have been made to develop a safe antidote or vaccine against this toxin [24].

9.6.1 Use of Unconjugated RIPs in Experimental Therapy

Interest in type 2 RIPs as anticancer agents began as early as 1970, when it was shown that ricin and abrin were more toxic to tumor cells than to normal cells [25]. Mistletoe extracts, based on their presumed immunostimulatory and antineoplastic effects, have been used in the complementary treatment of cancer patients, for more than 80 years. The active compound of these extracts is the type 2 RIP viscumin (MLs) which possess apoptosis-inducing effects on many types of cancers [26, 27]. Recently a recombinantly engineered rViscumin has been developed for application in cancer treatment and is being tested in phase I/II clinical studies [28]. Another type 2 RIP from *Ximenia americana*, termed riproximin, which is the active component of the plant material used in African traditional medicine to treat some forms of cancer, has shown potent antitumor activity in a rat metastasis model [29]. Therefore, it seems that type 2 RIPs such as viscumin, abrin or ricin bear apoptosis-inducing activities toward cancer cells by targeting different stages of apoptotic pathways [26].

Plants from Cucurbitaceae family, mainly belonging to the genus *Trichosanthes* and *Momordica*, have a long history of being used in old traditional Chinese medicine. Trichosanthin is a type 1 RIP purified from *T. kirilowii* that has been known for around 30 years. Trichosanthin is known to possess a broad spectrum of biological and pharmacological activities, including abortifacient which terminates early pregnancy, anti-tumor, immunomodulatory, nuclease, and anti-human immunodeficiency virus activities. Recently, trichosanthin has been found to induce apoptosis, enhance the action of chemokines and inhibit HIV-1 integrase [30]. Also of interest as a medicinal candidate, alpha-momorcharin, a type 1 RIP, purified from *M. charantia* has demonstrated strong anti-tumor growth activity and anti-HIV function besides abortifacient activity [31] (see Chap. 2 in this book). In addition, recombinant luffin, a type 1 RIP from *Luffa cylindrica*, displays in vitro cytotoxicity against various tumor cell lines [32].

9.6.2 Use of RIPs for the Construction of Immunotoxins

Another application of RIPs is the linking of them to monoclonal antibodies or their fragments to obtain immunotoxins that are specifically toxic to target cells [33–36]. The antibody (or antibody fragment) [37] binds specifically to an antigen on the surface of the cell to be killed and the immunotoxin internalizes through endocytosis following the antigen intracellular pathway. Finally the toxin translocates to the cytosol and kills the cell by inhibiting protein synthesis or inducing apoptosis.

These kinds of therapeutic agents were first postulated by the German immunologist Paul Ehrlich in 1897 who envisaged them as "magic bullets" [38]. However, they did not become a reality until the development of the monoclonal antibody, protein purification and protein cross-linking technologies. In fact the first immunotoxin containing a RIP was reported by Youle and Neville in 1980 [39]. It consisted of a monoclonal antibody directed against the murine T cells Thy 1.2 antigen, covalently linked to ricin using the crosslinking agent m-maleimidobenzoyl-N-hydroxysuecinimide ester (MBS). The immunotoxin showed cell type specificity when the unwanted binding via the B chain was blocked with lactose. Unfortunately, the necessary use of lactose limited its use to ex vivo therapy [40].

Immunotoxins have been constructed with selected potent microbial and plant proteins. Among microbial proteins, *Pseudomonas* exotoxin or diphtheria toxin have been used, and among plant proteins, both type 1 and type 2 RIPs have been used.

Since the construction of the first immunotoxin, which was prepared by chemically linking the whole molecule of an antibody to the toxin, several kinds of these chimeras have been designed in order to improve several important aspects such as specificity, efficacy, in vivo stability, immunogenicity and unwanted side effects. From a strategic point of view, several types of immunotoxins have been designed [1]: direct immunotoxins, indirect immunotoxins (directed to a primary antibody) and bispecific antibodies (directed to the toxin and the target simultaneously). In order to prevent unwanted side effects immunotoxins have been constructed with antibody fragments such as Fab' (obtained by pepsin digestion and mild reduction), Fab (obtained by papain digestion), Fv (containing only the two antibody variable domains, VH and VL, linked by a disulphide bridge) and scFv (single-chain variable fragment, containing the two antibody variable domains connected with a short linker peptide).

The choice of the crosslinker agent is also very important. Efforts have been made in order to optimize the efficient release of the toxic moiety when the target is reached. This question has been excellently reviewed recently by Dosio et al. [36]. The most used RIP for the construction of immunotoxins is ricin, but as discussed, it is extremely toxic. For this reason attempts have been made to reduce the non-specificity caused by the B-chain. Approaches are based on the blockage of the B-chain lectin binding ability [36] or the use of the deglycosylated ricin A-chain (the glycosylated ricin A-chain is toxic to macrophages and Kupffer cells) [36]. Another approach is to use type 1 RIPs instead of ricin. The most commonly used are saporin, PAP and gelonin [34]. Recombinant immunotoxins have also been designed consisting of single-chain variable fragments (scFv) genetically fused to a RIP such as gelonin [41].

Immunotoxins cause several undesirable side effects and toxicities. Some of these adverse effects are mild or moderate like fever, nausea, vomiting, diarrhoea, myalgia, edema and hypoalbuminemia. Other effects are severe and could pose a limitation to the therapeutic use of protein immunotoxins. An increased toxicity may result from immunogenicity as a result of the formation of human antimouse antibodies (HAMA) or anti-toxin-antibodies (HATA). These antibodies may prevent repeated cycles of therapy. The development of immunotoxins containing humanized antibodies or choosing smaller antibody formats containing only the variable domains may in part resolve these problems. Many efforts are being made also to decrease immunogenicity of the toxin moiety.

The most common toxicity in patients treated with immunotoxins, in particular ricin A chain-based ones, is vascular leak syndrome (VLS) characterized by an increase in vascular permeability caused by the weak binding of the immunotoxin to normal endothelial cells. Another typical toxicity related to immunotoxin administration is hepatotoxicity that is attributed to the binding of basic residues on the targeting Fv to negatively charged hepatic cells. Renal toxicity is also observed following treatment with immunotoxins. Moreover, successful treatment of solid tumors faces several obstacles including poor penetration into tumor masses and the immune response to the conjugate. Tumor progression is characterized by the formation of a neovasculature, which supplies tumor cells with oxygen and nutrients. The formation of new blood vessels (angiogenesis) is necessary for the growth and metastatic spread of solid tumor. Immunoconjugates containing RIPs targeting the tumor neovasculature have been designed and some of them have considerable potential for use in cancer therapy [7].

Immunotoxins have been included in experimental therapies against various malignancies, often achieving promising results but also often causing severe side effects in patients. Among RIPs, ricin, saporin and gelonin have been widely used to construct anti-cancer immunotoxins. The subject has been dealt with in a number of recent reviews [33–37] and for this reason only some selected examples are shown in Table 9.2.

Briefly, clinical trials with RIP-containing immunotoxins have been conducted against hematologic malignancies and solid tumors. Clinical studies have been carried out using both blocked ricin and its A-chain (glycosylated or deglycosylated) linked to several monoclonal antibodies for the treatment of acute lymphoblastic leukemia, B cell non-Hodgkin's lymphomas, Hodgkin's disease, cutaneous T cell lymphomas, graft-versus-host disease, haploidentical stem cell transplantation and T cell non-Hodgkin's lymphoma. Regarding solid tumors clinical applications of ricin-based immunotoxins include melanoma, colorectal cancer, small cell lung cancer, leptomeningeal neoplasia and breast cancer. Additionally, type 1 RIPs such as PAP, saporin or gelonin have been used in immunotoxins tested in clinical trials in patients with acute lymphoblastic leukaemia, Hodgkin's disease and myeloid malignancies.

9.6.3 Use of Other RIP Conjugates

Targeting can also be performed using tumor-specific ligands. Cancer cells overexpress cell-surface receptors including growth factor, transferrin and interleukin receptors. Several conjugates consisting of a tumor-specific ligand (cytokine, growth factor, transferrin or peptide hormone) coupled to a toxin by genetic fusion or by chemical ligation have been produced. The targeting protein or ligand binds to cell

Table 9.2 Selected examples of clinically evaluated/under evaluation immunotoxins

Immunotoxin	Target	RIP	Target diseases	Phases
Combotox	CD19/CD22	dgA	NHL, ALL	I
IMTOX-25	CD25	dgA	HD, CTCL, GVHD, HSCT, melanoma	I,II
Anti-B4-bR	CD19	Blocked ricin	NHL	II,III
$F(ab')_2BsAb$	CD22	Saporin	NHL	I
BER-H_2-Saporin	CD30	Saporin	HD	I/II
Hum-195/rGel	CD33	Gelonin	AML, CML	I
B43-PAP	CD19	PAP	ALL	I
XomaZyme-Mel	Melanoma	Ricin A-chain	Melanoma	I,II,I/II
N901-bR	CD56	Blocked ricin	SCLC	I,II
Anti-CEA-bR	CEA	Blocked ricin	Colorectal cancer	I/II

The data were obtained from the references cited in the text
Abbreviations ALL (acute lymphoblastic leukemia); *AML* (acute myelogenous leukemia); *CEA* (carcinoembryonic antigen); *CML* (chronic myelogenous leucemia); *CTCL* (cutaneous T cell lymphoma); *dgA* (deglycosylated ricin A-chain); *GVHD* (graft-versus-host disease); *HD* (Hodgkin's disease); *NHL* (non-Hodgkin's lymphoma); *HSCT* (haploidentical stem cell transplantation); *SCLC* (small cell lung cancer)

surface receptors and are subsequently internalized by endocytosis, resulting in cell death. The toxin moiety requires internalization and translocation to the cytosol to achieve the cytotoxic effect by inactivating cytosolic protein synthesis and inducing apoptosis [35, 42].

The transferrin receptor, a cell membrane-associated glycoprotein involved in iron homeostasis and cell growth, has been explored as a target to deliver therapeutics into cancer cells due to its increased expression on malignant cells. Conjugates have been made using the plasma protein transferrin as a carrier fused to several RIPs such as ricin A-chain [43], saporin [43], nigrin b [7] and ebulin 1 [7]. These conjugates have demonstrated anti-tumor activity in several in vitro and in vivo models [43].

Epidermal growth factor receptor (EGFR) is overexpressed in many different types of solid tumors and is associated with metastasis and poor prognosis. Several anti-cancer conjugates targeting the EFGR have been developed. Recombinant saporin has been linked to EGF both directly and via an adapter. Both conjugates were very effective in inhibiting an oral squamous cell carcinoma cell line. More importantly, the authors observe a significant reduction in the number of colonies formed [44].

Fibroblast growth factor receptors (FGFRs) are a family of at least 12 different proteins. Many solid tumors express receptors binding FGF2. FGF2-saporin was the first recombinant fusion chimera based on saporin expressed in *E. coli* shown to be highly selective and cytotoxic towards FGF2 receptor-expressing cells [45]. The conjugate also showed significant anti-proliferation activity when tested in animal models of human ovarian teratocarcinoma or melanomas in a combination therapy [46].

Another antitumor therapy approach involves the vascular endothelial growth factor (VEGF). VEGF-gelonin fusions contain a highly selective carrier, targeting tumor endothelial cells. This fusion protein is able to inhibit tumor growth and metastasis dissemination [47].

The therapeutic efficacy of natural ligand-toxin conjugates can be limited by the intracellular trafficking pathway followed by endogenous ligands. To improve the therapeutic efficacy of ligand-drug conjugates, research groups are developing new approaches for engineering the ligands to be more effective drug carriers [42].

Finally, the use of pure lectins for targeting has also been proposed [7]. Conjugates containing nigrin b as the toxic portion and the mucin-binding lectins SELld or SELfd as the targeting molecule, proved to be effective in killing COLO 320 and HeLa cells. In contrast, the free lectins had a very small or no effect on cell viability. SELld and SELfd are type 2 RIPs-related dimeric lectins isolated from the leaves and fruits of *Sambucus ebulus* respectively. These results open the possibility of using pure lectins as targeting molecules for cancer therapy.

9.7 Conclusions and Perspectives

Since their discovery, RIPs have been the subject of extensive investigation due to their strong enzymatic activity and toxicity. These enzymes share common properties such as conserved active site residues and reaction mechanisms. RIPs are

N-glycosidases capable of inhibiting protein synthesis by depurinating rRNA. However, the diversity among RIPs and their activities toward different targets make it difficult to extrapolate results in attempting to characterize the biological role of RIPs in plants. Discovery of multiple enzymatic activities in some RIPs, which can depurinate not only ribosomal substrates but may also damage DNA or RNA of pathogens and host cells, makes the picture more complicated. RIPs are broadly distributed in plants; however there are no systematic screening studies to generalize their occurrence. Results from such studies might provide new information about the phylogenetic distribution, structure and functions of RIPs. Also solving the three-dimensional structure of more RIPs will facilitate the elucidation of RIP structure–function relationships and will add valuable information regarding the structural homology of the RIPs. Plants produce RIPs that are able to kill mammalian cells once delivered to the cytosol. In mammalian cells, both type 1 and type 2 RIPs have been related to apoptosis. The mechanisms by which apoptosis is activated by a particular RIP may differ and may be independent of protein synthesis inhibition. Therefore, the process of RIP uptake and transport in cells needs more investigation in order to understand the mechanism by which RIPs lead to cell death. Knowing how they target cells will make it possible to use them to our advantage in medicine. RIPs have a potential as therapeutic agents if the toxicity can be specifically directed. RIP-containing conjugates have been used in many experimental strategies against cancer cells, often showing excellent clinical activity, for example immunotoxins targeting hematological malignancies. A major limitation of immunotoxins is the development of neutralizing antibodies to both the toxic and the carrier portions of the conjugate together with the development of VLS and hepatotoxicity. However, there have been significant advancements in the design of new immunotoxins that reduce side effects on normal cells. From the progress over the past twenty years it is apparent that immunotoxins are having an increasing impact in experimental therapy.

Acknowledgments This work was supported by grants FISPI04/1279 to J.M.F. and BIO39/VA42/10 to L.C. We thank Judy Callaghan (Monash University, Melbourne, Australia) for correcting the manuscript.

References

1. Barbieri L, Battelli MG, Stirpe F (1993) Ribosome-inactivating proteins from plants. Biochim Biophys Acta 1154:237–282
2. Stirpe F, Barbieri L (1986) Ribosome-inactivating proteins up to date. FEBS Lett 195:1–8
3. Girbes T, Ferreras JM, Arias FJ, Stirpe F (2004) Description, distribution, activity and phylogenetic relationship of ribosome-inactivating proteins in plants, fungi and bacteria. Mini Rev Med Chem 4:461–476
4. Obrig TG (1997) Shiga toxin mode of action in E. coli O157:H7 disease. Front Biosci 2:d635–d642
5. Barbieri L, Valbonesi P, Bonora E, Gorini P, Bolognesi A, Stirpe F (1997) Polynucleotide: adenosine glycosidase activity of ribosome-inactivating proteins: effect on DNA, RNA and poly(A). Nucleic Acids Res 25:518–522

6. Polito L, Bortolotti M, Farini V, Battelli MG, Barbieri L, Bolognesi A (2009) Saporin induces multiple death pathways in lymphoma cells with different intensity and timing as compared to ricin. Int J Biochem Cell Biol 41:1055–1061
7. Ferreras JM, Citores L, Iglesias R, Jimenez P, Girbes T (2011) Use of ribosome-inactivating proteins from *Sambucus* for the construction of immunotoxins and conjugates for cancer therapy. Toxins (Basel) 3:420–441
8. Ferreras JM, Citores L, Iglesias R, Jiménez P, Girbés T (2010) *Sambucus* ribosome-inactivating proteins and lectins. In: Lord JM, Hartley MR (eds) Toxic plant proteins-series plant cell monographs, vol 18. Springer, Berlin, pp 107–131
9. Mak AN, Wong YT, An YJ, Cha SS, Sze KH, Au SW, Wong KB, Shaw PC (2007) Structure-function study of maize ribosome-inactivating protein: implications for the internal inactivation region and the sole glutamate in the active site. Nucleic Acids Res 35:6259–6267
10. Wang HX, Ng TB (2001) Isolation of pleuturegin, a novel ribosome-inactivating protein from fresh sclerotia of the edible mushroom *Pleurotus tuber-regium*. Biochem Biophys Res Commun 288:718–721
11. Liu RS, Yang JH, Liu WY (2002) Isolation and enzymatic characterization of lamjapin, the first ribosome-inactivating protein from cryptogamic algal plant (*Laminaria japonica* A). Eur J Biochem 269:4746–4752
12. Endo Y, Tsurugi K (1988) The RNA N-glycosidase activity of ricin A-chain. The characteristics of the enzymatic activity of ricin A-chain with ribosomes and with rRNA. J Biol Chem 263:8735–8739
13. Shi X, Khade PK, Sanbonmatsu KY, Joseph S (2012) Functional role of the sarcin-ricin loop of the 23S rRNA in the elongation cycle of protein synthesis. J Mol Biol 419:125–138
14. Barbieri L, Ferreras JM, Barraco A, Ricci P, Stirpe F (1992) Some ribosome-inactivating proteins depurinate ribosomal RNA at multiple sites. Biochem J 286(Pt 1):1–4
15. Barbieri L, Brigotti M, Perocco P, Carnicelli D, Ciani M, Mercatali L, Stirpe F (2003) Ribosome-inactivating proteins depurinate poly(ADP-ribosyl)ated poly(ADP-ribose) polymerase and have transforming activity for 3T3 fibroblasts. FEBS Lett 538:178–182
16. Sestili P, Alfieri R, Carnicelli D, Martinelli C, Barbieri L, Stirpe F, Bonelli M, Petronini PG, Brigotti M (2005) Shiga toxin 1 and ricin inhibit the repair of H_2O_2-induced DNA single strand breaks in cultured mammalian cells. DNA Repair (Amst) 4:271–277
17. Sikriwal D, Batra JK (2010) Ribosome inactivating proteins and apoptosis. In: Lord JM, Hartley MR (eds) Toxic plant proteins-series plant cell monographs, vol 18. Springer, Berlin pp 167–189
18. Das MK, Sharma RS, Mishra V (2012) Induction of apoptosis by ribosome inactivating proteins: importance of N-glycosidase activity. Appl Biochem Biotechnol 166:1552–1561
19. Pascal JM, Day PJ, Monzingo AF, Ernst SR, Robertus JD, Iglesias R, Perez Y, Ferreras JM, Citores L, Girbes T (2001) 2.8-A crystal structure of a nontoxic type-II ribosome-inactivating protein, ebulin l. Proteins 43:319–326
20. van Deurs B, Sandvig K, Petersen OW, Olsnes S, Simons K, Griffiths G (1988) Estimation of the amount of internalized ricin that reaches the trans-Golgi network. J Cell Biol 106:253–267
21. Lord JM, Spooner RA (2011) Ricin trafficking in plant and mammalian cells. Toxins (Basel) 3:787–801
22. Battelli MG, Musiani S, Buonamici L, Santi S, Riccio M, Maraldi NM, Girbes T, Stirpe F (2004) Interaction of volkensin with HeLa cells: binding, uptake, intracellular localization, degradation and exocytosis. Cell Mol Life Sci 61:1975–1984
23. Vago R, Marsden CJ, Lord JM, Ippoliti R, Flavell DJ, Flavell SU, Ceriotti A, Fabbrini MS (2005) Saporin and ricin A chain follow different intracellular routes to enter the cytosol of intoxicated cells. FEBS J 272:4983–4995
24. Reisler RB, Smith LA (2012) The need for continued development of ricin countermeasures. Adv Prev Med 2012:149737
25. Lin JY, Tserng KY, Chen CC, Lin LT, Tung TC (1970) Abrin and ricin: new anti-tumour substances. Nature 227:292–293
26. Fu LL, Zhou CC, Yao S, Yu JY, Liu B, Bao JK (2011) Plant lectins: targeting programmed cell death pathways as antitumor agents. Int J Biochem Cell Biol 43:1442–1449

27. Pryme IF, Bardocz S, Pusztai A, Ewen SW (2006) Suppression of growth of tumour cell lines in vitro and tumours in vivo by mistletoe lectins. Histol Histopathol 21:285–299
28. Zwierzina H, Bergmann L, Fiebig H, Aamdal S, Schoffski P, Witthohn K, Lentzen H (2011) The preclinical and clinical activity of aviscumine: a potential anticancer drug. Eur J Cancer 47:1450–1457
29. Voss C, Eyol E, Frank M, von der Lieth CW, Berger MR (2006) Identification and characterization of riproximin, a new type II ribosome-inactivating protein with antineoplastic activity from *Ximenia americana*. FASEB J 20:1194–1196
30. Shaw PC, Lee KM, Wong KB (2005) Recent advances in trichosanthin, a ribosome-inactivating protein with multiple pharmacological properties. Toxicon 45:683–689
31. Fang EF, Ng TB (2011) Bitter gourd (*Momordica charantia*) is a cornucopia of health: a review of its credited antidiabetic, anti-HIV, and antitumor properties. Curr Mol Med 11:417–436
32. Liu L, Wang R, He W, He F, Huang G (2010) Cloning and soluble expression of mature alpha-luffin from *Luffa cylindrica* and its antitumor activities in vitro. Acta Biochim Biophys Sin (Shanghai) 42:585–592
33. Becker N, Benhar I (2012) Antibody-based immunotoxins for the treatment of cancer. Antibodies 1:39–69
34. Polito L, Bortolotti M, Pedrazzi M, Bolognesi A (2011) Immunotoxins and other conjugates containing saporin-s6 for cancer therapy. Toxins (Basel) 3:697–720
35. de Virgilio M, Lombardi A, Caliandro R, Fabbrini MS (2010) Ribosome-inactivating proteins: from plant defense to tumor attack. Toxins (Basel) 2:2699–2737
36. Dosio F, Brusa P, Cattel L (2011) Immunotoxins and anticancer drug conjugate assemblies: the role of the linkage between components. Toxins (Basel) 3:848–883
37. Madhumathi J, Verma RS (2012) Therapeutic targets and recent advances in protein immunotoxins. Curr Opin Microbiol 15:300–309
38. Gensini GF, Conti AA, Lippi D (2007) The contributions of Paul Ehrlich to infectious disease. J Infect 54:221–224
39. Youle RJ, Neville DM Jr (1980) Anti-Thy 1.2 monoclonal antibody linked to ricin is a potent cell-type-specific toxin. Proc Natl Acad Sci U S A 77:5483–5486
40. Filipovich AH, Vallera DA, Youle RJ, Quinones RR, Neville DM Jr, Kersey JH (1984) *Ex-vivo* treatment of donor bone marrow with anti-T-cell immunotoxins for prevention of graft-versus-host disease. Lancet 1:469–472
41. Cao Y, Marks JD, Huang Q, Rudnick SI, Xiong C, Hittelman WN, Wen X, Marks JW, Cheung LH, Boland K, Li C, Adams GP, Rosenblum MG (2012) Single-chain antibody-based immunotoxins targeting Her2/neu: design optimization and impact of affinity on antitumor efficacy and off-target toxicity. Mol Cancer Ther 11:143–153
42. Yoon DJ, Liu CT, Quinlan DS, Nafisi PM, Kamei DT (2011) Intracellular trafficking considerations in the development of natural ligand-drug molecular conjugates for cancer. Ann Biomed Eng 39:1235–1251
43. Daniels TR, Delgado T, Helguera G, Penichet ML (2006) The transferrin receptor part II: targeted delivery of therapeutic agents into cancer cells. Clin Immunol 121:159–176
44. Heisler I, Sutherland M, Bachran C, Hebestreit P, Schnitger A, Melzig MF, Fuchs H (2005) Combined application of saponin and chimeric toxins drastically enhances the targeted cytotoxicity on tumor cells. J Control Release 106:123–137
45. Lappi DA, Ying W, Barthelemy I, Martineau D, Prieto I, Benatti L, Soria M, Baird A (1994) Expression and activities of a recombinant basic fibroblast growth factor-saporin fusion protein. J Biol Chem 269:12552–12558
46. Davol P, Beitz JG, Mohler M, Ying W, Cook J, Lappi DA, Frackelton AR Jr (1995) Saporin toxins directed to basic fibroblast growth factor receptors effectively target human ovarian teratocarcinoma in an animal model. Cancer 76:79–85
47. Smagur A, Boyko MM, Biront NV, Cichon T, Szala S (2009) Chimeric protein ABRaA-VEGF121 is cytotoxic towards VEGFR-2-expressing PAE cells and inhibits B16–F10 melanoma growth. Acta Biochim Pol 56:115–124

Lucía Citores studied Chemistry at the University of Valladolid, Spain, and completed her PhD at the same University in 1996. From 1997 to 2004 she worked at the Institute for Cancer Research in Oslo, Norway, as a postdoctoral researcher. During this time she became interested in the endocytosis and intracellular transport of fibroblast growth factors and their receptors. She returned to Spain in 2004, and for five years she stayed at the University of Valladolid as a postdoctoral researcher. In 2009, she was appointed as Non-civil servant tenured Lecturer in Biochemistry and Molecular Biology at the University of Valladolid and in 2011 she was appointed as Associated Professor in the same area. Her research interests have focused on distribution, activity and structure of ribosome-inactivating proteins and how they enter and intoxicate mammalian cells.

Maria Rosario Iglesias studied Biology at the University of León, Spain, and completed her PhD at the University of Valladolid, Spain, in 1989. From 1991 to 1992 she worked at the University of Windsor, Canada, as a postdoctoral researcher in several aspects of genetic alterations of plasma lipoproteins, focused on the study of the transciptional regulation of Apo B gen in HepG2 cells. She returned to Valladolid in 1992 and she was appointed as Associated Professor in Biochemistry and Molecular Biology there in 1998. Her research interests have focused on the activity, molecular cloning and expression of ribosome-inactivating proteins, as well as the role played by RIPs in plant defense response to biotic and environmental stress.

José Miguel Ferreras studied Biology at the University of León, Spain, and completed his PhD at the University of Valladolid, Spain, in 1989. From 1990 to 1991 he worked at the University of Bologna, Italy, as a postdoctoral researcher. He returned to Valladolid in 1991 and he was appointed as Associated Professor in Biochemistry and Molecular Biology. His research interests have focused on the distribution, activity and structure of ribosome-inactivating proteins and their use for the construction of immunotoxins and other conjugates for cancer therapy.

Chapter 10
Current Status and Biomedical Applications of Ribosome-Inactivating Proteins

Riccardo Vago, Rodolfo Ippoliti and Maria Serena Fabbrini

Abstract Toxin domains from plants or bacterial origin (such as Diphtheria toxin (DT) or Pseudomonas exotoxin A (PEA), which are endowed with ADP-ribosylation activity of the Eukaryotic elongation factor-2) have been extensively exploited for research and therapeutic purposes. Denileukin diftitox is the first FDA approved recombinant fusion toxin between a truncated diphtheria toxin and human interleukin-2 for the treatment of cutaneous T cell lymphomas. Ribosome-inactivating proteins (RIPs) are also a class of potent inhibitors of protein synthesis that act differently, by catalytically depurinating an adenine residue (A4324 in rat) exposed in an universally conserved stem-loop region of 23/26/28S ribosomal RNA (rRNA), also known as the "α-sarcin loop". This causes an irreversible block in protein synthesis, leading to apoptotic cell death of mammalian target cells. RIP-containing conjugates have been used in the therapy of cancer and other incurable diseases with potential promising results and failures. In the last decade, many research efforts have been dedicated to the development of more efficient and less immunogenic chimeric fusions. In this chapter we will mainly focus on plant RIPs with the aim to report some of the most promising biomedical applications currently under investigation and further discuss their future perspectives.

R. Vago (✉)
Dibit-HSR Scientific Institute and Università Vita-Salute San Raffaele,
via Olgettina 58, 20132 Milan, Italy
e-mail: vago.riccardo@hsr.it

R. Ippoliti
Department of Life, Health and Environmental Sciences, University of L'Aquila,
Via Vetoio snc. Coppito, 67010 L'Aquila, Italy
e-mail: rodolfo.ippoliti@univaq.it

M. S. Fabbrini (✉)
Istituto Nazionale di Genetica Molecolare (INGM),
Via Francesco Sforza 28, 20122 Milan, Italy
e-mail: msfabbrini@gmail.com

E. F. Fang and T. B. Ng (eds.), *Antitumor Potential and Other Emerging Medicinal Properties of Natural Compounds*, DOI: 10.1007/978-94-007-6214-5_10,

10.1 Introduction

Ribosome-inactivating proteins (RIPs) belong to the N-glycosidase family of toxins (EC3.2.32.22) able to specifically and irreversibly inactivate the large 60S ribosomal subunits and thereby inhibit protein synthesis. Most plants and bacterial RIPs, such as Shiga and Shiga-like toxins from the bacteria Shigella dysenteriae and the Shigatoxigenic group of *Escherichia coli* (which include other enterohemorrhagic *E.coli* strains) in fact exert their toxic effects by depurinating a specific adenine base (A4324 in the rat 28S ribosomal rRNA) located in the universally conserved GAGA-tetraloop, also known as the α-sarcin/ricin loop, present in 23/26/28S rRNA (Fig. 10.1). This results in the inability of the ribosome to bind elongation factor 2, thus irreversibly blocks translation [1, 2]. RIPs are widely distributed in nature but are found predominantly in plants, bacteria and fungi. Since plants cannot run away from their predators, some have developed defense systems based on toxin accumulation in leaves, seeds and other tissues where RIPs

(a)

```
Rat 28S       4301 CCATGGTAATCCTGCTCAGTACGAGAGGAACCGCAGGTTCAGACATTTGGTGTA 4354
Yeast 25S     3001 CAATAGTAATTGAACTTAGTACGAGAGGAACAGTTCATTCGGATAATTGGTTTT 3054
E. Coli 23S   2637 TGAGGGGGGCTGCTCCTAGTACGAGAGGACCGGAGTGGACGCATCACTGGTGTT 2690
Rice 25S      2994 CGATAGTAATTCAACCTAGTACGAGAGGAACCGTTGATTCACA--ATTGGTCAT 3045
Tomato 25S    3001 CAATAGTAATTCAACCTAGTACGAGAGGAACCGTTGATTCACACAATTGGCCAT 3054
                     *  *        *  ************ * *      *  *    ***
```

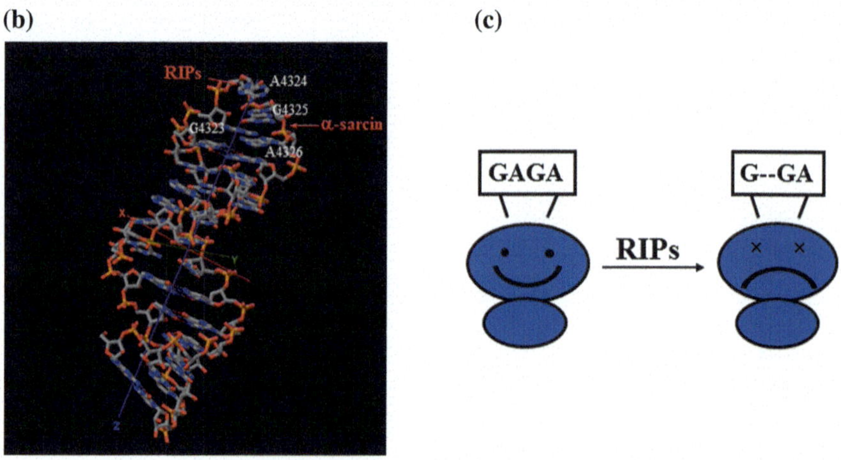

Fig. 10.1 RIPs act on a conserved loop in 23/25/28S rRNA. Alignment of 23/25/28S rRNA sequences containing the GAGA conserved tetraloop targeted by RIPs (*in yellow*) (**a**). Three-dimensional structure of the rat tetraloop; the GAGA bases and the targets of RIPs and α-sarcin are indicated. **b** Depurination of the adenine A4324 by RIPs irreversibly blocks the ribosomal activity (**c**)

may represent effective weapons plants used to ensure the maintenance of the species. Most RIP-expressing plants present multigene families as in the case of saporins [3], PAP [4] or ricin gene family [5], that seem to be under a clear selective gene pressure [6]. Indeed, all the isoforms of saporin are coded by a small gene family [7] with limited sequence variations, allowing to suppose a duplication of a common ancestor gene. Plant RIPs have been classified into three main types: Type I is composed of a single polypeptide chain of approximately 30 kDa, Type II is a heterodimer consisting of an A chain, functionally equivalent to the Type I polypeptide (Fig. 10.2) [8], linked to a B subunit which is endowed with lectin-binding properties [9] while Type III are synthesized as inactive precursors (ProRIPs) that require proteolytic processing events to form an active RIP and are not in use for therapeutic purposes. Different enzymatic activities besides the N-glycosidase one have been ascribed to the RIPs from plants [10], although they are still controversial. Indeed, the deoxyribonuclease activity attributed to ricin was clearly demonstrated to depend upon a nuclear contamination, [11] similarly to the one previously reported for saporin [12].

The ricin AB dimer is the prototype plant Type II RIP whose biochemical features, catalytic activity, biosynthetic pathway, and intracellular transport in intoxicated cells have been studied in greatest details so far [13, 14]. Ricin is among the most potent poisons known in nature showing potency in the low picomolar range thanks to its galactose-B-lectin domain that binds to cell surface exposed glycans and mediates toxin entry into most mammalian cells. Once endocytosed, the ricin heterodimer is able to be retrogradelly transported along the secretory route where in the endoplasmic reticulum the SS bridge linking A to B domain is reductively broken [15], leaving free unfolded A chain to undergo retrotranslocation into the cytosol where it rapidly refolds to inactivate target ribosomes. Conversely, type I RIPs have been demonstrated to follow different intoxication routes [16–18].

The most promising ways to exploit plant RIPs as weapons is to mimic ricin action either by designing molecules in which the toxic RIP domain(s) are linked to selective tumor targeting domains or directed against the desired cellular targets *via* ligand domains or simply by directly delivering complexed DNAs, as suicide genes for cancer gene therapy approaches.

10.2 Biochemical and Structural Characteristics of Plant RIPs

Many plant species (at least 50 plant species belonging to 17 different families, among them Cucurbitaceae, Euphorbiaceae, Poaceae, and families belonging to the superorder Caryophyllales) produce RIPs (for a more detailed description, please refer here to Chap. 11 by Ferreras and collaborators), although they are not found in many others. Type I RIPs usually show a high content in basic amino acids (almost 10 % Lysine in saporin) that confers to these toxins a high IP (>9) [19], greatly differing from Type II RIPs that are instead neutral proteins, very poor in lysines. These crucial differences may also reflect the different intoxication pathways followed by Type I and Type II RIPs in mammalian cells. The low lysine content of Ricin A chain (RTA) allows the toxin to avoid proteasomal degradation, most likely by hampering efficient cytosolic ubiquitination [20]. The sequence identities between RTA and Type I RIPs (Fig. 10.2) are low and found to be respectively: Saporin 22 %, Gelonin 30 %, PAP 29 %, Thricosanthin 35 %, Dianthin 19 %, Bouganin 29 %, Momordin 33 % and Momorcharin, 33 %. Several residues are highly conserved among RIPs, including amino acids found within the catalytic cleft and few others surrounding it are also highly conserved among Type I and Type II RIPs and belong to the "N-glycosidase signature" which includes Tyr80, Tyr123, and the key active site residues Glu177, Arg180, and Trp211 in RTA. However, as can be seen in the protein alignments between some Type I RIPs used for targeted therapies with the toxic A chain of ricin, amino acids and secondary structures may vary consistently [21]. Glu176 and Arg179 in saporin-6 (as Glu177 and Arg180 in RTA) are thought to be directly involved in catalysis. However, while a RTA Glu176 mutant was 20-fold less active than wild-type A chain in inhibiting translation in a reticulocyte lysate, the Arg179 saporin mutant was 200-fold less active [22]. A complete loss of *in vitro* and *in vivo* saporin cytotoxicity can be achieved when Glu176 and Arg179 are both mutated to lysine and glutamine residues, respectively. This double mutant (termed KQ) is, indeed, devoid of all the detrimental effects associated with RIP expression in several hosts [12, 23, 24]. Interestingly, mutation of Trp208 in saporin did not impair its *in vitro* enzymatic activity and cytotoxicity [22], but this same residue has been demonstrated to be crucial for the structural integrity of PAP [25]. Despite differences in amino acid sequences, the overall three-dimensional fold is well conserved among RIPs as estimated by the superimposition of the 3D structures of several Type I RIPs with the one of RTA (Fig. 10.3), which demonstrates that RTA and Type I RIPs all share a common "RIP fold" characterized by the presence of two major domains: an N-terminal domain, which is mainly beta-stranded, and a C-terminal domain that is predominantly alpha–helical. Although within the primary sequence comparisons certain tyrosine residues would appear conserved and aligned, saporin Y120 would instead appear to superimpose with F119 of PAP in 3D structures. (Fabbrini and Ippoliti unpublished observations). Insertions and deletions, as compared to PAP, momordin from *Momordica charantia L.* and RTA were found to lay mainly in random coil regions [26].

10.3 Heterologous Expression of RIPs and Chimeric Fusions

To avoid self-intoxication, most of the plants synthesize RIPs as pre-pro precursor proteins. *Ricinus communis* cells protect themselves from ricin activity, by producing an inactive single chain precursor which is then transported to the storage vacuoles and activated into SS linked A and B chains [27]. To date bacterial,

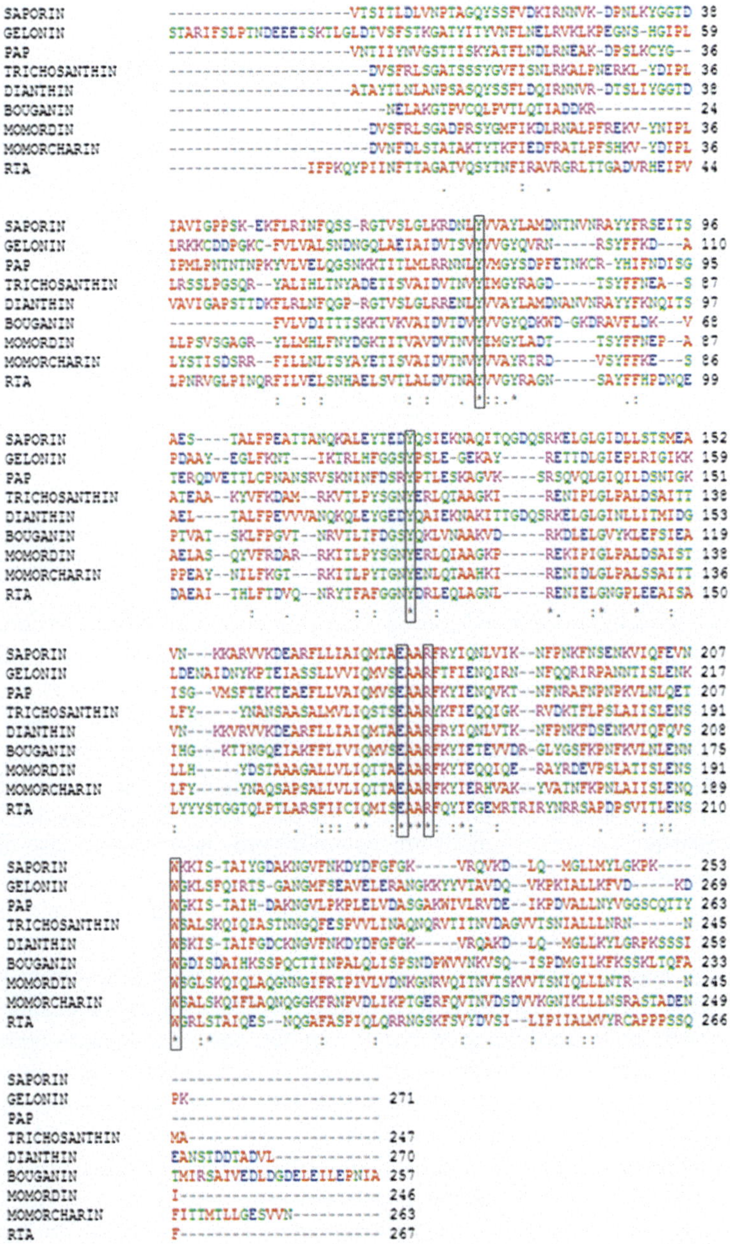

Fig. 10.2 Amino acid sequence alignment of different type I RIPs compared to Ricin A chain (RTA). Conserved residues crucial for the catalytic activity are boxed: Tyr 83, Tyr123, Glu177, Arg180 and Trp211 in RTA

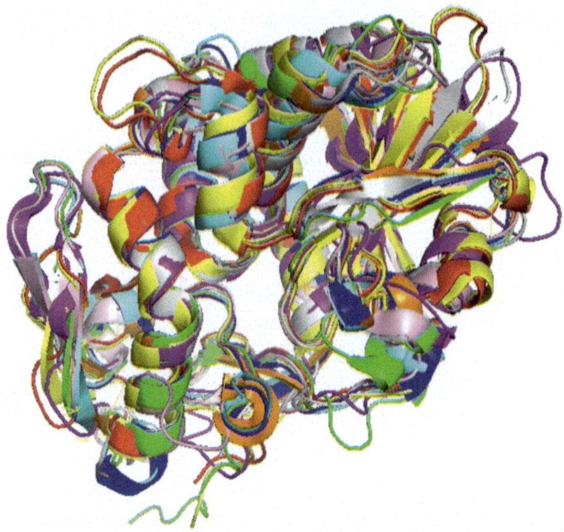

Fig. 10.3 Three-dimensional structure of different Type I RIPs and ricin A chain (RTA). Superimposition of secondary structure elements of Saporin (*red*, PDB code 1QI7), Gelonin (*pink*, 3KU0), PAP (*magenta*, 1GIK), Trichosanthin (*cyan*, 1QD2), Dianthin (*yellow*, 1RL0), Bouganin (*grey*, 3CTK), Momordin (*orange*, 1 MOM), Momorcharin (*blue*, 1AHA), RTA (*green*, 1J1 M)

yeast and model plant cells have been used to produce RIPs or RIP-based chimeric fusions. One common problem faced during the production of recombinant RIPs resides in their intrinsic toxicity toward the host ribosomes. Initial attempts to express recombinant Type I RIPs in *E. coli* were problematic, because, upon induction of RIPs, the bacterial growth rate was significantly impaired, as reported in the case of Mirabilis antiviral protein [28], PAP [29], as well as saporin [7]. Both *in vitro* depurination assays [2], with the observation that *E. coli* ribosomes were found depurinated *in vivo* [29], confirmed that host toxicity was due to RIP catalytic activity. Although the toxin expression can be tightly regulated by employing the *E. coli* strain BL21(λDE3)pLysS to get a satisfactory yield in PAP [29], in dianthin [30], and saporin (or saporin fusions) [21, 31–34] in some cases the protein may accumulate inside the cell as an insoluble fraction from which the full RIP activity is not easily recovered [35]. Endotoxin contamination [36] together with inefficient folding of certain secretory targeting domains appear as the main disadvantages of the bacterial expression systems, prompting the recent development of eukaryotic expression systems. The ATF-saporin fusion provides a typical example of a chimeric toxin difficult to express at high levels and in soluble form in bacterial hosts. A model system for the production of the ATF-saporin chimera was developed consisting in RNA micro-injection of *Xenopus laevis* oocytes that were protected with neutralizing anti-saporin antibodies injected into the cytoplasm in order to preserve oocyte's full translation capability [37].

In the last years, the methylotrophic yeast *Pichia pastoris* has been demonstrated to be a suitable platform for the expression of recombinant proteins including some fusion toxins, allowing protein post-translation modifications and a several-fold yield improvement in product with respect to bacteria [38]. Neville's group has been one of the first pathfinders in the expression of recombinant

Diphtheria toxin-based immunotoxin fusions in *P. pastoris*: they found GS115 strain particularly tolerant to this bacterial toxin [39] and could express several chimaeras successfully. *P. pastoris* has been exploited also to produce large amounts of secreted recombinant PAP showing same enzymatic activity of native PAP purified from *Phytolacca americana* L. Toxicity was likely prevented through rapid and efficient secretion of the toxin into the culture medium. [40]. Using different *Pichia* strains, (GS115 (His4)) was found more tolerant towards saporin and both active saporin as well as ATF-Saporin were successfully produced in *P. pastoris* following codon-usage optimization which greatly increased their expression levels [12]. This expression system could become a great resource in chimeric RIP's expression due to its limited costs, ease of manipulation, ability to perform complex post-translational modifications, and high expression levels in high scale in fermentors.

Finally, another way to produce recombinant plant RIPs has been explored by the use of plant tobacco protoplasts. However, saporin precursor was found very toxic towards this expression system [24]. Interestingly, the native preproricin construct could be expressed in tobacco protoplasts and its processing, glycosylation, and targeting to the vacuole occurred normally with no toxicity detected [41]. In contrast, the expression of an orphan secretory ricin toxic A chain (RTA) polypeptide resulted in the retro-translocation of RTA to the cytosol followed by inhibition of protein synthesis [41]. This suggesting once again that RIP toxicity could be detrimental also for the tobacco protoplast expression system. Indeed, among the various constructs expressed in transgenic tobacco plants, preprotrichosanthin was properly processed and the protein translation inhibitory activity of purified toxin was similar to native one only when the expressed precursor also included the C-terminal propeptide [42], suggesting in the latter an efficient segregation of this RIP.

10.4 IT-Therapeutic Potential and Applications

10.4.1 Toxins-Based Therapy of Hematological Tumors

Over one century ago, Paul Ehrlich elaborated the concept of "magic bullet", a drug that selectively targets a pathogen/cellular target, thus being innocuous for the surrounding healthy tissues. This concept was to some extent realized by the discovery of monoclonal antibodies, providing antigen-specific binding affinity [43]. The identification of specific antigens over-expressed at the surface of malignant cells, together with the development and the large-scale production of monoclonal antibodies, allowed the first generation of immunotoxins (ITs) to be created (Fig. 10.4). ITs are composed of a toxic domain chemically conjugated to an antibody or genetically fused to antibody fragment(s) which confer selective specificity to the IT target. More recently, ligand targeted toxins have also been developed, where a specific targeting ligand (usually a growth factor domain) replaces the antibody domains (Figure 10.4). ITs made of domains of the bacterial

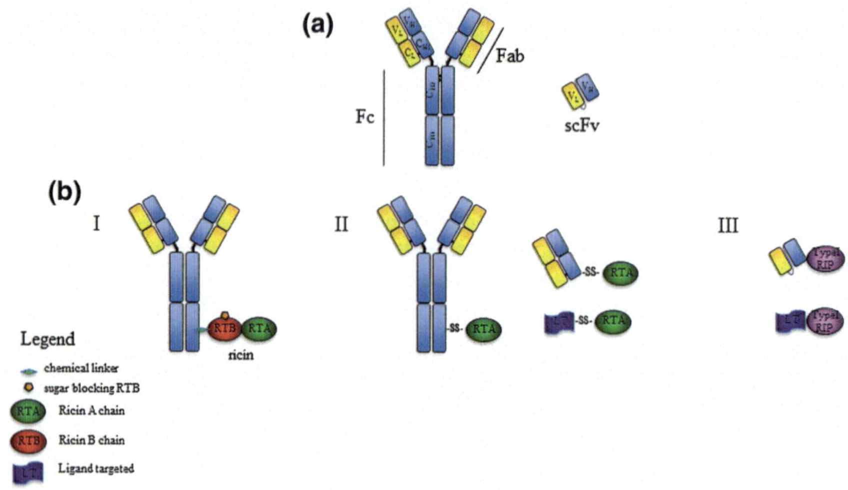

Fig. 10.4 Plant RIPs have been employed to prepare immunotoxins (IT) and ligand targeted (LT) toxins. Schematic representation of the antibody and its fragments used in the IT preparation (**a**). First (I), second (II) and third (III) generation of plant RIP-based ITs. LT toxins have been developed in parallel (**b**). Several mature type I RIPs have been employed in place of RTA to generate II generation ITs. Fc: fragment crystallizable; Fab: fragment antigen-binding; scFv: single-chain variable fragment; VL: Variable fragment of antibody light chain; VH: Variable fragment of antibody heavy chain; CL: light chain variable domain; CH: heavy chain constant domain. Light chains are in *yellow* and heavy chains in *blue*

Pseudomonas exotoxin A (PEA) and the diphtheria toxin (DT) together with the plant RIP ricin are the most studied toxins employed for therapeutic purposes [44], but several others are currently undergoing clinical evaluation, mostly in the context of oncological diseases. Toxins are potent, natural weapons since their toxicity has been increased by natural selection over millions of years, such as only minute amounts of molecules are needed to exert devastating effects.

The first-generation of ricin ITs were made by blocking the binding activity of the toxin (sugars in the case of ricin B chain). Among drawbacks were their poor stability, lack of specificity and heterogeneous composition(s), determining high variability among batches. The second-generation ITs were prepared with toxins that lack non-specific binding domains, conjugated with intact antibody or with a fragment antigen binding (Fab). The Fab construct gives a more uniform and smaller product which also potentially enables greater tumor penetration. The third-generation ITs are wholly recombinant molecules transcribed from a single DNA: such single-chain IT incorporates single-chain variable fragment (scFv) of the antibody (Fig. 10.4) [45].

Hematologic tumors are good candidates for ITs and ligand toxin treatments: they show distinct cell lineages with stages of differentiation expressing well-characterized surface target molecules, with the malignant cells being easily accessible by the ITs in the bloodstream [46]. One critical aspect in using mouse hybridoma derived antibodies for the IT construction is the recognition by the human immune system with induction of human anti-mouse antibody (HAMA) responses, resulting in short half-life, reduced efficacy, and in some cases increased toxicity in repeatedly treated patients [47].

Denileukin diftitox (Ontak) was the first IT-like drug approved by the Food and Drug Administration (FDA) for the treatment of cutaneous T cell lymphoma (CTCL). It is a genetic fusion between the catalytic domain of DT and the cytokine interleukin-2 (IL-2) [48]. In the pivotal phase III trial in 71 CTCL patients, a 30 % response rate was achieved, including 10 % complete remissions [49]. This approved recombinant fusion toxin represents a milestone in the cancer therapy field. IL-2 receptors have been targeted also by PEA and ricin based ITs. A single-chain IT including the Fv of an antibody to the CD25 subunit of the IL-2 receptor was fused to a 38-kDa fragment of the PEA (PE38) and was administered to 35 patients with chemotherapy-resistant leukemia, lymphoma, and Hodgkin's Disease (HD), resulting in a response rate of 40 % in the 20 patients receiving upper dose levels [50]. A phase I/II trial for treatment of CD25 positive CTCL and Chronic Lymphocytic Leukemia (CLL) were recently completed and others have now been initiated in adult T cell leukemia/lymphoma (ATL) and Hairy-Cell Leukemia (HCL). Treatment with an IT containing the anti-CD25 antibody chemically conjugated to deglycosylated ricin A-chain (dgA) demonstrated a more slowly clearance from blood stream [51] and resulted in a response rate of 11 % in refractory HD patients in a phase I/II study [52]. A Phase II study evaluating the side effects and efficiency of anti-CD25 dgA treatment in patients with CTCL was recently completed while a phase II study of ATL is currently recruiting participants. DgA has been conjugated to several other monoclonal antibodies, namely, CD19 [53], CD22 [54] and CD30 [55] to be tested in phase I trials showing moderate efficacy accompanied by systemic toxicity: vascular leak syndrome (VLS) was a common dose-related toxicity in these studies. Concerning Type I RIPs, an anti-CD19-PAP IT was tested in a phase I dose escalation study conducted on 17 relapsed patients with B-lineage acute lymphoblastic leukemia (ALL), resulting in four complete remissions, one partial remission and in a rapid reductions in numbers of leukemic blasts in circulation in additional five patients [56]. Anti-CD30 saporin IT was administrated to four patients with advanced refractory HD, resulting in a rapid and substantial reduction in tumor mass (50 % to >75 %) in three of them. Unfortunately, the clinical responses were transient (6–10 weeks) with antibodies developed in all subjects [57].

In animal models, a 3-day treatment with nontoxic doses of the same IT induced complete remissions in 80 % of mice carrying anaplastic large-cell lymphoma, suggesting its possible role as adjuvant therapy in patients with $CD30^+$ ALCL refractory to standard treatments [58]. Similar results were obtained by using IT containing anti-CD30 covalently linked to momordin or PAP and the sequential administration of two anti-CD30 ITs (momordin and saporin) was well tolerated and did not result in formation of antibodies cross-reacting with the two plant toxins [59]. Pursuing this strategy, combined administrations of ITs with different toxic or targeting moieties have been proposed. Treatment with momordin, saporin and PAP conjugated to anti-CD22 monoclonal antibodies, alone or in combination, significantly extended the survival time of mice bearing transplanted Daudi (human Burkitt's lymphoma) cells [60]. A combination of anti-CD19, -CD22, and -CD38-saporin ITs (3BIT) cured severe combined immunodeficient mice injected with the human B cell lymphoma cell line Ramos, resulting in 100 % disease-free survivors at 300 days [61]. This approach could likely overcome the problems arising from heterogeneity of target antigen expression that can limit the therapeutic efficacy of a single-IT therapy. To address this point, polyclonal antibodies, which recognize multiple antigens on a wide range of cells, thus reducing the escape of neoplastic clones during immunotherapy, have been used. Anti-thymocyte globulins (ATG), consisting of a mixture of polyclonal antibodies specific for cell surface proteins, have been conjugated to saporin, showing a strong cytotoxic effect on five lymphoma- and leukemia-derived cell lines, as well as on activated lymphocytes while sparing non-hematological cell lines [62]. Saporin, in particular, has been extensively studied to construct anti-cancer conjugates because of its high enzymatic activity, stability and resistance to conjugation procedures, resulting in efficient killing of target cells [63].

Gelonin is one of the most promising plant toxins used to construct several recombinant ITs. rGel (recombinant Gelonin)/BlyS (B-lymphocyte stimulator) is a toxin-cytokine fusion protein used for selective killing of malignant B cells expressing receptors for B-cell-activating factor. In an immunodeficient transplant model for human acute lymphoblastic leukemia it showed prolongs survival of ALL-bearing mice. The co-administration of AMD3100, a CXCR4 antagonist, with the aim to mobilize the leukemic cells protected in the bone marrow (BM) microenvironment, resulted in a significant reduction of the tumor load in the BM and complete eradication of ALL cells from the circulation [64].

A recent frontier in the in hemato-oncology approaches consists in the targeted elimination of leukemia stem cells (LSC), exploiting selective expression of LSC-associated surface antigens. This approach allows the eradication of therapy-resistant malignant cells with self-renewal capacity as well. Anti-CD123-PE38 toxin was proven to be cytotoxic in leukemic cell lines and the efficacy of various CD123-targeted immunoconjugates is currently under investigation [65].

The overall results achieved so far in the treatment of hematological tumors suggest that RIP may have still a potential for being an useful tool in cancer therapy, in addition to the traditional surgery, radio or chemotherapy, especially for relapsed patients.

10.4.2 Bone Marrow Purging

Conventional chemotherapeutics for acute myelogenous leukemia (AML) and acute lymphoblastic leukemia (ALL) results in 60–80 % remission, but unfortunately patients are prone to relapse with around 18 months median survival time without heterologous bone marrow transplantation which can be applied only to a subset of patients (about 40 %) finding HLA-matched donors [66]. Autologous bone marrow transplantation has been shown to be potentially effective, but hematopoietic recovery is slow and infectious or bleeding complications are also common. Another experimental approach is to use "toxin purging" in selected cases. A pioneering study showed an anti-CD5-saporin immunotoxin was employed to purge *ex vivo* bone marrow from a patient with T cell ALL in second remission who was treated with high doses of chemotherapy and radiotherapy, however, 22 days after the autologous transplantation the disease relapsed in the bone marrow with the same phenotype as at the onset. Retrospective analysis of the transplanted marrow cells revealed a graft contamination of a small share of malignant T cells [67]. The complete eradication of the very aggressive and potentially lethal cancer cells remains an open challenge to be addressed.

Over the years, different targets and toxins have been studied and proposed for the purging [68, 69]. As potential candidate for therapeutic use, an IT formed by gelonin conjugated to anti-CD33 antibody directed to an antigen presented in leukemic blasts of most patients, but absent in the earliest progenitor bone marrow cells has been proven to effectively purge HL60 leukemia cells from an excess of normal bone marrow cells *in vitro* [70]. A phase I/II clinical study aimed to determine the optimal dose of gelonin based IT for stem cell purging in patients with acute myelogenous and leukemia myelodysplastic syndrome to allow for rapid engraftment with a lower relapse rate and to therefore improve the therapeutic outcomes has been recently terminated (NCT00043810). Briefly, autologous stem cells, that were used for the transplant, were collected from patients and treated in the laboratory with the anti-CD33-gelonin conjugate. This procedure eliminated most leukemia cells without damaging the normal stem cells. After chemotherapic treatment, patients received their previously stored stem cells through a blood transfusion. At present, an analogous phase II study is ongoing where a conjugate combining IL-2 and diphtheria toxin has been used for stem cells purging in peripheral T cell lymphoma patients (NCT00632827).

The cell-surface glycolipid CD77 is another potentially interesting target, as its expression is restricted to a subset of activated B cells and derived cancers, such as Burkitt's lymphoma. CD77 is the natural ligand of the Shiga-like toxin-l which was proven to be very effective in animal model purging of bone marrows obtained from SCID mice, seeded with Burkitt's lymphoma cells, leading to substantial increase in disease-free survival animals [71].

Efficiently purging stem cells or bone marrow can be considered a relevant step towards the achievement of a cure but this has necessarily to go through a great improvement in the detection techniques of residual malignant cells among the purged cells in order to avoid the relapses, as in the initial pioneering studies.

10.4.3 Toxin-Based Therapy of Autoimmune Disorders

Autoimmune diseases such as type 1 *diabetes mellitus* or multiple sclerosis origin from chronic organ-specific damage mediated by autoreactive T cells.

Self-perpetuating T cell responses make ineffective most of the therapies and may have pathologic consequences in allotransplantation leading to the rejection of the implanted organ. Since the long-term administration of immunosuppressive agents after transplantation displayed risks of side toxicity and infections [72, 73], toxin-based approaches have been assayed. Almost 30 years ago, Rennie et al. showed that ricin conjugated to the autoantigen thyroglobulin specifically suppressed the auto-antibody response of lymphocytes from patients with Hashimoto's thyroiditis [74]. In the following three decades, various toxins have been used to counteract the T-cell response in autoimmune disorders. Saporin conjugated to MHC class I tetramer has been proven to selectively kill antigen-specific $CD8^+$ T cells *in vitro* [75]. A fusion protein consisting of gelonin and an acetylcholine receptor fragment prevented the onset symptoms of myasthenia gravis in a rat model [76]. RTA conjugated to anti-CD64, an antigen overexpressed in rheumatoid arthritis synovial macrophages, showed effective clearance of activated synovial inflammatory macrophages *in vitro* [77]. Along the same line of research, the treatment of SCID mice engrafted with rheumatoid arthritis synovial tissue with a PEA based IT reduced the numbers of macrophages, activated fibroblast-like cells, endothelial cells, and proliferating cells [78]. Recently, a panel of ITs made by fusion to the B cell motif c-myc epitope, as a model targeting sequence, has been tested *ex vivo* on native splenocytes [79]. Since the screened toxins so far showed variable responses, this novel approach could be expanded to plant RIPs, as well.

10.4.4 Toxins-Based Therapy of Solid Tumors

ITs have been mainly developed against lymphomas and leukemia, since they could be injected into the blood stream and would more easily find their path against target tumor cells. However, interesting results could be obtained in certain solid tumors, as well. Conjugates between plant RIPs and different growth factors whose receptors are overexpressed on the cell surface of malignant cells have been developed and tested *in vitro* and some *in vivo*.

Saporin linked to epidermal growth factor (EGF) has been assayed in mouse models of human breast carcinoma [80], adenocarcinoma [36] and cervical cancers [81] overexpressing EGF receptor (EGFR), leading in all cases to a consistent tumor growth reduction. Among them, a saporin fusion to epidermal growth factor via a cleavable adapter (SA2E) was investigated for its efficacy in inhibiting tumor growth and reduced side effects *in vivo*. Indeed, a lethal dose for BALB/c mice was three-times less for the adapter-containing toxin than for the adapter-free construct (SE) [36]. Furthermore, SE only reduced the average weight of induced tumors by 33 % whereas SA2E-treated mice exhibited a 71 % reduction showing an almost complete suppression in 60 % of the cases. Finally, the combined application of non permeabilizing concentrations of triterpenoid natural saponins increased cytotoxicity of SA2E on murine tumor cells transfected with human EGFR by almost 20,000-fold [82]. A fusion protein of anti-EGFR single-chain variable fragment and gelonin was assessed on EGFR-overexpressing non-small cell lung cancer-derived cell lines and xenografts in nude mice: the IT inhibited cell proliferation and induced a significant suppression of tumor growth [83].

Mice bearing human melanoma, teratocarcinoma or neuroblastoma cells were treated with basic fibroblast growth factor (bFGF) conjugated saporin. Significant tumor inhibition was seen after 1 month treatment and complete tumor regression was observed in 20 % of the neuroblastoma xenografts [84]. The same conjugate was also tested in murine melanoma leading to inhibition of tumor growth and metastatization and increased survival time [85]. The bFGF-saporin conjugate induced dramatic tumor reduction in nude mice bearing a prostatic carcinoma cell line [86]. Prostate cancer was also targeted by ITs: in mice bearing prostate carcinoma cells saporin was coupled to the prostate specific marker PSMA [87] and to anti-CUB domain-containing protein 1 [88], widely expressed in tumors. In the first case the IT inhibited metastasis, but not primary tumor; in the latter it significantly inhibited tumor growth. Mice bearing human pancreatic carcinoma cells treated with anti-carcinoembryonic antigen-related cell adhesion molecule 6 (CEACAM6) antibody followed by secondary SAP-conjugated immunoglobulin showed a marked reduction in mean tumor volume and in some cases tumor regression [89]. The receptor protein tyrosine phosphatase β (RPTPβ), primarily overexpressed in astrocytic tumors, facilitates tumor cell adhesion and migration through interactions with the extracellular matrix. SAP-conjugated antibodies were assayed in athymic nude mice bearing glioblastoma cells showing variable tumor growth delay [90].

A specific antibody (TP3) directed against a 80 kDa surface antigen with a unique expression pattern in osteosarcomas [91] has been conjugated to PAP and the resulting IT has been demonstrated to kill clonogenic human osteosarcoma cells *in vitro* and shows significant antitumor activity in a murine soft tissue sarcoma model [92].

The administration of an IT formed by gelonin and a monoclonal antibody against inducible fibroblast growth factor, overexpressed in multiple solid tumor types relative to matched adjacent normal tissue, promoted long-term tumor growth suppression in nude mice bearing T-24 human bladder cancer cell xenografts [93].

Epithelial cell adhesion molecule (EpCAM) is overexpressed on the cell surface of many tumors but shows relatively low levels of expression on normal epithelia [94]. An IT comprised of an anti-EpCAM Fab fragment linked to bouganin selectively killed EpCAM-positive cell lines with a greater potency than many commonly used chemotherapeutic agents. *in vivo* efficacy was demonstrated using an EpCAM positive human ovarian carcinoma xenograft model in SCID mice: at highest doses, 12 out of 15 mice were found tumor free at the end of this study [95].

The less successful results obtained with solid tumor treatments using ITs as compared to hematological malignancies, are not only due to the poor penetration of the ITs within the tumor mass. Major concerns have been raised regarding use of plant toxins in oncologic patients with solid tumor masses since they may require prolonged treatment regimens [45]. In fact, it must be underscored that IT approaches would seem to be better suited for fighting hematologic malignancies, especially because patients with such malignancies are often immunocompromised [45, 96], besides the different physiological/biologic barriers for local toxin concentration. In fact, treatment of solid tumors may present potential bottlenecks due to the high interstitial fluid pressure gradient within the tumor generated by the hastily formed vascular architecture that limits the diffusion of large macromolecules, such as full-length antibodies conjugates within the tumoral mass. These problems could be in part counteracted by using smaller antibody formats (single chain variable fragments, scFv) in fully recombinant molecules, and indeed, the scFv toxin fusions have been recently termed the "modern" ITs [96]. Additionally, to bypass these hurdles, many strategies involving the use of selective adjuvants (e.g., saponins for the saporin chimaeras) or photochemical internalization (PCI) approaches have been investigated to improve toxin cytosolic delivery and thus, the anti-tumor efficacy of chimeric molecules [25, 82, 97]. Despite some concerns, still these approaches for solid tumors may provide an added value in association to classical anti-tumor therapies and specifically for cancers that may develop drug resistance to chemotherapy, as well as for targeting small size metastases which could be easily reached and destroyed by ITs or for the minimal residual therapy approaches after resection of a primary tumor mass.

A Plasminogen activator inhibitor-saporin conjugate was evaluated in a fibrosarcoma model in SCID mice and injected intra-tumorally, showing an effective antitumor response on mice with no systemic toxicity, revealing also the presence of necrotic foci at the tumor site following administration [98]. The intriguing observation was also made at the injection site of B16 melanomas when using the native wild-type saporin DNA complex. In fact, additionally, for some accessible tumors such as those in head or neck, *in vivo* delivery of genes encoding saporin or other Type I RIPs might also have great therapeutic potential. We have demonstrated that the intra-tumoral delivery of as low as 50 ng of a plasmid DNA encoding cytosolic saporin in an animal model, bearing an aggressive tumor such as B16 melanoma, could significantly retard tumor growth [23]. These straightforward, easy and cost-effective cancer gene therapy approaches should deserve further investigations, namely using also other Type I RIP genes and accessible solid tumor models.

10.4.5 Targeting Neoangiogenesis

Neoangiogenesis is a key event in the promotion of cancer, that involves numerous soluble cytokines, growth factors and their receptors. Not only it is essential for tumor growth, but it is also implicated in the initial progression from a pre-malignant lesion to a fully invasive cancer, and in the subsequent growth of dormant micrometastases into clinically detectable metastatic lesions [99]. The progressive identification of molecules playing a relevant role in this process has prompted the development of a variety of new specific anti-cancer agents [100]. Several angiogenic factors, such as MHC class II [101, 102], CD44 [103], NCAM [104] and VEGF [105, 106] have been proposed as therapeutic targets and some derived toxin based conjugates have been produced and tested *in vitro* and *in vivo*.

RIPs were used for the first time in tumor vascular targeting by Burrows and Thorpe who targeted MHC class II antigen, not expressed in vascular endothelial cells in normal mouse tissues, unless activated by IFN-γ. The injection of a ricin-conjugated antibody against the MHC class II antigen caused complete occlusion of the tumor vasculature and dramatic regressions of large solid tumors. The treatment did not completely eradicate the tumor, since a small population of malignant cells at the tumor-host interface survived and proliferated to cause a relapse a week later. The addition of another RTA-based IT against MHC class I antigen able to target the tumor cells themselves gave a synergistic effect, obtaining a lasting complete remission of the tumors [101]. Subsequent studies focused on endoglin, a proliferation-linked endothelial cell marker up-regulated on vascular endothelial cells in solid human tumors, demonstrated that an anti-endoglin-RTA IT was selectively toxic for the dividing endothelial cells [102]. The same targeting strategy was used in the treatment of breast tumor xenografts and an RTA-based IT induced a complete tumor regression in the majority of the mice [107].

CD44 is an activation antigen on human endothelial cells which is up-regulated in the vasculature of tumors as a result of exposure to angiogenic factors. Specific anti-CD44 antibodies have been demonstrated to specifically deliver saporin to endothelial cells *in vitro* [103]. Moreover, the antibiotic with anti-tumoral activity neocarzinostatin conjugated to anti-CD44 monoclonal antibody (TES-23) showed a marked anti-tumor effects in rats bearing KMT-17 fibrosarcomas [108].

The neural cell adhesion molecule (NCAM), expressed during embryogenesis, is down-regulated during tissue differentiation but re-expressed during tumor associated neoangiogenesis, as in case of renal carcinomas. Indeed, targeting NCAM with a saporin-conjugated synthetic peptide ligand, C3d, induced a cytotoxic effect on tumor derived endothelial cells (TEC) but not on normal ones [104]. The same research group exploited a mouse model of human tumor angiogenesis, in which TEC were injected in SCID mice and organized in vascular structures connected with the mouse circulation to identify a cyclic peptides (BB1) showing specific binding to TEC and not to normal human endothelial cells or to murine tumor endothelial cells. Notably, BB1-saporin conjugate induced disruption of the TEC vessel network, without damaging any other murine organs [109].

Unlike antiangiogenic drugs that inhibit formation of new blood vessels, vascular targeting agents (VTA) may occlude pre-existing blood vessels of tumors to cause tumor cell death through ischemia and extensive hemorrhagic necrosis. Among ligand-based VTAs, we may find vascular endothelial growth factor fusions (VEGF) to gelonin, ITs made of monoclonal antibodies to endoglin conjugated to ricin A, as described above as well as antibodies linked to cytokines, liposomally encapsulated drugs or more recently the gene therapy approaches [110].

VEGF-gelonin is a highly selective carrier, targeting tumor endothelial cells, which show promising applications. VEGF-gelonin fusions are able to inhibit tumor growth and metastasis dissemination, and recently, by a fed-batch cultivation method the investigators could achieve high cell densities, comparable to those obtained with the methylotrophic yeast strains (of about 40 OD/mL) with an overall production of a biologically active histidine-tagged VEGF-gelonin of around 40 mg/L as compared to 1.5 mg/L produced using conventional flask cultivation methods [111].

The VEGF receptors are overexpressed on the endothelium of tumor vasculature but are almost undetectable in the adjacent normal tissue, making it an excellent target. A gelonin fusion protein containing the lowest molecular weight VEGF isoform, ($VEGF_{121}$) demonstrated a reduction in tumor volume down to 16 % of untreated controls in human melanoma (A-375) or human prostate (PC-3) xenografts. This fusion construct localized selectively to PC-3 tumor vessels causing thrombotic damage with extravasation of red blood cells into the tumor bed [105]. The recombinant fusion toxin was also used to treat mice bearing metastatic breast cancer cell lines. It targeted selectively the endothelium of the tumor vessels and could cut in half both the number of colonies per lung and the size of the metastatic foci present within the lung [106]. Furthermore, vascularization studies demonstrated that VEGF-gelonin inhibits bFGF-stimulated neovascular growth [112], as well as, it could induce regression of retinal neovascularization [113]. Noteworthy, VEGF121-gelonin is being also under development for the treatment of bone metastases, multiple solid tumors and ocular diseases including age-related macular degeneration (AMD) and diabetic retinopathy. Preclinical animal studies using VEGF/rGel have been recently completed for both these ocular disorders by the Clayton Biotechnologies company, owner of the US patent protecting the targeting of ocular neovascularization.

A key hallmark of antitumor agents targeting neoangiogienesis is a concomitant disruption of the tumor mass coupled to the inhibition of metastatic spread through the inhibition of the new blood vessel formation. New techniques allowing direct visualization of this peculiar effect have been very recently employed making use of nanoparticles (NP). To monitor the efficacy of the anti-angiogenic toxin-based therapy, VEGF-gelonin has been conjugated to manganese ferrite ($MnFe_2O_4$) NPs and then used as a contrast agent [114]. Since NPs exhibit superparamagnetic behavior, they can be traced by magnetic resonance imaging (MRI). Mice bearing human bladder cancer xenografts were injected with VEGF-gelonin NPs and indeed, MRI could detect a strong signal in the intratumoral vessels [115].

Another successful approach was aimed to isolate peptides which specifically could home to tumor blood vessels and made use of *in vivo* selection of phage display libraries. Through this screening, two main peptide motifs were identified that targeted the phage particles into the tumors. One motif contained the sequence Arg-Gly-Asp (RGD), that can bind selectively to a_vb_3 and a_vb_5 integrins while the other sequence Asn-Gly-Arg (NGR) was identified as a cell adhesion motif [116]. Since these peptide motifs can recognize the endothelium of neoangiogenic vessels, NGR and RGD have been tested to deliver a variety of chemotherapeutic drugs, proapoptotic peptides, and cytokines to tumor blood vessels [117].

This targeting approach resulted more effective in combination with chemotherapic agents [118]. RGD/NGR-mediated vascular targeting may thus represent an additional and effective strategy to targeted delivery of RIPs against tumor vessels, to ablate feeding of the surrounding tumors.

10.4.6 Immunolesioning

Functional neuroanatomy research has long relied on the analysis of the effects by selective lesions to infer the function(s) of distinct neural structures. Type II RIPs ricin, abrin, modeccin and volkensin have been used to ablate neurons projecting through a particular nerve to investigate neuronal plasticity and neurotransmission mechanisms [119, 120]. An important limitation of this approach relies on it's being based on retrograde axonal transport of the Type II toxins and included the unavoidable and indiscriminate destruction of different cell types at the injection site. To circumvent this drawback, in the last decades, ITs conjugated to selected monoclonal antibodies directed against neuronal-specific surface antigens have been exploited in order to selectively destroy well-defined sub-types of neurons. Saporin conjugated to the monoclonal antibody 192IgG against the nerve growth factor receptor was first employed to produce specific lesions of cholinergic neurons in rat basal forebrain with the aim to study the hippocampal and cerebral cortical cholinergic neurotransmission. Cholinergic cells in the basal forebrain abundantly express the $p75^{NGFr}$ and as such they are almost completely destroyed after intracerebroventricular administrations of saporin [120, 121]. In addition, cholinergic depletion by 192IgG-saporin represented a good animal model of Alzheimer disease (AD) and has been utilized to evaluate the efficacy of radioligands for the functional imaging of AD brains [122]. Given that 192IgG is specific for rat NGF receptor being not active towards any other species, a similar IT, which contains a monoclonal antibody to the human receptor, has been shown to be as effective in primates [123]. 192IgG-saporin IT was also used to investigate the role of the cholinergic neurotransmission in the synaptic plasticity of visual cortex [124], in the behaviorally-mediated cortical map plasticity [125] and to define the sources of cholinergic basal forebrain input to primary auditory cortex [126]. In addition, selective depletion of cholinergic neurons in the nucleus basalis by 192IgG-saporin, a valid model of sensorimotor gating deficits in the psychotic spectrum, has been used to test the cholinesterase inhibitor rivastigmine in immunolesioned rats [127]. Saporin-based IT for immunolesioning have been employed for studying of other types of neurotransmission: an IT constructed using a monoclonal antibody against the enzyme dopamine beta-hydroxylase coupled via a disulfide bond to saporin selectively destroys noradrenergic neurons in rats after systemic injection [128] and a saporin-containing IT directed to the dopamine transporter produces significant loss of dopaminergic neurons in the substantia nigra and ventral tegmental area, reproducing the neural degeneration characteristics of the Parkinson's disease [129]. Other plant RIPs have been also tested [130, 131] in order to increase the cartridges available to better understand the CNS complexity, to generate animal models for the study of human neuronal pathologies and to test the efficacy of new therapeutic drugs.

A neuropeptide-toxin conjugate warrants a special consideration as it is aimed at selectively destroying neurons expressing the pain related receptor for substance P neuropeptide, best known as substance P-saporin (SP-SAP) and currently under preclinical investigation for human neuropathic pain therapy. Substance P is a 11-amino acid neuropeptide selectively internalized by neurons expressing the neurokinin-1 receptor (NK-1R), which is widely distributed in the brain and has been implicated in signaling nociception in the spinal cord and in the modulation of cortical and striatal cell function [132]. Even if substance P is not the only player involved in nociception and its mechanism(s) of action have not yet been completely elucidated, SP-SAP has been demonstrated to selectively ablate NK-1R expressing spinal neurons in rats, resulting in a reduction of thermal hyperalgesia and mechanical allodynia, associated with persistent neuropathic and inflammatory pain states [133]. An important step forward was achieved by investigating the ability of the SP-SAP conjugate to diffuse along the radial axis of the spinal cord to reach the target dorsal horn neurons in higher species. The lumbar delivery of SP-SAP, but nor SAP alone in dogs resulted in a specific, dose-dependent reduction in superficial NK-1R bearing lumbar neurons and dendrites. Remarkably, no adverse effects upon animal behavior over the 1–3-month survival time were identified; motor functions were not chronically disturbed; changes in blood pressure or heart rate were not detected, there was no progression of the injury or any evidence of additional loss of neurons or function. Furthermore, the pharmacokinetic studies indicated that SP-SAP is cleared relatively rapidly (clearance of over 50 % of the SP-SAP occurring within 30–60 min.) [134]. These encouraging results led to an increased interest in SP-SAP for the treatment-of resistant pain status in humans. NIH has recently approved a Phase I trial granting $3 million award to "Advanced targeting Systems" to evaluate efficacy and safety of this novel pain molecular therapy (Targeting Trends Newsletter Q4, December 2010). The mechanism of action of SP-SAP has been well characterized, a rarity in the pain therapeutics field: a small number of neuronal cells that process the pathological pain signals can be selectively removed causing a permanent relief whereby the normal acute pain response remains unaffected after the SP-SAP treatment.

10.4.7 Adverse Results and Improvement of Safety and Efficacy

Plant and bacterial toxins have been extensively studied for their potential use in cancer as immunotoxins becoming part of our current arsenal of anti-cancer weapons. Frustrations were alternated with successes and investigators learned from defeats to try correcting previous errors and improve the clinical protocols.

A major drawback when using murine targeting domains, as well as plant or bacterial toxins as therapeutic agents, is represented by their potential immunogenicity after repeated administrations.

ITs containing chimeric/humanized antibodies were developed since the '90s. Development of ITs containing fully humanized antibodies or antibody fragment lacking Fc portions may in part resolve immunogenicity problems by avoiding the induction of human anti-mouse antibodies (HAMA) directed to the Fc murine fragments in treated patients [135–137]. Humanization of the variable antibody domains produces a further decrease in immunogenicity [138]. The latter involve grafting loops from both heavy and light variable domains parental antibody into a human acceptor antibody. Fully human antibodies were successfully created either recombinantly by phage display techniques or by the selection of variable regions of human origin [139] and can now be also produced as monoclonal antibodies in human immunoglobulin transgenic mice [140]. For instance, an IT formed by recombinant gelonin and a fully humanized version of anti-CD33 demonstrated low immunogenicity in a Phase 1 trial and dose limiting toxicity was not encountered after repeated administrations [141].

The presence of a bacterial or plant toxin can trigger the formation of neutralizing antibodies, hindering their efficacy. On one hand, for patients with B- or T cell malignancies, formation of neutralizing antibodies may be less frequent essentially because of their immuno-suppressed state, on the other hand, in patients with solid tumors, antibody responses are frequently detected as early as a few days after the first treatment regimen, thus preventing re-administration of the ITs [50, 142]. This is a critical issue to successful treatment of cancers, which limits the number of therapeutic administrations and may be crucial in the treatment of solid tumors, compromising the clinical efficacy [142–144]. Many efforts are being made to decrease immunogenicity of the toxin moiety; one possibility explored is masking of the therapeutic molecules by use of high-molecular-weight polyethylenglycol (PEG) derivatization techniques which shield antigenic determinants detection by the immune system. One major challenge in pegylation techniques is to avoid also the masking of the active sites that are essential for drug efficacy [145]. Indeed, PEGylated Type I RIPs, gelonin [146], trichosanthin [147] and momorcharin [148] as well as RTA [149] showed a consistent reduction of immunogenicity with an increased stability, but also suffered a significant decrease in activity. Likewise, coupling trichosanthin to dextrans was demonstrated to significantly decrease both IgG and IgE responses with concurrently halving the activity of the toxin [150].

Among most promising strategies to overcome these limitations, several were used for modifying *Pseudomonas aeruginosa* exotoxin A (PEA) and concerned the identification and substitution of epitopes recognized by the human immune system, without compromising the catalytic activity [151, 152]. Impressive results were obtained by Onda and coworkers, who identified the seven major immunogenic B cell epitopes in a truncated form PE38 and obtained a new anti-CD22 immunotoxin (HA22–8X) which resulted significantly less immunogenic in mice, yet retaining full cytotoxic and anti-tumor activities [151, 152]. PEA based bispecific anti CD22 and anti CD19 IT construct carrying an efficient ER-retrieval motif Lys-Asp-Glu-Leu (KDEL) instead of the "physiological" C-terminus present in PE, interestingly, showed reduced neutralizing anti-toxin antibodies by about 80 % with no apparent loss in *in vitro* or *in vivo* activity in mouse models [153]. Moreover, the conversion of two structurally adjacent residues mutated to cysteines introduced a disulfide bond, resulting in a remarkable increase in thermal stability and an enhanced resistance to trypsin degradation of an anti-CD22 PE38 IT, which retained full anti-tumor activity, while exhibiting significantly lower immunogenicity in mice [154].

In this regard, it is of primary importance checking the stability of the biological material to be administrated by a careful and safe preparation. It is worth to mention the Phase I study using "Combotox", a mixture of anti-CD19 and anti-CD22 RTA-based ITs, where unfortunately two deaths occurred following drug treatment, due to unexpected toxicity presumably derived by undetected protein aggregation [53]. This seriously adverse event reminds us that every precaution must be taken to ensure the stability of these therapeutic molecules. RTA is aggregation-prone being protein precipitation very likely caused by exposure of hydrophobic surfaces on RTA when used in the absence of its natural B-chain partner: the removal of the RTA C-terminal sequence led to much less precipitation under physiological conditions, but also to a concomitant reduction in RIP activity [155].

Another dangerous side-effect observed during IT clinical administration regards the unspecific binding of the toxin domains to the vascular endothelial cells, causing the so-called "vascular leak syndrome" (VLS), a side effect characterized by interstitial edema, hypoalbuminemia, weight gain, and in most severe cases, pulmonary edema and hypotension. The mechanism(s) causing VLS is not completely understood, but it appears that a consensus sequence "X-Asp-Y" (X = Leu, Ile, Gly or Val; Y = Val, Leu or Ser) present in RTA and in several Type I RIPs, also found in IL-2, binds to integrin receptors, inducing vascular damage to human endothelial cells [156–158]. Notably, this motif is common to viral disintegrins, which disrupt the function of integrin receptors [159]. To carry on clinical studies with RTA-based ITs, a panel of RTA mutants was produced and the Asn97 to Ala substitution in a region flanking the VLS-responsible motif exhibited a significant less vascular damage in mice models and consequently the mutated IT could be injected as CD22 conjugate in a five-fold higher concentration, as compared to the parental IT [160].

Plant Type I RIPs have been engineered to limit their antigenicity, as well: to define minimum size of saporin active variants, Colombatti's group prepared a series of mutated proteins with deletions at the N-terminus or at the C-terminus. An N-terminal (sapΔ1–20) deletion mutant of saporin displayed defective catalytic activity, drastically reduced cytotoxicity while a C-terminal (sapΔ239–253) deletion mutant showed instead only a moderate reduction in cytotoxic activity [161]. However, C-terminal domain deletions of trichosanthin which contains a putative antigenic site resulted, unfortunately, in a loss of the RIP activity in the truncated protein by about 10-fold [162]. By computer modeling, another group identified two potential antigenic sites which could instead reduce the immunogenicity of trichosanthin, maintaining the same activity of the natural protein [163].

Bouganin has been also mutated to remove the major T cell epitopes: a fusion construct containing an anti-epithelial cell adhesion molecule (EpCAM) Fab moiety with the epitope-depleted bouganin (VB6-845) was successfully tested on EpCAM-positive human tumor xenograft model in SCID mice [95]. Additional preclinical studies were carried out on rats and monkeys where in the latter VB6-845 was well tolerated and only minimally immunogenic [164].

A comprehensive series of pharmacology and toxicology studies have been conducted using VB6-845, including a phase I trial in patients bearing advanced squamous cell carcinomas of the head and neck. The results of this phase I trial showed that VB6-845 could reduce or stabilize tumors in 71 % of patients with a maximum tolerated dose of 280 mg of IT, administered daily for 5 days. The IT was well tolerated with pain as adverse effects, due to the intratumoral injections and reversibly elevated liver enzymes. These promising results warrant further clinical development of this plant RIP-based IT [165, 166]. Further studies aimed at limiting the IT immunogenicity, modifying the toxin domain or the antibody epitopes are expected to prompt the development of novel molecules, safe enough to allow repeated cycles of administrations and thus increasing their therapeutic potential.

10.4.8 Gene Delivery Approaches

More recently with respect to the development of protein based ITs or conjugates, DNA based suicide therapies have been developed. Gene therapy has been investigated as an alternative or complementary strategy to overcome the limitations of conventional anticancer therapy or the approaches relying on gene-directed enzyme prodrug therapy.

The plasmid DNA holds many positive aspects compared to proteins: it avoids costly and time-consuming production and purification even at large scale; it does not cause the side effects like immune system response or vascular leak syndrome, even after repeated administrations; low doses are active since very few molecules of the toxin generated in the tumor cells are sufficient to kill them; tumor cells do not develop resistance to the drug. Furthermore, acting in a cell cycle independent way, toxin genes can target quiescent and rapidly dividing tumor cells. This property makes them suited both to contrast aggressively growing tumors (e.g., melanoma) and for other tumor types with slower progression (e.g., prostate tumors). Working at DNA level, allows the direct screening of a wide variety of different constructs, employing diverse promoters and transcription regulatory elements, as well as exploring different delivery approaches [167, 168]. In the case of systemic delivery, to prevent detrimental expression of toxin in inappropriate tissues, DNA constructs could easily be improved with tissue and/or tumor specific promoters [169–171]. In addition, pharmacologically controlled gene expression could be ensured using inducible promoters such as the Tet-on [172] or mifepristone [173] system.

Two major gene delivery agents have been developed: either recombinant disarmed viruses or non-viral vectors, namely DNA-complexes. Both approaches have some advantages and limitations, but as long as an ideal vector for all-purposes is not available, the choice of the delivery vector should be determined by the requirements of the specific disease to be treated [174]. Replication-defective viruses, where non-essential viral genes were partly replaced by the therapeutic genes, represent the first generation of viral vectors earlier applied in gene therapy approaches [175]. Their promises as therapeutic agents are limited by their side effects such as toxicity and immunogenicity, their broad tropism as well as the cost of large-scale formulations. Although viral vectors could be modified to remove immunogenic components, these revised viruses are often difficult to produce in high titer. Given the importance of viral vectors for gene therapy of certain tumors, many efforts have been performed to improve their safety and efficiency [176, 177].

Non-viral vectors are a promising alternative as they offer improved safety profiles compared to the viruses counterparts, are much less expensive to produce, but suffer from significantly lower transfection efficiency than viruses, severely limiting their utility [178]. Cationic liposomes and cationic polymers represent the most commonly used non-viral gene delivery methods in transfection protocols of different cell lines, animal models and patients submitted to gene therapy clinical trials [179].

To date, a restricted panel of toxins has been used in cancer suicide gene therapy; DT is one of the best-studied therapeutic agents and it is a landmark for the toxin gene therapy. DT has been reported to be delivered through viral and non-viral vectors; to be effective in many cellular and animal models [180] and to be currently under investigation in clinical trials. Recently, two phase I/II clinical trials involving DT complexed with the synthetic polycation polyethylenimine (PEI) have been completed. The first aimed to assess the safety, tolerability, pharmacokinetics and preliminary efficacy of the H19 tumor promoter driven DT administered intratumorally in patients with unresectable, locally advanced pancreatic cancer (NCT00711997). In the second, the same construct was administrated as intravesical infusions into the bladder of patients with superficial bladder cancer who have failed intravesical therapy with the FDA approved Bacille Calmette-Guérin (NCT00393809).

Retroviral vectors bearing RTA or PEA resulted highly toxic towards rat (C6) and human (U-373-MG) glioblastoma cells. Remarkably, the two toxins induced a total remission of C6-derived brain tumor in Wistar rats [168].

Direct injection of saporin encoding plasmid DNA complexed with cationic lipids has been demonstrated to be highly cytotoxic: as few as 10 ng of plasmid DNA carrying the gene encoding wt saporin were sufficient to drastically inhibit protein synthesis *in vitro*, whereas the gene for an inactive saporin KQ mutant had no effects. In addition, in the B16 mouse melanoma model, direct intratumoral injection of saporin gene DNA complexes significantly attenuated tumor growth, with an effect that was improved upon repeated administrations. To demonstrate that the antitumor effect was specifically owing to the intrinsic RIP activity of the toxin, a plasmid encoding the catalytic inactive mutant SAP-KQ was shown to neither kill cultured cells, nor delay tumor growth in mice [23]. An intriguing observation we made was the presence of necrotic lesions at the injection site of B16 melanomas when using the native wild-type saporin DNA but not the KQ mutant, suggesting secondary effects could partly be due to bystander effects mediated by T-Killer cells.

The direct delivery approach using cationic lipids is in general applicable to accessible solid tumors, thus limiting the field of action to a few malignancies, for example, melanoma, head and neck tumors, mesotheliomas that, however, are sometimes resistant to conventional therapies or whose tumor-specific better defined surface marker antigens have not been identified yet. Thus, toxin-mediated gene therapy could be exploited in combination with surgery or classical radio- or chemotherapy to eradicate tumor mass.

Cationic liposomes therapeutical use was until now restricted to intratumoral administration, due to the considerable reduction in their delivery activity by serum components and to some observed toxicity following systemic administration [181]. To solve those problems, hydrophilic components, such as polyethylene glycol (PEG) which could mask the net positive charge of the complexes, have been added, increasing circulation time in blood stream and reducing liposome toxicity. In addition, the association of proteins, antibodies, peptides or other agents that could enhance the biological activity and specificity to target cells have been proposed [182]. In an original study performed by Hoganson and collaborators, targeting of FGF-2 receptor-bearing cells was obtained including basic FGF either on a DNA complex encoding cytosolic saporin or, for comparison, with the one encoding herpes simplex virus thymidine kinase gene (which remains, so far, the most widely used suicide gene in cancer gene therapy) able to efficiently kill target cells by activating the ganciclovir prodrug. When the investigators compared FGF2-mediated delivery of a saporin optimized gene to thymidine kinase DNA (followed by ganciclovir treatment) in tumor target cells, a comparable decrease in cell numbers was observed, while the specificity of gene delivery could be demonstrated in competition assays with free FGF2 [183].

In the last years, several new unconventional vectors, including bacteria, bacteriophages, virus-like particles, erythrocyte ghosts, and exosomes have been developed as carriers for the suicide gene therapy [184]. Among them, bacterial vectors have been the most extensively studied, given that anaerobic bacteria such as strains of Clostridia, Bifido bacteria and Salmonellae have the potential for therapy of solid tumors, being able to selectively colonize the hypoxic areas of tumors and destroy the tumor cells [185]. Furthermore, in the absence of natural tropism, modification of bacteria surface proteins by conjugation to a cell-type specific ligand can also alter tissue specificity to suit the application [186]. The intrinsic toxicity of bacteria, especially arising following systemic administration, varies depending on the specific strain. In clinical gene transfer trials of direct intratumoral [187] and intravenous [188] injection of attenuated *Salmonella* bacteria, no significant side effects were observed. The potential of this system has already been demonstrated with plasmids encoding prodrug convertases [189–191] and it is desirable that modified bacteria become an additional opportunity to selectively deliver RIPs to tumors, extending their applications.

Due to the unique properties of nano-scale matter, the diversity of available materials and infinite design schemes, nanoparticle-based delivery systems have also emerged as potential gene carriers [192]. These agents, which comprise polymeric and lipid systems, accumulate in solid tumors after intravenous administration, owing to an increased permeability of tumor blood vessels and a decreased rate of clearance within the tumor (the so-called Enhanced Permeability and Retention (EPR) effect) [193] could be further exploited for solid tumor targeting of therapeutic genes.

10.5 Conclusions and Perspectives

Over the past 30 years, RIPs have attracted great interest in the scientific community for their therapeutic potential and have been extensively investigated in order to exploit them as potent and versatile weapons against cancers or other debilitating human diseases. In combination with antibodies or ligands to specifically target malignant or pathogenic cells, RIPs have been widely used in animal models with encouraging results and some chimeric constructs have already undergone clinical trials. Adverse effects may be limited by an increasing knowledge about RIP's structure and mechanism of action, leading to protein optimization approaches and removal of potential antigenic epitopes. In parallel, the possibility to employ humanized antibody domains to limit adverse immune responses together with the improvement of heterologous expression systems allowing a cost-effective, high production in RIP containing chimeric fusions may contribute to advance the toxin-based therapies both in terms of efficacy and safety, making them more suitable and eligible in the next future for clinical applications in patients, also in combination cancer therapies along with surgery, chemo or radiotherapy. Of great importance will be the identification of early markers of the disease to increase the IT specificity and efficacy. In the last few years, methods have been developed to enhance the cytotoxicity of the ITs, by specifically acting on their internalization and delivery routes. Examples are, Photochemical internalization approaches [97, 194] or the use of saponins as adjuvant co-treatment [25, 82, 195] which have demonstrated to favor IT cytosolic escape, allowing the broadening of the therapeutic windows together with a simultaneous IT dose lowering.

Last but not least, advanced nano-platforms may offer a great opportunity for design and use of multi-functionalized nano carriers, where core material is combined with encapsulated biologically active agents and targeting/surface properties. This will allow for strategies that not only enable targeted stimuli-responsive delivery, but also a simultaneous delivery of multiple pharmacological agents, aimed at enhancing overall therapeutic effects and overcoming drug resistance in cancer.

To conclude, even if RIP-based chimeric constructs have not yet fulfilled the expectations, they may be a step less far from the Paul Ehrlich's magic bullet, if all our efforts will synergize.

References

1. Endo Y, Tsurugi K, Lambert JM (1988) The site of action of six different ribosome-inactivating proteins from plants on eukaryotic ribosomes: the RNA N-glycosidase activity of the proteins. Biochem Biophys Res Commun 150:1032–1036
2. Hartley MR, Legname G, Osborn R, Chen Z, Lord JM (1991) Single-chain ribosome inactivating proteins from plants depurinate Escherichia coli 23S ribosomal RNA. FEBS Lett 290:65–68
3. Stirpe F, Gasperi-Campani A, Barbieri L, Falasca A, Abbondanza A, Stevens WA (1983) Ribosome-inactivating proteins from the seeds of Saponaria officinalis L. (soapwort), of Agrostemma githago L. (corn cockle) and of Asparagus officinalis L. (asparagus), and from the latex of Hura crepitans L. (sandbox tree). Biochem J 216:617–625
4. Irvin JD, Uckun FM (1992) Pokeweed antiviral protein: ribosome inactivation and therapeutic applications. Pharmacol Ther 55:279–302
5. Despeyroux D, Walker N, Pearce M, Fisher M, McDonnell M, Bailey SC et al (2000) Characterization of ricin heterogeneity by electrospray mass spectrometry, capillary electrophoresis, and resonant mirror. Anal Biochem 279:23–36
6. Chan AP, Crabtree J, Zhao Q, Lorenzi H, Orvis J, Puiu D et al (2010) Draft genome sequence of the oilseed species Ricinus communis. Nat Biotechnol 28:951–956
7. Barthelemy I, Martineau D, Ong M, Matsunami R, Ling N, Benatti L et al (1993) The expression of saporin, a ribosome-inactivating protein from the plant Saponaria officinalis, in Escherichia coli. J Biol Chem 268:6541–6548
8. Olsnes S, Pihl A (1973) Different biological properties of the two constituent peptide chains of ricin, a toxic protein inhibiting protein synthesis. Biochemistry 12:3121–3126
9. Lord JM, Roberts LM, Robertus JD (1994) Ricin: structure, mode of action, and some current applications. Faseb J 8:201–208
10. Peumans WJ, Hao Q, Van Damme EJ (2001) Ribosome-inactivating proteins from plants: more than RNA N-glycosidases? Faseb J 15:1493–1506
11. Day PJ, Lord JM, Roberts LM (1998) The deoxyribonuclease activity attributed to ribosome-inactivating proteins is due to contamination. Eur J Biochem 258:540–545
12. Lombardi A, Bursomanno S, Lopardo T, Traini R, Colombatti M, Ippoliti R et al (2010) Pichia pastoris as a host for secretion of toxic saporin chimeras. Faseb J 24:253–265
13. Sandvig K, van Deurs B (2000) Entry of ricin and Shiga toxin into cells: molecular mechanisms and medical perspectives. EMBO J 19:5943–5950
14. Lord JM, Deeks E, Marsden CJ, Moore K, Pateman C, Smith DC et al (2003) Retrograde transport of toxins across the endoplasmic reticulum membrane. Biochem Soc Trans 31:1260–1262
15. Spooner RA, Watson PD, Marsden CJ, Smith DC, Moore KA, Cook JP et al (2004) Protein disulphide-isomerase reduces ricin to its A and B chains in the endoplasmic reticulum. Biochem J 383:285–293
16. Madan S, Ghosh PC (1992) Interaction of gelonin with macrophages: effect of lysosomotropic amines. Exp Cell Res 198:52–58
17. Chan WY, Huang H, Tam SC (2003) Receptor-mediated endocytosis of trichosanthin in choriocarcinoma cells. Toxicology 186:191–203
18. Vago R, Marsden CJ, Lord JM, Ippoliti R, Flavell DJ, Flavell SU et al (2005) Saporin and ricin A chain follow different intracellular routes to enter the cytosol of intoxicated cells. FEBS J 272:4983–4995

19. Stirpe F, Barbieri L (1986) Ribosome-inactivating proteins up to date. FEBS Lett 195:1–8
20. Deeks ED, Cook JP, Day PJ, Smith DC, Roberts LM, Lord JM (2002) The low lysine content of ricin A chain reduces the risk of proteolytic degradation after translocation from the endoplasmic reticulum to the cytosol. Biochemistry 41:3405–3413
21. Fabbrini MS, Rappocciolo E, Carpani D, Solinas M, Valsasina B, Breme U et al (1997) Characterization of a saporin isoform with lower ribosome-inhibiting activity. Biochem J 322(Pt 3):719–727
22. Bagga S, Seth D, Batra JK (2003) The cytotoxic activity of ribosome-inactivating protein saporin-6 is attributed to its rRNA N-glycosidase and internucleosomal DNA fragmentation activities. J Biol Chem 278:4813–4820
23. Zarovni N, Vago R, Solda T, Monaco L, Fabbrini MS (2007) Saporin as a novel suicide gene in anticancer gene therapy. Cancer Gene Ther 14:165–173
24. Marshall RS, D'Avila F, Di Cola A, Traini R, Spano L, Fabbrini MS et al (2011) Signal peptide-regulated toxicity of a plant ribosome-inactivating protein during cell stress. Plant J 65:218–229
25. Rajamohan F, Pugmire MJ, Kurinov IV, Uckun FM (2000) Modeling and alanine scanning mutagenesis studies of recombinant pokeweed antiviral protein. J Biol Chem 275:3382–3390
26. de Virgilio M, Lombardi A, Caliandro R, Fabbrini MS (2010) Ribosome-inactivating proteins: from plant defense to tumor attack. Toxins (Basel) 2:2699–2737
27. Hartley MR, Lord JM (2004) Cytotoxic ribosome-inactivating lectins from plants. Biochim Biophys Acta 1701:1–14
28. Habuka N, Murakami Y, Noma M, Kudo T, Horikoshi K (1989) Amino acid sequence of Mirabilis antiviral protein, total synthesis of its gene and expression in Escherichia coli. J Biol Chem 264:6629–6637
29. Chaddock JA, Lord JM, Hartley MR, Roberts LM (1994) Pokeweed antiviral protein (PAP) mutations which permit E.coli growth do not eliminate catalytic activity towards prokaryotic ribosomes. Nucleic Acids Res 22:1536–1540
30. Legname G, Fossati G, Monzini N, Gromo G, Marcucci F, Mascagni P et al (1995) Heterologous expression, purification, activity and conformational studies of different forms of dianthin 30. Biomed Pept Proteins Nucleic Acids 1:61–68
31. Lappi DA, Ying W, Barthelemy I, Martineau D, Prieto I, Benatti L et al (1994) Expression and activities of a recombinant basic fibroblast growth factor-saporin fusion protein. J Biol Chem 269:12552–12558
32. Fabbrini MS, Carpani D, Bello-Rivero I, Soria MR (1997) The amino-terminal fragment of human urokinase directs a recombinant chimeric toxin to target cells: internalization is toxin mediated. Faseb J 11:1169–1176
33. Heisler I, Keller J, Tauber R, Sutherland M, Fuchs H (2003) A cleavable adapter to reduce nonspecific cytotoxicity of recombinant immunotoxins. Int J Cancer 103:277–282
34. Giansanti F, Di Leandro L, Koutris I, Pitari G, Fabbrini MS, Lombardi A et al (2010) Engineering a switchable toxin: the potential use of PDZ domains in the expression, targeting and activation of modified saporin variants. Protein Eng Des Sel 23:61–68
35. Rajamohan F, Engstrom CR, Denton TJ, Engen LA, Kourinov I, Uckun FM (1999) High-level expression and purification of biologically active recombinant pokeweed antiviral protein. Protein Expr Purif 16:359–368
36. Fuchs H, Bachran C, Li T, Heisler I, Durkop H, Sutherland M (2007) A cleavable molecular adapter reduces side effects and concomitantly enhances efficacy in tumor treatment by targeted toxins in mice. J Control Release 117:342–350
37. Fabbrini MS, Carpani D, Soria MR, Ceriotti A (2000) Cytosolic immunization allows the expression of preATF-saporin chimeric toxin in eukaryotic cells. Faseb J 14:391–398
38. Cereghino JL, Cregg JM (2000) Heterologous protein expression in the methylotrophic yeast Pichia pastoris. FEMS Microbiol Rev 24:45–66
39. Woo JH, Liu YY, Mathias A, Stavrou S, Wang Z, Thompson J et al (2002) Gene optimization is necessary to express a bivalent anti-human anti-T cell immunotoxin in Pichia pastoris. Protein Expr Purif 25:270–282

40. Rajamohan F, Doumbia SO, Engstrom CR, Pendergras SL, Maher DL, Uckun FM (2000) Expression of biologically active recombinant pokeweed antiviral protein in methylotrophic yeast Pichia pastoris. Protein Expr Purif 18:193–201
41. Frigerio L, Vitale A, Lord JM, Ceriotti A, Roberts LM (1998) Free ricin A chain, proricin, and native toxin have different cellular fates when expressed in tobacco protoplasts. J Biol Chem 273:14194–14199
42. Krishnan R, McDonald KA, Dandekar AM, Jackman AP, Falk B (2002) Expression of recombinant trichosanthin, a ribosome-inactivating protein, in transgenic tobacco. J Biotechnol 97:69–88
43. Strebhardt K, Ullrich A (2008) Paul Ehrlich's magic bullet concept: 100 years of progress. Nat Rev Cancer 8:473–480
44. Madhumathi J, Verma RS (2012) Therapeutic targets and recent advances in protein immunotoxins. Curr Opin Microbiol 15:300–309
45. Pastan I, Hassan R, FitzGerald DJ, Kreitman RJ (2007) Immunotoxin treatment of cancer. Annu Rev Med 58:221–237
46. Foon KA, Todd RF 3rd (1986) Immunologic classification of leukemia and lymphoma. Blood 68:1–31
47. Zhang XW, Yan XJ, Zhou ZR, Yang FF, Wu ZY, Sun HB et al (2010) Arsenic trioxide controls the fate of the PML-RARalpha oncoprotein by directly binding PML. Science 328:240–243
48. Bacha P, Williams DP, Waters C, Williams JM, Murphy JR, Strom TB (1988) Interleukin 2 receptor-targeted cytotoxicity. Interleukin 2 receptor-mediated action of a diphtheria toxin-related interleukin 2 fusion protein. J Exp Med 167:612–622
49. Olsen E, Duvic M, Frankel A, Kim Y, Martin A, Vonderheid E et al (2001) Pivotal phase III trial of two dose levels of denileukin diftitox for the treatment of cutaneous T-cell lymphoma. J Clin Oncol 19:376–388
50. Kreitman RJ, Wilson WH, White JD, Stetler-Stevenson M, Jaffe ES, Giardina S et al (2000) Phase I trial of recombinant immunotoxin anti-Tac(Fv)-PE38 (LMB-2) in patients with hematologic malignancies. J Clin Oncol 18:1622–1636
51. Blakey DC, Watson GJ, Knowles PP, Thorpe PE (1987) Effect of chemical deglycosylation of ricin A chain on the in vivo fate and cytotoxic activity of an immunotoxin composed of ricin A chain and anti-Thy 1.1 antibody. Cancer Res 47:947–952
52. Schnell R, Borchmann P, Staak JO, Schindler J, Ghetie V, Vitetta ES et al (2003) Clinical evaluation of ricin A-chain immunotoxins in patients with Hodgkin's lymphoma. Ann Oncol 14:729–736
53. Messmann RA, Vitetta ES, Headlee D, Senderowicz AM, Figg WD, Schindler J et al (2000) A phase I study of combination therapy with immunotoxins IgG-HD37-deglycosylated ricin A chain (dgA) and IgG-RFB4-dgA (Combotox) in patients with refractory CD19(+), CD22(+) B cell lymphoma. Clin Cancer Res 6:1302–1313
54. Amlot PL, Stone MJ, Cunningham D, Fay J, Newman J, Collins R et al (1993) A phase I study of an anti-CD22-deglycosylated ricin A chain immunotoxin in the treatment of B-cell lymphomas resistant to conventional therapy. Blood 82:2624–2633
55. Schnell R, Staak O, Borchmann P, Schwartz C, Matthey B, Hansen H et al (2002) A Phase I study with an anti-CD30 ricin A-chain immunotoxin (Ki-4.dgA) in patients with refractory CD30 + Hodgkin's and non-Hodgkin's lymphoma. Clin Cancer Res 8:1779–1786
56. Uckun FM (1993) Immunotoxins for the treatment of leukaemia. Br J Haematol 85:435–438
57. Falini B, Bolognesi A, Flenghi L, Tazzari PL, Broe MK, Stein H et al (1992) Response of refractory Hodgkin's disease to monoclonal anti-CD30 immunotoxin. Lancet 339:1195–1196
58. Pasqualucci L, Wasik M, Teicher BA, Flenghi L, Bolognesi A, Stirpe F et al (1995) Antitumor activity of anti-CD30 immunotoxin (Ber-H2/saporin) in vitro and in severe combined immunodeficiency disease mice xenografted with human CD30 + anaplastic large-cell lymphoma. Blood 85:2139–2146
59. Terenzi A, Bolognesi A, Pasqualucci L, Flenghi L, Pileri S, Stein H et al (1996) Anti-CD30 (BER = H2) immunotoxins containing the type-1 ribosome-inactivating proteins momordin

and PAP-S (pokeweed antiviral protein from seeds) display powerful antitumour activity against CD30 + tumour cells in vitro and in SCID mice. Br J Haematol 92:872–879
60. Bolognesi A, Tazzari PL, Olivieri F, Polito L, Lemoli R, Terenzi A et al (1998) Evaluation of immunotoxins containing single-chain ribosome-inactivating proteins and an anti-CD22 monoclonal antibody (OM124): in vitro and in vivo studies. Br J Haematol 101:179–188
61. Flavell DJ, Noss A, Pulford KA, Ling N, Flavell SU (1997) Systemic therapy with 3BIT, a triple combination cocktail of anti-CD19, -CD22, and -CD38-saporin immunotoxins, is curative of human B-cell lymphoma in severe combined immunodeficient mice. Cancer Res 57:4824–4829
62. Polito L, Bortolotti M, Farini V, Pedrazzi M, Tazzari PL, Bolognesi A (2009) ATG-saporin-S6 immunotoxin: a new potent and selective drug to eliminate activated lymphocytes and lymphoma cells. Br J Haematol 147:710–718
63. Polito L, Bortolotti M, Pedrazzi M, Bolognesi A (2011) Immunotoxins and other conjugates containing saporin-s6 for cancer therapy. Toxins (Basel) 3:697–720
64. Parameswaran R, Yu M, Lyu MA, Lim M, Rosenblum MG, Groffen J et al (2012) Treatment of acute lymphoblastic leukemia with an rGel/BLyS fusion toxin. Leukemia 26:1786–1796
65. ten Cate B, de Bruyn M, Wei Y, Bremer E, Helfrich W (2010) Targeted elimination of leukemia stem cells; a new therapeutic approach in hemato-oncology. Curr Drug Targets 11:95–110
66. Appelbaum FR (1997) Allogeneic hematopoietic stem cell transplantation for acute leukemia. Semin Oncol 24:114–123
67. Bregni M, Siena S, Subar M, Villa S, Bonadonna G, Dalla-Favera R et al (1989) Detection of occult leukemic cells in the autologous bone marrow graft of a patient with T-cell acute lymphoblastic leukemia by a highly specific and sensitive assay. Haematologica 74:11–14
68. Lemoli RM, Tazzari PL, Fortuna A, Bolognesi A, Gulati SC, Stirpe F et al (1994) Positive selection of hematopoietic CD34 + stem cells provides 'indirect purging' of CD34- lymphoid cells and the purging efficiency is increased by anti-CD2 and anti-CD30 immunotoxins. Bone Marrow Transpl 13:465–471
69. Dinota A, Barbieri L, Gobbi M, Tazzari PL, Rizzi S, Bontadini A et al (1989) An immunotoxin containing momordin suitable for bone marrow purging in multiple myeloma patients. Br J Cancer 60:315–319
70. McGraw KJ, Rosenblum MG, Cheung L, Scheinberg DA (1994) Characterization of murine and humanized anti-CD33, gelonin immunotoxins reactive against myeloid leukemias. Cancer Immunol Immunother 39:367–374
71. LaCasse EC, Saleh MT, Patterson B, Minden MD, Gariepy J (1996) Shiga-like toxin purges human lymphoma from bone marrow of severe combined immunodeficient mice. Blood 88:1561–1567
72. Kelly DA (2005) Long-term challenges of immunosuppression in pediatric patients. Transpl Proc 37:1657–1662
73. Fishman JA, Rubin RH (1998) Infection in organ-transplant recipients. N Engl J Med 338:1741–1751
74. Rennie DP, McGregor AM, Wright J, Weetman AP, Hall R, Thorpe P (1983) An immunotoxin of ricin A chain conjugated to thyroglobulin selectively suppresses the antithyroglobulin autoantibody response. Lancet 2:1338–1340
75. Hess PR, Barnes C, Woolard MD, Johnson MD, Cullen JM, Collins EJ et al (2007) Selective deletion of antigen-specific CD8 + T cells by MHC class I tetramers coupled to the type I ribosome-inactivating protein saporin. Blood 109:3300–3307
76. Hossann M, Li Z, Shi Y, Kreilinger U, Buttner J, Vogel PD et al (2006) Novel immunotoxin: a fusion protein consisting of gelonin and an acetylcholine receptor fragment as a potential immunotherapeutic agent for the treatment of Myasthenia gravis. Protein Expr Purif 46:73–84
77. van Roon JA, van Vuuren AJ, Wijngaarden S, Jacobs KM, Bijlsma JW, Lafeber FP et al (2003) Selective elimination of synovial inflammatory macrophages in rheumatoid arthritis by an Fcgamma receptor I-directed immunotoxin. Arthritis Rheum 48:1229–1238

78. Nagai T, Tanaka M, Tsuneyoshi Y, Matsushita K, Sunahara N, Matsuda T et al (2006) In vitro and in vivo efficacy of a recombinant immunotoxin against folate receptor beta on the activation and proliferation of rheumatoid arthritis synovial cells. Arthritis Rheum 54:3126–3134
79. Stepanov AV, Belogurov AA Jr, Ponomarenko NA, Stremovskiy OA, Kozlov LV, Bichucher AM et al (2011) Design of targeted B cell killing agents. PLoS ONE 6:e20991
80. Chandler LA, Sosnowski BA, McDonald JR, Price JE, Aukerman SL, Baird A et al (1998) Targeting tumor cells via EGF receptors: selective toxicity of an HBEGF-toxin fusion protein. Int J Cancer 78:106–111
81. Hoffmann C, Bachran C, Stanke J, Elezkurtaj S, Kaufmann AM, Fuchs H et al (2010) Creation and characterization of a xenograft model for human cervical cancer. Gynecol Oncol 118:76–80
82. Bachran D, Schneider S, Bachran C, Urban R, Weng A, Melzig MF et al (2010) Epidermal growth factor receptor expression affects the efficacy of the combined application of saponin and a targeted toxin on human cervical carcinoma cells. Int J Cancer 127:1453–1461
83. Zhou X, Qiu J, Wang Z, Huang N, Li X, Li Q et al (2012) In vitro and in vivo anti-tumor activities of anti-EGFR single-chain variable fragment fused with recombinant gelonin toxin. J Cancer Res Clin Oncol 138:1081–1090
84. Beitz JG, Davol P, Clark JW, Kato J, Medina M, Frackelton AR Jr et al (1992) Antitumor activity of basic fibroblast growth factor-saporin mitotoxin in vitro and in vivo. Cancer Res 52:227–230
85. Ying W, Martineau D, Beitz J, Lappi DA, Baird A (1994) Anti-B16-F10 melanoma activity of a basic fibroblast growth factor-saporin mitotoxin. Cancer 74:848–853
86. Davol P, Frackelton AR Jr (1996) The mitotoxin, basic fibroblast growth factor-saporin, effectively targets human prostatic carcinoma in an animal model. J Urol 156:1174–1179
87. Kuroda K, Liu H, Kim S, Guo M, Navarro V, Bander NH (2010) Saporin toxin-conjugated monoclonal antibody targeting prostate-specific membrane antigen has potent anticancer activity. Prostate 70:1286–1294
88. Siva AC, Wild MA, Kirkland RE, Nolan MJ, Lin B, Maruyama T et al (2008) Targeting CUB domain-containing protein 1 with a monoclonal antibody inhibits metastasis in a prostate cancer model. Cancer Res 68:3759–3766
89. Duxbury MS, Ito H, Ashley SW, Whang EE (2004) CEACAM6 as a novel target for indirect type 1 immunotoxin-based therapy in pancreatic adenocarcinoma. Biochem Biophys Res Commun 317:837–843
90. Foehr ED, Lorente G, Kuo J, Ram R, Nikolich K, Urfer R (2006) Targeting of the receptor protein tyrosine phosphatase beta with a monoclonal antibody delays tumor growth in a glioblastoma model. Cancer Res 66:2271–2278
91. Bruland OS, Fodstad O, Stenwig AE, Pihl A (1988) Expression and characteristics of a novel human osteosarcoma-associated cell surface antigen. Cancer Res 48:5302–5309
92. Anderson PM, Meyers DE, Hasz DE, Covalcuic K, Saltzman D, Khanna C et al (1995) In vitro and in vivo cytotoxicity of an anti-osteosarcoma immunotoxin containing pokeweed antiviral protein. Cancer Res 55:1321–1327
93. Zhou H, Marks JW, Hittelman WN, Yagita H, Cheung LH, Rosenblum MG et al (2011) Development and characterization of a potent immunoconjugate targeting the Fn14 receptor on solid tumor cells. Mol Cancer Ther 10:1276–1288
94. Momburg F, Moldenhauer G, Hammerling GJ, Moller P (1987) Immunohistochemical study of the expression of a Mr 34,000 human epithelium-specific surface glycoprotein in normal and malignant tissues. Cancer Res 47:2883–2891
95. Cizeau J, Grenkow DM, Brown JG, Entwistle J, MacDonald GC (2009) Engineering and biological characterization of VB6-845, an anti-EpCAM immunotoxin containing a T-cell epitope-depleted variant of the plant toxin bouganin. J Immunother 32:574–584
96. Kreitman RJ (2009) Recombinant immunotoxins containing truncated bacterial toxins for the treatment of hematologic malignancies. BioDrugs 23:1–13
97. Selbo PK, Weyergang A, Hogset A, Norum OJ, Berstad MB, Vikdal M et al (2010) Photochemical internalization provides time- and space-controlled endolysosomal escape of therapeutic molecules. J Control Release 148:2–12

98. Jankun J, Hart R (1997) Method of delivery of a medicament to a cancer cell using a pathway of plasminogen activator material. US Patent 5,679,350, 21 Oct 1997
99. Skobe M, Rockwell P, Goldstein N, Vosseler S, Fusenig NE (1997) Halting angiogenesis suppresses carcinoma cell invasion. Nat Med 3:1222–1227
100. Carmeliet P, Jain RK (2011) Molecular mechanisms and clinical applications of angiogenesis. Nature 473:298–307
101. Burrows FJ, Thorpe PE (1993) Eradication of large solid tumors in mice with an immunotoxin directed against tumor vasculature. Proc Natl Acad Sci U S A 90:8996–9000
102. Burrows FJ, Derbyshire EJ, Tazzari PL, Amlot P, Gazdar AF, King SW et al (1995) Up-regulation of endoglin on vascular endothelial cells in human solid tumors: implications for diagnosis and therapy. Clin Cancer Res 1:1623–1634
103. Griffioen AW, Coenen MJ, Damen CA, Hellwig SM, van Weering DH, Vooys W et al (1997) CD44 is involved in tumor angiogenesis; an activation antigen on human endothelial cells. Blood 90:1150–1159
104. Bussolati B, Grange C, Bruno S, Buttiglieri S, Deregibus MC, Tei L et al (2006) Neural-cell adhesion molecule (NCAM) expression by immature and tumor-derived endothelial cells favors cell organization into capillary-like structures. Exp Cell Res 312:913–924
105. Veenendaal LM, Jin H, Ran S, Cheung L, Navone N, Marks JW et al (2002) In vitro and in vivo studies of a VEGF121/rGelonin chimeric fusion toxin targeting the neovasculature of solid tumors. Proc Natl Acad Sci U S A 99:7866–7871
106. Ran S, Mohamedali KA, Luster TA, Thorpe PE, Rosenblum MG (2005) The vascular-ablative agent VEGF(121)/rGel inhibits pulmonary metastases of MDA-MB-231 breast tumors. Neoplasia 7:486–496
107. Matsuno F, Haruta Y, Kondo M, Tsai H, Barcos M, Seon BK (1999) Induction of lasting complete regression of preformed distinct solid tumors by targeting the tumor vasculature using two new anti-endoglin monoclonal antibodies. Clin Cancer Res 5:371–382
108. Tsunoda S, Ohizumi I, Matsui J, Koizumi K, Wakai Y, Makimoto H et al (1999) Specific binding of TES-23 antibody to tumour vascular endothelium in mice, rats and human cancer tissue: a novel drug carrier for cancer targeting therapy. Br J Cancer 81:1155–1161
109. Bussolati B, Grange C, Tei L, Deregibus MC, Ercolani M, Aime S et al (2007) Targeting of human renal tumor-derived endothelial cells with peptides obtained by phage display. J Mol Med (Berl) 85:897–906
110. Thorpe PE (2004) Vascular targeting agents as cancer therapeutics. Clin Cancer Res 10:415–427
111. Kim S, Mohamedali KA, Cheung LH, Rosenblum MG (2007) Overexpression of biologically active VEGF121 fusion proteins in Escherichia coli. J Biotechnol 128:638–647
112. Mohamedali KA, Ran S, Gomez-Manzano C, Ramdas L, Xu J, Kim S et al (2011) Cytotoxicity of VEGF(121)/rGel on vascular endothelial cells resulting in inhibition of angiogenesis is mediated via VEGFR-2. BMC Cancer 11:358
113. Akiyama H, Mohamedali KA, E Silva RL, Kachi S, Shen J, Hatara C et al (2005) Vascular targeting of ocular neovascularization with a vascular endothelial growth factor121/gelonin chimeric protein. Mol Pharmacol 68:1543–1550
114. Cho EJ, Yang J, Mohamedali KA, Lim EK, Kim EJ, Farhangfar CJ et al (2011) Sensitive angiogenesis imaging of orthotopic bladder tumors in mice using a selective magnetic resonance imaging contrast agent containing VEGF121/rGel. Invest Radiol 46:441–449
115. Shan L (2011) Vascular endothelial growth factor A isoform 121-gelonin fusion protein-conjugated manganese ferrite nanoparticles. Molecular Imaging and Contrast Agent Database (MICAD)
116. Arap W, Pasqualini R, Ruoslahti E (1998) Cancer treatment by targeted drug delivery to tumor vasculature in a mouse model. Science 279:377–380
117. Corti A, Curnis F (2011) Tumor vasculature targeting through NGR peptide-based drug delivery systems. Curr Pharm Biotechnol 12:1128–1134
118. Zarovni N, Monaco L, Corti A (2004) Inhibition of tumor growth by intramuscular injection of cDNA encoding tumor necrosis factor alpha coupled to NGR and RGD tumor-homing peptides. Hum Gene Ther 15:373–382

119. Wiley RG, Blessing WW, Reis DJ (1982) Suicide transport: destruction of neurons by retrograde transport of ricin, abrin, and modeccin. Science 216:889–890
120. Wiley RG, Stirpe F (1987) Neuronotoxicity of axonally transported toxic lectins, abrin, modeccin and volkensin in rat peripheral nervous system. Neuropathol Appl Neurobiol 13:39–53
121. Rossner S, Schliebs R, Hartig W, Bigl V (1995) 192IGG-saporin-induced selective lesion of cholinergic basal forebrain system: neurochemical effects on cholinergic neurotransmission in rat cerebral cortex and hippocampus. Brain Res Bull 38:371–381
122. Quinlivan M, Chalon S, Vergote J, Henderson J, Katsifis A, Kassiou M et al (2007) Decreased vesicular acetylcholine transporter and alpha(4)beta(2) nicotinic receptor density in the rat brain following 192 IgG-saporin immunolesioning. Neurosci Lett 415:97–101
123. Fine A, Hoyle C, Maclean CJ, Levatte TL, Baker HF, Ridley RM (1997) Learning impairments following injection of a selective cholinergic immunotoxin, ME20.4 IgG-saporin, into the basal nucleus of Meynert in monkeys. Neuroscience 81:331–343
124. Kuczewski N, Aztiria E, Leanza G, Domenici L (2005) Selective cholinergic immunolesioning affects synaptic plasticity in developing visual cortex. Eur J Neurosci 21:1807–1814
125. Ramanathan D, Tuszynski MH, Conner JM (2009) The basal forebrain cholinergic system is required specifically for behaviorally mediated cortical map plasticity. J Neurosci 29:5992–6000
126. Kamke MR, Brown M, Irvine DR (2005) Origin and immunolesioning of cholinergic basal forebrain innervation of cat primary auditory cortex. Hear Res 206:89–106
127. Ballmaier M, Casamenti F, Scali C, Mazzoncini R, Zoli M, Pepeu G et al (2002) Rivastigmine antagonizes deficits in prepulse inhibition induced by selective immunolesioning of cholinergic neurons in nucleus basalis magnocellularis. Neuroscience 114:91–98
128. Picklo MJ, Wiley RG, Lappi DA, Robertson D (1994) Noradrenergic lesioning with an anti-dopamine beta-hydroxylase immunotoxin. Brain Res 666:195–200
129. Wiley RG, Harrison MB, Levey AI, Lappi DA (2003) Destruction of midbrain dopaminergic neurons by using immunotoxin to dopamine transporter. Cell Mol Neurobiol 23:839–850
130. Kwok KH, Law KB, Wong RN, Yung KK (2000) Immunolesioning of glutamate receptor GluR1-containing neurons in the rat neostriatum using a novel immunotoxin. Cell Mol Neurobiol 20:483–496
131. Monti B, D'Alessandro C, Farini V, Bolognesi A, Polazzi E, Contestabile A et al (2007) In vitro and in vivo toxicity of type 2 ribosome-inactivating proteins lanceolin and stenodactylin on glial and neuronal cells. Neurotoxicology 28:637–644
132. Mantyh PW, Allen CJ, Ghilardi JR, Rogers SD, Mantyh CR, Liu H et al (1995) Rapid endocytosis of a G protein-coupled receptor: substance P evoked internalization of its receptor in the rat striatum in vivo. Proc Natl Acad Sci U S A 92:2622–2626
133. Nichols ML, Allen BJ, Rogers SD, Ghilardi JR, Honore P, Luger NM et al (1999) Transmission of chronic nociception by spinal neurons expressing the substance P receptor. Science 286:1558–1561
134. Allen JW, Mantyh PW, Horais K, Tozier N, Rogers SD, Ghilardi JR et al (2006) Safety evaluation of intrathecal substance P-saporin, a targeted neurotoxin, in dogs. Toxicol Sci 91:286–298
135. Holmes MA, Foote J (1997) Structural consequences of humanizing an antibody. J Immunol 158:2192–2201
136. Presta LG (2006) Engineering of therapeutic antibodies to minimize immunogenicity and optimize function. Adv Drug Deliv Rev 58:640–656
137. Sharkey RM, Goldenberg DM (2008) Use of antibodies and immunoconjugates for the therapy of more accessible cancers. Adv Drug Deliv Rev 60:1407–1420
138. Hwang WY, Foote J (2005) Immunogenicity of engineered antibodies. Methods 36:3–10
139. Winter G, Griffiths AD, Hawkins RE, Hoogenboom HR (1994) Making antibodies by phage display technology. Annu Rev Immunol 12:433–455
140. Bruggemann M, Spicer C, Buluwela L, Rosewell I, Barton S, Surani MA et al (1991) Human antibody production in transgenic mice: expression from 100 kb of the human IgH locus. Eur J Immunol 21:1323–1326

141. Borthakur G, Rosenblum M, Talpaz M, Daver N, Ravandi F, Faderl S et al (2012) Phase 1 study of anti-CD33 immunotoxin HUM-195/rGEL in patients withadvanced myeloid malignancies. Haematologica
142. Hassan R, Bullock S, Premkumar A, Kreitman RJ, Kindler H, Willingham MC et al (2007) Phase I study of SS1P, a recombinant anti-mesothelin immunotoxin given as a bolus I.V. infusion to patients with mesothelin-expressing mesothelioma, ovarian, and pancreatic cancers. Clin Cancer Res 13:5144–5149
143. Wayne AS, Kreitman RJ, Findley HW, Lew G, Delbrook C, Steinberg SM et al (2010) Anti-CD22 immunotoxin RFB4(dsFv)-PE38 (BL22) for CD22-positive hematologic malignancies of childhood: preclinical studies and phase I clinical trial. Clin Cancer Res 16:1894–1903
144. Dinndorf P, Krailo M, Liu-Mares W, Frierdich S, Sondel P, Reaman G (2001) Phase I trial of anti-B4-blocked ricin in pediatric patients with leukemia and lymphoma. J Immunother 24:511–516
145. Molineux G (2003) Pegylation: engineering improved biopharmaceuticals for oncology. Pharmacotherapy 23:3S–8S
146. Arpicco S, Dosio F, Bolognesi A, Lubelli C, Brusa P, Stella B et al (2002) Novel poly(ethylene glycol) derivatives for preparation of ribosome-inactivating protein conjugates. Bioconjug Chem 13:757–765
147. An Q, Lei Y, Jia N, Zhang X, Bai Y, Yi J et al (2007) Effect of site-directed PEGylation of trichosanthin on its biological activity, immunogenicity, and pharmacokinetics. Biomol Eng 24:643–649
148. Meng Y, Liu S, Li J, Meng Y, Zhao X (2012) Preparation of an antitumor and antivirus agent: chemical modification of alpha-MMC and MAP30 from Momordica Charantia L. with covalent conjugation of polyethyelene glycol. Int J Nanomedicine 7:3133–3142
149. Hu RG, Zhai QW, He WJ, Mei L, Liu WY (2002) Bioactivities of ricin retained and its immunoreactivity to anti-ricin polyclonal antibodies alleviated through pegylation. Int J Biochem Cell Biol 34:396–402
150. Chan WL, Shaw PC, Li XB, Xu QF, He XH, Tam SC (1999) Lowering of trichosanthin immunogenicity by site-specific coupling to dextran. Biochem Pharmacol 57:927–934
151. Onda M, Nagata S, FitzGerald DJ, Beers R, Fisher RJ, Vincent JJ et al (2006) Characterization of the B cell epitopes associated with a truncated form of Pseudomonas exotoxin (PE38) used to make immunotoxins for the treatment of cancer patients. J Immunol 177:8822–8834
152. Onda M, Beers R, Xiang L, Nagata S, Wang QC, Pastan I (2008) An immunotoxin with greatly reduced immunogenicity by identification and removal of B cell epitopes. Proc Natl Acad Sci U S A 105:11311–11316
153. Vallera DA, Oh S, Chen H, Shu Y, Frankel AE (2010) Bioengineering a unique deimmunized bispecific targeted toxin that simultaneously recognizes human CD22 and CD19 receptors in a mouse model of B-cell metastases. Mol Cancer Ther 9:1872–1883
154. Liu W, Onda M, Kim C, Xiang L, Weldon JE, Lee B et al (2012) A recombinant immunotoxin engineered for increased stability by adding a disulfide bond has decreased immunogenicity. Protein Eng Des Sel 25:1–6
155. Olson MA, Carra JH, Roxas-Duncan V, Wannemacher RW, Smith LA, Millard CB (2004) Finding a new vaccine in the ricin protein fold. Protein Eng Des Sel 17:391–397
156. Baluna R, Rizo J, Gordon BE, Ghetie V, Vitetta ES (1999) Evidence for a structural motif in toxins and interleukin-2 that may be responsible for binding to endothelial cells and initiating vascular leak syndrome. Proc Natl Acad Sci U S A 96:3957–3962
157. Baluna R, Coleman E, Jones C, Ghetie V, Vitetta ES (2000) The effect of a monoclonal antibody coupled to ricin A chain-derived peptides on endothelial cells in vitro: insights into toxin-mediated vascular damage. Exp Cell Res 258:417–424
158. Vitetta ES (2000) Immunotoxins and vascular leak syndrome. Cancer J 6(Suppl 3):S218–S224
159. Coulson BS, Londrigan SL, Lee DJ (1997) Rotavirus contains integrin ligand sequences and a disintegrin-like domain that are implicated in virus entry into cells. Proc Natl Acad Sci U S A 94:5389–5394

160. Smallshaw JE, Ghetie V, Rizo J, Fulmer JR, Trahan LL, Ghetie MA et al (2003) Genetic engineering of an immunotoxin to eliminate pulmonary vascular leak in mice. Nat Biotechnol 21:387–391
161. Bonini F, Traini R, Comper F, Fracasso G, Tomazzolli R, Dalla Serra M et al (2006) N-terminal deletion affects catalytic activity of saporin toxin. J Cell Biochem 98:1130–1139
162. Chan SH, Shaw PC, Mulot SF, Xu LH, Chan WL, Tam SC et al (2000) Engineering of a mini-trichosanthin that has lower antigenicity by deleting its C-terminal amino acid residues. Biochem Biophys Res Commun 270:279–285
163. An Q, Wei S, Mu S, Zhang X, Lei Y, Zhang W et al (2006) Mapping the antigenic determinants and reducing the immunogenicity of trichosanthin by site-directed mutagenesis. J Biomed Sci 13:637–643
164. Entwistle J, Brown JG, Chooniedass S, Cizeau J, Macdonald GC (2012) Preclinical evaluation of VB6-845: an anti-EpCAM immunotoxin with reduced immunogenic potential. Cancer Biother Radiopharm
165. MacDonald GC, Rasamoelisolo M, Entwistle J, Cizeau J, Bosc D, Cuthbert W et al (2009) A phase I clinical study of VB4-845: weekly intratumoral administration of an anti-EpCAM recombinant fusion protein in patients with squamous cell carcinoma of the head and neck. Drug Des Devel Ther 2:105–114
166. MacDonald GC, Rasamoelisolo M, Entwistle J, Cuthbert W, Kowalski M, Spearman MA et al (2009) A phase I clinical study of intratumorally administered VB4-845, an anti-epithelial cell adhesion molecule recombinant fusion protein, in patients with squamous cell carcinoma of the head and neck. Med Oncol 26:257–264
167. Zarovni N, Vago R, Fabbrini MS (2009) Saporin suicide gene therapy. Methods Mol Biol 542:261–283
168. Martin V, Cortes ML, de Felipe P, Farsetti A, Calcaterra NB, Izquierdo M (2000) Cancer gene therapy by thyroid hormone-mediated expression of toxin genes. Cancer Res 60:3218–3224
169. Gommans WM, van Eert SJ, McLaughlin PM, Harmsen MC, Yamamoto M, Curiel DT et al (2006) The carcinoma-specific epithelial glycoprotein-2 promoter controls efficient and selective gene expression in an adenoviral context. Cancer Gene Ther 13:150–158
170. Lipinski KS, Djeha HA, Gawn J, Cliffe S, Maitland NJ, Palmer DH et al (2004) Optimization of a synthetic beta-catenin-dependent promoter for tumor-specific cancer gene therapy. Mol Ther 10:150–161
171. Li Y, McCadden J, Ferrer F, Kruszewski M, Carducci M, Simons J et al (2002) Prostate-specific expression of the diphtheria toxin A chain (DT-A): studies of inducibility and specificity of expression of prostate-specific antigen promoter-driven DT-A adenoviral-mediated gene transfer. Cancer Res 62:2576–2582
172. Zabala M, Wang L, Hernandez-Alcoceba R, Hillen W, Qian C, Prieto J et al (2004) Optimization of the Tet-on system to regulate interleukin 12 expression in the liver for the treatment of hepatic tumors. Cancer Res 64:2799–2804
173. Schmidt M, Gruensfelder P, Roller J, Hagen R (2011) Suicide gene therapy in head and neck carcinoma cells: an in vitro study. Int J Mol Med 27:591–597
174. Beltinger C, Uckert W, Debatin KM (2001) Suicide gene therapy for pediatric tumors. J Mol Med (Berl) 78:598–612
175. Boeckle S, Wagner E (2006) Optimizing targeted gene delivery: chemical modification of viral vectors and synthesis of artificial virus vector systems. Aaps J 8:E731–E742
176. Bourbeau D, Lau CJ, Jaime J, Koty Z, Zehntner SP, Lavoie G et al (2007) Improvement of antitumor activity by gene amplification with a replicating but nondisseminating adenovirus. Cancer Res 67:3387–3395
177. Ye X, Liang M, Meng X, Ren X, Chen H, Li ZY et al (2003) Insulation from viral transcriptional regulatory elements enables improvement to hepatoma-specific gene expression from adenovirus vectors. Biochem Biophys Res Commun 307:759–764
178. Bergen JM, Park IK, Horner PJ, Pun SH (2008) Nonviral approaches for neuronal delivery of nucleic acids. Pharm Res 25:983–998

179. Al-Dosari MS, Gao X (2009) Nonviral gene delivery: principle, limitations, and recent progress. Aaps J 11:671–681
180. Glinka EM (2012) Eukaryotic expression vectors bearing genes encoding cytotoxic proteins for cancer gene therapy. Plasmid 68:69–85
181. Patil SD, Rhodes DG, Burgess DJ (2005) DNA-based therapeutics and DNA delivery systems: a comprehensive review. Aaps J 7:E61–E77
182. Duarte S, Carle G, Faneca H, Lima MC, Pierrefite-Carle V (2012) Suicide gene therapy in cancer: where do we stand now? Cancer Lett 324:160–170
183. Hoganson DK, Chandler LA, Fleurbaaij GA, Ying W, Black ME, Doukas J et al (1998) Targeted delivery of DNA encoding cytotoxic proteins through high-affinity fibroblast growth factor receptors. Hum Gene Ther 9:2565–2575
184. Seow Y, Wood MJ (2009) Biological gene delivery vehicles: beyond viral vectors. Mol Ther 17:767–777
185. Wei MQ, Ellem KA, Dunn P, West MJ, Bai CX, Vogelstein B (2007) Facultative or obligate anaerobic bacteria have the potential for multimodality therapy of solid tumours. Eur J Cancer 43:490–496
186. Akin D, Sturgis J, Ragheb K, Sherman D, Burkholder K, Robinson JP et al (2007) Bacteria-mediated delivery of nanoparticles and cargo into cells. Nat Nanotechnol 2:441–449
187. Nemunaitis J, Cunningham C, Senzer N, Kuhn J, Cramm J, Litz C et al (2003) Pilot trial of genetically modified, attenuated Salmonella expressing the E. coli cytosine deaminase gene in refractory cancer patients. Cancer Gene Ther 10:737–744
188. Toso JF, Gill VJ, Hwu P, Marincola FM, Restifo NP, Schwartzentruber DJ et al (2002) Phase I study of the intravenous administration of attenuated Salmonella typhimurium to patients with metastatic melanoma. J Clin Oncol 20:142–152
189. King I, Bermudes D, Lin S, Belcourt M, Pike J, Troy K et al (2002) Tumor-targeted Salmonella expressing cytosine deaminase as an anticancer agent. Hum Gene Ther 13:1225–1233
190. Friedlos F, Lehouritis P, Ogilvie L, Hedley D, Davies L, Bermudes D et al (2008) Attenuated Salmonella targets prodrug activating enzyme carboxypeptidase G2 to mouse melanoma and human breast and colon carcinomas for effective suicide gene therapy. Clin Cancer Res 14:4259–4266
191. Cheng CM, Lu YL, Chuang KH, Hung WC, Shiea J, Su YC et al (2008) Tumor-targeting prodrug-activating bacteria for cancer therapy. Cancer Gene Ther 15:393–401
192. Jabr-Milane L, van Vlerken L, Devalapally H, Shenoy D, Komareddy S, Bhavsar M et al (2008) Multi-functional nanocarriers for targeted delivery of drugs and genes. J Control Release 130:121–128
193. Iyer AK, Khaled G, Fang J, Maeda H (2006) Exploiting the enhanced permeability and retention effect for tumor targeting. Drug Discov Today 11:812–818
194. Yip WL, Weyergang A, Berg K, Tonnesen HH, Selbo PK (2007) Targeted delivery and enhanced cytotoxicity of cetuximab-saporin by photochemical internalization in EGFR-positive cancer cells. Mol Pharm 4:241–251
195. Fuchs H, Bachran D, Panjideh H, Schellmann N, Weng A, Melzig MF et al (2009) Saponins as tool for improved targeted tumor therapies. Curr Drug Targets 10:140–151

Author Biography

Riccardo Vago graduated in Pharmaceutical Chemistry and Technology and completed his PhD in 2004 at the University of Milan. From 2000 to 2004 in UK, Italy and Switzerland he investigated the RIP's intoxication route(s) and worked on the development of saporin protein/DNA as therapeutic agent. From 2005 to 2007 as postdoctoral researcher at the Institute of Molecular Genetics, Pavia, he investigated the DNA replication/repair regulation; since 2007 he acts as project leader at Dibit-HSR Scientific Institute, working on a genetic form of Parkinson's disease. In 2011 he was appointed as Assistant Professor of Biochemistry at Università Vita-Salute San Raffaele, Milan. His interests focus on molecular and cellular mechanisms underlying complex pathologies aiming at development of novel strategies and therapeutics.

Rodolfo Ippoliti is Full Professor of Molecular Biology at the University of L'Aquila. He got his MS degree in Biology in 1985 and his PhD in Biochemistry in 1990 at the University of Rome La Sapienza. He has worked at the Department of Biochemical Sciences at the University La Sapienza since 1984 to 2000, with his main scientific interests being in the characterization of structure/function relationships among selected proteins, particularly focusing on respiratory proteins and plant toxins. He got an Associate Professor position in Biochemistry at the University of L'Aquila in 2000, pursuing his work on protein toxins by focusing on the characterization of intracellular delivery of toxins using biochemical, molecular biology and cellular biology methods. 2011-present: Coordinator of the PhD School of Cellular and Molecular Biology. He is member of the Scientific Board of the Chemical-Pharmaceutical Innovation Pole of the Region Abruzzo.

Maria Serena Fabbrini got her degree in plant cell biology in Milano. (1988–1992) Fellow at European Molecular Biology Lab (Heidelberg, D) and at Insitute for Molecular Genetics of Maize (I). Consultant for Farmitalia company on the processing of human Endothelin-1.
Since 1990-member of ABCD (Association of Cellular Biology) & European Cell Biologists organization (ECBO, now ELSO). (1992–2004) Staff Scientist at Dibit-HSR Institute. Tutor for Molecular & Cell Biology Courses at Università Vita-Salute for the Medical School students where she supervised several experimental student's thesis and PhD thesis.
(2005–2009) Visiting Scientist at CNR (national research council).
Biotechnology Consultant for different institutions (Telethon(I); Leukaemia Busters(UK); INGM(I)).
2011-present teaching duties for MIUR (Italian Ministery of instruction and research).
Her research interests focussed on cell biology studies on plant and disease proteins and on recombinant expression of chimaeras including Saporin from *Saponaria officinalis*.

Part IV
The Research History of Protease Inhibitors and their Health Implications

Chapter 11
The Health Benefits of the Bowman-Birk Inhibitor

Ann R. Kennedy

Abstract There have been numerous reviews on the beneficial effects and health benefits of the soybean protease inhibitor known as the Bowman-Birk Inhibitor (BBI) [1–15]. The early in vitro studies were so promising that in vivo studies were performed at a very early stage of BBI development. The results of animal carcinogenicity studies were sufficiently promising that permission to perform BBI human trials was sought. It was planned that human trials using BBI would be performed with BBI Concentrate (BBIC), a soybean extract enriched in BBI. BBIC achieved Investigational New Drug (IND) Status with the FDA in 1992, and human trials began at that time. There have been six INDs involving BBI human trial work for the following endpoints: cancer prevention (regression of a pre-malignant lesion known as oral leukoplakia), treatment of benign prostatic hyperplasia, prostate cancer detection and treatment (with measurements of prostate specific antigen levels and prostate volume, etc.), treatment of ulcerative colitis, gingivitis, or esophagitis (and/or alleviation of adverse side effects of lung cancer treatment), as described elsewhere [15]. There are numerous publications on the results from these trial areas [16–21]. In this "expert opinion", the current status of BBI experimentation and human trials will be discussed.

11.1 Studies on the Beneficial Effects of BBI/BBIC

The Bowman-Birk inhibitor (BBI) is a soybean-derived protease inhibitor, which is known by the names of the scientists who identified and characterized it. BBI was initially discovered by Dr. Donald Bowman in 1946 [22] and was subsequently

A. R. Kennedy (✉)
Department of Radiation Oncology, Perelman School of Medicine,
University of Pennsylvania, Philadelphia, PA 19104, USA
e-mail: akennedy@mail.med.upenn.edu

E. F. Fang and T. B. Ng (eds.), *Antitumor Potential and Other Emerging Medicinal Properties of Natural Compounds*, DOI: 10.1007/978-94-007-6214-5_11,

further characterized by Dr. Yehudith Birk over the past several decades [23]. It has been evaluated by those in the Kennedy laboratory for well over three decades for its numerous health benefits, as reviewed in references [1–5]. Most of the publications about BBI relate to its potent anti-inflammatory and anticarcinogenic activities.

The known beneficial health effects of BBI/BBIC include: (1) anticarcinogenic and cancer preventive activities, (2) anti-inflammatory activities, (3) prevention of hair and weight loss in animals with cancer, (4) prevention of a radiation-induced birth defect (exencephaly), (5) increased life-span, (5) enhancement in cytotoxicity of *cis*-platinum as a chemotherapeutic agent, and (6) actions as a radioprotective agent for normal cells in vitro (specific references for each of these effects are given in reference (15). Recent studies have indicated that BBI/BBIC can prevent muscle atrophy in animals exposed to simulated microgravity [24, 25] and in mdx mice, which represent an animal model of Duchenne muscular dystrophy [26] and can have beneficial effects in animals with experimental autoimmune encephalomyelitis (EAE), an animal model system for multiple sclerosis [27–30], as well as in animals with experimental neuritis, an animal model system for Guillain-Barré syndrome (Abdolmohamad Rostami, personal communication).

The results from the studies involving animal models of multiple sclerosis and muscular dystrophy have been particularly striking, but human trials of BBI in patients with these diseases are not possible at this time because there is no current supplier of BBI/BBIC for human clinical trial work. Central Soya, Inc (Ft. Wayne, Indiana). Manufactured BBIC for decades for use in cells, animals and human trials, but Central Soya, Inc. ceased to exist in the recent past when the business was taken over by Solae (As of May 1, 2012, DuPont assumed full ownership of Solae, which was previously a joint venture between DuPont and Bunge). As Solae was not interested in producing a drug product, they were no longer able to supply BBI/BBIC for use in human trial work. At this point, we have begun to search for a new company to manufacture BBI/BBIC as an IND drug that could be used in human trials focused on the ability of BBI to prevent/mitigate the adverse health effects associated with muscle atrophy occurring in aging populations or those with non-weightbearing disorders (e.g. persons with spinal cord injury, etc.), or patients with multiple sclerosis or muscular dystrophy.

References

1. Kennedy AR, Szuhaj BF, Newberne PM, Billings PC (1993) Preparation and production of a cancer chemopreventive agent, Bowman-Birk Inhibitor Concentrate. Nutr Cancer 19:281–302
2. Kennedy AR (1993) Overview: anticarcinogenic activity of protease inhibitors. In: Troll W, Kennedy AR (eds) Protease inhibitors as cancer chemopreventive agents. Plenum publishing corporation, New York, pp 9–64 (Chapter)
3. Kennedy AR (1993) In vitro studies of anticarcinogenic protease inhibitors In: Troll W, Kennedy AR (eds) Protease inhibitors as cancer chemopreventive agents. Plenum publishing corporation, New York, pp 65–91 (Chapter)
4. Chang JD, Kennedy AR (1993) Suppression of c-myc by anticarcinogenic protease inhibitors. In: Troll W, Kennedy AR (eds) Protease inhibitors as cancer chemopreventive agents. Plenum Publishing corporation, New York, pp 265–280 (Chapter)

5. Kennedy AR (1994) Prevention of carcinogenesis by protease inhibitors. Cancer Res 54:1999s–2005s
6. Kennedy AR (1993) Cancer prevention by protease inhibitors. Prev Med 22:796–811
7. Kennedy AR (1995) The evidence for soybean products as cancer preventive agents. J Nutr 125:733s–743s
8. Kennedy AR (1998) Cancer prevention by Bowman-Birk Inhibitor Concentrate (BBIC). In: Prasad KN, Cole WC (eds) Cancer and Nutrition. IOS Press, Amsterdam, pp 93–98
9. Kennedy AR (1998) The Bowman-Birk inhibitor from soybeans as an anticarcinogenic agent. Am J Clin Nut 68(suppl):1406S–1412S
10. Kennedy AR (1998) Chemopreventive agents: protease inhibitors. Pharmacol Ther 78:167–209
11. Kennedy AR, Wan XS (2002) Biological effects of a soybean-derived protease inhibitor, the Bowman-Birk inhibitor. In: Descheemaeker K, Debruyne I (eds) Soy and Health 2002—Clinical evidence, dietetic applications. Garant Publishers, Apeldoorn, p 129–136
12. Kennedy AR (2007) Status of current human trials utilizing Bowman Birk Inhibitor Concentrate. In: Descheemaeker K. Debruyne I (eds) Proceedings of a Symposium, "Soy and Health 2006—Clinical evidence, dietetic applications", held in Dusseldorf, Germany, October 2006; Garant Publishers, Apeldoorn, pp 73–79
13. Kennedy AR (2009) Protease Inhibitors. In: Knasmüller S, DeMarini DM, Johnson I, Gerhäuser C (eds) Chemoprevention of cancer and DNA damage by dietary factors, Chapter 42, Wiley, GmbH and Co, KGaA, Weinheim, pp 761–767
14. Kennedy AR, Wan XS (2011) Countermeasures for space radiation induced adverse effects. Adv Space Res 48:1460–1479
15. Kennedy AR (2006) The status of human trials utilizing Bowman-Birk Inhibitor Concentrate from soybeans. In: Sugano M (ed) Soy in health and disease prevention. CRC Press, Taylor and Francis Group LLC, Chapter 12, Boca Raton, Florida, USA, pp 207–223 (Chapter)
16. Armstrong WB, Kennedy AR, Wan XS, Atiba J, McLaren CE, Meyskens FL Jr (2000) Single dose administration of Bowman-Birk Inhibitor Concentrate (BBIC) in patients with oral leukoplakia. Cancer Epidemiol Biomark Prev 9:43–47
17. Armstrong WB, Kennedy AR, Wan XS, Taylor TH, Nguyen QA, Jensen J, Thompson W, Lagerberg W, Meyskens FL Jr (2000) Clinical modulation of oral leukoplakia and protease activity by Bowman-Birk Inhibitor Concentrate in a phase IIa chemoprevention trial. Clin Cancer Res 6:4684–4691
18. Malkowicz SB, McKenna WG, Vaughn DJ, Wan XS, Propert KJ, Rockwell K, Marks SHF, Wein AJ, Kennedy AR (2001) Effects of Bowman-Birk Inhibitor Concentrate in patients with benign prostatic hyperplasia. Prostate 48:16–28
19. Armstrong WB, Wan XS, Kennedy AR, Taylor TH, Meyskens FL (2003) Development of the Bowman-Birk Inhibitor for oral cancer chemoprevention and analysis of neu immunohistochemical staining intensity with Bowman-Birk Inhibitor Concentrate treatment. Laryngoscope 113(10):1687–1702
20. Lichtenstein GR, Deren JJ, Katz S, Lewis JD, Kennedy AR, Ware JH (2008) Bowman-Birk Inhibitor Concentrate (BBIC): a novel therapeutic agent for patients with active ulcerative colitis. Dig Dis Sci 53(1):175–180
21. Lin L, Mick M, Ware J, Metz J, Lustig R, Vapiwala N, Rengan R, Kennedy AR (2009) Phase I randomized double-blind placebo controlled single dose safety studies of Bowman Birk Inhibitor Concentrate. (submitted)
22. Bowman DE (1046) Differentiation of soybean antitryptic factors. Proc Soc Exp Biol Med 63:547–550
23. Birk Y (1961) Purification and some properties of a highly active inhibitor of trypsin and α-chymotryotrypsin from soybeans. Biochim Biophys Acta 54:378–381
24. Morris CA, Morris LD, Kennedy AR, Sweeney HL (2005) Attenuation of skeletal muscle atrophy via protease inhibition. J Appl Physiol 99(5):1719–1727
25. Arbogast S, Smith J, Matuszczak Y, Hardin B, Moylan J, Smith JD, Ware J, Kennedy AR, Reid MB (2007) Bowman-Birk Inhibitor Concentrate prevents atrophy, weakness and oxidative stress in soleus muscle of lindlimb-unloaded mice. J Appl Physiol 102(3):956–964

26. Morris CA, Selsby JT, Morris LD, Pendrak K, Sweeney HL (2010) Bowman-Birk inhibitor attenuates dystrophic pathology in mdx mice. J Appl Physiol 109(5):1492–1499
27. Gran B, Tabibzadeh N, Martin A, Ventura ES, Ware JH, Zhang G-X, Parr JL, Kennedy, AR, Rostami A (2006) The protease inhibitor, Bowman-Birk inhibitor, suppresses experimental autoimmune encephalomyelitis: a potential oral therapy for multiple sclerosis. Multiple Scler 12(6):688–697
28. Touil T, Ciric B, Ventura E, Shindler KS, Gran B, Rostami A (2008) Bowman-Birk inhibitor suppresses autoimmune inflammation and neuronal loss in a mouse model of multiple sclerosis. J Neurol Sci 27(1–2):191–202
29. Dai H, Ciric B, Zhang GX, Rostami A (2011) Bowman-Birk inhibitor attenuates experimental autoimmune encephalomyelitis by delaying infiltration of inflammatory cells into the CNS. Immunol Res 51(2–3):145–152
30. Dai H, Ciric B, Zhang GX, Rostami A (2012) Interleukin-10 plays a crucial role in suppression of experimental autoimmune encephalomyelitis by Bowman-Birk inhibitor. J Neuroimmunol 245(1–2):1–7

Author Biography

Ann R. Kennedy is a Professor in the Department of Radiation Oncology at the University of Pennsylvania School of Medicine and the Richard Chamberlain Professor of Research Oncology at the University of Pennsylvania. She received her doctoral degree in Radiation Biology from Harvard University and remained at Harvard as a faculty member for many years before joining the faculty at the University of Pennsylvania. She has been performing research on mechanisms of carcinogenesis and cancer prevention for her entire career, with research investigations including in vitro studies on molecular mechanisms, animal studies and human trials. Much of the recent work in the Kennedy laboratory has focused on the soybean-derived protease inhibitor known as the Bowman-Birk inhibitor (BBI). BBI was originally identified as a cancer preventive agent from in vitro studies, and was then shown to prevent the development of cancer in many different models of animal carcinogenesis. BBI, as Bowman Birk Inhibitor Concentrate (BBIC), is now being evaluated as a human cancer chemopreventive agent and as a therapeutic agent for several different human diseases. Much of her recent effort has been aimed at developing countermeasures for radiation induced adverse health effects which could occur in astronauts during space travel. She is currently the Team Leader for the Radiation Effects Team of the National Space Biomedical Research Institute (NSBRI), as well as a Co-Team Lead, along with Dr. Francis Cucinotta, of the Space Radiation Element of the Human Research Program at the National Aeronautics and Space Administration (NASA). Dr. Kennedy is the author of more than 250 scientific papers.

Chapter 12
Peptidic Inhibitors of Serine Proteinases of Plant Origin

Krzysztof Rolka, Adam Lesner, Anna Łęgowska and Magdalena Wysocka

Abstract Serine proteinases play important roles in many physiological processes and in consequence, when unbalanced, are responsible for numerous severe diseases. The most predominant mechanism of their control is the ubiquitous presence of their inhibitors. On the basis of their inhibition mode, serine proteinase inhibitors are classified into canonical (standard mechanism) inhibitors, serpins and non-canonical inhibitors. The best studied are inhibitors assigned to the first group. At least 18 canonical inhibitor families have been recognized. Inhibitors isolated from the seeds of Cucurbitaceae are small (around 30 amino acid residues), containing three disulfide brides. Some of them are characterized by a cyclic polypeptide backbone. Head-to-tail cyclization is also present in the smallest (14 amino acid residues) trypsin inhibitor isolated from sunflower seeds. All these potent inhibitors display well-defined, rigid structures and, unlike most proteins, are also resistant to denaturizing agents. Modifications introduced into the molecules of these low-molecular-mass inhibitors are well tolerated, retaining their tertiary structure and inhibitory activity. They are able to cross cell membranes and are the first examples of cyclic cell-penetrating peptides. Recent results obtained on plant peptidic inhibitors and discussed in this mini-review have proved that they are promising molecules for drug design.

K. Rolka (✉) · A. Lesner · A. Łęgowska · M. Wysocka
Faculty of Chemistry, University of Gdansk, Sobieskiego 18, 80-952 Gdansk, Poland
e-mail: rolka@chem.univ.gda.pl

A. Lesner
e-mail: adas@chem.univ.gda.pl

A. Łęgowska
e-mail: legowska@chem.univ.gda.pl

M. Wysocka
e-mail: mwysocka@chem.univ.gda.pl

E. F. Fang and T. B. Ng (eds.), *Antitumor Potential and Other Emerging Medicinal Properties of Natural Compounds*, DOI: 10.1007/978-94-007-6214-5_12,

12.1 Introduction

Proteolysis is probably the most widespread mechanism of biological regulation. By controlling protein synthesis, turnover and activity, it is involved in fundamental physiological processes including cell differentiation, growth and signaling, apoptosis, fertilization, blood coagulation, immune response and digestion. Yet, uncontrolled proteolysis can be harmful for organisms, causing amongst others such diseases as emphysema, inflammation, cancer, neurodegenerative, immunological, and cardiovascular disorders. Proteolytic enzymes can be divided into at least seven families: serine, aspartic, cysteine, glutamic, metallo, asparagine and threonine proteases. Serine proteases (SPs) are among the most widely studied proteins; they are widespread in nature and are involved in numerous and diverse physiological and pathological processes. Over 82 thousand sequences of SPs have been collected in the MEROPS database (http://merops.sanger.ac.uk/), whereas the second largest family is the family of metalloproteinases (over 75 thousands). SPs originate from extracellular matrix, mast cells, cytotoxic cells, and neutrophils. Some of the most studied SPs are those involved in blood coagulation. There are several mechanisms found in mammalian systems that strictly control the activities of proteases. Like other proteins, their ribosomal synthesis can be controlled at several stages, followed in many cases by post-translational modifications. Compartmentalization and the manner in which proteinases are released in the pericellular environment are additional important mechanisms for the regulation of their activity. Release and folding of the active enzyme from its inactive precursor is one of the predominant methods of protease control. A second method is effected by the ubiquitous presence of protease inhibitors. Naturally occurring inhibitors are proteins or peptides except in the case of non-proteinaceous ones isolated from microorganisms.

According to recent data brought together in the MEROPS database almost 21 thousand inhibitor amino acid sequences have been identified. Based on their sequence homology, they have been grouped into 91 families and further assigned to 36 clans according to their similar tertiary structures. There are also a large number of small-molecule inhibitors, both naturally occurring and synthetic. These peptidomiometics, or simply small organic molecules, are outside the scope of this mini-review. Since more than one third of identified proteases are serine proteinases, it is no surprise that their inhibitors (SPIs) also represent the largest group. They are found in all living organisms. In fact, the first described were trypsin inhibitors. In 1936 Kunitz and Northrop isolated and crystallized a bovine pancreatic trypsin inhibitor (BPTI). Later, Kunitz identified another, named STI, from soybean. In the meantime, another inhibitor, the Bowman-Birk inhibitor (BBI), was identified from the same source and characterized. These proteins inhibited trypsin and at neutral pH formed stable complexes which were dissociated at low pH. Progress in bioanalytical techniques has confirmed earlier suggestions that these inhibitors interact with enzymes in a substrate-like manner. They are probably the best studied proteins and their complexes with cognate enzymes are among the best known examples of protein—protein interactions. This was also one of the reasons that knowledge about inhibitors of other proteolytic enzymes was, until the last decade, significantly less well advanced.

12.2 Inhibition Mode of Serine Proteinase Inhibitors

On the basis of their mechanism of action, SPIs are classified as: canonical (standard mechanism), serpins and non-canonical. Except for the last group, SPIs interact with target enzymes in a substrate-like manner. Canonical inhibitors are small proteins or peptides with molecular weights varying from 1.5 to 2.1 kD, corresponding to 14–200 amino acid residues. They are characterized by rather compact, thermally resistant structures, and although they possess different folding motives, they are usually stabilized by several disulfide bridges. These native structures are also stable in denaturizing agents. A common structural element in canonical inhibitors is a short fragment comprising less than 10 amino acid residues which is directly involved in interaction with cognate enzymes. This exposed fragment, named the "binding loop" or the "primary binding segment", has an extended conformation with a similar anti-parallel β-strand. Bode and Huber [1] proposed to call it "canonical". Some inhibitors contain a single binding loop, some others consist of multiple domains where such elements are located. Thus, such inhibitors interact with proteinases with stoichiometry other than 1:1. Alternatively, they are able to inhibit more than one enzyme simultaneously. It transpires that all standard mechanism inhibitors exhibit this canonical conformation of binding loop, and are therefore named "canonical inhibitors". This mode of inhibition was studied in detail by the group of Laskowski [2]. Upon enzyme—inhibitor interaction, a stable complex C is formed rapidly and then (unlike in substrates) very slowly dissociated into a free enzyme and modified inhibitor. As indicated by X-ray data, the majority of contacts with enzymes are made by the inhibitor binging loop [1] that fit into the active site of a cognate enzyme. The intermolecular contact area is relatively small, restricted to 6–9 nm^2. In the middle of this loop is located a peptide bond P_1-P_1' (reactive site). According to the Schechter and Berger nomenclature [3], P_1-P_1' is the scissile peptide bond and both amino acid residues P_1 and P_1' interact with the corresponding enzyme—S_1 and S_1'- subsites, respectively. The hyper-exposed amino acid residue in position P_1 interacts with the S_1 cavity of the enzyme accounting for the largest part (up to 50 %) of the total association energy released during complex formation with the target enzyme. Therefore, P_1 residue determines inhibitor specificity and is often referred to as the primary specificity residue. As in substrates, the reactive site P_1-P_1' is selectively cleaved by proteinase yielded modified inhibitor. After sufficient incubation time, an equilibrium between intact and modified inhibitor is formed. The determined values of the hydrolysis constant ($K_{hyd} = [I]/[I^*]$) for most canonical inhibitors are, in neutral pH, close to unity. It is worth mentioning that in a number of 3D structures of proteinase-inhibitor complexes determined by X-ray crystallography the reactive site P_1-P_1' remains intact. The mechanism of inhibition of canonical inhibitors with cognate enzymes is shown in Fig. 12.1. One of the most eminent researchers in the field of serine proteinase inhibitors was undoubtedly Michael Laskowski Jr. who devoted his whole scientific life to the elucidation of this mode of inhibition. In recognition of his contribution, MEROPS suggests the adoption of the new name, the "Laskowski mechanism", for what has been accepted as the standard mechanism.

The acronym "Serpin" (serine proteinase inhibitors) was introduced in the mid 1980s [4] to describe a superfamily of relatively large proteins (45–55 kD); some of them are serine and cysteine protease inhibitors, others play non-inhibitory functions. Despite a low sequence homology, they adopt a similar metastable conformation comprising three β-sheets, eight to nine α-helical segments and an exposed binding loop, also called the "reactive center loop" (RCL). This large family (I4) consists of over 2500 members, found in all kingdoms of organisms including 44 in humans. Serpins interact with target proteinases based on a unique mechanism that is described as the suicide substrate-like inhibition. Inhibition by serpins is irreversible. Proteinase is inactivated, kinetically trapped in the complex and can be cleared from the circulation by binding to members of the low density lipoprotein receptor family. Irreversibility is also reinforced by the induced proteolytic susceptibility of proteases in complex with serpins.

The first non-canonical inhibitor, hirudin, was isolated from the saliva of the medicinal leech in the 1950s and a decade later its primary structure was determined. This polypeptide, consisting of 65 amino acid residues, displays anticoagulant activity and is a very potent and specific thrombin inhibitor. Hirudin and its variants isolated from leech species share a common 3D structure, *N*-terminal globular domain cross-linked by three disulfide bridges and an acidic *C*-terminal fragment, which displays flexible high conformational freedom. Both structural elements are directly involved in the interaction with proteinase. Negatively charged *C*-terminal segment directs hirudin towards a positively charged thrombin exosite I (fibrinogen recognition exosite), followed by their tight binging. This induces conformational changes in the enzyme allowing the interaction of *N*-terminal tripeptide of the inhibitor with the thrombin active site. In contrast to canonical inhibitors, this fragment does not form an anti-parallel but rather a parallel β-sheet with side chains of a thrombin Ser^{24}—Gly^{219} segment and a negatively

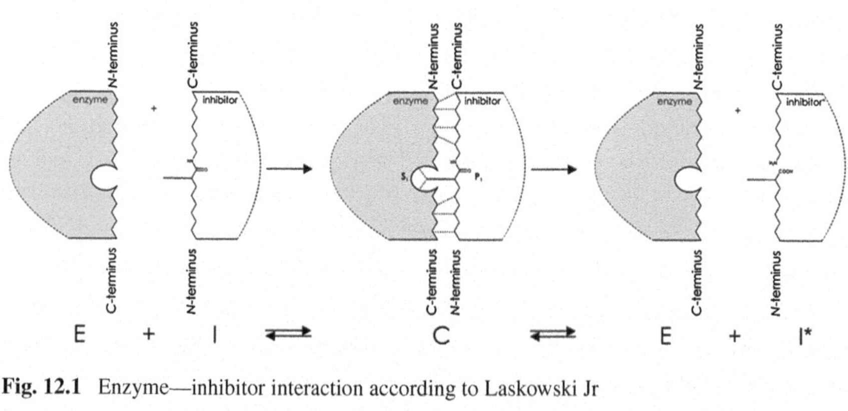

Fig. 12.1 Enzyme—inhibitor interaction according to Laskowski Jr

charged side chain of Asp^{189} located in the enzyme S_1 subsite does not interact with the inhibitor. Another thrombin inhibitor isolated from leeches (*Hirudinaria manillensis*)—named haemadin, although it does not share sequence homology with hirudin, displays a similar tertiary structure and similar mode of thrombin inhibition. This was the reason that hirudin and haemadin were assigned to the same family (I14). Nevertheless, the analysis of its complex with thrombin revealed several differences. One is the direct contact of the positively charged guanidine group of inhibitor's Arg2 with carboxylate of enzyme's Asp189. This resembles the P_1—S_1 interaction of canonical inhibitors. The second is the interaction of the *C*-terminal segment not with exosite I but with exosite II. Several Cys-rich inhibitors including antistatin, therostatin, tessulin, bdellastatin, guamerin and the most potent thrombin inhibitor theromin were assigned to this family (I15) and this is collectively called the antistatin family. These inhibitors display a common folding motif consisting of several repeats of *N*- and *C*-domains. On the basis of limited structural data, it is postulated that inhibitors of the I15 family interact with target enzymes in a canonical (substrate-like) manner with P_1 position located in the *C*-domain, while (in at least some inhibitors) the highly acidic *N*-domain may interact with thrombin exosites. A more detailed mechanism of interaction between non-canonical inhibitors and proteinases has been presented in a recently published review [5].

12.3 Classification of the Canonical Serine Proteinase Inhibitors

Extensive studies on the isolation, characterization and determination of the 3D structures of inhibitors obeying a common, so-called standard mechanism (later named canonical) led to the conclusion that not all inhibitors are homologous. They were found in almost all tissues where they were looked for—animal, plant and microbial. Laskowski and Kato proposed grouping them into families. Originally, nine families were recognized, but information on three of them was rather marginal. Subsequently, with progress in the identification and characterization of new canonical inhibitors more members of each family were found and, in addition, more families were added to the list. Finally, based on results obtained over the following two decades, Laskowski Jr's group [2] recognized at least 16 different families (Table 12.1). More recently, Rawlings et al. [6] proposed a new classification system for the protease inhibitors that allows them to be added to the MEROPS database. They introduced the term "inhibitor unit" which is a segment of the inhibitor containing a single reactive site. An inhibitor containing a single inhibitor unit is named a simple inhibitor, whereas one containing multiple units is termed a compound inhibitor. Based on sequence homology, inhibitors were grouped into 48 families (currently, there are 79 distinct families). In addition, inhibitors which display the same type of protein fold were assigned to the same clan. To date 39 clans have been recognized, and this means that some clans contain representatives of more than one family. In the case of the canonical inhibitors of serine proteinases 18 families have been recognized and grouped in

Table 12.1 Families and clans of canonical serine proteinase inhibitors

Family				Number	
According to Laskowski [2]	According to MEROPS	Clan	Representative	Entries/ sequences	PDB structures
Kazal	I1	IA	OMTKY	1,413	13
Kunitz (BPTI)	I2	IB	BPTI	1,448	16
Kunitz (STI), Arrowhead	I3	IC	STI	231	10
Squash	I7	IE	CMTI-I	32	4
Ascaris	I8	IE	BTI	397	5
Marinostatin	I10	Unassigned	Marinostatin	23	0
Ecotin	I11	IN	Ecotin	107	1
BBI	I12	IF	BBI, SFTI-1	138	11
Potato I	I13	IG	PPI	131	3
Antistasin	I15	IM	Antistasin	36	3
SSI	I16	IY	S–SI	69	1
Chelonianin	I17	IP	SLPI	443	3
Rapeseed	I18	JD	MTI	18	2
Grasshopper	I19	IW	LMPI	14	2
Potato II	I20	JO	PPI-II	57	5
Bombyx subtilisin inhibitor	I40	Unassigned	BSI	1	0

OMTKY turkey ovomucoid trypsin inhibitor; *BPTI* bovine pancreatic trypsin inhibitor; *STI* soybean trypsin inhibitor; *CMTI-I Cucurbita maxima* trypsin inhibitor; *BTI* barley trypsin inhibitor; *BBI* Bowman-Birk inhibitor, *SFTI-1* Sunflower trypsin inhibitor-1; *S–SI Streptomyces* subtlisin inhibitor, *SLPI* secretory leukocyte peptidase inhibitor; *MTI* mustard trypsin inhibitor-2; *LMPI Locusta migratoria* trypsin inhibitor; *PPI* potato peptidase inhibitor II; *BSI* Bombyx subtilisin inhibitor; *VTI Veronica* trypsin inhibitor

14 clans. Table 12.1 presents a summary of the Laskowski and MEROPS classifications of these inhibitors. The size of these proteins varies from 14 to 200 amino acid residues. As shown in Table 12.1, canonical inhibitors are widely distributed in the plant kingdom. To date, at least eight families have been classified as being of plant origin. They are especially ubiquitous in tissues responsible for storage (seeds and drupes). It has been proved that the biosynthesis of inhibitors in plants is stimulated by infection or damage of vegetative tissue. Thus, they are important components of plant defense systems targeting exogenous proteinases.

12.4 Squash Family of Serine Proteinase Inhibitors

The first members of this family were isolated from squash (*Cucurbita maxima*) seeds and characterized at the beginning of the 1980s independently by Polanowski et al. from University of Wroclaw (Poland) and Hojima et al. from NIH, Bethesda

(USA). These polypeptides inhibiting several serine proteinases (such as bovine β-trypsin, human Hageman factor XIIa and kallikreins) were named ITD I, ITD III and PHFI, respectively. It turned out that ITD III is identical to PHFI. Subsequently, more homologues inhibitors were isolated from *Cucurbitaceae*, and in 1985 Wieczorek et al. [7] postulated that these low-molecular mass inhibitors could be assigned to a new family of serine proteinase inhibitors (I7). In addition, the Polish group introduced a new nomenclature using the scientific names of the plant which is consequently applied to all members of the squash family. Thus, the acronym ITD was replaced by CMTI (*Cucurbita maxima* trypsin inhibitor). To date over 40 inhibitors assigned to this family have been identified in seeds of Cucurbitaceae. Their amino acid sequences, sources and years of discovery are summarized in Table 12.2. A unique feature of the discussed family of inhibitors is its unusually small size, compact structure and high resistance to denaturizing conditions. Squash inhibitors consisting of 28–34 amino acid residues and their structure are stabilized by three disulfide bridges. Two disulfide bridges form a ring that is threaded by a third one. Such a structural motive is called the cystine knot (CK) [8]. Interestingly, two members of the discussed family called MCoTI-I and MCoTI-II isolated from *Momordica cochinchinensis* possess an additional cyclic element—head-to-tail cyclization and are the first macrocyclic inhibitors of the squash family [9]. At present over 160 such cyclic peptides found in nature are collectively named cyclotides [10]. They display a characteristic folding path where the cystine knot is embedded within a macrocyclic backbone, defining a motif referred to as a cyclic cystine knot (CCK). The topology of disulfide bridges, cystine knots and CCK motives is illustrated in Fig. 12.2.

12.4.1 Cystine Knot Inhibitors

In squash inhibitors, the P_1-P_1' corresponds to positions 5 and 6 in CMTI inhibitors. In substrate specificity P_1 position Arg or Lys is present; therefore, they are

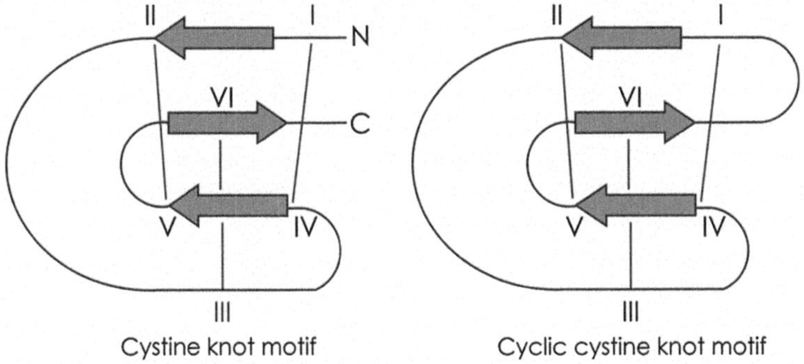

Fig. 12.2 The cystine knot motifs

Table 12.2 Amino acid sequences of squash family inhibitors

Sequence	Name	Source	Year of discovery							
RV	CPRILME	CKKDSD	CLAE	CV	CLEH	GY	CG	CMTI-I	*Cucurbita maxima*	1980, 1983
RV	CPRILMK	CKKDSD	CLAE	CV	CLEH	GY	CG	CMTI-III	*Cucurbita maxima*	1980, 1983
HEERV	CPRILMK	CKKDSD	CLAE	CV	CLEH	GY	CG	CMTI-IV	*Cucurbita maxima*	1980, 1983
HEERV	CPKILME	CKKDSD	CLAE	CI	CLEH	GY	CG	CPTI-III	*Cucurbita pepo*	1983
RV	CPKILME	CKKDSD	CLAE	CI	CLEH	GY	CG	CPTI-II	*Cucurbita pepo*	1983
MV	CPKILMK	CKHDSD	CLLD	CV	CLEDI	GY	CGVS	CSTI-IIb	*Cucumis sativus*	1983
MM	CPRILMK	CKHDSD	CLPG	CV	CLEHI	EY	CG	CSTI-IV	*Cucumis sativus*	1983
GI	CPRILME	CKRDSD	CLAQ	CV	CKRQ	GY	CG	MRTI-I	*Momordica repens*	1984
AI	CPRILME	CKRDSD	CLAQ	CV	CKRQ	GY	CG	MRTI-III	*Momordica repens*	1984
	CPRILMP	CKVNDD	CLRG	CK	CLSN	GY	CG	TTI I	*Trichosanthes kirilowii*	1984
	CPRILMP	CQVNDD	CLRG	CK	CLSN	GY	CG	TTI II	*Trichosanthes kirilowii*	1984
RV	CPKILMK	CKKDSD	CLAE	CI	CLEH	GY	CG	CPGTI-I	*Cucurbita pepo var Giromontia*	1985
GRR	CPRIYME	CKRDAD	CLAD	CV	CLQH	GI	CG	CVTI-I	*Citrullus vulgaris*	1987
<ERG	CPRILMR	CKRDSD	CLAG	CV	CQKN	GY	CG	BDTI-I	*Bryonia dioica*	1987
RG	CPRILMR	CKRDSD	CLAG	CV	CQKN	GY	CG	BDTI-II	*Bryonia dioica*	1987
G	CPRILMR	CKQDSD	CLAG	CV	CGPN	GF	CGSP	EETI-II	*Ecballium elaterium*	1989
ERR	CPRILKQ	CKRDSD	CPGE	CI	CMAH	GF	CG	MCTI-I	*Momordica charantia*	1989
RI	CPRIWME	CKRDSD	CMAQ	CI	CVD	GH	CG	MCTI-II	*Momordica charantia*	1989
RI	CPLIWME	CKRDSD	CLAQ	CI	CVD	GH	CG	MCEI-I	*Momordica charantia*	1989
RI	CPRILME	CSSDSD	CLAE	CI	CLEQ	GF	CG	LCTI-1	*Luffa cylindrical*	1990
RI	CPRILME	CSSDSD	CLAE	CI	CLEQD	GF	CG	LCTI-2	*Luffa cylindrical*	1990
RS	CPRIWME	CTRDSD	CMAK	CI	CVA	GH	CG	MCTI-A	*Momordica charantia*	1992
M	CPKILMK	CKQDSD	CLLD	CV	CLKE	GF	CG	CMCTI-I	*Cucumis melo*	1992
RM	CPKILMK	CKQDSD	CLLD	CV	CLKE	GF	CG	CMCTI-II	*Cucumis melo*	1992
<ERM	CPKILMK	CKQDSD	CLLD	CV	CLKE	GF	CG	CMCTI-III	*Cucumis melo*	1992
<ERR	CPRIYME	CKHDSD	CLAD	CV	CLEH	GI	CGG	LLTI-I	*Langeria leucantha*	1992
RR	CPRIYME	CKHDSD	CLAD	CV	CLEH	GI	CG	LLTI-II	*Langeria leucantha*	1992
ERR	CPRIYME	CKHDSD	CLAD	CV	CLEH	GI	CG	LLTI-III	*Langeria leucantha*	1992
R	CPRIYME	CKHDSD	CLAD	CV	CLPQ	GI	CG	BHIT-I	*Benicasa hispida*	1992

Table 12.2 (continued)

Sequence	Name	Source	Year of discovery							
RR	CPRIYME	CKHDSD	CLAD	CV	CLPQ	GI	CG	BHIT-III	*Benicasa hispida*	1992
QRM	CPKILMK	CKQDSD	CLLD	CV	CLKE	GF	CG	CMTI-II	*Cucumis melo*	1992, 1995
VG	CPRILMK	CKTDDD	CLLG	CK	CLSN	GY	CG	HMTI-I	*Hami melon*	1993
I	CPRILMP	CSSDSD	CLAE	CI	CLEN	GF	CG	TGTI-I	*Luffa cylindrical*	1993
MASVAESSGVVEVIELISDGGNDLPRKIMSGRHGGI	CPRILMP	CKTDDD	CMLD	CR	CLSN	G		TGTI-II	*Luffa cylindrical*	1993
MAAFVESARAGAGAEVIQLVSDGVNEYSEKMME GVVA	CPRILMP	CKVNDD	CLRG	CK	CLS			TTII	*Trichosanthes kirilowii*	1994
RI	CPRILME	CSYDSD	CFGE	CI	CLPS	GY	CG	LCTI II	*Luffa cylindrical*	1994
RI	CPRILME	CSSDSD	CLAE	CI	CLEN	GF	CG	LCTI III	*Luffa cylindrical*	1994
RM	CPRILMK	CKQDSD	CTTD	CV	CKKE	GF	CG	CMeTI-A	*Cucumis melo*	1995
VG	CPRILMK	CKTDRD	CLTG	CT	CKRN	GY	CG	CMeTI-B	*Cucumis melo*	1995
ERG	CPRILKQ	CKQDSD	CPGE	CI	CMAH	GF	CG	MCTI-III	*Momordica charantia*	1995
ERG	CPLILKQ	CKQDSD	CPGE	CI	CVD	GF	CG	MCEI-II	*Momordica charantia*	1995
EERG	CPLILKQ	CKQDSD	CPGE	CI	CVD	GF	CG	MCEI-III	*Momordica charantia*	1995
EEERG	CPLILKQ	CKQDSD	CPGE	CI	CVD	GF	CG	MCEI-IV	*Momordica charantia*	1995
KEEQRV	CPRILMR	CKRDSD	CLAQ	CT	CQQS	GF	CG	ELTI-I	*Echinocystis lobata*	1996
RV	CPRILMR	CKRDSD	CLAQ	CT	CQQS	GF	CG	ELTI-II	*Echinocystis lobata*	1996
I	CPRILME	CSHDSD	CFGE	CI	CLSS	GY	CG	LA I	*Luffa acutangula*	1996
IR	CPRIYME	CSHDSD	CLGE	CI	CLES	GF	CG	LA II	*Luffa acutangula*	1996
RV	CPRILMR	CKRDSD	CLAE	CT	CQGS	GY	CG	SATI-I	*Sicyos australis*	1996
GRI	CPRILMR	CKRDSD	CLAE	CT	CQS	GY	CG	SATI-II	*Sicyos australis*	1996
<ERGRI	CPRILMR	CKRDSD	CLAE	CT	CQS	GY	CG	SATI-III	*Sicyos australis*	1996
&SGSDGGV	CPKILQR	CRRDSD	CPGA	CI	CRGN	GY	CG&	MCoTI-I	*Momordica cochinchinensis*	2000
&SGSDGGV	CPKILKK	CRRDSD	CPGA	CI	CRGN	GY	CG&	MCoTI-II	*Momordica cochinchinensis*	2000
<ERA	CPRILKK	CRRDSD	CPGE	CI	CKEN	GY	CG	MCoTI-III	*Momordica cochinchinensis*	2000
EDRK	CPKILMR	CKRDSD	CLAK	CT	CQES	GY	CG	SETI-IIa	*Sechium edule*	2006
EEDRK	CPKILMR	CKRDSD	CLAK	CT	CQES	GY	CG	SETI-IIb	*Sechium edule*	2006
	CPKILMK	CKLDTD	CFPT	CT	CRPS	GF	CG	SETI-V	*Sechium edule*	2006

(continued)

Table 12.2 (continued)

Sequence	Name	Source	Year of discov-ery							
I	CPRILME	CKADSD	CLAQ	CI	CEES	GF	CG	CyPTI I	*Cyclanthera pedata*	2006
I	CPRILME	CKADSD	CLAQ	CI	CQES	GF	CG	CyPTI II	*Cyclanthera pedata*	2006
RI	CPRILME	CKADSD	CLAQ	CI	CEES	GF	CG	CyPTI III	*Cyclanthera pedata*	2006
RI	CPRILME	CKADSD	CLAQ	CI	CQEN	GF	CG	CyPTI IV	*Cyclanthera pedata*	2006
RI	CPRILME	CKADSD	CLAQ	CI	CQES	GF	CG	CyPTI V	*Cyclanthera pedata*	2006
ARI	CPRILMK	CKKDSD	CLAE	CI	CEEH	GF	CG	CyPTI VI	*Cyclanthera pedata*	2006
ARI	CPRILMK	CKKDSD	CLAE	CI	CQEH	GF	CG	CyPTI VII	*Cyclanthera pedata*	2006

potent inhibitors of trypsin and other serine proteinases with a negatively charged S_1 subside. The exceptions are MCEI inhibitors with Leu5 which therefore inhibit elastase. Also, P_1' position is conservative and is occupied in all in squash inhibitors by Ile residue.

The unusually small size of squash inhibitors as compared with other *proteinaceous* serine proteinase inhibitors makes them readily accessible by means of chemical synthesis. In contrast to biological methods of synthesis where introduction into the protein molecule other than by proteinogenic amino acids is rather difficult, chemical methods allow practically all modifications to be achieved. CMTI-III and -I, the first members of the family were synthesized by us in the mid 1980s [11]. This definitely confirmed the correctness of their sequences. A few years later, the French group isolated, synthesized and determined solution structure [12] homologues EETI-II from seeds of *Ecballium elaterium*. The authors showed that deletion of its *C*-terminal dipeptide Ser^{29}-Pro^{30} retained its inhibitory activity. Both CMTI-III and EETI-II (shortened 28 amino acid residue variant) were used for structure- activity-relationships. Since the first synthesis, we have synthesized over 60 analogues of CMTI/EETI analogues. The obvious modification introduced by several groups into the squash inhibitors was the substitution of Arg5 by aromatic or aliphatic amino acid residues. This yielded analogues with specific chymotrypsin and elastase inhibitory activity, respectively. On the other hand, we have shown that the requirements for P_1 position are very strict. The introduction of Arg or Lys derivatives in this position dramatically lowered trypsin inhibitory activity. It is worth mentioning that derivatives differ from proteiongenic amino acids by one methylene group in the side chains (L-homoarginine or L-homolysine) introduced in the P_1 position reducing the affinity of such analogues towards trypsin by 2–3 fold. We were also able to considerably simplify the structure of CMTI. The analogue with a truncated *N*-terminal dipeptide, and *C*-terminal Gly and contains 7 amino acid residues replaced by Gly or Pro residues retaining high inhibitory activity. All three disulfide bridges are important in maintaining the tertiary structure of squash inhibitors. On the other hand, we have recently shown [13] the substitution of each disulfide bridge by diselenide yielded potent trypsin inhibitors and more redox-resistant EETI-II analogues. Le-Nguyen et al. [14] reported the synthesis of hybrids of EETI-II and carboxypeptidase inhibitors. The rationale of their work was the fact that both peptides display the same fold as the cystine knot motive. It turned out that such an analogue inhibited both proteases used. Analogues of EETI-II were also synthesized by Hilpert et al. [15]. They replaced an *N*-terminal heptapeptide in EETI-II by an optimized binding loop of the turkey ovomucoid inhibitor (OMTKY3). The hybrid inhibitor HEI-TOE I retained inhibitory activity against porcine pancreatic elastase with considerably higher proteolytic stability. Inhibitors of this family were also successfully obtained using biological methods. Sato et al. [16] obtained recombinants of CMTI-II by expression of its gene in silkworms, whereas Grzesiak et al. [17] expressed series of CMTI-I analogues modified within a binding loop in *E. coli*. Since plant inhibitors play an important role in defense systems against insects or micro-organism predators, Milner et al. [18] fused CMTI-1 to the coat protein of potato potyvirus Y and bacterial β-glucuronidase in order to check whether such an insert could protect a protein against proteolysis. They showed that inhibitory activity was retained when the inhibitor was fused to the first protein only. All the above mentioned synthetic studies on squash inhibitors proved that their analogues with possible therapeutic applications can be accessible for both chemists and biochemists.

12.4.2 Cyclic Cystine Knot Inhibitors

It has already been mentioned that MCoTI-I/II are members of a family of miniproteins named cyclotides. Peptides displaying the CCK motive possess a diverse range of biological activities including antimicrobial, insecticidal and pesticidal activity. Cyclotides represent attractive scaffolds for the design of peptide-based drugs. This was the main reason that MCoTI-I/II became the most studied inhibitors of the squash family. Both were isolated from the seeds of the squash *Momordica cochinchinensis* in 2000, and in the following year the tertiary structure of MCoTI-II was determined. It was independently shown by two groups that its structure is similar to that of the other members of the squash family. Cysteine-rich proteins are attractive models to study oxidative folding pathways. Bearing in mind that the discussed inhibitors contain 6 Cys residues, 15 different sets of disulfide bridges can be formed. It turned out that the pathways of oxidation folding of the reduced precursors of cyclic and acyclic squash inhibitors (EETI-II and MCoTI-II were studied) are similar, yielding the native species. In both cases, only one major intermediate with two native disulfide bridges was present [19].

In contrast to other inhibitors of the discussed family, MCoTIs contain macrocycle formed by a polypeptide chain. Therefore, chemical synthesis of such inhibitors requires an additional step. One of the options is generation of a thioester at the *C*-terminus of the fully protected precursor followed by removal of the side chain protecting groups and a ring-closing macrolactam formation by native chemical ligation. Oxidative refolding of homodetic peptides gives the desired inhibitor [20]. MCoTIs were also biosynthesized in *E. coli* cells using a biomimetic approach that involves the use of a modified protein splicing unit in combination with an in-cell intramolecular native chemical ligation reaction [21]. Briefly, a MCoTI-I linear precursor was inserted into a modified intein yielding Met-Cys dipeptide at the *N*-terminus. After the enzymatic removal of Met and generation of thiester at the *C*-terminus of the MCoTI precursor, native chemical ligation was used to accomplish the main chain cyclization. In the final step, oxidative folding produced MCoTI. Finally, the inhibitor was purified on HPLC or using trypsin-immobilized Sepharose beads. A biochemical approach was applied to obtain the library of MCoTI-I analogues and affinity on trypsin-immobilized Sepharose beads was utilized to select active ones. Leatherbarrow's group [22] reported on modification (including deletion of fragment 3–6) within the binding loop of MCoTI-II that gave potent inhibitors of physiologically important tryptase and human leukocyte elastase (HLE). Substitution of Lys10 located in substrate specificity P_1 position by Arg increased inhibitory activity towards the first enzyme, whereas the introduction in this position of Val or Ala gave analogues possessing potent inhibitory activity against HLE. Moreover, considering the fact that a shortened analogue (28 amino acid residues) preserved tryptase inhibitory activity without significantly altering the three-dimensional structure, this proved the impressive adaptability of this cyclotide scaffold. These results correlate well with high proteolytic stability in the stomach and in plasma of cyclic hybrid of MCoTI-II(6-21) and EETI-II(16-28) named McoEeTI [23]. It was pointed out that this inhibitor can be considered a novel pharmacophoric carrier for oral peptide drug delivery.

Perhaps the most promising feature of MCoTIs is their membrane permeability. MCoTI-II biotin labeled at all three Lys residues was internalized into macrophages RAW264.7 and breast cancer MCF-7 cells at a non-toxic concentration [24]. This was the first reported cyclic cell-penetrating peptide. In this first report Craik's group has shown that translocation into cells is mediated through macropinocytosis. By contrast, another cyclotide—kalata B1 analogue (with Thr16 replaced by the biotin labeled Lys)—is not only significantly more toxic but is also not able to cross the cell membrane. This finding opens a new possibility to use MCoTI-II as a carrier of bioactive peptides to intracellular targets. The above mentioned internalization of MCoTI-II was analyzed in fixed cells. Recently, this process was studied on MCoTI-I in live HeLa cells by Camarero's group [25]. Since this inhibitor contains one Lys residue only (corresponding to inhibitor P_1 position), its acylation by AlexaFluor488 N-hydroxysuccinimide yielded a mono-labeled analogue. With the use of various endocytic markers and inhibitors, the authors proved the uptake of AF488-MCoTI-I in HeLa cells following multiple endocytic pathways. Moreover, a significant amount of MCoTI-I was found in lysosomes, and also in late endosomes or other types of acidic organelles. These results were confirmed very recently by Craik's group who re-evaluated their studies; this time on live cells [26]. They observed that Alexa-labeled MCoT-II and also kalata B1 analogue (Thr20 replaced by labeled Lys) were present inside macrophages and breast cancer cells. The first peptide penetrated cells mainly by macropinocytosis, whereas the second did through direct membrane interaction. Interestingly, uptake in MCF-7 cells of the mono-labeled analogue of MCoT-II (at Lys10 only with the remaining two replaced by Ala) was increased by 20 % as compared with the native sequence.

12.5 Sunflower Trypsin Inhibitor 1 (SFTI-1)

Since its discovery in 1999, trypsin inhibitor isolated from sunflower seeds (SFTI-1) has become the object of study for several groups. To date, SFTI-1 is the smallest naturally occurring circular proteinase inhibitor. SFTI-1 is a member of the Bowman-Birk group of inhibitors (I12) retaining the nonapepide binding loop of other inhibitors of this family, spanning the disulfide bridge. It consists of 14 amino acid residues and, in addition to the circular polypeptide chain, contains the second cycle—the disulfide bridge formed by Cys residues. The P_1-P_1' reactive site is located at Lys^5-Ser^6. Despite the fact that this is much smaller than other BBIs which are composed of 60–90 amino acids, SFTI-1 is the most potent trypsin inhibitor of this family. The elimination of one of the cycles retained inhibitory activity and the three dimensional structures of such analogues. Structure-activity-relationships studies on SFTI-1 have shown that substitution in the P_1 position caused dramatic changes in specificity, opening the possibility to design inhibitors of physiologically important proteinases. Several groups have shown that modifications introduced within the SFTI-1 reactive site and its vicinity led to potent physiologically important enzymes including: chymotrypsin, human neutrophil elastase, cathepsin G, matriptase, β-tryptase, proteinase K, and human kallikrein-related peptidase 4. Nevertheless, none of the analogues displaced inhibitory activity higher than the wild inhibitor. This might indicate that the structure of SFTI-1 is already optimized by nature. Most of the results obtained for SFTI-1 in the first decade of its study are already summarized in review papers [27, 28]; therefore, here we wish to discuss only those published recently. Pereira et al. [29] proposed the use of SFTI-1 analogues with different specificity immobilized on Sepharose as tools to identify proteinases in biological samples. Authors covalently attached SFTI-1 analogues through hydrazone bonds formed by the hydrazine group of Sepharose and the aldehyde moiety of a side chain of *N*-terminal amino acid of the inhibitor. The aldehyde group was introduced by the oxidation of the hydroxyl group of additional Ser coupled to *N*-termini of inhibitors. Such affinity resins display high capacity and stability under experimental conditions. An interesting issue discussed in recent papers was the cyclization mechanism of SFTI-1. Craik's group [30] determined that in the case of SFTI-1, the ratio between intact and modified inhibitors reached 9:1, regardless of whether or not intact or modified inhibitor was incubated with trypsin. More recently, the same group studied cyclization and degradation of series SFTI-1 analogues with cleaved peptides bonds at Arg2, Lys5 [31]. In the case of substrates, proteolysis of a single peptide bond usually gives two products, whereas in these cyclic inhibitors a single product is formed and re-synthesis of cleaved bonds is observed. Interestingly, when an analogue lacking a disulfide bridge was incubated with an enzyme, two peptide bonds were cleaved (after Arg2 and Lys5), tripeptide Arg-Abu-Lys was removed and peptide bond Arg^2-Ser^6 was formed yielding a 11 amino acid residue cyclic analogue. This corresponds with our results on proteolysis of double-sequence of SFTI-1 analogues. Following this mechanism, incubation of these analogues with trypsin or chymotrypsin yielded monocyclic SFTI-1 or [Phe^5]SFTI-1, depending on the amino acid residue present in P_1 position. We proposed to name this re-arrangement of the peptide chain peptide splicing [32].

Recently, Cascales et al. [26] reported that fluorescent Alexa-labeled SFTI-1 penetrates MCF-7 cells and is located in endosomal compartments. The cellular uptake of SFTI-1 is comparable with that determined for Alexa-labeled MCoTI-II, but in contrast to the latter inhibitor, no affinity of SFTI-1 to phospholipids was observed. Taking into consideration that all three cyclic peptides tested (MCoTI, kalata B1 and SFTI-1) enter cells, the authors proposed that they constitute a new family of cyclic cell-penetrating peptides. Moreover, in our recent work with the help of new fluorescent labels, we have shown that also monocyclic SFTI-1 analogues (with disulfide bridge only) are able to enter cancer cells. The work is in progress, and only preliminary results have been published [28].

12.6 Biosynthesis of Plant Cyclic Peptides

Cyclic peptides discussed herein were found in the plant kingdom. All are composed of proteinogenic amino acids. Their precursors are synthesized on ribosomes that undergo post-translational processing to generate cyclized inhibitors. It was observed that Asp or Asn is a conservative *C*-terminal residue present in cyclotides (including MCoTIs) and also in SFTI-1 precursors. This led to the suggestion that an asparaginyl endoprotease is involved in head-to tail-cyclization in plants. According to the proposed mechanism, in the first step a peptide bond after Asp (or Asn) is cleaved, followed by the formation of the peptide bond with *N*-terminal residue. This mechanism is also supported by experiments showing in vitro cyclization of linear precursors. Although other cyclization mechanisms are not excluded (e.g. the peptide splicing discussed above) the one mediated by asparaginyl endoprotease seems to be the most accepted [33].

12.7 Conclusions and Future Perspectives

Inhibitors of serine proteinase isolated from the seeds of *Cucurbitaceae* and sunflowers are a very exciting group of peptides. They are the smallest naturally occurring proteinase inhibitors composed of proteinogenic amino acids. They display well-defined, compact structures stabilized by disulfide bridges. Some of them are characterized by a head-to-tail cyclic backbone. These cyclic peptides are able to cross cell membranes. Unlike most peptides discussed here, low-molecular plant inhibitors are exceptionally resistant to denaturizing conditions and proteolysis. The observed rearrangement of the polypeptide chain (peptide splicing) upon incubation of some SFTI-1 analogues with serine proteinases is another feature worth considering for further studies. It might be possible to utilize such compounds for transfer and release inside the cell of certain peptidic or peptidomimetic drugs. All the information discussed above makes these plant serine proteinase inhibitors valuable scaffolds for designing compounds with potential commercial possibilities for medical and agrochemical applications. The size of these inhibitors makes them readily accessible by chemical synthesis. In contrast to biological methods, the possible modifications introduced to the chemically synthesized compounds are almost unlimited.

Acknowledgments This work was supported by the National Science Center under No. UMO-2011/01/B/ST5/03772.

References

1. Bode W, Huber R (1992) Natural protein proteinase inhibitors and their interaction with proteinases. Eur J Biochem 204:433–451
2. Laskowski M Jr (2000) Qasim MA, What can the structures of enzyme-inhibitor complexes tell us about the structures of enzyme substrate complexes? BBA 1477:324–337
3. Schechter I, Berger A (1967) On the size of the active site in proteases. I. Papain. Biochem Bophys Res Commun 27:157–162
4. Carrell RW, Travis J (1985) Trends α1-antitrypsin and the serpins: Variation and countervariation. Biochem Sci 10:20–24
5. Corral-Rodriguez MA, Macedo-Ribeiro S, Pereira PJB, Fuentes-Prior P (2010) Leech-derived thrombin inhibitors: from structures to mechanisms to clinical applications. J Med Chem 53:3847–3861
6. Rowlings ND, Tolle DP, Barret AJ (2004) Evolutionary families of peptidase inhibitors. Biochem J 378:705–716
7. Wieczorek M, Otlewski J, Cook J, Parks K, Leluk J, Wilimowska-Pelc A et al (1985) The squash family of serine proteinase inhibitors: amino acid sequences and association equilibrium constants of inhibitors from squash, summer squash, zucchini, and cucumber seeds. Biochem Biophys Res Co 126:646–652
8. Pallahgy PK, Norton RS, Nielsen KJ, Craik DJ (1994) A common structural motif incorporating a cystine knot and a triple-stranded beta-sheet in toxic and inhibitory polypeptides. Protein Sci 3:1833–1839
9. Hernandez J-F, Gagnon J, Chiche L, Nguyen TM, Andrieu J-P, Heitz A et al (2000) Squash trypsin inhibitors from *Momordica cochinchinensis* exhibit an atypical macrocyclic structure. Biochemistry 39:5722–5730
10. Ireland DC, Clark RJ, Daly NL, Craik DJ (2010) Isolation, sequencing, and structure-activity relationships of cyclotides. J Nat Prod 73:1610–1622
11. Kupryszewski G, Ragnarsson U, Rolka K, Wilusz T (1985) Synthesis of trypsin inhibitor CMTI III from squash seeds (*Cucurbita maxima*). Experientia 41:1422–1423
12. Favel A, Mattras H, Coletti-Previero MA, Zwilling R, Robinson EA, Castro B (1989) Protease inhibitors from *Ecballium elaterium* seeds. Int J Pept Protein Res 33:202–208
13. Walewska A, Jaśkiewicz A, Bulaj G, Rolka K (2011) Selenopeptide analogs of EETI-II retain potent trypsin inhibitory activities. Chem Biol Drug Des 77:93–97
14. Le-Nguyen D, Mattras H, Coletti-Previero MA, Castro B (1989) Design and chemical synthesis of a 32 residues chimeric microprotein inhibiting both trypsin and carboxypeptidase A. Biochem Biophys Res Commun 162:1425–1430
15. Hilpert K, Wessner H, Schneider-Mergener J, Welfle K, Misselwitz R, Welfle H et al (2003) Design and characterization of a hybrid miniprotein that specifically inhibits porcine pancreatic elastase. J Biol Chem 278:24986–24993
16. Sato S, Kamei K, Taniguchi M, Sato H, Takano R, Mori H et al (2000) Cloning and expression of the Momordica charantia trypsin inhibitor II gene in silkworm by using a baculovirus vector. Biosci Biotechnol Biochem 64:393–398
17. Grzesiak A, Buczek O, Petry I, Szewczuk Z, Otlewski J (2000) Inhibition of serine proteinases from human blood clotting system by squash inhibitor mutants. Biochim Biophys Acta 1478:318–324
18. Milner M, Chroboczek J, Zagorski-Ostoja W (2007) Engineered resistance against proteinases. Acta Biochim Pol 54:523–536
19. Cemazar M, Daly NL, Haggblad S, Lo LP, Yulyaningsih E, Craik DJ (2006) Knots in rings. The circular knotted protein Momordica cochinchinensis trypsin inhibitor-II folds via a stable two-disulfide intermediate. J Biol Chem 281:8224–8232
20. Thongyoo P, Tate EW, Leatherbarrow RJ (2006) Total synthesis of the macrocyclic cysteine knot microprotein MCoTI-II. Chem Commun 27:2848–2850

21. Austin J, Wang W, Puttamadappa S, Shekhtman A, Camarero JA (2009) Biosynthesis and biological screening of a genetically-encoded library based on the cyclotide MCoTI-I. ChemBioChem 10:663–670
22. Thongyoo P, Bonomelli C, Leatherbarrow RJ, Tate EW (2009) Potent Inhibitors of β-Tryptase and Human Leukocyte Elastase Based on the MCoTI-II Scaffold. J Med Chem 52:6197–00
23. Werle M, Kafedjiiski K, Kolmar H, Bernkop-Schnurch A (2007) Evaluation and improvement of the properties of the novel cystine-knot microprotein McoEeTI for oral administration. Int J Pharm 332:72–79
24. Greenwood KP, Daly NL, Brown DL, Stow JL, Craik DJ (2007) The cyclic cystine knot miniprotein MCoTI-II is internalized into cells by macropinocytosis. Int J Biochem Cell B 39:2252–2264
25. Contreras J, Elnagar AZO, Hamm-Alvarez SF, Camarero JA (2011) Cellular uptake of cyclotide MCoTI-I follows multiple endocytic pathways. J Control Relase 155:134–143
26. Cascales L, Henriques ST, Kerr MC, Huang Y-H, Sweet MJ, Daly NL et al (2011) Identification and characterization of a new family of cell-penetrating peptides. J Biol Chem 286:36932–36943
27. Korsinczky MLJ, Schirra HJ, Craik DJ (2004) Sunflower trypsin inhibitor-1. Curr Protein Pept Sci 5:351–364
28. Lesner A, Łęgowska A, Wysocka M, Rolka K (2011) Sunflower trypsin inhibitor 1 as a molecular scaffold for drug discovery. Curr Pharm Design 38:4308–4317
29. Pereira HJV, Salgado MCO, Oliveira OB (2009) Immobilized analogues of sunflower trypsin inhibitor-1 constitute a versatile group of affinity sorbents for selective isolation of serine proteases. J Chrom B 877:2039–2044
30. Marx UC, Korsinczky MJL, Schirra HJ, Jones A, Condie B, Otvos L et al (2003) Enzymatic cyclization of potent Bowman-Birk protease inhibitor, sunflower trypsin inhibitor-1, and solution structure of an acyclic precursor peptide. J Biol Chem 278:21782–21789
31. Colgrave ML, Korsinczky MJL, Clark RJ, Foley F, Craik DJ (2010) Sunflower trypsin inhibitor-1 proteolytic studies on a trypsin inhibitor peptide and its analogs. Biopolymers 94:665–672
32. Łęgowska A, Lesner A, Bulak E, Jaśkiewicz A, Sieradzan A, Cydzik M et al (2010) Inhibitory activity of double-sequence analogues of trypsin inhibitor SFTI-1 from sunflower seeds: an example of peptide splicing. FEBS J 277:2351–2359
33. Gillon AD, Saska I, Jennings CV, Guarino RF, Craik DJ, Anderson MA (2008) Biosynthesis of circular proteins in plants. Plant J 53:505–515

About Biography

Krzysztof Rolka achieved in 1980 his MS in Chemistry at the University of Gdansk, Poland, and completed his PhD at the same University, in 1985. From 1985 to 1986 he worked at the Institute of Molecular Biology and Biophysics, Swiss Federal Institute (ETH) Zurich, Switzerland as a postdoctoral researcher. He returned to Gdansk in 1986 and was appointed as Assistant Professor in the Bioorganic Chemistry Department at the University of Gdansk. In 1991 he received his DSc. From 1991 to 1992 he worked as Lecturer at the School of Pharmacy, University of London. In 1992 he was appointed as Associate Professor at the Faculty of Chemistry, University of Gdansk and in 1996 he was appointed as a full Professor. He is co-author over 110 original papers and more than 150 conference papers. His research centres on peptide chemistry, combinatorial chemistry, structure–activity-relationships of serine proteinase inhibitors and chromo- and fluorogenic substrates.

Adam Lesner achieved in 1994 his MS in Chemistry at University of Gdansk, Gdansk, Poland, and completed his PhD at the same University in 2001. From 2001 to 2004 worked in Virology Department of Columbia University NYC (USA) as a postdoctoral researcher. He returned to Gdansk in 2004 and was appointed as Assistant Professor in the Bioorganic Chemistry Department at the University of Gdansk. In 2009 he received his DSc. He is co-author of 40 original papers and more than 50 conference papers. His major interest centers on peptide synthesis, combinatorial chemistry and proteolysis, along with substrates and inhibitors of selected proteases. Right now he is appointed as Associate Professor at University of Gdansk.

Anna Łęgowska obtained her MS in 1984 at University of Gdansk, Gdansk, Poland, and completed her PhD at the same University, in 1993. From 1996 she has been working in the Bioorganic Chemistry Department at Faculty of Chemistry, University of Gdansk as Assistant Professor. In 1998 she spent 5 months in Menarini Ricerche S.P.A in Florence, Italy, where she completed postdoctoral fellowship. In 2011 she received her DSc. She is a co-author of 43 original papers and more than 60 conference papers. Her main interests are peptide and peptoid synthesis, combinatorial chemistry, inhibitors of serine proteases. Right now she is appointed as Associate Professor at University of Gdansk.

Magdalena Wysocka achieved her MS in 2001 at University of Gdansk, Gdansk, Poland, and completed her PhD at the same University, in 2008. Right now she is appointed as Assistant Professor at Bioorganic Chemistry Dept., Faculty of Chemistry, University of Gdansk. She is co-author of over 20 original papers and more than 40 conference papers. Her major interests are combinatorial chemistry, protease mapping, peptide synthesis, along with substrates and inhibitors of selected proteases.

Chapter 13
Anti-Tumor Effect of Synthetic Serine Protease Inhibitor

Kenei Furukawa, Tadashi Uwagawa and Katsuhiko Yanaga

Abstract Synthetic serine protease inhibitors, such as nafamostat mesilate, gabexate mesilate and ulinastatin inhibit various kinds of plasma proteinases and are widely used for treatment of pancreatitis, disseminated intravascular coagulation and as an anticoagulant for hemolysis. Several reports describe in preclinical cancer models that synthetic serine protease inhibitors induce apoptosis, prevent tumor invasion or metastasis, and sensitize chemotherapy by inhibiting NF-κB activity, proteinases such as urokinase-type plasminogen activator (u-PA) and matrix metalloproteinases (MMPs) and trypsin combined with protease-activated receptor-2 (PAR-2). Nafamostat mesilate, the first synthetic serine protease inhibitor that underwent clinical testing, has showed impressive anti-tumor effect and manageable toxicities in Phase I and II trials by combination chemotherapy of gemcitabine with nafamostat mesilate for unresectable pancreatic cancer. Synthetic serine protease inhibitors have minimal adverse effects and will have a potential to become a new therapeutic option for cancer patients. Below we discuss the rationale behind targeting the serine protease inhibitor for cancer therapy, and review the preclinical and clinical data.

13.1 Introduction

13.1.1 Serine Protease Inhibitor

Serine protease inhibitors inhibit serine proteases which are enzymes that cleave peptide bonds in proteins, in which serine serves as the nucleophilic amino acid at the active site [1] by acting as 'suicide substrates' and inactivating serine proteases through the formation of a 1:1 complex [2]. Serine protease inhibitors can

---

K. Furukawa (✉) · T. Uwagawa · K. Yanaga
Department of Surgery, Jikei University School of Medicine, 3-25-8,
Nishi-Shinbashi, Minato-ku, Tokyo 105-8461, Japan
e-mail: k-furukawa@jikei.ac.jp

E. F. Fang and T. B. Ng (eds.), *Antitumor Potential and Other Emerging Medicinal Properties of Natural Compounds*, DOI: 10.1007/978-94-007-6214-5_13,

be categorized into physiologic inhibitors that are naturally present in tissues and nonphysiologic inhibitors that are either produced by microorganisms or synthesized chemically. Physiologic serine protease inhibitors belong to the superfamily of serpins, which include, for example, antithrombin III and plasminogen activator inhibitor [3]. Nonphysiologic synthetic serine protease inhibitors include nafamostat mesilate, gabexate mesilate and ulinastatin.

13.1.2 Synthetic Serine Protease Inhibitors

Nafamostat mesilate [4] is widely used for treatment of pancreatitis [5], disseminated intravascular coagulation [6], and anticoagulation in hemodialysis [7] in Japan, and has only minimal adverse effects such as hyperkalemia or hyponatremia [8]. Gabexate mesilate [9] has been used for disseminated intravascular coagulation and acute pancreatitis in the clinical field in Japan and Korea [10]. Ulinastatin has been used in treatment of inflammatory diseases, including disseminated intravascular coagulation, shock and pancreatitis [11].

13.2 Molecular Targets of Serine Protease Inhibitors

Tumor cell proliferation, invasion, metastasis and chemoresistance are regarded as multistep phenomena involving many pathways. There are several important molecular targets for anti-tumor effect regulated by serine protease inhibitors (Table 13.1).

Table 13.1 A list of synthetic serine protease inhibitors, mechanism of action and ongoing clinical trials

Drug	Mechanism	Clinical trials
NM	NF-κB inhibitor for pancreatic and gastric cancer Tumor-associated trypsinogen inhibitor into PAR-2 for pancreatic cancer	Phase 1 and 2 clinical trial in combination chemotherapy of gemcitabine with nafamostat mesilate for unresectable pancreatic cancer at the Jikei University School of Medicine
GM	NF-κB inhibitor for pancreatic cancer u-PA and MMPs inhibitor for pancreatic and colon cancer Tumor-associated trypsinogen inhibitor into PAR-2 for cholangiocarcinoma	None ongoing
Ulinastatin	NF-κB inhibitor for breast cancer	None ongoing

NM Nafamostat mesilate; *GM* Gabexate mesilate; *PAR-2* protease-activated receptor-2; *u-PA* urokinase-type plasminogen activator; *MMPs* matrix metalloproteinases

13.2.1 NF-κB

NF-κB transcription factor family plays an important role in the regulation of cell apoptosis, inflammation, and oncogenesis [12, 13]. Inhibition of NF-κB is considered as one of new treatment strategies for cancer patients [14, 15]. We have reported that nafamostat mesilate inhibits activation of NF-κB and induces apoptosis, prevents adhesion, invasion and peritoneal dissemination for pancreatic cancer in vitro and in vivo [16–18]. We also have reported in an experimental study that better outcome of combination chemotherapy can be achieved with gemcitabine and nafamostat mesilate for pancreatic cancer and paclitaxel and nafamostat mesilate for gastric cancer in comparison with mono-drug therapy by inhibition of NF-κB activation induced by gemcitabine or paclitaxel [19, 20]. Also, we applied nafamostat mesilate, an NF-κB inhibitor, to clinical trials [21, 22] and gene therapy [23].

NF-κB is typically a heterodimer that consists of the p65 (RelA) and p50 proteins. In an inactive form of NF-κB, proteins are sequestered in the cytoplasm with IκBα. Following IκBα phosphorylation by IκB kinase complex (IKK), NF-κB proteins are released and translocated into the nucleus, where they activate transcription of target genes involved in inflammatory cytokines, adhesion molecules and anti-apoptotic proteins. We have demonstrated that mechanism of nafamostat mesilate-mediated NF-κB inhibition is blocking IKK-mediated phosphorylation of IκBα [16].

In another study, Takahashi et al. reported that gabexate mesilate inhibited TNF-α-induced NF-κB activation and enhanced apoptosis in human pancreatic cancer cell lines [24]. Also, Wang et al. reported that ulinastatin inhibited expression of NF-κB in breast carcinoma cells [25]. However, it is unclear how gabexate mesilate and ulinastatin inhibit activation and expression of NF-κB in cancer cells.

13.2.2 Proteinases

Proteolytic degradation of extracellular matrix (ECM) components is a process essential to tumor invasion and metastasis. Many proteinases are capable of degrading ECM components of which two families of enzymes appear to be particularly important for ECM degradation, namely, the urokinase-type plasminogen activator (u-PA) and the matrix metalloproteinases (MMPs) [26]. Inhibitors of u-PA and MMPs can block these enzymes and subsequently inhibit invasion and metastasis [27, 28].

The u-PA is a serine protease that converts the zymogen plasminogen to active plasmin. Of the fibrinolytic components, thrombin, a serine protease, has been identified as an activator of pro-MMP-2 via a membrane-type 1-MMP-independent pathway [29]. Moreover, plasmin, a serine protease, converts pro-MMP-3 to activate MMP-3, which becomes a potent activator of MMP-9 [30]. Therefore, activation pathways of some MMPs are closely related to serine proteases.

Uchima et al. reported that gabexate mesilate prevented the invasive and liver metastatic potential of pancreatic cancer cells by antagonizing the activities of u-PA and MMPs [31]. Wan et al. reported inhibition of MMPs gabexate mesilate reduced the invasion and metastasis of colon cancer cells [32].

Also, we showed that nafamostat mesilate reduced peritoneal metastasis in pancreatic cancer by inhibiting NF-κB activation, resulting in reduced expression of MMPs because MMPs are downstream target genes of NF-κB [18].

13.2.3 Trypsin and PAR-2

Trypsin, the activated form of trypsinogen, is a digestive enzyme produced by pancreatic acinar cells, and is one of serine proteases. This enzyme is expressed in a variety of cancer cell lines and cancer tissues [33]. Also, trypsin activates protease-activated receptor-2 (PAR-2), as one of the G protein-coupled receptors [34]. PAR-2 activation induces cell proliferation [35], and PAR-2 expression has been found in human solids tumors, such as gastric, colon and pancreatic cancer [35–37]. Therefore, several studies have suggested that trypsin plays an important role in proliferation and serves as a potent mitogen in PAR-2-positive cancer cells [37, 38]. Outa et al. reported that nafamostat mesilate inhibited growth and invasion of pancreatic cancer cells by blocking tumor-associated trypsinogen combined with PAR-2 [37]. Also, Nakanuma et al. reported that gabexate mesilate inhibited cell growth by inhibition of trypsin in PAR-2-positive intrahepatic cholangiocarcinoma cells [38].

13.3 Clinical Data

Although several reports have showed anti-tumor effect of serine protease inhibitors in a variety of cancers, clinical trials are limited to ours [21, 22]. We applied our strategy to the clinical trials by combination chemotherapy of gemcitabine with nafamostat mesilate for unresectable pancreatic cancer In Phase I/II study. Our previous data showed that a dose of 1×10^{-6} M of nafamostat mesilate is necessary to inhibit apoptosis of pancreatic cancer cells [26]. As the acceptable maximum clinical dose of nafamostat mesilate could reach 1.8×10^{-7} M (under the condition of continuous infusion), this concentration of nafamostat mesilate is insufficient to inhibit NF-κB activation. Therefore, we started this clinical trials of i.v. gemcitabine combined with regional arterial infusion of nafamostat mesilate because continuous regional arterial infusion to achieve adequate concentration of nafamostat mesilate in the pancreas as compared with i.v. infusion. The eligibility of this study is patients diagnosed as unresectable pancreatic cancer, aged between 20 and 80 years and exhibiting adequate bone marrow, renal and hepatic functions. The treatment schedule of this study is gemcitabine with intravenous administration at a fixed dose of 1,000 mg/m^2 for 30 min, and nafamostat mesilate was administered one hour before the infusion of gemcitabine via a port-catheter system for 24 h through the common hepatic artery and celiac artery on days 1, 8 and 15 of each 28-day cycle. This regimen was repeated at 28-day intervals. To determine the dose-limiting toxic effects (DLTs), a standard '3 + 3' phase I dose-escalation protocol was used. The initial dose of nafamostat mesilate was 2.4 mg/kg and was escalated in increments of 1.2 mg/kg until a dose of 4.8 mg/kg was reached. The phase I study showed that none of the patients experienced DLTs, and nafamostat mesilate was well tolerated at doses up to 4.8 mg/kg [21]. Therefore, the recommended phase I dose of nafamostat mesilate in combination with gemcitabine was 4.8 mg/kg. The phase II included 35 patients and the median survival time was 10.0 months, while the response rate and disease control rate were 17.1 % and 88.6 %, respectively [22].

13.4 Conclusions

Synthetic serine protease inhibitors have anti-tumor effect in a broad spectrum of tumor cell lines, and in vivo xenograft models. As to the mechanism of their anti-tumor effects, NF-κB activity, proteinases such as u-PA, MMPs and trypsin, and PAR-2 likely play important roles, but other factors seem to be involved. Since synthetic serine protease inhibitors are used clinically with minimal adverse effects in patients with poor performance status and organ dysfunction due to pancreatitis, disseminated intravascular coagulation or chronic renal failure. Synthetic serine protease inhibitors have a potential to become a new therapeutic option to augment the effect of chemotherapeutic agents for cancer patients.

References

1. Hedstrom L (2002) Serine protease mechanism and specificity. Chem Rev 102:4501–4524
2. Silverman GA, Bird PI, Carrell RW, Church FC, Coughlin PB, Gettins PG, Irving JA, Lomas DA, Luke CJ, Moyer RW, Pemberton PA, Remold-O'Donnell E, Salvesen GS, Travis J, Whisstock JC (2001) The serpins are an expanding superfamily of structurally similar but functionally diverse proteins. Evolution, mechanism of inhibition, novel functions, and a revised nomenclature. J Biol Chem 276:33293–33296
3. DeClerck YA, Imren S (1994) Protease inhibitors: role and potential therapeutic use in human cancer. Eur J Cancer 30A:2170–2180
4. Fujii S, Hitomi Y (1981) New synthetic inhibitors of C1r, C1 esterase, thrombin, kallikrein and trypsin. Biochim Biophys Acta 661:342–345
5. Iwaki M, Ino Y, Motoyoshi A, Ozeki M, Sato T, Kurumi M, Aoyama T. (1986) Pharmacological studies of FUT-175, nafamostat mesilate. V. Effects on the pancreatic enzymes and experimental acute pancreatitis in rats. Jpn J Phamacol 41: 155–162
6. Takahashi H, Takizawa S, Tatewaki W, Nagai K, Wada K, Hanano M, Shibata A (1989) Nafamostat mesilate (FUT-175) in the treatment of patients with disseminated intravascular coagulations. Thomb Haemost 62:372
7. Ohtake Y, Hirasawa H, Sugai T, Oda S, Shiga H, Matsuda K, Kitamura N (1991) Nafamostat mesilate as anticoagulant in continuous hemofiltration and continuous hemodiafiltration. Contr Nephrol 93:215–217

8. Kitagawa H, Chang H, Fujita T (1995) Hyperkalemia Due to Nafamostat Mesylate. N Engl J Med 332:687
9. Ohno H, Kosaki G, Kambayashi J, Imaoka S, Hirata F (1980) FOY: [ethyl P-(6-guanidinohexanoyloxy) benzoate] methanesulfonate as a serine proteinase inhibitor. I. Inhibition of thrombin and factor Xa in vitro. Thromb Res 19:579–588
10. Chen HM, Chen JC, Hwang TL, Jan YY, Chen MF (2000) Prospective and randomized study of gabexate mesilate for the treatment of severe acute pancreatitis with organ dysfunction. Hepatogastroenterology 47:1147–1150
11. Inoue K, Takano H (2010) Urinary trypsin inhibitor as a therapeutic option for endotoxin-related inflammatory disorders. Expert Opin Investig Drugs 19:513–520
12. Amer BA, David B (1996) An essential role for NF-κB in preventing TNF-alpha-induced cell death. Science 274:782–784
13. Karin M, Lin A (2002) NF-κB at the crossroads of life and death. Nature Immunol 3:221–227
14. Matsumoto G, Namekawa J, Muta M, Nakamura T, Bando H, Tohyama K, Toi M, Umezawa K (2005) Targeting of nuclear factor κB pathways by Dehydroxymethyleoxyquinomicin, a novel inhibitor of breast carcinomas: antitumor and antiangiogenic potential in vivo. Clin Cancer Res 11:1287–1293
15. Huang S, Pettaway CA, Uehara H, Bucana CD, Fidler IJ (2001) Blockade of NF-κB activity in human prostate cancer cells is associated with suppression of angiogenesis, invasion, and metastasis. Oncogene 20:4188–4197
16. Uwagawa T, Li Z (2007) Chang Zhe, Xia Q, Peng B, Sclabas GM, Ishiyama S, Hung MC, Evans DB, Abbruzzese JL, Chiano PJ. Mechanisms of synthetic serine protease inhibitor (FUT-175)-mediated cell death. Cancer 109:2142–2153
17. Furukawa K, Iida T, Shiba H, Fujiwara Y, Uwagawa T, Shimada Y, Misawa T, Ohashi T, Yanaga K (2010) Anti-tumor effect by inhibition of NF-κB activation using nafamostat mesilate for pancreatic cancer in a mouse model. Oncol Rep 24:843–850
18. Fujiwara Y, Furukawa K, Haruki K, Shimada Y, Iida T, Shiba H, Uwagawa T, Ohashi T, Yanaga K (2011) Nafamostat mesilate can prevent adhesion, invasion and peritoneal dissemination of pancreatic cancer thorough nuclear factor kappa-B inhibition. J Hepatobiliary Pancreat Sci 18:731–739
19. Uwagawa T, Chiano PJ, Gocyo T, Hirohara S, Misawa T, Yanaga K (2009) Combination chemotherapy of nafamostat mesilate with gemcitabine for pancreatic cancer targeting NF-κB activation. Anticancer Res 29:3173–3178
20. Haruki K, Shiba H, Fujiwara Y, Furukawa K, Iwase R, Uwagawa T, Misawa T, Ohashi T, Yanaga K (2012) Inhibition of nuclear factor-κB enhances the antitumor effect of paclitaxel against gastric cancer with peritoneal dissemination in mice. Dig Dis Sci [Epub ahead of print]
21. Uwagawa T, Misawa T, Sakamoto T, Ito R, Gocho T, Shiba H, Wakiyama S, Hirohara S, Sadaoka S, Yanaga K (2009) A phase I study of full-dose gemcitabine and regional arterial infusion of nafamostat mesilate for advanced pancreatic cancer. Ann Oncol 20:239–243
22. Uwagawa T, Misawa T, Tsutsui N, Ito R, Gocho T, Hirohara S, Sadaoka S, Yanaga K (2011) Phase II study of gemcitabine in combination with regional arterial infusion of nafamostat mesilate for advanced pancreatic cancer. Am J Clin Oncol [Epub ahead of print]
23. Furukawa K, Ohashi T, Haruki K, Fujiwara Y, Iida T, Shiba H, Uwagawa T, Kobayashi H, Yanaga K (2011) Combination treatment using adenovirus vector-mediated tumor necrosis factor-alpha gene transfer and a NF-κB inhibitor for pancreatic cancer in mice. Cancer Lett 306:92–98
24. Takahashi H, Funahashi H, Sawai H, Matsuo Y, Yamamoto M, Okada Y, Takeyama H, Manabe T (2007) Synthetic serine protease inhibitor, gabexate mesilate, prevents nuclear factor-kappaB activation and increases TNF-alpha-mediated apoptosis in human pancreatic cancer cells. Dig Dis Sci 52:2646–2652

25. Wang H, Sun X, Gao F, Zhong B, Zhang YH, Sun Z (2012) Effect of ulinastatin on growth inhibition, apoptosis of breast carcinoma cells is related to a decrease in signal conduction of JNk-2 and NF-κB. J Exp Clin Cancer Res 31:2
26. Liotta LA, Stetler-Stevenson WG (1991) Tumor invasion and metastasis: an imbalance of positive and negative regulation. Cancer Res 15:5054s–5059s
27. Naito K, Kanbayashi N, Nakajima S, Murai T, Arakawa K, Nishimura S, Okuyama A (1994) Inhibition of growth of human tumor cells in nude mice by a metalloproteinase inhibitor. Int J Cancer 58:730–735
28. Reich R, Thompson EW, Iwamoto Y, Martin GR, Deason JR, Fuller GC, Miskin R (1988) Effects of inhibitors of plasminogen activator, serine proteinases, and collagenase IV on the invasion of basement membranes by metastatic cells. Cancer Res 48:3307–3312
29. Nguyen M, Arkell J, Jackson CJ (1999) Thrombin rapidly and efficiently activates gelatinase A in human microvascular endothelial cells via a mechanism independent of active MT1 matrix metalloproteinase. Lab Invest 79:467–475
30. Ramos-DeSimone N, Hahn-Dantona E, Sipley J, Nagase H, French DL, Quigley JP (1999) Activation of matrix metalloproteinase-9 (MMP-9) via a converging plasmin/stromelysin-1 cascade enhances tumor cell invasion. J Biol Chem 274:13066–13076
31. Uchima Y, Sawada T, Nishihara T, Maeda K, Ohira M, Hirakawa K (2004) Inhibition and mechanism of a protease inhibitor in human pancreatic cancer cells. Pancreas 29:123–131
32. Yoon WH, Jung YJ, Kim TD, Li G, Park BJ, Kim JY, Lee YC, Kim JM, Park JI, Park HD, No ZS, Lim K, Hwang SD, Kim YS (2004) Gabexate mesilate inhibits colon cancer growth, invasion, and metastasis by reducing matrix matalloproteinases and angiogenesis. Clin Cancer Res 10:4517–4526
33. Koivunen E, Saksela O, Itkonen O, Osman S, Huhtala ML, Stenman UH (1991) Human colon carcinoma, fibrosarcoma and leukemia cell lines produce tumor-associated trypsinogen. Int J Cancer 47:592–596
34. Nystedt S, Emilsson K, Larsson AK, Strömbeck B, Sundelin J (1995) Molecular cloning and functional expression of the gene encoding the human proteinase-activated receptor 2. Eur J Biochem 232:84–89
35. Darmoul D, Gratio V, Devaud H, Laburthe M (2004) Protease-activated receptor 2 in colon cancer: trypsin-induced MAPK phosphorylation and cell proliferation are mediated by epidermal growth factor receptor transactivation. J Biol Chem 279:20927–22094
36. Miyata S, Koshikawa N, Yasumitsu H, Miyazaki K (2000) Trypsin stimulates integrin alpha(5)beta(1)-dependent adhesion to fibronectin and proliferation of human gastric carcinoma cells through activation of proteinase-activated receptor-2. J Biol Chem 275:4592–4598
37. Ohta T, Shimizu K, Yi S, Takamura H, Amaya K, Kitagawa H, Kayahara M, Ninomiya I, Fushida S, Fujimura T, Nishimura G, Miwa K (2003) Protease-activated receptor-2 expression and the role of trypsin in cell proliferation in human pancreatic cancers. Int J Oncol 23:61–66
38. Nakanuma S, Tajima H, Okamoto K, Hayashi H, Nakagawara H, Onishi I, Takamura H, Kitagawa H, Fushida S, Tani T, Fujimura T, Kayahara M, Ohta T, Wakayama T, Iseki S, Harada S (2010) Tumor-derived trypsin enhances proliferation of intrahepatic cholangiocarcinoma cells by activating protease-activated receptor-2. Int J Oncol 36:793–800

Author Biography

Kenei Furukawa

2006 Jikei University School of Medicine, Tokyo, Japan, MD, Medicine

2008 Department of Surgery, Jikei University School of Medicine, Tokyo, Japan

2011 ~ present Department of Surgery, Jikei University School of Medicine, Tokyo, Japan, MD PhD

Tadashi Uwagawa

1990 Yamaguchi University, Yamaguchi, Japan, MD, Medicine
1992 Department of Surgery, Jikei University School of Medicine, Tokyo, Japan
2001 Department of Surgery, Jikei University School of Medicine, Tokyo, Japan, MD PhD
2003 Research fellowship, Surgical Oncology, University of Texas, MD Anderson Cancer Center, Houston, USA
2011 ~ present Associate professor, Department of Surgery, Jikei University school of Medicine, Tokyo, Japan

Katsuhiko Yanaga

1979 Kyushu University, Fukuoka, Japan, MD, Medicine
1988 Assistant Professor, Department of Surgery, University of Pittsburgh, PA, USA
1989 Assistant Professor, Department of Surgery II, Kyushu University, Fukuoka, Japan
1991 Kyushu University, Fukuoka, Japan, MD PhD
1997 Associate Professor, Department of Surgery II, Kyushu University, Fukuoka, Japan
2000 Associate Professor, Department of Surgery II, Nagasaki University, Nagasaki, Japan
2003 ~ present Professor, Department of Surgery, Jikei University school of Medicine, Tokyo, Japan

Chapter 14
A Landscape of the Health Benefits of Different Natural Protease Inhibitors

Ho Him Leung, Evandro Fei Fang and Tzi Bun Ng

Abstract Protease inhibitors are important since the deregulation of proteolytic activities is tightly linked to different pathological conditions including severe inflammatory disorders and cancer. There are around 85 families of protease inhibitors according to the MEROPS database. In this article we have reviewed the medicinal applications of different types of protease inhibitors, in particular the serine protease inhibitors (exemplified as Kunitz type and Bowman-Birk type inhibitors) which are the most extensively studied. Although the main protease inhibitors cited here are of plant and animal origin, antibody and engineered protease inhibitors are also discussed, albeit lesser extensively. By binding to serine proteases or other targets, many protease inhibitors display a vast range of properties that are beneficial for humans. These include anti-fungal, anti-apoptotic, anti-inflammatory, anti-metastatic, anti-proliferative, and anti-HIV activities. The multipotency of such inhibitors has made them strong candidates for treating diseases and disorders. The current limitation mainly lies on whether these properties could be transferred from bench-top level to clinical trials and eventually employed as substitutes for existing drugs.

14.1 Introduction

Protease inhibitors are molecules that inhibit the activity of proteases. They are mainly found in plants such as legumes, potato and soybean, but are also found in animals and humans [1, 2]. Protease inhibitors display several characteristics:

H. H. Leung
Department of Medicine, Imperial College London, London, UK

E. F. Fang (✉)
Laboratory of Molecular Gerontology, National Institute on Aging, National Institutes of Health, Baltimore, MD 21224, USA
e-mail: fange@mail.nih.gov; evandro.fang@nih.gov

E. F. Fang · T. B. Ng (✉)
Faculty of Medicine, School of Biomedical Sciences, The Chinese University of Hong Kong, Hong Kong, China
e-mail: b021770@mailserv.cuhk.edu.hk

E. F. Fang and T. B. Ng (eds.), *Antitumor Potential and Other Emerging Medicinal Properties of Natural Compounds*, DOI: 10.1007/978-94-007-6214-5_14,

They are small in size, and typically utilize a large number of disulphide bonds to stabilize the inhibitory domains. In view of the significant relationship between different human diseases and the imbalance of proteases and protease inhibitors, much research has recently focused on the screening and development of therapeutic protease inhibitors from both natural and laboratory setting.

14.2 Classification and Modes of Inhibition of Protease Inhibitors

There are different methods of categorizing plant protease inhibitors. One of these is to classify them depending on the class of protease they inhibit, which include serine proteases, cysteine proteases, aspartic proteases and metallocarboxy-proteases [3–5]. Serine protease inhibitors are subdivided into canonical (standard mechanism) inhibitors, serpins and non-canonical inhibitors, on the basis of their inhibition mechanisms. Canonical inhibitors contain a rigid reactive-loop complementary to the substrate binding site of the protease. They are considered good inhibitors due to their low K_m and K_{cat} values, representing a tight binding and slow rate of hydrolysis. The tight binding is achieved by maintaining a stable conformation of the reactive-site loop, facilitated by other secondary contacts between the inhibitor and the protease, whereas the slow hydrolysis rate is accomplished by hydrogen bonding and disulphide bonds which stabilize the complex and increase the activation energy barrier. Different from canonical inhibitors, non-canonical inhibitors such as potato carboxypeptidase inhibitors (CPI) and cystatin are less common and use a less common mechanism to inhibit protease [3]. In CPI, the rigid reactive-site loop is only present in the bound form, and it forms more stable interaction with the protease than the enzyme to prevent dissociation. Cystatin, on the other hand, does not contain any sites for the cysteine protease to cleave, and thus blocks the active site of the enzyme [3]. Please see Chap. 12 by Prof. Krzysztof Rolka for more information on classification and inhibition mechanism of serine protease inhibitors.

Among all protease inhibitors, Bowman-Birk and Kunitz inhibitors are the most well-studied group, and both belong to canonical serine protease inhibitors. Kunitz inhibitors are approximately 200 amino acid residues in size, and consist of 12 anti-parallel β-strands, linked by 13 loops [3]. The reactive-site binding loop is generally located at loop 5 but can be found in different locations, such as in loop 6 and loop 10 for *Sagittaria sagittifolia* [3]. The overall structure of Kunitz inhibitors is stabilized by disulphide bonds and hydrogen bonds. The number of disulphide bonds varies between species, but the most frequently observed disulphide bonds are found to bridge loop 3 and loop 6, and loop 10 to β-strand 9. Kunitz inhibitors can be further subdivided into three groups based on the number of disulphide bonds they contain. Those with more than 4 cysteine residues, forming 2 intra or intermolecular bondings, are grouped as group 1, Examples include *Acacia confusa, Pithecellobium dulce* and *Acacia plumosa,* which contain

1 intrachain disulphide bond within the α-helix chain and an interchain disulphide bond with the β-chain, *or* DMTI-I and DMTI-II from *Dimorphandra mollis*, which only contain 2 intrachain disulphide bonds. Group 2 inhibitors are those containing only 2–3 cysteine residues, and are only able to form one disulphide bond, such as *Entada Scandens* and *Inga laurina* trypsin inhibitors. Inhibitors with one or no cysteine residue, and therefore unable to form disulphide bridges, are classified as group 3, an example being *Bauhinia bauhiniodes* Kallikrein inhibitor [6, 7].

Bowman-Birk inhibitors (BBI) vary in size between species, containing 33 and 133 amino acid residues. They are commonly found in monocotyledonous grass species and dicotyledonous legume species (including soybean) [8]. BBI found in grass family or legume species are slightly different. The former can be subdivided into two groups, depending on the number of active sites. The latter are small in size (6–9 kDa), and consist of two reactive loops which are able to inhibit two proteases independently [8]. The loops are highly conserved in sequence, which allows them to inhibit specific proteases. For example, loop 1 provides specificity for serine proteases, whereas loop 2 is specific towards trypsin- or chymotrypsin-like proteases. Similar to Kunitz inhibitors, the seven disulphide bonds of soybean BBI (C9–C24, C14–C22, C8–C62, C12–C58, C36–C51, C32–C39, C41–C49) are fundamental in providing structural stability and tolerance towards extreme conditions. From the X-ray crystal structure, hydrophobic patches are observed, which allow BBI to self-associate to form a dimer, trimer, tetramer and hexamer as well as monomer [8, 9].

As previously mentioned, many other protease inhibitors exist in the nature, such as Potato I and II inhibitors, sunflower trypsin inhibitor, and inhibitors from the cereal and rapseed families. Their three-dimensional structures have been determined, along with many others, including a novel Kazel-type trypsin inhibitor from *Agalychnis callidryas* [1].

14.3 Medicinal Applications of Protease Inhibitors of Plant Origins

14.3.1 The Health Benefits of Kunitz-Type Protease Inhibitors

The Kunitz-type protease inhibitors have continued to be exploited because they elicit many responses that are beneficial to mankind, e.g. for combatting inflammation and tumour progression, in addition to their natural physiological functions. Bovine pancreatic trypsin inhibitor (BPTI) is one of the most extensively studied animal protease inhibitors. It is a monomeric protein with 58 amino acids connected by 3 disulphide bonds and containing one Kunitz-type protease inhibitory domain, is able to form a more stable and higher affinity complex with trypsin and different serine proteases in close proximity, which allows assays to be conducted under extreme conditions (Fig. 14.1) [10, 11]. This BPTI-trypsin detection system is found useful in several aspects, for example, in detection of

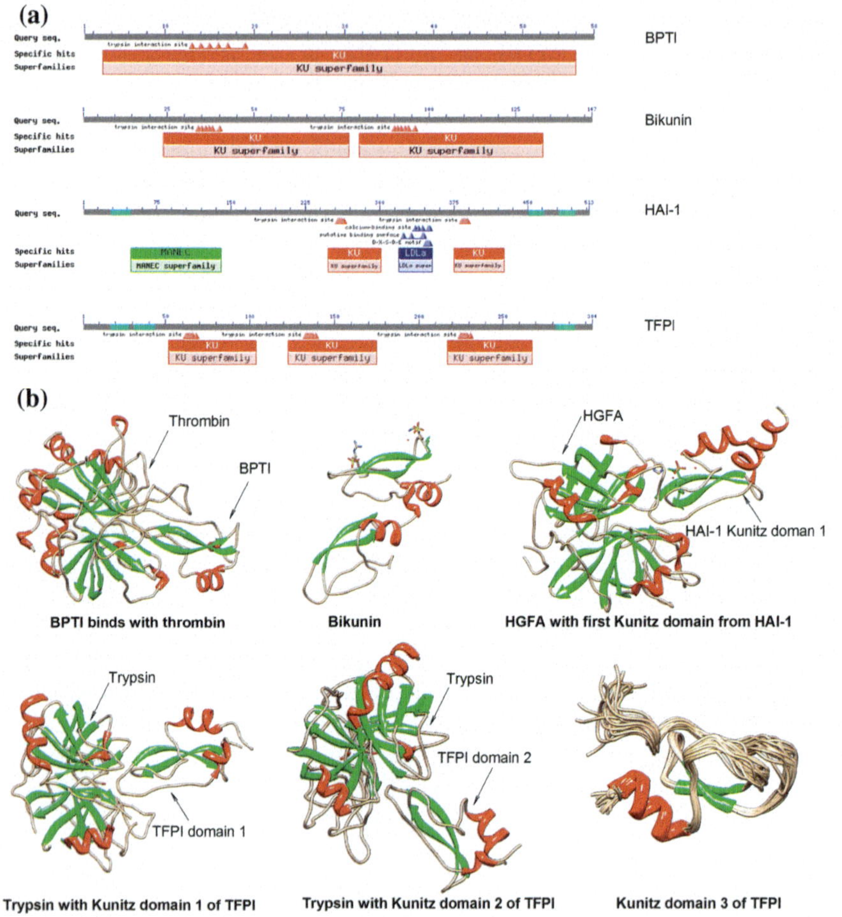

Fig. 14.1 Some representative Kunitz-type trypsin inhibitors. **a** Different Kunitz domains among Kunitz type protease inhibitors. Abbreviations: Bovine pancreatic trypsin inhibitor (*BPTI*), Hepatocyte growth factor activator (*HGFA*) inhibitor 1B (*HAI-1*), Tissue factor pathway inhibitor (*TFPI*). Identification of conserved domains was executed in the NCBI website. **b** Ribbon diagram of the crystal structures of those Kunitz-type trypsin inhibitors. Crystal data were from RCSB protein data bank and visualized by the UCSF chimera software as used previously [34]: BPTI (PDB ID: 1BTH), bikunin (1BIK), HAI-1 (1YC0), TFPI Kunitz domain 1 (1ZR0), TFPI Kunitz domain 2 (1TFX), TFPI Kunitz domain 3 (1IRH). The helix and strand are shown in *red* and *green*, respectively

receptor targets of trypsin, in affinity chromatography and in identifying protein topography [10]. More recently, anti-inflammatory properties of BPTI have also been discovered. When activated by trypsin, protease-activated receptor 1 (PAR1) produces proinflammatory cytokines and chemokines as well as platelet-derived growth factor (PDGF). As BPTI binds and inhibits trypsin, PAR1 would subsequently be inhibited [12]. Although the use of BPTI to reduce the inflammatory response has undergone clinical trials and has been proven to be effective and safe, concerns have arisen over its association with the development of renal failure and anaphylaxis after repeated administration.

In contrast to BPTI, there are many other mammalian protease inhibitors that contain more than one Kunitz-type protease inhibitory domain. Bikunin (bis Kunitz inhibitor) contains two Kunitz-type protease inhibitory domains (Fig. 14.1), and is released from the precursor Inter-alpha Inhibitor (IaI) upon substitution of hyaluronan (HA) with chondroitin sulfate to form a complex with the heavy chain of IaI [13]. The Kunitz 1 domain located closer to the N-terminus is required to bind bikunin receptors or the link protein, whereas Kunitz 2 domain is a protease inhibitor domain responsible for binding proteases. In humans, bikunin is found predominantly in amniotic fluid, as well as in serum and urine [13]. The physiological function is not limited to inhibiting serine proteases, as reports have shown that bikunin, together with anti-cancer drugs, suppresses metastasis by inhibiting expression of urokinase-type plasminogen activator (uPA), achieved by negatively regulating the ERK1/2 phosphorylation or MAP kinase pathway upstream. Suzuki and his colleagues suggested this inhibition of metastasis is due to prevention of oligomerization of CD44, a key player in tumour progression. They proposed that when the bikunin binds to its receptor, it physically interacts with CD44 and prevents dimerization of CD44. As a result the monomeric CD44 cannot form a complex with hyaluronan (HA), and therefore is not recognized by the cells [13]. In view of this, researchers hoped to apply this for anti-metastatic therapy, and the experiments have been performed on rodents which showed significant inhibition of metastasis two weeks after inoculation of tumor cells. Clinical trials have also been conducted, where a dose of 150 mg/day bikunin was used in addition to chemotherapy, and the results showed that it prolonged the survival of patients with advanced ovarian cancer [13].

Both hepatocyte growth factor (HGF) activator inhibitors (HAI) 1 and 2 contain two Kunitz-type protease inhibitor domains (Fig. 14.1), with the first domain inhibiting HGF and the second domain inhibiting serine proteases. Tissue factor pathway inhibitor (TFPI) is another example of a serine protease inhibitor. It contains three Kunitz-domains (Fig. 14.1). All of the Kunitz-domain containing protease inhibitors manifest anti-inflammatory activity. And in some cases, tumor progression is suppressed and other diseases are prevented.

In addition to Kunitz protease inhibitors from mammals, there are numerous medicinal Kunitz protease inhibitors purified from plants. Soybean is one of the main sources of nutrients in Asia with some cultivars being used as Chinese traditional medicine. In 2009 and 2011, our group identified the presence of protease inhibitors in Korean Black soybean and Chinese mini-black soybean by chromatographic purification [14, 15]. By further determining the molecular weights (MW) and amino acid sequences, they were both classified as Kunitz type protease inhibitors (KTPI) with similar MW and homologous N-terminal sequences. Different from most of the protease inhibitors extracted from soybeans, there was no antifungal activity detected. Instead, both Korean and Chinese mini- black soybean protease inhibitors displayed potent anti-tumour and HIV-1 reverse transcriptase inhibitory activities as evidenced by the low IC_{50} value (<10 μM) [14, 15]. To be specific, Korean black soybean protease inhibitor suppressed the proliferation of breast cancer MCF-7 cells and hepatoma HepG2 cells, whereas Chinese min-black soybean protease inhibitor was able to inhibit proliferation of nasopharyngeal carcinoma cells CNE-2 and HNE-2 in addition to the previous two types of cancer cells. HIV-1 reverse transcriptase is an important factor as HIV replication relies on the viral RNA to be reverse transcribed into DNA that could enter into the host genome. Together with the tolerance of low pH values, these protease inhibitors are potential therapeutic agents for HIV infections and cancers.

The Bauhinia inhibitors are in the family of Kunitz inhibitors, but they differ both in structure and inhibitory specificity. *Bauhinia bauhiniodes* kallikrein inhibitor (BbKI) and *Bauhinia bauhinioides* cruzipain inhibitor (BbCI) for example, do not contain any disulphide bridges. BbKI only has one cysteine residue whereas BbCI does not have any cysteine residue. Strikingly, these inhibitors are still able to hold a similar tertiary structure by interaction between Tyr_{125}, H_2O and Gly_{132} [7]. In 2011, our group isolated *B. purpurea* trypsin inhibitor (BPLTI) from the seeds of *B. purpurea* L. by ion-exchange chromatography and size-exclusion. We named it BPLTI is to discriminate it from the common bovine pancreatic trypsin inhibitor, abbreviated as BPTI. BPLTI shows homology at the N-terminus with other members of the family, but when treated with increasing concentrations of DTT a gradual reduction of inhibitory activity was seen, signifying the presence and importance of disulphide bonds [16]. By interfering with different mechanisms (including formation of apoptotic bodies, stimulating chromatin condensation and mitochondrial depolarization), BPLTI successfully inhibited the proliferation of human carcinoma Hep G2 cells as well as other tumour cells to a lesser extent. Similar to many Kunitz domain-containing protease inhibitors, BPLTI has the ability to induce cytokine production. Cytokines regulated by BPLTI include IL-1beta, IL-2 and TNF-alpha [16]. IL-1 alone is involved in immunomodulation, inflammation, wound healing, hematopoiesis, metabolism and the endocrine system, and together with IL-2, improve cellular immunity. *B. pupurea and B. variegata* are both common Bauhinia trees species found in Hong Kong. From the seeds of *B. variegata*, a Kunitz type trypsin inhibitor (BvvTI) was isolated. It has the ability to induce the same cytokines as BPLTI, but instead of Hep G2 cells, it is most efficient against human nasopharyngeal carcinoma cell CNE-1 and can repress HIV-1-RT activity as shown by decreased values of IC_{50} [17].

14.3.2 Bowman-Birk Inhibitors Constitute a Source of Promising Drugs

The name of Bowman-Birk inhibitor is meant to give honor to Bowman, who was the first to isolate this inhibitor from soybeans in 1946, and Birk, who characterized this inhibitor in 1961. BBI is the second most abundant soybean protease inhibitor. It is most commonly found in soybean, but also in other dicotyledonous or monocotyledonous seeds. In the early years after its discovery, it was considered as anti-nutrients by nutritionists due to the potential harmful effects when consumed in milligram quantities [18]. But more studies have shown that small amount levels (nM) of BBI can give significant health benefits, such as suppression of X-ray induced transformation and slowing progression of some human diseases [18, 19].

BBI remains active under a wide range of pH values, implying that it survives the digestive processes occurring in the gastrointestinal tract (GIT). This allows the BBI to be absorbed and spread through the bloodstream and target different organs. To investigate the distribution of BBI, a study using rodents as a model was conducted. The animals were fed with labeled BBI, and it was found that 40–50 % of BBI was excreted in the feces, with the remainder distributed in different organs including esophagus, stomach, kidneys, liver and lungs, but not the brain. Once BBI reached an organ, it rapidly appeared in the corresponding epithelial cells [8]. Within 5 h, BBI traveled through the GIT to the colon tissue, feces or was taken up into the bloodstream [20]. The BBI found in feces or urine remained active and was still able to inhibit proteases, proving its resistance to pH and ability to survive in the digestive system.

BBI exhibits antitumor activities and other therapeutic effects against colorectal cancer (CRC), prostate cancer, oral leukoplakia, multiple sclerosis, and Duchenne muscular dystrophy (DMD). CRC is the second most common cancer that causes death, with roughly 100,000 new CRC cases per year. One of the methods for suppressing CRC development is via dietary uptake of BBI. When dimethyhydrazine (DMH)-treated mice were fed with BBI, the frequency of colon tumours drops, without interfering with animal growth or causing physiological changes [8]. Furthermore, in vivo studies show that the colon tumour cell line LoVo and Caco2 cells are suppressed in the presence of BBI, but when BBI is inactivated by chemical modification, the anti-proliferative activities are not expressed, hence proving the importance of BBI [8]. Prostate cancer is one of the high- mortality- rate cancers, and is liable to spread to other organs in the advanced stage. BBI exhibits different mechanisms that target prostate cancer cells: it has been shown to induce the expression of connexin 43 (Cx43) in prostate cancer LNCaP cells and in transgenic rats developing adenocarcinoma of the prostate (TRAP) which causes cell death [8]. The generation of reactive oxygen species was suppressed, whereas the tumour suppressor protein p53 (an important protein involved in DNA repair, inducing growth arrest and initiating apoptosis) was activated. The use of BBI as chemopreventive agent in other cancers and diseases has met with similar success, with oral leukoplakia already in the phase II clinical trial [21] and with the multiple sclerosis showing signs of improvement [22].

Multiple sclerosis is an autoimmune disease in which the immune cells infiltrate the central nervous system (CNS) and cause demyelination, and axonal and neuronal degeneration. An animal model of experimental autoimmune encephalomyelitis (EAE) is used to unravel the mechanisms and provide insights into this disease. In the animal model, although BBI cannot stop the disease from developing, it is able to slow down the progression of EAE by retarding the inflammatory cells from infiltrating to the CNS [23]. This retardation is dependent on the production of the anti-inflammatory cytokine IL-10. There are two roles of IL-10 regarding the retardation. First, it suppresses the antigen presentation ability of antigen presenting cells, and secondly, it inhibits the synthesis of proinflammatory cytokines and chemokines. In addition to IL-10, the shift from Th1 and Th17 to Treg also contributes to the protective effects of BBI.

For clinical studies, an orally available BBI concentrate (BBIC) was developed by Kennedy [18]. Other than BBI, BBIC also contains soybean cystatin, isoflavones, saponins and others. Duchenne muscular dystrophy (DMD) is a genetic disorder caused by the mutation of the gene in the X chromosome, and that predominantly affects males under the age of 5 [24]. As a result of mutation of the dystrophin gene, the protein dystrophin is suppressed. Due to the serious side effects caused by the conventional treatments, efforts to explore alternative therapies without side effects continue. BBIC have been demonstrated to inhibit proteases involved in the TGF-β1 and fibrosis signaling pathways via attenuation of Smad2/3, which are suspected to play a role in DMD. When dystrophic (mdx) mice were treated with BBIC, an improvement in tetanic force, increase in muscle mass and resistance in contraction-induced injury were observed, which together with reduction in fibrosis suggests a retardation in degeneration [25]. BBIC also attenuates the calpain activity which is increased in DMD patients and mdx mice. Calpain is involved in release of sarcomeric proteins which cause protein degradation and inhibits the Akt signaling pathway [25]. For the interesting story of BBIC, please refer to Chap. 11 by Prof. Ann R. Kennedy for details.

In 2010 and 2011, our group isolated two novel Bowman-Birk type trypsin inhibitors from the faba bean *Vicia faba*, termed VFTI-G1 and VFTI-E1, both with a molecular size of 15 kDa [26, 27]. They were both classified as Bowman-Birk type due to their small molecular size and homologous sequences at the N-terminus to other Bowman-Birk type inhibitors. VFTI-G1 exhibited a relatively weak trypsin inhibitory activity when compared with other trypsin inhibitors, and had no anti-fungal activity at all, due to differences in structure and the ability to bind enzyme [27]. The trypsin inhibitory activity might contribute to the suppression of cancers, as trypsinogen (the zymogen of trypsin) was found to be overexpressed in different tumours. The specific antiproliferative activity towards HepG2 hepatoma cells and ability to inhibit HIV-1 reverse transcriptase activity provide important information, as the drug resistance of VFTI-E1 is the main limitation in treating AIDS. Different to other trypsin inhibitors, VFTI-E1 displays various mechanisms of anti-fungal activity including chitin deposition, modification of hyphal morphology, permeabilization of the membrane and induction of a rapid reflux of cations [26].

14.3.3 Other Small Protease Inhibitors

Other than the Kunitz and Bowman-Birk protease inhibitors, there are many other protease inhibitors that are effective and are currently being used to treat diseases. Neutrophil elastase (NE) is one of the targeted proteins for the treatment of inflammatory disorders. Alpha-1 antitrypsin (A1AT) is an example of NE inhibitors [28]. As well as suppressing NE, it contains anti-apoptotic and immunomodulatory functions, which make it a promising therapeutic agent. The ability to suppress NE proves to be effective against cystic fibrosis (CF). However, the main drawback comes from the methionine residue it contains in the reactive site loops, which increases susceptibility to oxidative inactivation. Secretory leukocyte protease inhibitor (SLPI) is another protease inhibitor that is used for treating cystic fibrosis. In clinical trials, it was shown to be able to reduce the amount of neutrophil and neutrophil elastase, as well as modulating the level of pro-inflammatory cytokines [28]. Similar to other protease inhibitors, it has a small molecular weight and several properties, including triggering of anti-inflammatory response, inhibition of HIV replication and mediation of non-redundant functions necessary for normal wound healing [29].

Kazal-type proteinase inhibitors (KPIs) are serine proteinase inhibitors. Depending on the species, they can contain one or more Kazal inhibitory domains. The Kazal inhibitory domain is composed of around 50 amino acid residues, with vertebrate species possessing a larger domain than invertebrate species. The inhibitory domain is linked by 6 intra-disulphide bonds and inhibits proteinase by binding to the active site of the cognate proteins via its reactive site loop in a non-covalent manner, with a high association constant. Although there is only one contact position that is responsible for the specificity, 11 additional contact positions facilitate its binding and recognition. By mutation of the 11 contact positions, the Ki value is significantly reduced. Most recently, a novel Kazal-type trypsin inhibitor was discovered from the skin of the red-eyed leaf frog *Agalychnis callidryas* [1]. This discovery of a new Kazal-type trypsin would prove useful as it displays various biological and physiological functions that could be used as pharmacological tools. Kazal1 from *Hydra magnipapillata* prevents excessive autophagy, and when this protein is silenced, it quickly causes disorganization and death of gland cells. If the protein is not restored, death eventually results. Bacteriostatic activity against *Bacillus subtilis* in *Penaeus monodon* and bacteriocidal activity against *Staphylococcus aureus* in *Hydra magnipapillata* have been reported, suggesting a mechanism with which KPI takes part in the host defense.

14.4 Biological Protease Inhibitors

In addition to natural protease inhibitors, there is great interest in the development and application of biological protease inhibitors which may circumvent the low specificity of some natural protease inhibitors, as well as increase protein quantity and purity [28, 30]. Examples include antibodies for different proteases and engineered protease inhibitors. Antibodies are one of the most prevailing biological agents, with 12 antibodies approved by US Food and Drug Administration, along with over 240 compounds under different stages of clinical trial [28, 30]. They can recognize target cells in an agonistic manner, for example, to promote apoptosis in tumour cells, or binding to receptors in an antagonistic manner and inhibiting the pathway or signaling. Apart from recognizing receptors or cells, they can also bind to proteases directly and act as protease inhibitors. Occasionally, the antibody recognizes and blocks the active site of the target protein, thereby preventing the substrate from entering. More commonly, it is able to bind the protease and induce conformational changes of the active site, thereby inhibiting the activity by preventing the recognition of substrate. Alternatively, a group has shown that by binding the antibody mab1696 to the conserved homologous N-terminal region between HIV-1 and HIV-2 proteases, it prevents the dimerization of the protease, which is required for activity [31]. Other than potential for treating AIDS, another antibody E2 has been reported to bind to a serine protease matriptase (MT-SP1) near its active site. MT-SP1 is upregulated in the presence of several tumours and accounts for the proliferation of tumour cells. Upon binding of E2, it blocks the entrance of substrate to the protease and therefore inhibits its ability to activate PI3 K-AKT pathway [28]. Another example is Fsn0503, an antagonistic antibody against cathepsin S, which is a cysteine protease and involves in tumor invasion. Burden and colleagues found that Fsn0505 blocked colorectal tumour invasion and angiogenesis in both in vitro and in vivo experiments [32].

The use of protease inhibitors as novel therapeutic agents by engineering different domains of inhibitors is considered to be an effective strategy, with the Kunitz domain being a promising candidate because of its low concentration requirement and its stability, which allows the structure to remain stable upon point mutation. An example of mutated Kunitz domains is the plasma kallikrein inhibitor DX-88 (Dyax-Genzyme LLC), also known as ecallantide and DX-1000. Ecallantide has the ability to inhibit plasma kallikrein, hence reducing the vascular permeability and relieve the symptoms of hereditary angioderma (HAE), and even most attractive as therapeutic agent for breast cancer [28, 33]. Most inhibitors are selected by phage display, including DX-1000. They are able to inhibit plasmin to trigger the anti-tumour effects via blockage of a type IV collagenase activation. Similarly, the neutrophil elastase (NE) could be inhibited by a phage display-derived BPTI-variant, and this inhibition brings about a lowering of the level of pro-inflammatory cytokine IL-8 and a reduction of NE activity in sputum [28].

14.5 Conclusion and Future Perspectives

In this review, we have explored the properties of different protease inhibitors with particular focus on the Bowman-Birk inhibitors and Kunitz type protease inhibitors. These bind to the serine protease with a canonical or non-canonical mechanism. By binding to serine proteases or other targets, they display a spectacular array of properties that are beneficial to humans encompassing anti-fungal, anti-apoptotic, anti-inflammatory, anti-metastatic, anti-proliferative, and anti-HIV activities. The multifarious activities of such inhibitors advocate their therapeutic application for diseases and disorders. The questions at the moment lie on whether the use of some of these protease inhibitors is confined to the bench top level, and whether they have the ability to target a specific organ or display a broad systemic effect when ingested orally? To date, most experiments are conducted in vitro, hence there is a need to conduct trials in vivo. The anti-HIV and HIV-reverse transcriptase inhibitory activities of protease inhibitors derived from soybean could be particularly useful. As soybean is a common food source and can be produced on a large scale at a relatively cheap price, soybean protease inhibitors could be used as a potential substitute for the current antiretroviral drugs.

References

1. Li R, Wang H, Jiang Y, Yu Y, Wang L, Zhou M et al (2012) A novel Kazal-type trypsin inhibitor from the skin secretion of the Central American red-eyed leaf frog *Agalychnis callidryas*. Biochimie 94:1376–1381
2. Liener IE (1986) Trypsin inhibitors: concern for human nutrition or not? J Nutr 116:920–923
3. Bateman KS, James MN (2011) Plant protein proteinase inhibitors: structure and mechanism of inhibition. Curr Protein Pept Sci 12:340–347
4. Volpicella M, Leoni C, Costanza A, De Leo F, Gallerani R, Ceci LR (2011) Cystatins, serpins and other families of protease inhibitors in plants. Curr Protein Pept Sci 12:386–398
5. Laskowski M, Qasim MA (2000) What can the structures of enzyme-inhibitor complexes tell us about the structures of enzyme substrate complexes? Biochim Biophys Acta 1477:324–337
6. Oliva ML, Ferreira Rda S, Ferreira JG, de Paula CA, Salas CE, Sampaio MU (2011) Structural and functional properties of Kunitz proteinase inhibitors from leguminosae: a mini review. Curr Protein Pept Sci 12:348–357
7. Oliva ML, Sampaio UM (2008) Bauhinia Kunitz-type proteinase inhibitors: structural characteristics and biological properties. Biol Chem 389:1007–1013
8. Clemente A, Sonnante G, Domoney C (2011) Bowman-Birk inhibitors from legumes and human gastrointestinal health: current status and perspectives. Curr Protein Pept Sci 12:358–373
9. Birk Y (1985) The Bowman-Birk inhibitor. Trypsin- and chymotrypsin-inhibitor from soybeans. Int J Pept Protein Res 25:113–131
10. Borjigin J, Nathans J (1993) Bovine pancreatic trypsin inhibitor-trypsin complex as a detection system for recombinant proteins. Proc Natl Acad Sci USA 90:337–341
11. Wlodawer A, Walter J, Huber R, Sjolin L (1984) Structure of bovine pancreatic trypsin inhibitor. Results of joint neutron and X-ray refinement of crystal form II. J Mol Biol 180:301–329
12. Shigetomi H, Onogi A, Kajiwara H, Yoshida S, Furukawa N, Haruta S et al (2010) Anti-inflammatory actions of serine protease inhibitors containing the Kunitz domain. Inflamm Res 59:679–687
13. Kobayashi H, Suzuki M, Hirashima Y, Terao T (2003) The protease inhibitor bikunin, a novel anti-metastatic agent. Biol Chem 384:749–754
14. Fang EF, Wong JH, Ng TB (2010) Thermostable Kunitz trypsin inhibitor with cytokine inducing, antitumor and HIV-1 reverse transcriptase inhibitory activities from Korean large black soybeans. J Biosci Bioeng 109:211–217
15. Ye XJ, Ng TB (2011) Antitumor and HIV-1 reverse transcriptase inhibitory activities of a hemagglutinin and a protease inhibitor from mini-black soybean. Evid Based Complement Altern Med 2011:851396
16. Fang EF, Bah CS, Wong JH, Pan WL, Chan YS, Ye XJ et al (2012) A potential human hepatocellular carcinoma inhibitor from *Bauhinia purpurea* L. seeds: from purification to mechanism exploration. Arch Toxicol 86:293–304

17. Fang EF, Wong JH, Bah CS, Lin P, Tsao SW, Ng TB (2010) *Bauhinia variegata* var. variegata trypsin inhibitor: from isolation to potential medicinal applications. Biochem Biophys Res Commun 396:806–811
18. Losso JN (2008) The biochemical and functional food properties of the bowman-birk inhibitor. Crit Rev Food Sci Nutr 48:94–118
19. Yavelow J, Collins M, Birk Y, Troll W, Kennedy AR (1985) Nanomolar concentrations of Bowman-Birk soybean protease inhibitor suppress X-ray-induced transformation in vitro. Proc Natl Acad Sci USA 82:5395–5399
20. Kennedy AR (1998) The Bowman-Birk inhibitor from soybeans as an anticarcinogenic agent. Am J Clin Nutr 68:1406S–1412S
21. Armstrong WB, Kennedy AR, Wan XS, Taylor TH, Nguyen QA, Jensen J et al (2000) Clinical modulation of oral leukoplakia and protease activity by Bowman-Birk inhibitor concentrate in a phase IIa chemoprevention trial. Clin Cancer Res: Off J Am Assoc Cancer Res 6:4684–4691
22. Gran B, Tabibzadeh N, Martin A, Ventura ES, Ware JH, Zhang GX et al (2006) The protease inhibitor, Bowman-Birk inhibitor, suppresses experimental autoimmune encephalomyelitis: a potential oral therapy for multiple sclerosis. Multiple Sclerosis 12:688–697
23. Dai H, Ciric B, Zhang GX, Rostami A (2012) Interleukin-10 plays a crucial role in suppression of experimental autoimmune encephalomyelitis by Bowman-Birk inhibitor. J Neuroimmunol 245:1–7
24. Ray PN, Belfall B, Duff C, Logan C, Kean V, Thompson MW et al (1985) Cloning of the breakpoint of an X; 21 translocation associated with Duchenne muscular dystrophy. Nature 318:672–675
25. Morris CA, Selsby JT, Morris LD, Pendrak K, Sweeney HL (2010) Bowman-Birk inhibitor attenuates dystrophic pathology in mdx mice. J Appl Physiol 109:1492–1499
26. Fang EF, Hassanien AA, Wong JH, Bah CS, Soliman SS, Ng TB (2010) Purification and modes of antifungal action by *Vicia faba* cv. Egypt trypsin inhibitor. J Agric Food Chem 58:10729–10735
27. Fang EF, Hassanien AA, Wong JH, Bah CS, Soliman SS, Ng TB (2011) Isolation of a new trypsin inhibitor from the Faba bean (*Vicia faba* cv. Giza 843) with potential medicinal applications. Protein Pept Lett 18:64–72
28. Scott CJ, Taggart CC (2010) Biologic protease inhibitors as novel therapeutic agents. Biochimie 92:1681–1688
29. Ashcroft GS, Lei K, Jin W, Longenecker G, Kulkarni AB, Greenwell-Wild T et al (2000) Secretory leukocyte protease inhibitor mediates non-redundant functions necessary for normal wound healing. Nat Med 6:1147–1153
30. Scott AM, Wolchok JD, Old LJ (2012) Antibody therapy of cancer. Nat Rev Cancer 12:278–287
31. Rezacova P, Brynda J, Lescar J, Fabry M, Horejsi M, Sieglova I et al (2005) Crystal structure of a cross-reaction complex between an anti-HIV-1 protease antibody and an HIV-2 protease peptide. J Struct Biol 149:332–337
32. Burden RE, Gormley JA, Jaquin TJ, Small DM, Quinn DJ, Hegarty SM et al (2009) Antibody-mediated inhibition of cathepsin S blocks colorectal tumor invasion and angiogenesis. Clin Cancer Res: Off J Am Assoc Cancer Res 15:6042–6051
33. Devy L, Rabbani SA, Stochl M, Ruskowski M, Mackie I, Naa L et al (2007) PEGylated DX-1000: pharmacokinetics and antineoplastic activity of a specific plasmin inhibitor. Neoplasia 9:927–937
34. Fang EF, Ng TB (1815) Ribonucleases of different origins with a wide spectrum of medicinal applications. Biochim Biophys Acta 2011:65–74

Part V
Plethora of Health Benefits from Small Molecules: From Bench-Top to Clinical Settings

Chapter 15
Autophagy Upregulation as a Therapeutic Strategy for Neurodegenerative Diseases

Mariana Pavel and David C. Rubinsztein

Abstract Autophagy is an evolutionarily conserved process, which enables the sequestration of intracytoplasmic long-lived proteins or damaged organelles in double-membraned autophagosomes followed by their degradation within the acidic lysosomal compartment. Autophagic failure at any step of the pathway perturbs cellular homeostasis with consequences for various pathologies, from cancer and infectious disorders to neurodegenerative diseases. Many neurodegenerative diseases are caused by gain-of-function mutations which favor the accumulation of intracytoplasmic aggregate-prone proteins. Thus, an effective strategy to clear the mutant proteins is to enhance autophagic flux. Here we highlight the recent progress in upregulating autophagy as a therapeutic approach in models of neurodegenerative diseases, which may guide further studies in the clinic.

15.1 Introduction

Autophagy is a highly conserved process that mediates the degradation of intracytoplasmic proteins, protein complexes and organelles by capturing them in double-membraned vesicles, called autophagosomes. Autophagosomes move bidirectionally along microtubules with a bias towards the microtubule organizing center to eventually fuse with lysosomes forming autolysosomes, in which the cargoes are degraded [1–3].

In mammalian cells, autophagosome generation is mediated by different ATG (autophagy) proteins. Some of these regulate autophagy by participating in two complexes (ULK1 and Vps34/Beclin1 complexes), while others regulate two ubiquitin-like reactions involved in the formation of Atg12-Atg5.Atg16 and lipidated LC3-II [2]. MAP1LC3 (microtubule-associated protein 1 light chain 3), or simply LC3, is the only mammalian protein that specifically associates with the

M. Pavel · D. C. Rubinsztein (✉)
Department of Medical Genetics, Cambridge Institute for Medical Research, Wellcome/MRC Building, Addenbrooke's Hospital, Hills Road, Cambridge CB2 0XY, UK
e-mail: dcr1000@hermes.cam.ac.uk

E. F. Fang and T. B. Ng (eds.), *Antitumor Potential and Other Emerging Medicinal Properties of Natural Compounds*, DOI: 10.1007/978-94-007-6214-5_15,

autophagosomal membranes and therefore the level of LC3-II correlates with the number of autophagosomes [4].

The physiological relevance of autophagy is related to the maintenance of both cellular quality control and energy balance [5]. Basal levels of autophagy mediate the normal turnover of cytoplasmic components and the clearance of misfolded proteins, or damaged organelles. For instance, mice specifically deficient in key autophagy genes in neuronal cells develop neurodegeneration with progressive motor function deficiency, accompanied by the accumulation of intraneuronal aggregates [6, 7]. Likewise, autophagy is stimulated by stress conditions, such as starvation or oxidative stress, to break down macromolecules into basic components required for biosynthetic processes, therefore restoring the nutrient balance [8, 9]. Newborn mice undergo a period of relative starvation before receiving energetic supplies through breastfeeding, and autophagy impairment at this stage prevents them from recycling nutrients, and causes early postnatal lethality [10]. In addition, autophagy protects against pro-apoptotic insults by upregulating mitophagy (degradation of mitochondria by autophagy) and consequently reducing ROS levels and caspase activation [11]. Apart from the roles of autophagy in cellular homeostasis in both basal and stress conditions, there are data in model organisms suggesting that it may have beneficial roles in lifespan extension, aging, senescence, immunity, and defense against pathogens [12–14].

Perturbations of the autophagic balance (excessive or insufficient autophagy) have been broadly associated with cancer, heart disease, gastrointestinal disorders, type 2 diabetes and with neurodegenerative diseases [2, 15]. Many neurodegenerative diseases are caused by gain-of-function mutations which make the target proteins aggregate-prone and toxic [16]. As many of these proteins are found in the cytoplasm and are autophagy substrates, impaired autophagosome formation or degradation reduces their clearance and contributes to the onset and progression of the disease [17]. The clearance of mutant proteins, such as mutant huntingtin (Huntington's disease, HD), mutant α-synuclein (Parkinson's disease), mutant tau (frontotemporal dementia), or mutant ataxin 3 (spinocerebellar ataxia type 3, SCA3), is impaired when autophagy is inhibited, compared to their wild-type counterparts [17–19]. On the other hand, autophagy upregulation slows disease onset in cell, *Drosophila* and zebrafish models of conditions like Huntington's disease and this is associated with enhanced clearance of the mutant protein [17, 20–22]. These results demonstrate the importance of autophagy in continuously maintaining the homeostatic turnover of cytosolic proteins and organelles in neurons.

15.2 Enhancing Autophagic Clearance of Aggregate-Prone Proteins by mTORC1 Inhibition

The evolutionary conserved serine/threonine protein kinase mTOR is a negative regulator of autophagy and drugs that inhibit its activity enhance autophagic flux [23]. mTOR participates in signaling pathways, which are responsible for cell growth, proliferation, transcription, ribosome biogenesis and cytoskeleton organization, in two functional complexes: mTORC1 and mTORC2 [24, 25]. The mTORC1 complex containing the raptor subunit is regulated by various signals (amino acids, glucose, growth factors, energy status, and various stress factors) and impacts on autophagy. Rapamycin and starvation stabilize the raptor-mTOR association, inhibiting its kinase activity, and therefore enhance autophagy [26]. Recent studies showed that mTORC1 is not completely inhibited by rapamycin due to its rapamycin-insensitive properties. Torin 1, a new inhibitor of both mTORC1 and mTORC2, has been reported to increase autophagy more potently than rapamycin due to the inhibition of both the rapamycin-sensitive and -insensitive functions of mTORC1 [27]. However, the autophagy-inducing effects of Torin 1 are mTORC2-independent.

Rapamycin was first shown to enhance mutant huntingtin clearance and to reduce its toxicity in cell, *Drosophila* and mouse models of HD [17, 21]. Furthermore, treatment with rapamycin reduced the toxicity and the levels of isolated polyglutamine expansions (which cause 10 neurodegenerative diseases, including HD and SCA3), and both wild-type and mutant tau in cells and flies, suggesting that this strategy may have utility in a range of neurodegenerative conditions. Similarly, mutant forms of α-synuclein, which cause familial Parkinson's disease (PD) are also removed more rapidly after autophagy induction [20–22] and subsequent work has shown that rapamycin reduces the accumulation of α-synuclein in mice [28]. Likewise, temsirolimus (a rapamycin ester) reduced the number of aggregates and the levels of soluble mutant ataxin-3, another polyglutamine-expanded protein, in the brains of transgenic SCA3 mice and improved their motor performance [29]. In the context of Alzheimer's disease (AD), rapamycin reduced the levels of amyloid-beta and ameliorated the cognitive deficits, therefore slowing or blocking the progress of the disease in transgenic and triple transgenic mouse models [30, 31]. In an AD model, rapamycin was beneficial only when administrated prophylactically, before the formation of plaques and tangles which are known to contribute to the neuronal impairment [30]. This contrasts with the SCA3 situation, where the temsirolimus treatment had great benefits when initiated after the onset of early signs of disease [29]. The benefits of autophagy upregulation have also been seen in a form of motor neuron disease characterized by the accumulation of the nuclear protein TDP-43 in ubiquitin-positive inclusions [32], as well as in a mouse model of the prion disease Gerstmann-Straussler-Scheinker [33].

The protective effects of the treatment with rapamycin appear to be autophagy-dependent in *Drosophila* models of such diseases, since no amelioration was seen when the activities of different autophagy genes were reduced [20, 34]. Rapamycin administration in neurodegenerative diseases may also have additional positive effects, as it prolongs life-span and delays aging, as revealed in "normal" *Drosophila* and mice [14, 35]. Some of these benefits may also be associated with the anti-apoptotic effects of autophagy [11].

15.3 mTOR Independent Therapeutic Strategies

Since mTORC1 inhibition by rapamycin affects many processes independently of autophagy and has side-effects like poor wound healing and immunosuppression, there have been efforts to identify safer ways of inducing autophagy for long-term treatment of neurodegenerative diseases. New classes of drugs that act independently of mTOR have been identified and characterized. For instance, intracellular inositol and inositol-1,4,5-triphosphate (IP3) levels are negative regulators of autophagy and drugs like lithium, sodium valproate, and carbamazepine, which reduce inositol recycling or synthesis, enhance the clearance of autophagic substrates [36, 37]. In transgenic mice overexpressing human mutant tau, treatment with lithium increased autophagy and this was associated with a reduction in the motor abnormalities [38].

Other recent screens of FDA-approved drugs identified autophagy inducers including imidazoline receptor agonists, L-type Ca^{2+} channel antagonists and calpain inhibitors [36, 39]. The first class acts by decreasing the levels of cAMP and they were shown to enhance autophagy and increase the clearance of mutant huntingtin protein [39], raising the possibility for their use in the treatment of polyglutamine diseases. L-type Ca^{2+} channel antagonists (loperamide, verapamil, amiodarone, nitrendipine), K+ ATP channel openers (minoxidil), and calpain inhibitors (calpastatin, calpeptin) reduce intracellular Ca^{2+} levels and enhance autophagy, while increased levels of Ca^{2+} inhibit autophagic flux [39, 40]. Many of these compounds are already used in the clinic for the treatment of other diseases (hypertension, cardiac arrhythmias) and have well characterized safety profiles, and thus may be valuable in diseases where autophagy induction is beneficial.

15.4 Additional mTOR Independent Drugs

A number of other mTOR-independent autophagy-inducing drugs have been identified for which the targets are less clear, or are currently unknown. One of those compounds is the disaccharide trehalose. It has been shown to induce autophagy and to enhance the clearance of mutant proteins both in vivo and in cell culture [36, 41]. In a mouse model of tauopathy with parkinsonism with both overexpressed human mutant tau protein and a parkin deletion, trehalose ameliorated the dropout of the dopaminergic neurons, reduced the numbers of phosphorylated tau positive neuritic plaques, improved motor function and increased life expectancy [42]. Trehalose was also beneficial in another tau transgenic mouse model [43].

Other screens in yeast and mammalian cells identified small molecule enhancers (SMERs) or inhibitors (SMIRs) of autophagy. Three SMERs (SMER10, SMER18, SMER28) were shown to positively regulate autophagosome synthesis and to enhance the clearance of mutant huntingtin and A53T α-synuclein proteins in mammalian cells and *Drosophila* [44]. A number of other small molecule modulators of autophagy have been identified to enhance the clearance of

Table 15.1 Selected publications describing benefits of autophagy upregulating drugs in rodent models of neurodegenerative diseases

Type of disease	Rodent model	Drug/therapeutic strategy	Features
Huntington's disease	Mice expressing a fragment of mutant huntingtin [21]	CCI-779 (a rapamycin ester) administration up to 5–6 months starting at around 4 weeks of age	Increased clearance of mutant huntingtin protein Reduced aggregate formation Improved behavioral tasks: rotarod test, wire maneuver test, grip strength test and tremors
Alzheimer's disease	APP transgenic mice [50]	Lentiviral vector expressing of Beclin 1	Reduced both intracellular and extracellular amyloid pathology
	PDAPP transgenic mice [31]	Rapamycin fed for 13 weeks starting at 4 months of age	Delayed or blocked AD progression
	TgCRND8 mice [51]	Latrepirdine administration for 1 month, one i.p. injection per day starting at 3 months of age	Improved learning behavior Reduced levels of Aβ42 and α-synuclein
	TgCRND8 mice [49]	Genetic deletion of the lysosomal cysteine protease inhibitor, cystatin B (CstB)	Increased clearance of Aβ and autophagic substrates Increased cathepsin activity and restored lysosomal proteolysis Improved learning and memory deficits
	3xTg-AD mice [52]	Rapamycin fed for 10 weeks starting at 6 month of age	Improved cognitive functions Reduced levels of Aβ
	3xTg-AD mice [30]	Rapamycin fed for 16 month starting at 2 months of age	Reduced number of amyloid plaques and neurofibrillary tangles Increased cognitive function
		Rapamycin fed for 3 month starting at 15 months of age	No effect on improving AD pathology

(continued)

Table 15.1 (contniued)

Type of disease	Rodent model	Drug/therapeutic strategy	Features
Parkinson's disease and Lewy body disease	α-synuclein transgenic mice [53]	Lentiviral vector expressing Beclin 1	Increased clearance of α-synuclein Enhanced lysosomal activity Ameliorated synaptic dysfunctions
	α-synuclein transgenic mice [28]	Intracerebral infusions with Rapamycin for 2 weeks starting at 9 months of age Lentiviral vector expressing Atg7 starting at 9 months of age	Increased clearance of α-synuclein Enhanced lysosomal activity
	Wild type mice [47]	Latrepirdine for 3 weeks starting at 4 months of age	Enhanced degradation of α-synuclein Increased levels of autophagic markers
Spinocerebellar ataxia type 3	Transgenic mouse model of SCA3 [29]	Temsirolimus (a rapamycin ester)—i.p. injections for 2 months or 21 weeks starting at 6 weeks of age	Improved motor performances Increased clearance of mutant ataxin-3 Reduced aggregate formation
	Transgenic mouse model and rat model of SCA3 [54]	Lentiviral vector expressing Beclin 1	Increased clearance of mutant ataxin-3 Enhanced autophagic flux Neuroprotective effects
Tauopathies	Mouse model of tauopathy with parkinsonism—$PK^{-/-}/Tau^{VLW}$ [42]	Trehalose fed for 2.5 months starting at 3 months of age Trehalose fed for 3 weeks starting at 14 months of age	Improved dropout of dopamine neurons Increased levels of dopamine-related proteins Improved motor deficit Reduced phosphorylated tau levels Reduced number of Aβ plaques
	Human mutant P301S tau transgenic mice [43]	Trehalose administration until 20 weeks of age	Enhanced clearance of tau inclusions Improved neuronal survival (brainstem and cerebral cortex) No effect on motor dysfunction
	Human mutant P301L tau transgenic mice [38]	Oral Lithium Chloride for 4 months starting at 5 months of age	Improved motor deficit Reduced phosphorylated tau levels Reduced number of neurofibrillary tangles Increased levels of autophagic markers

(continued)

Table 15.1 (contniued)

Type of disease	Rodent model	Drug/therapeutic strategy	Features
Frontotemporal lobar dementia and Amyotrophic lateral sclerosis	TDP43 mouse model [32]	Rapamycin fed for 4 months starting at 2 months of age	Increased learning and memory behavior
		Rapamycin, Tamoxifen, Spermidine, Carbamazepine for 1 month starting at 6 months of age	Improved motor deficit Reduced neuronal loss Decreased caspase-3 levels
Prion diseases	Tg(PrP-A116V) mouse model of Gerstmann-Straussler-Scheinker Disease [33]	Rapamycin i.p. injections starting at 6 weeks of age until killed for analysis (~5 months)	Delayed onset of disease Improved symptomathology Improved survival Reduced levels of insoluble PrP-A116V Reduced number of PrP amyloid plaques

aggregate-prone proteins [36, 44–46], including latrepirdine, which enhances the clearance of α-synuclein and reduces its toxicity in both neuronal cells and mice brains [47].

15.5 Other Considerations

For many neurodegenerative diseases, upregulation of autophagy may be an attractive therapeutic strategy. However, diseases with impairment in the late-stages of autophagy (e.g. autophagosome-lysosome fusion) may not benefit from this therapy, since increasing autophagosome formation would not necessarily result in more substrate delivery to the lysosomes. Thus, the most likely successful therapy for some complex diseases may be to increase the autophagic flux by enhancing all autophagic steps. A novel target for drug development may be the modulation of TFEB activity, which impacts on both autophagosome formation and lysosomal biogenesis by simply controlling its phosphorylation state [48]. Another strategy may be to develop drugs targeting endogenous lysosomal protease inhibitors, as genetic depletion of cystatin B, a typical lysosomal cysteine protease inhibitor, rescues the AD phenotype in transgenic mice [49].

To date, many studies on rodent models of neurodegenerative diseases have confirmed the relevance of earlier results in various cell lines and primary neuron cultures (reviewed in Table 15.1). Therefore, this justifies continuous efforts to identify new and better autophagy-modulating drugs, some of which will hopefully move into the clinic.

Acknowledgments The work in DCR's lab is supported by the Wellcome Trust (Principal Fellowship), a Wellcome Trust/Medical Research Council Strategic Award on Neurodegenerative Diseases, Medical Research Council, and the National Institute for Health Research Biomedical Research Unit in Dementia at Addenbrooke's Hospital. MP is funded by Marie Curie Actions through the European Community's Seventh Framework Programme FP7/2010 under grant agreement no. 264508.

References

1. Jahreiss L, Menzies FM, Rubinsztein DC (2008) The itinerary of autophagosomes: from peripheral formation to kiss-and-run fusion with lysosomes. Traffic 9:574–587
2. Ravikumar B, Sarkar S, Davies JE, Futter M, Garcia-Arencibia M, Green-Thompson ZW, Jimenez-Sanchez M, Korolchuk VI, Lichtenberg M, Luo S et al (2010) Regulation of mammalian autophagy in physiology and pathophysiology. Physiol Rev 90:1383–1435
3. Xie Z, Klionsky DJ (2007) Autophagosome formation: core machinery and adaptations. Nat Cell Biol 9:1102–1109
4. Kabeya Y, Mizushima N, Ueno T, Yamamoto A, Kirisako T, Noda T, Kominami E, Ohsumi Y, Yoshimori T (2000) LC3, a mammalian homologue of yeast Apg8p, is localized in autophagosome membranes after processing. EMBO J 19:5720–5728
5. Klionsky DJ (2010) The molecular machinery of autophagy and its role in physiology and disease. Semin Cell Dev Biol 21:663

6. Hara T, Nakamura K, Matsui M, Yamamoto A, Nakahara Y, Suzuki-Migishima R, Yokoyama M, Mishima K, Saito I, Okano H et al (2006) Suppression of basal autophagy in neural cells causes neurodegenerative disease in mice. Nature 441:885–889
7. Komatsu M, Waguri S, Chiba T, Murata S, Iwata J, Tanida I, Ueno T, Koike M, Uchiyama Y, Kominami E et al (2006) Loss of autophagy in the central nervous system causes neurodegeneration in mice. Nature 441:880–884
8. Singh R, Cuervo AM (2011) Autophagy in the cellular energetic balance. Cell Metab 13:495–504
9. Verhoef LG, Lindsten K, Masucci MG, Dantuma NP (2002) Aggregate formation inhibits proteasomal degradation of polyglutamine proteins. Hum Mol Genet 11:2689–2700
10. Kuma A, Hatano M, Matsui M, Yamamoto A, Nakaya H, Yoshimori T, Ohsumi Y, Tokuhisa T, Mizushima N (2004) The role of autophagy during the early neonatal starvation period. Nature 432:1032–1036
11. Ravikumar B, Berger Z, Vacher C, O'Kane CJ, Rubinsztein DC (2006) Rapamycin pre-treatment protects against apoptosis. Hum Mol Genet 15:1209–1216
12. Bandyopadhyay U, Cuervo AM (2007) Chaperone-mediated autophagy in aging and neurodegeneration: lessons from alpha-synuclein. Exp Gerontol 42:120–128
13. Massey AC, Kiffin R, Cuervo AM (2006) Autophagic defects in aging: looking for an "emergency exit"? Cell Cycle 5:1292–1296
14. Rubinsztein DC, Marino G, Kroemer G (2011) Autophagy and aging. Cell 146:682–695
15. Sridhar S, Botbol Y, Macian F, Cuervo AM (2012) Autophagy and disease: always two sides to a problem. J Pathol 226:255–273
16. Rubinsztein DC (2006) The roles of intracellular protein-degradation pathways in neurodegeneration. Nature 443:780–786
17. Ravikumar B, Duden R, Rubinsztein DC (2002) Aggregate-prone proteins with polyglutamine and polyalanine expansions are degraded by autophagy. Hum Mol Genet 11:1107–1117
18. Iwata A, Christianson JC, Bucci M, Ellerby LM, Nukina N, Forno LS, Kopito RR (2005) Increased susceptibility of cytoplasmic over nuclear polyglutamine aggregates to autophagic degradation. Proc Natl Acad Sci U S A 102:13135–13140
19. Shibata M, Lu T, Furuya T, Degterev A, Mizushima N, Yoshimori T, MacDonald M, Yankner B, Yuan J (2006) Regulation of intracellular accumulation of mutant Huntingtin by Beclin 1. J Biol Chem 281:14474–14485
20. Berger Z, Ravikumar B, Menzies FM, Oroz LG, Underwood BR, Pangalos MN, Schmitt I, Wullner U, Evert BO, O'Kane CJ et al (2006) Rapamycin alleviates toxicity of different aggregate-prone proteins. Hum Mol Genet 15:433–442
21. Ravikumar B, Vacher C, Berger Z, Davies JE, Luo S, Oroz LG, Scaravilli F, Easton DF, Duden R, O'Kane CJ et al (2004) Inhibition of mTOR induces autophagy and reduces toxicity of polyglutamine expansions in fly and mouse models of Huntington disease. Nat Genet 36:585–595
22. Webb JL, Ravikumar B, Atkins J, Skepper JN, Rubinsztein DC (2003) Alpha-Synuclein is degraded by both autophagy and the proteasome. J Biol Chem 278:25009–25013
23. Noda T, Ohsumi Y (1998) Tor, a phosphatidylinositol kinase homologue, controls autophagy in yeast. J Biol Chem 273:3963–3966
24. Laplante M, Sabatini DM (2012) mTOR signaling. Cold Spring Harb Perspect Biol 4. doi: pii: a011593. 10.1101/cshperspect.a011593
25. Wullschleger S, Loewith R, Hall MN (2006) TOR signaling in growth and metabolism. Cell 124:471–484
26. Kim DH, Sarbassov DD, Ali SM, King JE, Latek RR, Erdjument-Bromage H, Tempst P, Sabatini DM (2002) mTOR interacts with raptor to form a nutrient-sensitive complex that signals to the cell growth machinery. Cell 110:163–175
27. Thoreen CC, Kang SA, Chang JW, Liu Q, Zhang J, Gao Y, Reichling LJ, Sim T, Sabatini DM, Gray NS (2009) An ATP-competitive mammalian target of rapamycin inhibitor reveals rapamycin-resistant functions of mTORC1. J Biol Chem 284:8023–8032
28. Crews L, Spencer B, Desplats P, Patrick C, Paulino A, Rockenstein E, Hansen L, Adame A, Galasko D, Masliah E (2010) Selective molecular alterations in the autophagy pathway in patients with Lewy body disease and in models of alpha-synucleinopathy. PLoS ONE 5:e9313

29. Menzies FM, Huebener J, Renna M, Bonin M, Riess O, Rubinsztein DC (2010) Autophagy induction reduces mutant ataxin-3 levels and toxicity in a mouse model of spinocerebellar ataxia type 3. Brain 133:93–104
30. Majumder S, Richardson A, Strong R, Oddo S (2011) Inducing autophagy by rapamycin before, but not after, the formation of plaques and tangles ameliorates cognitive deficits. PLoS ONE 6:e25416
31. Spilman P, Podlutskaya N, Hart MJ, Debnath J, Gorostiza O, Bredesen D, Richardson A, Strong R, Galvan V (2010) Inhibition of mTOR by rapamycin abolishes cognitive deficits and reduces amyloid-beta levels in a mouse model of Alzheimer's disease. PLoS ONE 5:e9979
32. Wang IF, Guo BS, Liu YC, Wu CC, Yang CH, Tsai KJ, Shen CK (2012) Autophagy activators rescue and alleviate pathogenesis of a mouse model with proteinopathies of the TAR DNA-binding protein 43. Proc Natl Acad Sci U S A 109:15024–15029
33. Cortes CJ, Qin K, Cook J, Solanki A, Mastrianni JA (2012) Rapamycin delays disease onset and prevents PrP plaque deposition in a mouse model of Gerstmann-Straussler-Scheinker disease. J Neurosci 32:12396–12405
34. Pandey UB, Nie Z, Batlevi Y, McCray BA, Ritson GP, Nedelsky NB, Schwartz SL, DiProspero NA, Knight MA, Schuldiner O et al (2007) HDAC6 rescues neurodegeneration and provides an essential link between autophagy and the UPS. Nature 447:859–863
35. Harrison DE, Strong R, Sharp ZD, Nelson JF, Astle CM, Flurkey K, Nadon NL, Wilkinson JE, Frenkel K, Carter CS et al (2009) Rapamycin fed late in life extends lifespan in genetically heterogeneous mice. Nature 460:392–395
36. Renna M, Jimenez-Sanchez M, Sarkar S, Rubinsztein DC (2010) Chemical inducers of autophagy that enhance the clearance of mutant proteins in neurodegenerative diseases. J Biol Chem 285:11061–11067
37. Sarkar S, Floto RA, Berger Z, Imarisio S, Cordenier A, Pasco M, Cook LJ, Rubinsztein DC (2005) Lithium induces autophagy by inhibiting inositol monophosphatase. J Cell Biol 170:1101–1111
38. Shimada K, Motoi Y, Ishiguro K, Kambe T, Matsumoto SE, Itaya M, Kunichika M, Mori H, Shinohara A, Chiba M et al (2012) Long-term oral lithium treatment attenuates motor disturbance in tauopathy model mice: implications of autophagy promotion. Neurobiol Dis 46:101–108
39. Williams A, Sarkar S, Cuddon P, Ttofi EK, Saiki S, Siddiqi FH, Jahreiss L, Fleming A, Pask D, Goldsmith P et al (2008) Novel targets for Huntington's disease in an mTOR-independent autophagy pathway. Nat Chem Biol 4:295–305
40. Berridge MJ, Bootman MD, Roderick HL (2003) Calcium signalling: dynamics, homeostasis and remodelling. Nat Rev Mol Cell Biol 4:517–529
41. Sarkar S, Davies JE, Huang Z, Tunnacliffe A, Rubinsztein DC (2007) Trehalose, a novel mTOR-independent autophagy enhancer, accelerates the clearance of mutant huntingtin and alpha-synuclein. J Biol Chem 282:5641–5652
42. Rodriguez-Navarro JA, Rodriguez L, Casarejos MJ, Solano RM, Gomez A, Perucho J, Cuervo AM, Garcia de Yebenes J, Mena MA (2010) Trehalose ameliorates dopaminergic and tau pathology in parkin deleted/tau overexpressing mice through autophagy activation. Neurobiol Dis 39:423–438
43. Schaeffer V, Lavenir I, Ozcelik S, Tolnay M, Winkler DT, Goedert M (2012) Stimulation of autophagy reduces neurodegeneration in a mouse model of human tauopathy. Brain 135:2169–2177
44. Sarkar S, Perlstein EO, Imarisio S, Pineau S, Cordenier A, Maglathlin RL, Webster JA, Lewis TA, O'Kane CJ, Schreiber SL et al (2007) Small molecules enhance autophagy and reduce toxicity in Huntington's disease models. Nat Chem Biol 3:331–338
45. Fleming A, Noda T, Yoshimori T, Rubinsztein DC (2011) Chemical modulators of autophagy as biological probes and potential therapeutics. Nat Chem Biol 7:9–17
46. Rubinsztein DC, Codogno P, Levine B (2012) Autophagy modulation as a potential therapeutic target for diverse diseases. Nat Rev Drug Discov 11:709–730

47. Steele JW, Ju S, Lachenmayer ML, Liken J, Stock A, Kim SH, Delgado LM, Alfaro IE, Bernales S, Verdile G et al (2012) Latrepirdine stimulates autophagy and reduces accumulation of alpha-synuclein in cells and in mouse brain. Mol Psychiatry. doi: 10.1038/mp.2012.115
48. Settembre C, Di Malta C, Polito VA, Garcia Arencibia M, Vetrini F, Erdin S, Erdin SU, Huynh T, Medina D, Colella P et al (2011) TFEB links autophagy to lysosomal biogenesis. Science 332:1429–1433
49. Yang DS, Stavrides P, Mohan PS, Kaushik S, Kumar A, Ohno M, Schmidt SD, Wesson DW, Bandyopadhyay U, Jiang Y et al (2011) Therapeutic effects of remediating autophagy failure in a mouse model of Alzheimer disease by enhancing lysosomal proteolysis. Autophagy 7:788–789
50. Pickford F, Masliah E, Britschgi M, Lucin K, Narasimhan R, Jaeger PA, Small S, Spencer B, Rockenstein E, Levine B et al (2008) The autophagy-related protein beclin 1 shows reduced expression in early Alzheimer disease and regulates amyloid beta accumulation in mice. J Clin Invest 118:2190–2199
51. Steele JW, Lachenmayer ML, Ju S, Stock A, Liken J, Kim SH, Delgado LM, Alfaro IE, Bernales S, Verdile G et al (2012b) Latrepirdine improves cognition and arrests progression of neuropathology in an Alzheimer's mouse model. Mol Psychiatry
52. Caccamo A, Majumder S, Richardson A, Strong R, Oddo S (2010) Molecular interplay between mammalian target of rapamycin (mTOR), amyloid-beta, and Tau: effects on cognitive impairments. J Biol Chem 285:13107–13120
53. Spencer B, Potkar R, Trejo M, Rockenstein E, Patrick C, Gindi R, Adame A, Wyss-Coray T, Masliah E (2009) Beclin 1 gene transfer activates autophagy and ameliorates the neurodegenerative pathology in alpha-synuclein models of Parkinson's and Lewy body diseases. J Neurosci 29:13578–13588
54. Nascimento-Ferreira I, Santos-Ferreira T, Sousa-Ferreira L, Auregan G, Onofre I, Alves S, Dufour N, Colomer Gould VF, Koeppen A, Deglon N et al (2011) Overexpression of the autophagic beclin-1 protein clears mutant ataxin-3 and alleviates Machado-Joseph disease. Brain 134:1400–1415

Author Biography

Mariana Pavel is currently a PhD student at the University of Cambridge under the supervision of Prof. David Rubinsztein. She completed her medical studies in 2011 with a Bachelor Degree in Medicine from the "Gr. T. Popa" University of Medicine and Pharmacy of Iasi and she received a Bachelor Degree in Physics in 2008 and a Master Degree in Physics in 2010 from the "Alexandru Ioan Cuza" University of Iasi, Romania. In 2010, she participated as a summer student in the Amgen Scholars Programme-University of Cambridge. She was awarded the CARPATH Excellence Prize by the Centre for Applied Research in Physics and Advanced technologies in Romania.

David Rubinsztein is Professor of Molecular Neurogenetics and a Wellcome Trust Principal Research Fellow at the University of Cambridge, where he is deputy director of the Cambridge Institute for Medical Research. After completing his basic medical training and housejobs, David did a BSc(Med) Hons and PhD in the Medical Research Council/University of Cape Town Unit for the Cell Biology of Atherosclerosis. He came to Cambridge in 1993 as a senior registrar in Genetic Pathology. During this period, he started working on Huntington's disease and developed an independent research group. His current research has a major focus on the biology of autophagy and its relationship with neurodegenerative diseases. In 1997, David Rubinsztein acquired his Certificate of Completion of Specialist Training in Clinical Molecular Genetics and was awarded a 6 year Glaxo Wellcome Fellowship. In 2001 he was awarded Wellcome Trust

Senior Clinical Fellowship, which was renewed in 2006. In 2011, he was awarded a Wellcome Trust Principal Research Fellowship. David Rubinsztein was elected as a Fellow of the Academy of Medical Sciences in 2004, was awarded the Graham Bull Prize for Clinical Science by the Royal College of Physicians in 2007 and was elected as a member of EMBO in 2011.

Chapter 16
Rapamycin: Current and Future Uses

Morten Scheibye-Knudsen

Abstract The mTOR inhibitor rapamycin has received growing attention due to its immunosuppresive, antineoplastic and lifespan extending properties. The discovery of the drug and its target has had widespread implications for our understanding of the underlying metabolic processes in the cell. Indeed, rapamycin has opened up a new field of potential pharmacological targets. In the following chapter I will briefly review the molecular target of rapamycin and the current clinical applications for this compound

16.1 Introduction

Rapamycin (aka. Sirolimus) is an antifungal compound synthesized by the bacterium *Streptomyces hygroscopicus* that was initially discovered in the soil on Easter Island (Fig. 16.1). This compound has received growing interest due to its potential treatment of a long list of disorders. The drug was first characterized as an immunosuppressant and later shown to inhibit cellular proliferation. In this regard rapamycin was approved by the FDA in the late nineties for immunosuppression after organ transplantation. About a decade later a major breakthrough was achieved when it was shown that the drug had life extending properties in mammals even when fed late in life [1]. In this chapter I will attempt to describe the molecular target of rapamycin as well as the current and potential future applications for this drug in a clinical setting.

16.2 mTOR Pathway

16.2.1 mTOR

The target of rapamycin (TOR) was initially discovered in yeast, where the antifungal activity of rapamycin was found to occur through inhibition of TOR1 and 2. In mammals, it was later shown to inhibit the activity of the *m*echanistic

M. Scheibye-Knudsen (✉)
Laboratory of Molecular Gerontology, National Institute on Aging, National Institutes of Health, Baltimore, MD 21224, USA
e-mail: scheibyem@mail.nih.gov

E. F. Fang and T. B. Ng (eds.), *Antitumor Potential and Other Emerging Medicinal Properties of Natural Compounds*, DOI: 10.1007/978-94-007-6214-5_16,

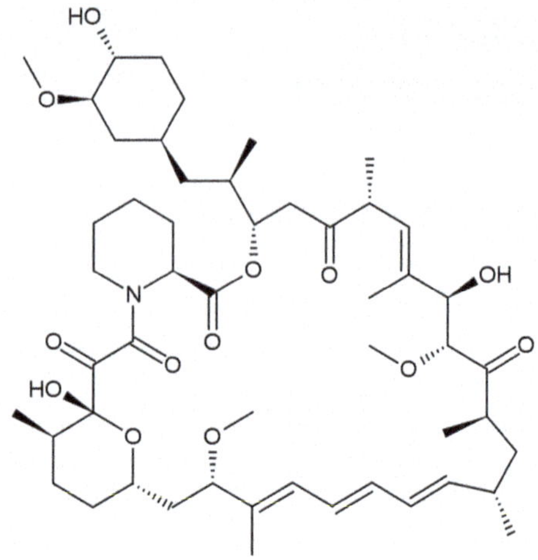

Fig. 16.1 Molecular structure of rapamycin

(previously known as mammalian) TOR (mTOR). MTOR is a member of the phosphoinositide-3-kinase related serine/threonine kinase (PIKK) family together with ataxia-telangiectasia mutated (ATM), ataxia and Rad-3 related (ATR), DNA-dependent protein kinase—catalytic subunit (DNA-PK_{cs}), suppressor of morphogenesis 1 (SMG1) and transformation/transcription domain-associated protein (TTRAP) (Fig. 16.2). Interestingly, all these proteins seem to be involved in the cellular adaptation to homeostatic alterations. In the case of ATM, ATR and

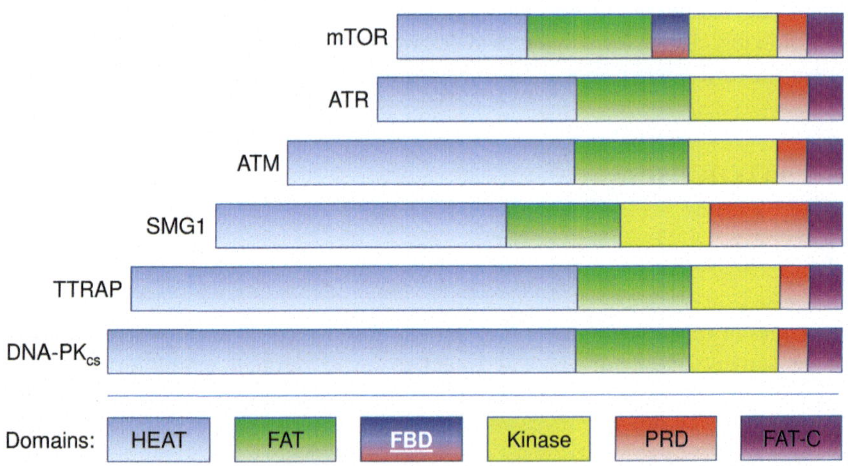

Fig. 16.2 The PIKK family of kinases. Rapamycin facilitates the association of FKBP12 with mTOR at the FKBP12 binding domain (FBD) that is specific to mTOR. *HEAT* Huntingtin, Elongation factor 3, A subunit of protein phosphatase 2A and TOR1. *FAT* FRAP, ATM and TTRAP. *PRD* Protein regulatory domain. *FAT-C* FAT C-terminal domain

DNA-PK_{cs} these proteins are among the early responders to DNA damage directing the recruitment of DNA metabolizing proteins while mTOR may be further downstream potentially orchestrating a broader long term response to damage. Indeed, ATM inhibits mTOR activity after oxidative stress by activating the endogenous mTOR inhibitor, TSC2 [2].

mTOR forms the catalytic core of two main complexes: mTOR complex 1 and 2 (mTORC1 and mTORC2). A growing number of proteins are believed to participate in these complexes with LST8, DEPTOR, tti1 and tel2 acting in both complexes while the proteins raptor and PRAS40 are only found in mTORC1 and rictor, protor1/2 and msin1 in mTORC2 [3]. Although there undoubtedly is crosstalk between these complexes, separate pathways have evolved that rely more heavily on one than the other and I will discuss this in more detail below. MTORC1 is the primary target of rapamycin although recent literature has shown possible mTORC2 dependent effects of this drug [4, 5]. Interestingly, the pharmacological inhibition does not happen directly through rapamycin but via binding of rapamycin to the protein FKBP12. This rapamycin-FKBP12-complex associates with the FRB-region on mTOR resulting in decreased kinase activity. Rapamycin does not seem to have an acute direct effect on mTORC2 although chronic stimulation may lead to disassembly of this complex [6]. Indeed, a recent paper shows that the rather mysterious diabetogenic action of rapamycin is dependent on mTORC2 [4]. In the following I will describe the two main mTOR complexes and their downstream signaling pathways.

16.2.2 mTORC1

As previously stated mTORC1 consists of mTOR, LST8, DEPTOR, tti1, tel2, raptor and PRAS40. In the last decade it has become obvious that this complex serves as a key player in the regulation of cellular metabolism in response to many endogenous and exogenous forms of stress. Indeed, alterations in mTORC1 activity have been shown after nutrient starvation, changes in the insulin/igf-1 axis, hypoxia, DNA damage, inflammation and others (Fig. 16.3). These stimuli alter the downstream targets of mTORC1 shifting the equilibrium from a catabolic state when mTOR is inhibited to an anabolic state upon activation. The most well studied target for mTORC1 is the translational machinery where activation of mTORC1 leads to phosphorylation of two key proteins 4E-BP1 and S6K1 and thereby initiation of translation and cell growth [3]. mTORC1 also stimulates lipid synthesis by phosphorylating lipin-1 leading to the activation of the master regulator of lipid synthesis SREBP-1 [7]. Further, mTORC1 stimulates HIF-1alpha thereby regulating mitochondrial and glycolytic energy metabolism [8]. Importantly, mTORC1 regulates autophagy by phosphorylating key proteins ULK-1 and ATG13 thereby inhibiting the initiation step of autophagy [9]. This role is particularly pertinent in light of the role of autophagy in aging. Caloric restriction is the only known behavioral alteration that can lead

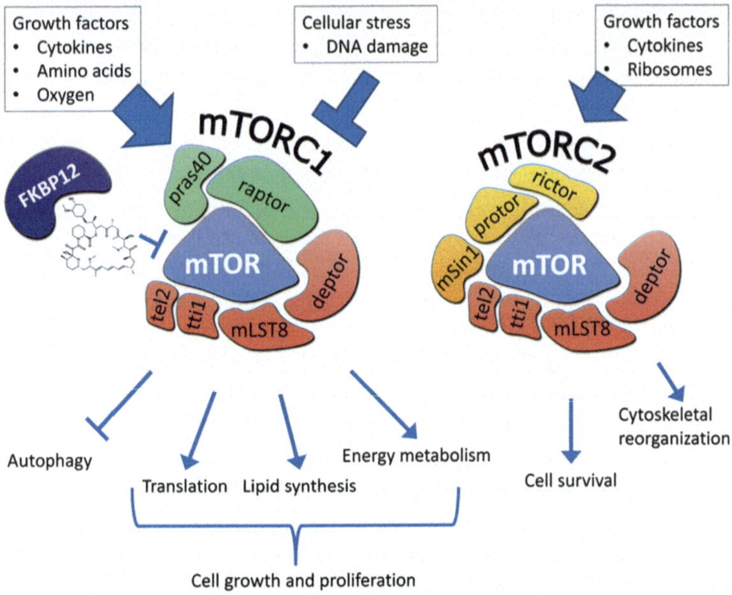

Fig. 16.3 mTORC1 and mTORC2. The mTORC1 complex consists of mTOR, pras40, raptor, deptor, mLST8, tti1 and tel2. This complex acts as a switch between catabolism and anabolism. Rapamycin interacts with FKBP12 and inhibits mTORC1. mTORC2 consists of mTOR, rictor, protor1/2, mSin1, deptor, mLST8, tti1 and tel2. Activation of this complex by growth factors and possibly through ribosomal interactions leads to cell survival through inhibition of apoptosis

to life span extension in a number of organisms and this effect is strictly dependent on autophagy.

16.2.3 mTORC2

mTORC2 consists of the core proteins LST8, DEPTOR, tti1 and tel2 and the mTORC2 specific rictor, protor1/2 and msin1. Much less research has been conducted regarding mTORC2 compared with mTORC1. One prominent downstream effector of mTORC2 is SGK-1. Stimulation of SGK-1 by mTORC2 leads to inhibition of apoptosis and cell survival possibly through a Foxo3a dependent pathway [10]. mTORC2 has also been described to influence cytoarchitectural alterations that are particularly important in cell migration and tumor invasion. Interestingly, autophagy is intimately regulated by the association of autophagosome with the microtubule network and mTORC2 could therefore regulate autophagy through cytoskeletal changes. From a pharmacological standpoint mTORC2 is interesting since the diabetogenic effect of rapamycin is mediated through this pathway [4]. This may be due to increased unchecked gluconeogenesis leading to hyperglycemia and hyperinsulinemia as recently shown in a liver specific rictor knockout mouse [5].

16.3 Medicinal Applications of Rapamycin

16.3.1 Rapamycin in Immunosuppression

The earliest and most common pharmacological use of rapamycin is as an immune suppressant. Rapamycin strongly inhibits T cell proliferation by repressing activation of T cells upon association with an antigen presenting cell. In this scenario, the Akt pathway is activated upon recognition of the antigen by the T-cell receptor leading to mTORC1 stimulation and initiation of cell growth and proliferation [11]. Secondarily, rapamycin also regulates the antigen presenting cells by altering dendritic cell maturation, migration, antigen presentation and apoptosis [12]. Rapamycin was initially a promising drug in this regard since it had a better adverse effect profile than traditional immunosuppressing drugs such as Ciclosporin A and FK506 that showed marked nephrotoxicity. This was, interestingly, not observed with rapamycin treatment. More recent data has, however, shown nephrotoxicity in the form of proteinuria as well as increased risk of early transplant rejection in patients treated primarily with rapamycin compared to calcineurin inhibitors [13]. Now rapamycin is a first choice in treatment of transplant patients that have developed secondary malignancies due to immune suppression. Here rapamycin has shown excellent performance as a result of its antiproliferative effects. In addition to the use of rapamycin in immune suppression in transplant patients the drug has received increasing attention in other disorders with hyperactivation of the immune system. Indeed, a large number of clinical trials are currently underway exploring the effect of rapamycin in systemic lupus erythematosus, uveitis, pemphigus, oral lichen planus and others (www.clinicaltrials.gov).

16.3.2 Rapamycin in Cardiovascular Protection

Cardiovascular disease will remain the leading cause of death in America and the western world in the coming decades and therefore represent a substantial healthcare and economic burden [14]. Rapamycin has shown promising potential in attenuating atherosclerosis in mouse models although the efficacy of this drug in prevention of human atherosclerosis remains unknown [15, 16]. However, the antiproliferative actions of rapamycin have been used for many years in stent treatment of thrombotic coronary artery disease. Here, the artery is balloon dilated and a stent is placed at the site of the atherosclerotic plaque to keep the artery from collapsing and to inhibit restenosis. One early observation in these otherwise successful treatments was the tendency for overgrowth of the stent and later restenosis of the artery. Rapamycin has been used with great success in these stents by inhibiting the attachment and proliferation of cells to the stent and rapamycin eluding stents remain a first choice in many hospitals. Recently, very long term restenosis have been suggested to be increased in rapamycin eluding stents [17] although another meta-analysis has cemented the anti-stenotic effect of rapamycin [18]. Rapamycin is now being investigated in peripheral artery disease with a number of clinical trials underway in this regards (www.clinicaltrials.gov). It is worth noting that one of the most significant risk factors of cardiovascular disease is diabetes and the diabetogenic effect of rapamycin may therefore hinder the systemic use of this drug. Development of rapamycin analogues that only target mTORC1 and potentially thereby avoiding this side effect may therefore be a promising future therapeutic avenue. Indeed, decreasing mTORC1 activation leads to increased insulin sensitivity in a mouse model [4].

16.3.3 Rapamycin in Proliferative Diseases

One of the main downstream effects of mTORC1 is stimulation of proliferation. Both S6K and 4E-BP1 phosphorylation are intimately involved in upregulation of translation, cell growth and initiation of cell divisions. In this regard rapamycin has shown excellent pharmacological properties for intervention in diseases with increased cell proliferation. The prime example of a proliferative disease is cancer where rapamycin may be effective in slowing down cell divisions. Indeed, rapamycin has shown significant efficacy in the treatment of cancer in many mouse model and literally hundreds of clinical trials are now underway to explore this promising effect. In these diseases proliferation and inflammation may be two interconnected phenomena since it has been shown that TNF-alpha and oxidative stress can induce cell cycle progression in model organisms and human cells. Notably, rapamycin will decrease both inflammation and oxidative stress through several pathways and may thereby attenuate both these pathological initiators. However, the use of rapamycin as a monotherapy has been disappointing so far. This is not surprising considering the idea that inhibition of mTOR by itself will not induce apoptosis but merely slow cell growth. As the tumor progresses, cells able to divide despite missing mTOR signaling will evolve and rapamycin will no longer be able to slow the cell growth.

16.3.4 Rapamycin in Neurodegenerative Disorders

Rapamycin has recently been suggested to be a promising pharmaceutical drug in several neurodegenerative disorders [19]. The underlying philosophy is the emerging realization that many neurodegenerative disorders show accumulation of protein aggregates, increased inflammation and neuronal apoptosis. Indeed, reentry of neurons into the cell cycle has been proposed as a possible pathway to apoptosis possible through mTOR activation [20]. However, only a few clinical trials have investigated the role of mTOR inhibitors in neurodegenerative disorders. Tuberous sclerosis complex (TSC) was early recognized as a possible rapamycin treatable disease. This inherited disease is caused by mutations in TSC1 or 2 leading to over-activation of mTOR through loss of rheb repression. In addition to autism like behavioral changes, learning disabilities and epilepsy, spontaneous tumorigenesis is a hallmark feature of this disease. Reflecting the hyperproliferative consequence of aberrant mTOR activation a recent study shows shrinkage of astrocytomas in response to rapamycin treatment in TSC [21]. In addition, rapamycin has been shown to decrease seizure frequency and has consequently been added as a viable treatment option in recent guidelines [22] although results from an ongoing clinical trial are still pending. Other neurodegenerative disorders have been suggested to be treatable by rapamycin. Alzheimer's disease is a common neurodegenerative disorder characterized by progressive dementia and accumulation of beta amyloid and hyperphosphorylated tau protein in the brain of affected individuals. Interestingly, defects in autophagy have been shown to be part of the pathology. Indeed, stimulating autophagy by inhibiting mTOR attenuates the phenotype in several mouse models of this disease [23]. Parkinson's disease is another common age related neurodegenerative disorder caused by progressive loss of dopaminergic neurons in the basal ganglia leading to motor dysfunction and later dementia. Here, defects in mitochondrial autophagy have been shown to underlie some cases of familial Parkinson's disease [24, 25]. Accordingly, rapamycin attenuates neuronal death in models of Parkinson's disease [26]. In addition to these pro-autophagic aspects of rapamycin the anti-inflammatory feature of this drug has been suggested to be useful in the treatment of multiple sclerosis. In this disease progressive autoimmune demyelination occurs in foci in the central nervous system. Rapamycin has been shown to be able to reduce the microglial activation and decrease neuronal inflammation in model systems and is a promising compound in this disease [27]. Indeed, we have recently shown that rapamycin is able to reverse the mitochondrial phenotype of cells from patients suffering from the accelerated aging disorder Cockayne syndrome [28]. This disorder is neuropathologically characterized by inflammation and microglial activation and rapamycin may therefore be useful in patients suffering from this disease. Although clinical trials are still missing, rapamycin shows great potential as a neuroprotective compound in many diseases. We will surely know more in the coming years.

16.4 The Future of Rapamycin

The potential of rapamycin seems almost limitless. Nothing gold can, however, stay and several trials have shown disappointing results. These failures may be due to compensatory responses, mTORC2 inhibition or other of target effects. The field is, nevertheless, rapidly progressing and a number of new refined rapamycin analogues are now in clinical trials. There will undoubtedly be disappointments, but the immense interest in this group of drugs from major pharmaceutical companies will certainly lead to new treatments for a number of incurable and hard to treat disorders.

Acknowledgments I would like to thank Dr. Vilhelm A. Bohr and Dr. Evandro Fei Fang for invaluable feedback and critically reading the manuscript. This research was supported entirely by the Intramural Research Program of the NIH, National Institute on Aging.

References

1. Harrison DE, Strong R, Sharp ZD, Nelson JF, Astle CM, Flurkey K, Nadon NL, Wilkinson JE, Frenkel K, Carter CS et al (2009) Nature 460:392–395
2. Alexander A, Cai SL, Kim J, Nanez A, Sahin M, MacLean KH, Inoki K, Guan KL, Shen J, Person MD et al (2010) Proc Natl Acad Sci U S A 107:4153–4158
3. Laplante M, Sabatini DM (2012) Cell 149:274–293
4. Lamming DW, Ye L, Katajisto P, Goncalves MD, Saitoh M, Stevens DM, Davis JG, Salmon AB, Richardson A, Ahima RS et al (2012) Science 335:1638–1643
5. Hagiwara A, Cornu M, Cybulski N, Polak P, Betz C, Trapani F, Terracciano L, Heim MH, Ruegg MA, Hall MN (2012) Cell Metab 15:725–738
6. Sarbassov DD, Ali SM, Sengupta S, Sheen JH, Hsu PP, Bagley AF, Markhard AL, Sabatini DM (2006) Mol Cell 22:159–168
7. Peterson TR, Sengupta SS, Harris TE, Carmack AE, Kang SA, Balderas E, Guertin DA, Madden KL, Carpenter AE, Finck BN et al (2011) Cell 146:408–420
8. Toschi A, Lee E, Gadir N, Ohh M, Foster DA (2008) J Biol Chem 283:34495–34499
9. Chan EY (2009) Sci Signal 2:e51
10. Brunet A, Park J, Tran H, Hu LS, Hemmings BA, Greenberg ME (2001) Mol Cell Biol 21:952–965
11. Chi H (2012) Nat Rev Immunol 12:325–338
12. Thomson AW, Turnquist HR, Raimondi G (2009) Nat Rev Immunol 9:324–337
13. Langone AJ, Helderman JH (2012) Chest 142:734–737
14. Fleg JL, Aronow WS, Frishman WH (2011) Nat Rev Cardiol 8:13–28
15. Castro C, Campistol JM, Sancho D, Sanchez-Madrid F, Casals E, Andres V (2004) Atherosclerosis 172:31–38
16. Elloso MM, Azrolan N, Sehgal SN, Hsu PL, Phiel KL, Kopec CA, Basso MD, Adelman SJ (2003) Am J Transplant 3:562–569
17. De LG, Dirksen MT, Spaulding C, Kelbaek H, Schalij M, Thuesen L, van der Hoeven B, Vink MA, Kaiser C, Musto C et al (2012) Arch Intern Med 172:611–621
18. Palmerini T, Biondi-Zoccai G, Della RD, Stettler C, Sangiorgi D, D'Ascenzo F, Kimura T, Briguori C, Sabate M, Kim HS et al (2012) Lancet 379:1393–1402
19. Bove J, Martinez-Vicente M, Vila M (2011) Nat Rev Neurosci 12:437–452
20. Modi PK, Komaravelli N, Singh N, Sharma P (2012) Mol Biol Cell 23:3722–3730
21. Krueger DA, Care MM, Holland K, Agricola K, Tudor C, Mangeshkar P, Wilson KA, Byars A, Sahmoud T, Franz DN (2010) N Engl J Med 363:1801–1811
22. Curatolo P, Jozwiak S, Nabbout R (2012) Eur J Paediatr Neurol 16:582–586
23. Cai Z, Zhao B, Li K, Zhang L, Li C, Quazi SH, Tan Y (2012) J Neurosci Res 90:1105–1118
24. Valente EM, Abou-Sleiman PM, Caputo V, Muqit MM, Harvey K, Gispert S, Ali Z, Del TD, Bentivoglio AR, Healy DG et al (2004) Science 304:1158–1160
25. Vives-Bauza C, Zhou C, Huang Y, Cui M, de Vries RL, Kim J, May J, Tocilescu MA, Liu W, Ko HS et al (2010) Proc Natl Acad Sci U S A 107:378–383
26. Malagelada C, Jin ZH, Jackson-Lewis V, Przedborski S, Greene LA (2010) J Neurosci 30:1166–1175
27. Dello RC, Lisi L, Feinstein DL, Navarra P (2012) Glia [Epub ahead of print]
28. Scheibye-Knudsen M, Ramamoorthy M, Sykora P, Maynard S, Lin PC, Minor RK, Wilson DM, III, Cooper M, Spencer R, de CR et al (2012) J Exp Med 209:855–869

Author Biography

Dr. Morten Scheibye-Knudsen first brush with science was during his M.D. studies were he received a scholarship to investigate mitochondrial bioenergetics in the lab of Professor Bjørn Quistorff at the University of Copenhagen. After getting his M.D. from the University of Copenhagen he worked briefly as a physician in hospitals in Denmark and Greenland before joining the lab of Professor Will Bohr where he has been since 2008. Here he has authored and co-authored a number of papers and received a several highly competitive awards. His focus in this lab has been on mitochondrial function in relationship to aging and accelerated aging disorders. Here he has found a substantial component of mitochondrial dysfunction in the accelerated aging disorder Cockayne Syndrome. Using rapamycin he has shown the reversal of the bioenergetics dysfunction in Cockayne syndrome perhaps indicating the first treatment in this incurable disorder. In addition to his scientific endeavors he has shown considerable entrepreneurship and founded or cofounded several companies. One of them, Forsoegsperson.dk, is the most successful recruiter of volunteers for clinical trials in his home country Denmark.

Chapter 17
Eating, Drinking, Smoking and Cancer Prevention: A Focus on Acetaldehyde

Roshanna Rajendram, Rajkumar Rajendram and Victor R Preedy

Abstract At room temperature acetaldehyde (ethanal) is a flammable, colourless gas with a fruity odour. In 2009, the International Agency for Research on Cancer of the world health organisation classified acetaldehyde as a Class 1 toxin (human carcinogen) [2]. Acetaldehyde is an aldehyde that is highly reactive and toxic, causing damage at the cellular and genomic levels. The main source of the exposure of the general public to this toxin is through consumption of alcohol. In vivo, ethanol is predominantly metabolised to acetaldehyde. However, there are many natural and manufactured sources of acetaldehyde and even those who are teetotal are exposed to this carcinogen. The development of cancer is a multifactorial process in which acetaldehyde has an important role. However, exposure to acetaldehyde and therefore risk of the associated cancers is affected by a complex matrix of behavioural, dietary and genetic factors. Despite the widespread prevalence of acetaldehyde, exposure to the toxin can be limited. For example, potential public health measures to reduce acetaldehyde exposure include reduction of smoking and alcohol consumption. For such measures to be effective it is important to concurrently sensitise the general public to the potential adverse effects of acetaldehyde.

17.1 Introduction

It is estimated that over 30 % of cancer related mortality is attributable to diet [1]. Indeed; the carcinogenic effects of various different dietary constituents are being discovered all the time. Recent interest has focused on acetaldehyde. In 2009,

R. Rajendram
School of Medicine, University of Birmingham, Edgbaston, Birmingham B15 2TT, UK

R. Rajendram
Departments of General Medicine and Intensive Care, John Radcliffe Hospital, Oxford OX3 9DU, UK

R. Rajendram (✉) · V. R. Preedy
Diabetes and Nutritional Sciences Research Division, School of Medicine, King's College London, Franklin–Wilkins Building, 150 Stamford Street, London SE1 8WA, UK
e-mail: rajkumarrajendram@doctors.org.uk

E. F. Fang and T. B. Ng (eds.), *Antitumor Potential and Other Emerging Medicinal Properties of Natural Compounds*, DOI: 10.1007/978-94-007-6214-5_17,

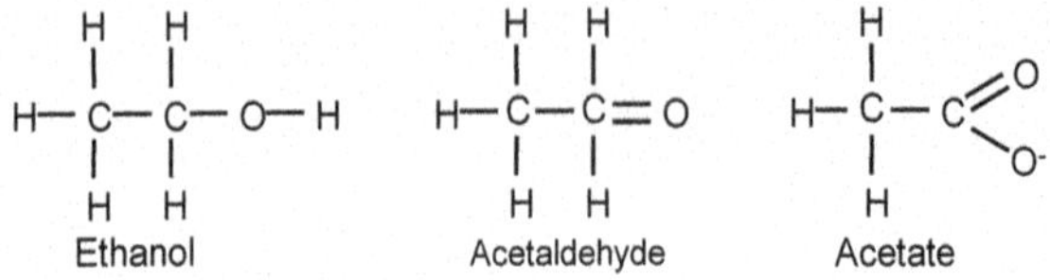

Fig. 17.1 The chemical structures of acetaldehyde, ethanol and acetate. Figure adapted from Rajendram et al. [5] with permission from Elsevier

the International Agency for Research on Cancer (IARC) of the world health organisation (WHO) reclassified acetaldehyde from Class 2B (probable human carcinogen) to Class 1 (human carcinogen) [2]. This chapter reviews the evidence for this reclassification, discusses the pathogenesis of malignancies related to acetaldehyde and outlines possible methods to prevent cancers induced by acetaldehyde.

17.1.1 Acetaldehyde

Acetaldehyde (Fig. 17.1), which is also known as ethanal, is part of the aldehyde family of compounds. At room temperature acetaldehyde is a colourless gas with a fruity odour that is flammable and unstable in air. It is highly reactive and toxic, and causes damage at the cellular and genomic levels. The chemical properties of acetaldehyde are listed in Table 17.1.*Sources* [a]HSDB 2009, [b] ChemIDplus 2009

17.1.2 Sources of Exposure to Acetaldehyde

The public is exposed to acetaldehyde through ingestion, inhalation, and dermal contact with the many natural and manmade sources of this toxin. The main source of this exposure is through consumption of alcoholic beverages [3]. However, there are many other sources of acetaldehyde so even those who are teetotal are exposed. The main use of acetaldehyde is as an intermediate in the production of vinegar (acetic acid), pyridine and pyridine bases, peracetic acid (an antimicrobial for use on indoor hard surfaces), pentaerythritol [used in the preparation of many compounds such as the explosive pentaerythritol tetranitrate (PETN)], butylene glycol (used in the resolution of carbonyl compounds in gas

Table 17.1 The chemical properties of acetaldehyde (ethanal)

Property	
Description	Colourless liquid with a fruity aroma
Chemical formula	CH_3CHO
Class	Aldehyde
Molecular weight	44.0

chromatography), and chloral (sedative) [3]. Acetaldehyde is also used as a preservative for fish and fruit and as a food flavouring that adds orange, apple, and buttery flavours [3]. Acetaldehyde is approved for use with phenolic resins in molded containers for contact with non-acidic foods [3].

Acetaldehyde is produced in most hydrocarbon oxidation reactions and is formed as an intermediate in the metabolism of most plants. It is found in trace amounts in many fruit and vegetable-based foods including apples, broccoli, coffee, grapefruit, grapes, lemons, mushrooms, onions, oranges, peaches, nectarines, pears, pineapples, raspberries, strawberries, cranberries, sour cherries, and mango. It has been detected in the essential oils of alfalfa, rosemary, balm, clary sage, daffodil, bitter orange, camphor, angelica, fennel, mustard, peppermint, and lychee, and in oak and tobacco leaves and cotton leaves and blossoms [3].

Acetaldehyde may be present in breast milk and dairy products, including cheese, yogurt, and milk [3]. Acetaldehyde has also been detected in cooked beef, chicken, and fish and is used as a synthetic flavouring ingredient in processed foods, especially margarine [3].

The United States Food and Drug Administration (FDA) state that acetaldehyde is safe for use as a flavouring agent and adjuvant. Acetaldehyde is often added to dairy products, baked goods, fruit juices, sweets, desserts, and soft drinks. Concentrations of acetaldehyde in food can be up to 0.047 % (470 mg/L) [3]. Unfortunately an acetaldehyde concentration as low as 100 μM (4.4 mg/L) can induce mutation of genetic material [4] and may therefore be carcinogenic.

Alcoholic beverages are the main source of the acetaldehyde to which the general public is exposed. The ethanol in these beverages is metabolised to acetaldehyde in vivo [5]. As acetaldehyde forms in alcoholic beverages after exposure to air, alcoholic beverages (including wines, beer, and spirits) also frequently contain acetaldehyde as a volatile component [3].

17.2 Alcohol and Acetaldehyde

After caffeine, ethanol (Fig. 17.1) is the most commonly used recreational drug worldwide. 'Alcohol' is synonymous with 'ethanol,' and 'drinking' often describes the consumption of beverages containing ethanol [5].

Alcoholism is associated with an increased incidence of cancers of the head and neck, gastrointestinal tract, breast, and liver (Table 17.2) [6, 7]. As ethanol itself is not carcinogenic, the pathogenesis of this phenomenon is unclear. Various theories have been proposed to explain this indirect carcinogenic effect of ethanol [6, 8]:

- Ethanol induces enzymes which promote tumour formation
- Ethanol affects metabolism or absorption of antioxidants
- Ethanol affects the hormonal balance in the body
- Ethanol increases exposure to harmful oxidants

Table 17.2 Increase in risk of cancer associated with 1 g of alcohol per day

Cancer type	Increase in risk/g ethanol per day
Oral cavity and pharynx	0.0185
Larynx	0.0136
Oesophagus	0.0129
Colorectal cancer	0.0080
Breast	0.0071
Liver	0.0059

Adapted with permission from Macmillan Publishers Ltd on behalf of Cancer Research UK: British Journal of Cancer (Parkin DM. Cancers attributable to consumption of alcohol in the UK in 2010. Br J Cancer 2011; 105: S14–S18, copyright (2011) [9]

- Ethanol suppresses the immune system
- Ethanol is mixed with other carcinogens in beverages
- Ethanol is a solvent for carcinogens

These theories can explain the systemic role of alcohol in carcinogenesis but not the local effects. For example, these theories do not explain the increased risk of oral and pharyngeal cancers associated with alcohol intake (see Table 17.2) [8, 9].

The acetaldehyde present naturally in alcoholic beverages arises as result of actions of yeasts and bacteria, and also by autooxidation. Both the local and systemic carcinogenic effects of alcohol consumption could be caused by acetaldehyde. The concentration of acetaldehyde varies between alcoholic beverages; high concentrations of acetaldehyde are found in spirits, lower concentrations are found in beers. See Table 17.3 for examples of the acetaldehyde content of various alcoholic beverages [10].

17.2.1 Metabolism of Ethanol to Acetaldehyde

The metabolism of ethanol to acetaldehyde can occur by one of three pathways (Fig. 17.2) [5]:

1. Alcohol dehydrogenase (ADH) in the cytosol
2. Microsomal ethanol oxidizing system (MEOS) in endoplasmic reticulum
3. Catalase in peroxisomes

Table 17.3 Concentrations of acetaldehyde in various alcoholic beverages

Alcoholic Beverage	Concentration (mM)
Calvados (French brandy) 40 % v/v ethanol	1.86
Shochu (Japanese liquor) 25 % v/v ethanol	1.16
Red wine 13 % v/v ethanol	0.25
Beer 5 % v/v ethanol	0.14

Adapted from: Yokoyama A, Tsutsumi E, Imazeki H, Suwa Y, Nakamura C, Mizukami T, Yokoyama T., Salivary Acetaldehyde Concentration According to Alcoholic Beverage Consumed and Aldehyde Dehydrogenase-2 Genotype. Alcohol Clin Exp Res 2008; 32: 1607–1614 [10]

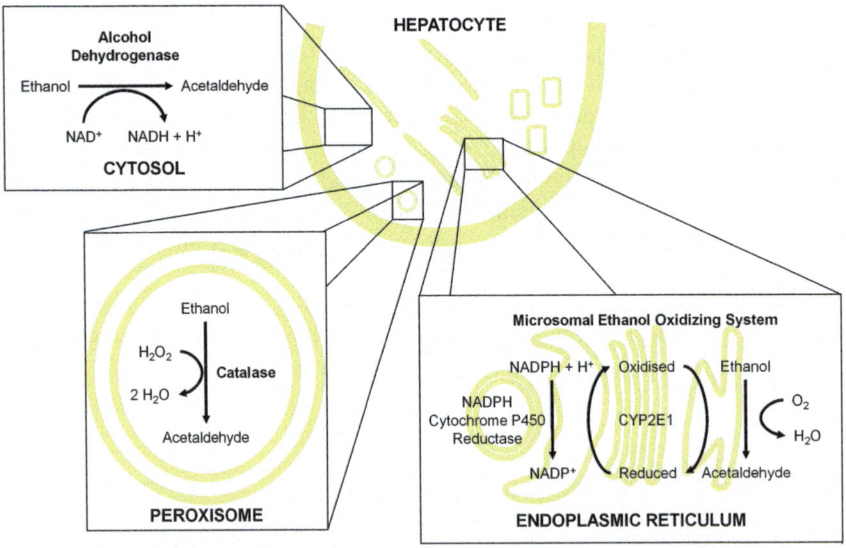

Fig. 17.2 The pathways of acetaldehyde production from ethanol metabolism. Figure adapted from Rajendram et al., [5] with permission from Elsevier

17.2.2 Alcohol Dehydrogenase

Alcohol dehydrogenase (ADH) is the main enzyme involved in the oxidation of ethanol to acetaldehyde [5]. It is a cytosolic enzyme that is mainly found in the liver. ADH is also present in the cells of the gastric mucosa and in various other tissues, including the lung. Alcohol dehydrogenase oxidises ethanol in the presence of NAD+ which is concomitantly converted to NADH. There are various polymorphisms of ADH and as a result individuals are classified as fast, intermediate or slow metabolisers of acetaldehyde (see section on genetic variation below). Some bacteria also possess ADH, which accounts for some of the oxidation of ethanol to acetaldehyde that occurs within the gastrointestinal tract [5].

17.2.3 Microsomal Ethanol Oxidising System

Cytochrome p450 2E1 (Cyp450 2E1) is another pathway which produces acetaldehyde from ethanol [5]. It metabolises approximately 20 % of ethanol [5]. This Cyp450 2E1 system is located in the endoplasmic reticulum microsomes of liver cells and it is involved in the oxidation of many compounds, including fatty acids, steroids, and exogenous drugs. Oxidation occurs via an electron transport chain: electrons are transferred from the substrate to electron carriers, one of which is Cyp450. In this reaction NADPH is oxidised and 2 molecules of water are formed [5].

In general, in those who do not misuse alcohol this microsomal ethanol oxidising system (MEOS) is usually much less active than the cytosolic ADH enzyme pathway [5]. However, consumption of large volumes of ethanol saturates the ADH enzyme pathway and increase Cyp450 2E1 activity. In addition, chronic alcohol consumption (over 40 g/day for at least a week) induces proliferation of the smooth endoplasmic reticulum and upregulates this pathway [5]. Upregulation of this MEOS pathway is one of the mechanisms that increase tolerance to ethanol. Thereafter, the pathway down regulates after a few days of abstinence [5].

Importantly, upregulation of this pathway also increases generation of reactive oxygen species which can damage DNA and lead to tumour formation [11]. The activation of dietary and environmental carcinogens also increase.

17.2.4 Catalase

Most cell types possess the enzyme catalase within peroxisomes. This enzyme oxidises ethanol to acetaldehyde. Catalase activity is dependent on the production of hydrogen peroxide by other reactions and is inhibited by the products of ADH enzyme activity; therefore it is responsible for less than 2 % of ethanol oxidation [5]. This pathway would only have a minor role if any at all in alcohol-acetaldehyde pathology.

17.3 Location of Ethanol, Acetaldehyde and Acetate Metabolism

17.3.1 Alcohol Metabolism

The main site of metabolism of ethanol is in the liver. However, extrahepatic metabolism also occurs in many tissues, including the gastrointestinal tract, and the lungs

17.3.1.1 Gastrointestinal Tract

In the gastrointestinal tract, ethanol oxidation is catalysed by mucosal enzymes and endogenous flora [5]. Importantly, this flora can also convert certain foods into alcohols. For example, glucose is fermented into ethanol, which is then rapidly oxidised to acetaldehyde. This not only affects the gastrointestinal mucosa, but also the lining of oropharynx and larynx (via transport in saliva). Rats administered acetaldehyde at doses to achieve concentrations of 120 mM, developed hyperplastic and hyperproliferative changes in the upper gastrointestinal tract mucosa [12].

Two other important risk factors for the development of gastric cancer are *Helicobacter pylori* (*H. pylori*) infection and atrophic gastritis [13, 14]. Indeed, some strains of H. pylori have ADH activity and can produce acetaldehyde, so it is possible that this factor may play a role in carcinogenesis. Furthermore, in individuals with achlorhydric atrophic gastritis bacteria or yeasts colonise the stomach. If there is excess bacterial or yeast proliferation, glucose may be fermented to produce ethanol, which can then be converted to acetaldehyde [15]. After consumption of alcohol the intragastric acetaldehyde concentrations in these individuals are over 6 times higher than the concentrations in those without this condition [15].

17.3.1.2 Lung

Cytochrome p450 is present in the lungs and Cyp450 activity is increased by smoking [16]. In the presence of ethanol this will further increase production of acetaldehyde in the lungs. This may lead to damage to genetic material and tumour development (see section on smoking).

17.3.2 Acetaldehyde Metabolism

Aldehyde dehydrogenase (ALDH) catalyses the conversion of acetaldehyde to acetate (Fig. 17.1) [5]. It is accompanied by the reduction of NAD^+. There are various isoenzymes of ALDH, the most important of which are ALDH2 (mitochondrial) which metabolises most of the acetaldehyde, and ALDH1 which is active in the cytosol. The presence of ALDH in tissues may reduce the toxic effects of acetaldehyde. Importantly, in alcoholics, when the oxidation of ethanol is increased by induction of MEOS, the activity of mitochondrial ALDH is reduced [5]. Hepatic acetaldehyde concentration therefore increases with chronic ethanol consumption; a finding reflected by the significant increase of acetaldehyde levels in hepatic venous blood. The carcinogenic potential of acetaldehyde may be increased in chronic alcoholics (Fig. 17.3).

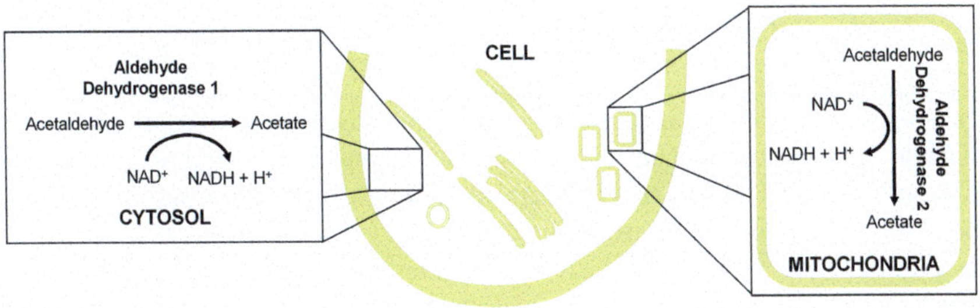

Fig. 17.3 The pathways of acetate production from acetaldehyde. Figure adapted from Rajendram et al. [5] with permission from Elsevier

17.3.3 Metabolism of Acetate

The fate of acetate derived from ethanol and acetaldehyde has not yet been fully elucidated [5]. However, some important principles have been established [5]:

1. The majority of absorbed ethanol is metabolized in the liver to acetaldehyde and then released as acetate. Acetate release from the liver increases 21 times after ethanol consumption [5].
2. Acetyl-CoA synthetase catalyses the conversion of acetate to acetyl-CoA. This reaction requires adenosine triphosphate (ATP) which is converted to adenosine monophosphate (AMP). The AMP is converted to adenosine in a reaction catalyzed by 5′-nucleosidase [5].
3. Acetyl-CoA may be converted to glycerol, glycogen, and lipid, particularly in the fed state. However, this only accounts for a small fraction of absorbed ethanol [5].
4. The acetyl-CoA generated from acetate may be used to generate ATP via the Krebs cycle [5].
5. Acetate readily crosses the blood–brain barrier and is actively metabolized in the brain. Acetylcholine, a neurotransmitter, is produced from acetyl-CoA [5].
6. Both cardiac and skeletal muscle are very important in the metabolism of acetate [5].

Based on these observations, future studies on the effects of acetaldehyde-acetate metabolism should focus on adipose tissue, cardiac and skeletal muscle, and the brain.

17.4 Genetic Variation in Ethanol and Acetaldehyde Metabolism

Genetic variation in metabolic rates provides a human model for different levels of acetaldehyde exposure. This allows us to examine the role of acetaldehyde in carcinogenesis.

17.4.1 Alcohol Dehydrogenase

There are three important genetic loci for Alcohol dehydrogenase (ADH); ADH1A, ADH1B and ADH1C. To date polymorphisms have only been found to occur at two of these loci (ADH1B and ADH1C). At the ADH1B locus, the ADH1B * 2 allele codes for an enzyme which oxidises ethanol at a higher rate than that generated from the ADH1B * 1 allele. In fact, homozygotes for the ADH1B * 1 allele (slow metabolisers) only have 1 % of the capacity to oxidise ethanol of heterozygotes (ADH1B * 1/ADH1B * 2; intermediate metabolisers) and 0.5 % of the oxidation capacity of homozygotes of the ADH1B * 2 allele, who are fast metabolisers [17].

As the faster metabolisers of ethanol produce more acetaldehyde, then in theory, if acetaldehyde is carcinogenic to humans, these individuals would be at higher risk of developing carcinomas. However, paradoxically, the slow metabolisers, theoretically exposed to less acetaldehyde, are at the highest risk of developing oesophageal cancer, and the fast metabolisers are at the lowest risk, with intermediate metabolisers having a level of risk between these two extremes [17]. This observation may be explained by work conducted by Rivera-Meza et al. [17] who created an animal model, in which rats selectively bred for traits of high alcohol consumption were made to express a rat analogue of the human gene ADH1B * 2. On administration of ethanol liver ADH activity increased by 90 % and was accompanied by an arterial surge in acetaldehyde concentrations 5 times greater than that of control mice without the allele. This was associated with reductions in ethanol consumption by the mice. This demonstrates that genotype interacts with behaviour to affect exposure to acetaldehyde, and probably risk of cancer.

For the other polymorphic ADH1C locus, the ADH1C * 1 allele codes for a protein which oxidises ethanol 2.5 times faster than the protein produced by the allele ADH1C * 2 [18]. Again, in theory, those who are fast metabolisers, and are exposed to higher levels of carcinogenic acetaldehyde should be at higher risk of developing cancers. However, a meta-analysis of seven case control studies with a combined total of 1,325 cases and 1,760 controls concluded that possessing the ADH1C * 1 allele was not associated with increased risk of head and neck cancers [18]. Nevertheless, smoking behaviour was not fully accounted for, and the participants were mainly mild or moderate drinkers [18]. It is possible that individuals with this genotype would be at higher risk of developing a cancer if they consume alcohol frequently. Indeed, in heavy drinkers the ADH1C * 1 allele was a risk factor for developing head and neck as well as oesophageal and hepatocellular carcinomas [18].

17.4.2 Aldehyde Dehydrogenase

The ALDH2 isoenzyme metabolises the majority of the acetaldehyde in the body. One particular allele (ALDH2 * 2), found more frequently in Orientals, has a point mutation which produces an inactive ALDH enzyme [19]. Whilst homozygotes cannot metabolise acetaldehyde, heterozygotes with one mutated allele will metabolise acetaldehyde at a much slower rate. Respectively, serum acetaldehyde concentrations will be up to nineteen and six times higher than controls with normal metabolism.

In homozygotes for the mutant ALDH2 * 2 allele, the toxin builds up causing an unpleasant reaction of flushing, tachycardia and nausea, which deters heavy alcohol consumption. This may therefore have a protective effect. A recent meta-analysis conducted by Fang et al. found a reduced risk for oesophageal cancer in ALDH2 * 2 homozygotes overall, although this effect was not statistically significant [20].

However, if homozygotes for the mutant ALDH2 * 2 allele choose to imbibe alcohol, they are at very high risk of cancer. The same meta-analysis, observed that the odds of moderate drinkers with the ALDH2 * 2/ALDH2 * 2 genotype contracting oesophageal cancer, was 8 times greater than the odds of controls (non- and moderate drinkers without the mutation) developing cancer. Notably, the odds of developing cancer in heavy drinkers with the normal ALDH2 * 2/ALDH2 * 2 genotype was 7.05 times greater than that of controls. This implies that a smaller amount of alcohol consumption in ALDH2 deficient individuals confers a greater risk of oesophageal cancer when compared to other drinkers. In addition, in individuals with liver cirrhosis, cases who had hepatocellular carcinoma, had a higher prevalence of ALDH2 * 2/ALDH2 * 2 genotype than their cancer free counterparts [21]. Thus it is possible that the acetaldehyde exposure may also be involved in the development of hepatocellular cancer [21].

Yokoyama et al. [22] investigated the association with other cancers in a case control study of alcoholics, of whom 487 had cancer and 237 did not. After adjusting for age, amount of alcohol consumed and cigarette smoking, the odds of head and neck, oesophageal, stomach, colon and lung cancer were higher in individuals who possessed 1 mutant ALDH2 allele. Interestingly, no evidence was found for increased odds of liver and other cancers. It is possible that other risk factors for liver cancer, such as hepatitis infection and cirrhosis, which were not adjusted for, may have had a greater influence on carcinogenesis and may have confounded any potential association [22].

17.5 Mechanism of Carcinogenesis

Acetaldehyde is highly toxic, mutagenic and carcinogenic. By affecting both DNA synthesis and repair (Fig. 17.4) acetaldehyde can induce the development of tumours. In dividing cells, acetaldehyde is converted to crotonaldehyde ($CH_3CH = CHCHO$) which binds co-valently to deoxyribonucleic acid (DNA) to produce stable carcinogenic DNA adducts [4].

In the presence of basic amino acids or histones, acetaldehyde reacts with deoxyguanosine in DNA to form the DNA adduct, 1,N^2-propano-dG (PdG) [11].

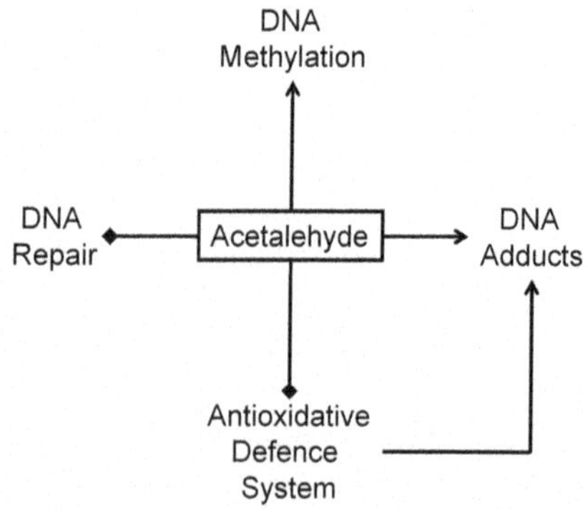

Fig. 17.4 Acetaldehyde causes DNA adducts, inhibits DNA repair and DNA methylation and damages the antioxidative defense system (AODS). These effects are toxic, mutagenic and carcinogenic

This induces point mutations in DNA, and chromosomal aberrations in cultures of mammalian cells in vivo at acetaldehyde concentrations from 40 to 1,000 μM [4, 11]. Adducts to DNA and other constituents of cells (e.g. proteins and lipids) cause intracellular protein retention, and impair antioxidant mechanisms [23]. This increases intracellular concentrations of free radicals causing lipid peroxidation and further damage to DNA. Adducts can thereby significantly affect the integrity of the genome.

The major stable DNA adduct, N^2-ethyl-2-deoxyguanosine (N^2-Et-dG) is present in DNA samples from white blood cells of human alcoholics. Although there is little evidence that this lesion is mutagenic or carcinogenic, it can be detected in human urine samples, and so could be useful as a marker of acetaldehyde induced DNA damage [11].

Acetaldehyde is highly mutagenic because it also affects DNA repair mechanisms. For example acetaldehyde inhibits O6-methyl-guanyltransferase enzyme activity [24]. This enzyme catalyses repair of DNA damaged by alkylating agents.

17.5.1 Smoking, Acetaldehyde and Carcinogenesis

Acetaldehyde is one of the most abundant carcinogens in tobacco smoke, with an average of 0.71 mg in each cigarette [25]. The main sources of acetaldehyde are the natural polymers present in tobacco leaf such as cellulose and lignin. A small amount is also produced from sugars added to the tobacco blend and casings of the cigarette, which are intended to enhance the flavouring and to act as humectants (i.e. help a product retain water) [26].

This exposure may contribute to the development of lung cancers: in animal models, inhalation of acetaldehyde vapour induces inflammation, metaplasia and tumours in the respiratory epithelium [19].

Smoking modifies the endogenous flora of the buccal cavity. This becomes rapidly colonised with Gram negative bacteria which produce more acetaldehyde than Gram positive bacteria. Indeed, Salaspuro observed acetaldehyde concentrations in smokers that were up to 60 % higher than in non-smokers [27]. This may explain the synergistic effect of smoking and drinking which increases risk of oral cancer far more than the risk associated with misuse of each substance independently [27].

17.5.2 Fermented Food Products

Acetaldehyde is also found in various fermented foods including vinegar, kimchi (a traditional fermented Korean dish made of vegetables), soybean pastes and soy sauce. Acetaldehyde is even present in bread. Furthermore, whilst fermented soy foods increase the risk of gastric cancer non-fermented soy foods (for example soy milk) do not [28]. The production of acetaldehyde by micro-organisms involved in the fermentation process may therefore play a role in carcinogenesis.

17.6 Recommendations

Despite the ubiquitous presence of acetaldehyde in foods, beverages and tobacco, it is possible to limit exposure to the toxin. Potential public health measures include [29]:

- Encouraging abstinence from or reduction of smoking and drinking behaviours.
- Promoting awareness of the carcinogenic effects of acetaldehyde.
- Strictly regulating acetaldehyde concentrations in foods.
- Banning of use of acetaldehyde as a food additive.
- Treating individuals exposed to high concentrations of acetaldehyde with L-cysteine (binds and inactivates acetaldehyde) [30].

17.7 Conclusion

Although the development of cancer is a multifactorial process, acetaldehyde has an important role. Selected evidence for the association of acetaldehyde with development of cancer has been described above. Exposure to acetaldehyde and therefore risk of the associated cancers is affected by behavioural, dietary and genetic factors. It is important to increase the awareness of the potential adverse effects of this compound, as well as to introduce protective public health measures in order to limit individual's exposure to its carcinogenic effects.

References

1. Doll R, Peto R (1981) The causes of cancer: quantitative estimate of avoidable risks of cancer in the United States today. J Natl Cancer Inst 66:1191–1308
2. Secretan B, Straif K, Baan R, Grosse Y, El Ghissassi F, Bouvard V et al (2009) A review of human carcinogens—part E: tobacco, areca nut, alcohol, coal smoke, and salted fish. Lancet Oncol 110:1033–1034
3. National Toxicology Program, U.S. Department of Health and Human Services (2011) Acetaldehyde. in national toxicology program. Report on carcinogens (12th ed). U.S. Department of Health and Human Services, USA, pp 21–24
4. Theruvathu JA, Jaruga P, Nath RG, Dizdorglu M, Brooks PJ (2005) Polyamines stimulate the formation of mutagenic 1 N2-propanodeoxyguanosine adducts from acetaldehyde. Nucleic Acid Res 33:3512–3520
5. Rajendram R, Hunter R, Preedy V, Peters T (2005) Alcohol: absorption, metabolism and physiological effects. In: Cabellero B, Allen L, Prentice A (eds) Encyclopedia of human nutrition (2nd ed). Elsevier, London, p 48–57
6. Baan R, Straif K, Grosse Y, Secretan B, El Ghissassi F, Bouvard V et al (2007) Carcinogenicity of alcoholic beverages. Lancet Oncol 8:292–293
7. Seitz HK, Pöschl G, Simanowski UA (1998) Alcohol and cancer. Recent Dev Alcohol 14:67–95

8. Homann N, Jousimies-Somer H, Jokelainen K, Heine R, Salaspuro M (1997) High acetaldehyde levels in saliva after ethanol consumption: methodological aspects and pathogenetic implications. Carcinogenesis 18:1739–1743
9. Parkin DM (2011) Cancers attributable to consumption of alcohol in the UK in 2010. Br J Cancer 105:S14–S18
10. Yokoyama A, Tsutsumi E, Imazeki H, Suwa Y, Nakamura C, Mizukami T, Yokoyama T (2008) Salivary acetaldehyde concentration according to alcoholic beverage consumed and aldehyde dehydrogenase-2 genotype. Alcohol Clin Exp Res 32:1607–1614
11. Seitz HK, Stickel F (2007) Molecular mechanisms of alcohol-mediated carcinogenesis. Nat Rev Cancer 7:599–612
12. Homann N, Koivisto R, Nosova T, Jokelainen K, Salaspuro M (1997) Effects of acetaldehyde on cell regeneration and differentiation of the upper gastrointestinal tract mucosa. J Natl Cancer Inst 89:1692–1697
13. Xue F-B, Xu Y–Y, Wan Y, Pan B-R, Ren J, Fan DM (2003) Association of *H pylori* infection with gastric carcinoma: a meta-analysis. World J Gastroenterol 7:801–804
14. Sipponen P, Kekki M, Haapakoski J, Ihamäki T, Siurala M (1985) Gastric cancer risk in chronic atrophic gastritis: statistical calculations and cross-sectional data. Int J Cancer 35:173–177
15. Väkeväinen S, Mentula S, Nuutinen H, Salmela KS, Jousimies-Somer H, Färkkilä M et al (2002) Ethanol-derived microbial production of carcinogenic acetaldehyde in achlorhydric atrophic gastritis. Scand J Gastroenterol 37:648–655
16. Villard P-H, Seree EM, Re J-L, Meo MD, Barra Y, Attolini L et al (1998) Effects of tobacco smoke on the gene expression of the Cyp1a, Cyp2b, Cyp2e and Cyp3a subfamilies in mouse liver and lung: relation to single strand breaks of DNA. Toxicol Appl Pharmacol 148:94–204
17. Rivera-Meza M, Quintanilla ME, Tampier L, Mura CV, Sapag A, Israel Y (2010) Mechanism of protection against alcoholism by an alcohol dehydrogenase polymorphism: development of an animal model. FASEB J 24:266–274
18. Homann N, Stickel F, König IR, Jacobs A, Junghanns K, Benesova M et al (2006) Alcohol dehydrogenase 1C*1 allele is a genetic marker for alcohol-associated cancer in heavy drinkers. Int J Cancer 118:1998–2002
19. Seitz HK, Matsuzaki S, Yokoyama A, Homann N, Väkeväinen S, Wang XD (2001) Alcohol and Cancer. Alcohol Clin Exp Res 25:137–143
20. Fang P, Jiao S, Zhang X, Liu Z, Wang H, Gao Y et al (2011) Meta-analysis of ALDH2 variants and esophageal cancer in Asians. Asian Pac J Cancer Prev 12:2623–2627
21. Kato S, Tajiri T, Matsukura N, Matsuda N, Taniai N, Mamada H et al (2003) Genetic polymorphisms of aldehyde dehydrogenase 2, cytochrome p450 2E1 for liver cancer risk in HCV antibody-positive Japanese patients and the variations of CYP2E1 mRNA expression levels in the liver due to its polymorphism. Scand J Gastroenterol 38:886–893
22. Yokoyama A, Muramatsu T, Ohmori T, Yokoyama T, Okuyama K, Takahshi H et al (1998) Alcohol-related cancers and aldehyde dehydrogenase-2 in Japanese alcoholics. Carcinogenesis 19:1383–1387
23. Holstege A, Bedossa P, Poynard T, Kollinger M, Chaput JC, Houglum K et al (1994) Acetaldehyde-modified epitopes in liver biopsy specimens of alcoholic and nonalcoholic patients: localization and association with progression of liver fibrosis. Hepatology 19:367–374
24. Espina N, Lima V, Lieber CS, Garro AJ (1988) In vitro and in vivo inhibitory effect of ethanol and acetaldehyde on O6-methylguanine transferase. Carcinogenesis 9:761–766
25. IARC (1997) IARC monographs on the evaluation of carcinogenic risks to humans. Overall evaluations for carcinogenicity: an updating of IARC monographs, vol 160. International Agency for research in cancer, Lyon
26. Seeman JI, Dixon M, Haussmann H-J (2002) Acetaldehyde in mainstream tobacco smoke: formation and occurrence in smoke and bioavailability in the smoker. Chemical Res Toxicol 15:1331–1350
27. Salaspuro V, Salaspuro M (2004) Synergistic effect of alcohol drinking and smoking on in vivo acetaldehyde concentration in saliva. Int J Cancer 111:480–483

28. Wu Ah, Yang D, Pike MC (2000) A meta-analysis of soyfoods and risk of stomach cancer: the problem of potential confounders. Cancer Epidemiol Biomark prev 9:1051–1058
29. Salaspuro M (2012) Interactions of alcohol and tobacco in gastrointestinal cancer. J Gastroenterol Hepatol 27:135–139
30. Salaspuro V, Hietala J, Kaihovaara P, Pihlajarinnne J, Marvola M, Salaspuro M (2002) Removal of acetaldehyde from saliva by a slow-release buccal tablet of L-cysteine. Int J Cancer 97:361–364

Author Biography

Dr Rajkumar Rajendram FRCA MRCP (UK) MBBS (Dist) BSc (Hons) AKC graduated in 2001 with a distinction from Guy's, King's and St. Thomas Medical School, in London. As an undergraduate he was awarded several prizes, merits and distinctions in pre-clinical and clinical subjects. This was followed by training in general medicine and intensive care in Oxford, during which period he attained membership of the Royal College of Physicians (London, MRCP) in 2004. Dr Rajendram went on to train in anaesthesia and intensive care in the Central School of Anaesthesia, London Deanery and became a fellow of the Royal College of Anaesthetists (FRCA) in 2009. He has completed higher training in regional anaesthesia and pain medicine and advanced training in intensive care. He returned to Oxford in his current position as a Locum Consultant in General Medicine at the John Radcliffe Hospital, Oxford. He co-authored the Oxford Case Histories in Cardiology which was published by the Oxford University Press in 2011. He is currently preparing the text for the Oxford Case Histories in Intensive Care. His unique training and experience has been tailored for a career in intensive care with a sub-specialty interest in peri-operative medicine. Dr Rajendram recognises that nutritional support is a fundamental aspect of critical care. He has therefore devoted significant time and effort into nutritional science research. He has investigated the effects of alcohol and acetaldehyde on skeletal muscle. As a visiting research fellow in the Nutritional Sciences Research Division of King's College London he has published over 50 textbook chapters, review articles, peer-reviewed papers and abstracts from his work with Professor Preedy.

Miss Roshanna Rajendram BMedSci is a final year medical student at the University of Birmingham. She has a strong interest in public health and population sciences and obtained an intercalated degree in this subject in 2012. She is also keenly interested in medical education and has co-authored a book about the application process to medical school.

Professor Victor R. Preedy BSc PhD DSc FSB FRCPath FRSPH is currently Professor of Nutritional Biochemistry in the Department of Nutrition and Dietetics, King's College London and Honorary Professor of Clinical Biochemistry in the Department of Clinical Biochemistry, King's College Hospital. He is also Director of the Genomics Centre, King's College London. Professor Preedy gained his PhD in 1981 and in 1992 he received his Membership of the Royal College of Pathologists (MRCPath). He was elected a Fellow of the Royal College of Pathologists (FRCPath) in 2000. In 1993 he gained his second doctorial degree (DSc) for his outstanding contribution to protein metabolism. In 2004 Professor Preedy was elected as a Fellow to both the Royal Society for the Promotion of Health (FRSH) and The Royal Institute of Public Health (FRIPHH). In 2009 he was elected as a Fellow of the Royal Society for Public Health (RSPH). He is also a Fellow of The Society of Biology (FSB). Professor Preedy has written or edited over 550 articles. His interests pertain to matters concerning pathology, biochemistry, food science, nutrition and health at the individual and societal levels.

Chapter 18
The Use of Edible Mushroom Water Soluble Polysaccharides in the Treatment and Prevention of Chronic Diseases: A Mechanistic Approach

Betty Schwartz, Yitzhak Hadar and Daniel Sliva

Abstract The use of dietary components having protective and/or preventive effects on chronic diseases is an important emerging field of research. Identifying new food supplements and understanding their mechanisms of action are some of the main challenges in using functional foods. In this manuscript we reviewed the sources, the chemical composition and medicinal properties of polysaccharides extracted from edible mushrooms. In addition we brought insights into the putative mechanisms of action behind each health-promoting activity of these interesting biomolecules.

18.1 Introduction

Many of the Basidiomycetes produce polysaccharides with very important advantageous medicinal properties. Advances in biochemical techniques have allowed the isolation, purification and characterization of polysaccharides from mushrooms that exhibit advantageous medicinal properties [1] that can be produced at the commercial level via a variety of technologies. The medicinal properties of these compounds have been shown to depend not only on the molecular weight of the polysaccharide but also on the structure of the polymeric backbone, the sugar composition and the degree of branching as well as the producing fungus [2–4].

Mushroom polysaccharides have been associated to prevention of cancer, diabetes, hyperlipidemia, arteriosclerosis, hepatitis, and other chronic diseases [5].

B. Schwartz (✉) · Y. Hadar
The Robert H. Smith Faculty of Agriculture, Food and Environment,
The Hebrew University of Jerusalem, Rehovot, Israel
e-mail: betty.schwartz@mail.huji.ac.il

D. Sliva
Cancer Research Laboratory, Methodist Research Institute,
Indiana University Health, Indianapolis, IN, USA

E. F. Fang and T. B. Ng (eds.), *Antitumor Potential and Other Emerging Medicinal Properties of Natural Compounds*, DOI: 10.1007/978-94-007-6214-5_18,

Biologically active water soluble polysaccharides harvested from edible mushroom are widely distributed among the medicinal mushrooms. The bioactive polysaccharides isolated from mushroom fruiting-bodies, submerged cultured fungal biomass, or liquid culture fermentation broths are either water-soluble α- and β-D-glucans, β-D-glucans with heterosaccharide chains of xylose, mannose, galactose, or uronic acid, or β-D-glucan-protein complexes, β-(1,3)-D-glucans with β-(1,6) branches in other words, proteoglycans [5–7]. Polysaccharides differ in their primary structure (type of basic sugar), type of linkage (α, β, etc.), degree of branching, and molecular weight, among other parameters [8]. We aim in this article to review some of the health-promoting activities of these polysaccharides and describe some of their possible mechanism of action.

18.2 Sources and Cultivation Methodology for Extraction of Polysaccharides from Medicinal Mushrooms

18.2.1 General

The medicinal fungi, belonging to the Basidiomycetes, show a vast potential for bioactive compounds. Wasser [3] summarized a list of 651 species belonging to 182 genera of hetero- and homobasidomycetes demonstrated to produce polysaccharides with antitumor or immunostimulating activity. Most polysaccharides are classified as nonspecific bioactive substances because their exact mode of action is not completely demonstrated. Within the basidiomycestes, homopolymers of D-glucose (glucans) are very common. Glucans are polysaccharide constituents very commonly found in the fungal kingdom. Their diversity results from the different bonds among monomer units. Condensation can take place with any hydroxylic group of any carbon atom and can result in the conformation of either α- or β-bonds, since there are at least eight different ways in which two glucose units can link. The diversity of glucans is further increased by substitutions of sugar rings and by branching of chains [2, 9]. Their molecular weight, chemical composition, the number of branches of side chains as well as spatial configuration determine the physical and therapeutic properties of mushroom glucans [10]. These glucans can be water soluble or insoluble. The insoluble fractions are usually structural components of the cell wall and cross linked to other polysaccharides like chitin or to proteins. The soluble fraction provides 20–50 % of the total glucans [9]. When the fungi are cultivated in submerged fermentation glucans can be secreted at levels that directly depend on chemical composition of the growth medium. In general nitrogen starvation will enhance soluble glucan secretion [11].

Heteroglycans are also common constituents found in the fungal kingdom. They contain xylose, galactose, mannose, arabinose, fructose and glucuronic acid as side chain components. Among the basidiomycetes, bioactive polysaccharides are produced by several mushrooms traditionally known for their medicinal properties, mostly in China, Korea and Japan. Out of over 2,000 edible fungi found worldwide, only 20 are cultivated on a commercial level. Among them, three species predominate: *Agaricus bisporus* (common mushroom) (approx. 32 % total weight of produced mushrooms), *Lentinula edodes* (approx. 25 %) and *Pleurotus spp.* (approx. 25 %) [10]. In addition to being popular culinary mushrooms, *Lentinula* and *Pleurotus* exhibit medicinal properties. In this review we focus on four fungi, which based on traditional medicine have become a target for extensive scientific clinical research during the last few decades, both in the east and west, as a source for searching bioactive polysaccharides with medicinal properties.

18.2.2 Lentinus Edodes

Lentinus. edodes also known as the "Shiitake Mushroom", is an edible medicinal mushroom, native to China and cultivated in East Asia. The mushrooms can be consumed fresh or dried as well as dried powder for complementary and alternative medicine purposes [12]. *L. edodes* is traditionally grown on wood logs. A new method of cultivation based on artificial logs, was developed utilizing heat-treated substrates based on sawdust enclosed in plastic bags. In this method the cultivation cycle is shorter and the yields are higher [13].

The commercial polysaccharide lentinan, thought to be responsible for many of the beneficial effects produced by the mushroom *L. edodes* is a β-glucan [1, 14]. Lentinan can be produced also in submerged, agitated liquid cultures. The polysaccharide obtained by this methodology has a similar infrared spectrum compared to the polysaccharide obtained from the fruiting bodies or from the extracellular culture fluids. However, some differences appear in the molecular composition of the intra- and extracellular lentinan. Lentinan produced from the extracellular fluids of the fungus is much more effective in stimulating antibody production in mice than that extracted either from fruiting bodies or from the biomass produced by fermentation methods [14].

18.2.3 Pleurotus spp.

Mushrooms of the genus *Pleurotus,* also known as the oyster mushrooms, are edible and among the most popular cultivated mushrooms worldwide. *Pleurotus* spp. is among the easiest mushrooms to cultivate. In nature, they grow on wood, usually on dead, standing trees or on fallen logs. Large variety of lignocellulosic substrates such as wood chips, corn wheat or rice straw, cotton stalks, waste hulls and other locally available agricultural wastes, can be used for *Pleurotus* cultivation [13, 15, 16].

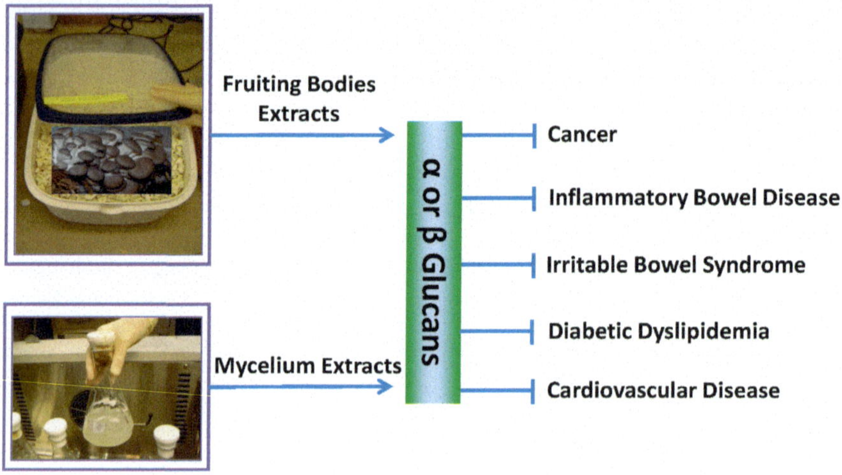

Fig. 18.1 Preparation of alpha and beta glucan polysaccharides and their effects on chronic diseases

The genome of *P. ostreatus* was sequenced (http://genome.jgi-psf.org/PleosPC9_1/PleosPC9_1.home.html) and the fungus became accessible to genetic engineering, thus it may be possible to improve its properties by gene replacement via a gene targeting system [17]. Studies from different regions of the world, reviewed by Khan and Tania [18] establish the medicinal importance of this popular and highly nutritious mushroom. Lavi et al. [19] compared the molecular weight and composition of the glycosyl residues, and the types of glycoside bonds, in polysaccharides harvested from *P. pulmonarius* under different growing conditions. Polysaccharides extracted from fruiting bodies produced by the fungus grown on straw (FBE) were compared to glucan extracted from mycelia produces in submerged culture (ME). These conditions represent commercial mushroom production that are technologically developed and common worldwide, on the one hand, and submerged fermentation of the fungal mycelium utilizing biotechnological practices, on the other (see Fig. 18.1). Growing mushrooms in the farm may take several weeks and it is difficult to control the quality of the final product while large scale fermentation is more sophisticated and controlled. The products of the different methodologies can also vary. For example, Lavi et al. [19] found that the glucan extracted from FBE contained 85 % glucose as compared to that extracted from ME contained only 64 % glucose. Both FBE and ME glucans contained significant and equal amounts of galactose (8.3 %). The ME polysaccharide also contained fucose and xylose, which were not found in the FBE glucan. The reason for the difference in carbohydrate content could be explained by the differences in carbon sources and growing conditions. ^{13}C and ^{1}H NMR analyses of the FBE preparation showed mixed α- linkages and β-anomeric carbon linkages, whereas the ME polysaccharide demonstrated only α-glucan linkages.

18.2.4 *Grifola frondosa*

Grifola. frondosa (maitake) produces a polysaccharide also known commercially as D-fraction. It is considered as a delicious edible mushroom. Production of *G. frondosa* is usually on a lignocelulosic substrate enclosed in bottles or bags. A common substrate used for production is composed of sawdust amended with rice bran or wheat bran. It has also been produced on a substrates consisting of oak sawdust, wheat bran, millet, and rye. Highest yields, best quality, and shortest crop cycle time (12 weeks) is usually the target for optimization. Due to its short history of cultivation only relatively little effort has been directed to improve culture conditions for *G. frondosa* fruiting bodies production on a farm level [20, 21]. For example Svagelj et al. [22] used solid-state substrate consisting of milled whole corn plant and olive press cake supplemented with mineral additives and olive oil keeping the moisture level above 70 %. In this study four fractions of pure extracellular β-D-glucans and four fractions of intracellular polysaccharides were isolated. Montoya et al. [23] investigated the enzymes that enable *G. frondosa*, to colonize and deconstruct two formulations based on industrial lignocellulosic by-products. *G. frondosa* degraded the substrates (oak-sawdust plus corn bran, and oak/corn bran supplemented with coffee spent-ground), however the coffee spent-ground amendment inhibited mushroom production [23]. Chen et al. [24] isolated polysaccharides from the fruiting bodies of *G.frondosa* and purified by DEAE cellulose-52 chromatography and Sephadex G-100 size-exclusion chromatography. They obtained three main polysaccharides with antioxidant activity. They concluded that polysaccharides isolated from *G. frondosa* can be used as a reliable source of natural antioxidants for the food processing industry.

18.2.5 *Ganoderma lucidum*

Ganoderma. lucidum, and other *Ganoderma* species, also known as lingzhi mushroom, reishi and 'the mushroom of immortality', is a white-rot basidiomycete and the best known and widely used medicinal mushroom in Asia. It is one of the oldest mushrooms known to have been used for medicinal purposes. Unlike the other fungi described here, *Ganoderma*, due to its bitter taste, is not a typical edible mushroom. Different processes of cultivation of *Ganoderma* spp. and development of their products were reviewed by Zhou et al. [25]. An important step towards large scale production of any crop is genetic improvement. Zhou et al. [25] described the use of several techniques such as selection, mutation, cross-breeding, cell fusion and genetic engineering. A simple and efficient transformation system for *G. lucidum* was developed recently [26]. These authors based their methodology on the relatively easy detection expression capability of "enhanced green fluorescent protein" and "β-glucuronidase" reporter genes in fungal transformants using glyceraldehyde-3-phosphate dehydrogenase promoter.

Several cultivation methods are used for *Ganoderma* polysaccharides production and are similar in principle to the other medicinal mushrooms. As *Ganoderma* in nature, a white rot wood degrading fungus, woody raw material or other lignocellulosic substrates are used in solid state cultivation for the fruiting bodies production. *Ganoderma* mycelium can be produced by liquid submerged fermentation in artificial growth media [25]. Chen et al. [24] sequenced the complete genome of monokaryotic *G. lucidum* strain 260125-1, and identified a large set of genes and potential gene clusters involved in secondary metabolism and its regulation. The genome sequence and its further investigation will promote the use of *G. lucidum* as a source of pharmacologically active compounds as well as industrial enzymes.

Production of polysaccharides from additional well studied mushrooms such as *Agaricus* spp., *Trametes versicolor*, *Schizophyllum commune* is not detailed here, as the principles are similar, but their medicinal properties are described below.

18.3 The Use of Mushroom Polysaccharides in the Treatment and Prevention of Chronic Diseases

18.3.1 General Description

Since ancient times mushrooms have been claimed to exert antiviral, anti-inflammatory, hypotensive, hypocholesterolemic, hypoglycemic effects [27]. Recent advances in biochemical techniques have allowed the partial isolation and purification of compounds from medicinal mushrooms—especially polysaccharides—that exhibit health-promoting activities [28], see Fig. 18.1.

18.3.2 Anticancer Activity

The most striking effects reported hitherto are the antitumor effect of mushroom polysaccharides especially glucans. Their immunomodulating ability, anti-inflammatory properties and anti-tumor effects in vitro and in vivo have been reported [29–33]. Their antitumor effects have been shown to depend not only on the molecular weight of the polysaccharide but also on the structure of the polymeric backbone and on the degree of branching [34]. Polysaccharides with antitumor properties have been screened mostly in the fruiting bodies, less so in liquid culture medium and mycelia, nonetheless we have characterized also one that is pro-apoptotic [35]. The main anticancer polysaccharides that have been subjected to early clinical trials are lentinan (*L. edodes*), schizophyllan (*S. commune*), PSK and PSP (*T. versicolor*), and Grifron-D (*G. frondosa*) [4, 36–40]. Grifron-D is a relatively new anticancer mushroom polysaccharide that has demonstrated promising pre-clinical activity in cell lines and xenografts [41]. The American Federal

Drug Administration (FDA) has approved Grifron-D for clinical trials under an investigational new drug application for patients with advanced solid cancers [42]. Most of the major medicinal mushroom polysaccharides have been subjected to extensive preclinical toxicological studies. The degree of safety testing on mushroom products is in general more advanced than that for most herbal products [5].

There are two proposed mechanisms by which polysaccharides extracted from mushroom exert their antitumor effect: an indirect activity through the regulation of the host's immune system or alternatively a direct activity on the tumor cells. The indirect pathway, in which the polysaccharide acts only as a mediator in the immune response, has been studied in detail [43], however the exact mechanism(s) involved in the indirect mechanism still remain to be completely understood; polysaccharides may up or down-regulate various aspects of the humoral or cellular arms of the immune system [44]. This functional heterogeneity is likely to be the explanation for the wide activity that has been reported against a number of different cancers [45]. The ability of mushroom polysaccharides to enhance or suppress the host immune response depends on a number of other factors, including dosage, route of administration, timing and frequency of administration, as well as their mechanism of action. We have recently demonstrated that soluble glucans prepared from the edible mushroom *P. pulmonarius* inhibit colitis-associated colon carcinogenesis induced in mice through modulation of cell proliferation, induction of apoptosis and inhibition of inflammation [46].

Rice et al. [47] showed that orally administered water-soluble glucans translocate from the gastrointestinal (GI) tract into the systemic circulation. The glucans are bound by GI epithelial and Gut-Associated Lymphoid Tissue (GALT) cells and they modulate the expression of pattern recognition receptors in the GALT, increase IL-12 expression, and induce protection against infectious challenge. Most medicinal mushroom polysaccharides have been shown to be multi-cytokine inducers, capable of modulating gene expression of various cytokines [48]. It is now believed that β-D-glucans induce many biological responses through binding to membrane complement receptor type three on immune effector cells (macrophages), leading to ligand-receptor complex internalization [49]. Little is known about the further intracellular events that occur after β-glucan-receptor binding.

As for the direct pathway, in which the polysaccharide itself inhibits the cancerous growth, only recently Chen et al. showed that exopolysaccharide from the mushroom *Fomes fomentarius* has a direct antiproliferative effect in vitro on human gastric cancer cells in a dose- and time-dependent manner [50]. Xie et al. showed that *G. lucidum* extract inhibits proliferation of SW480 human colorectal cancer cells [51]. We have detected proapoptotic and antiadhesive effects exerted by polysaccharides on human colon cancer cells and shown that the polysaccharide itself inhibits the cancer cell growth [19, 35, 50].

The hot water extract of the Basiodiomycete fungus *A. blazei Murill* has potent antitumor activity in sarcoma 180-bearing mice (3–6) and the antitumor activity was postulated to reside in the β-(1–6) glucan fraction [52]. The methanol-soluble fraction showed antitumor and antiangiogenic activities [53].

18.4 Effect on Inflammatory Bowel Disease

Few studies have been devoted to investigate the biological effects of fungal polysaccharides on inflammatory bowel disease (IBD), therefore only a relatively poor understanding of the mechanistic associated with the effects on this disease is available [48, 54–56]. IBDs are represented mainly by Crohn's disease (CD) and ulcerative colitis (UC) [57]. IBD is a chronic idiopathic inflammatory disease of the gastrointestinal tract affecting significant population numbers in the Western world and the frequency of which has increased considerably over the past few decades [58] however the existing therapies not only show limited benefits, they also have unwanted side effects [57]. Consequently, there is a need for alternative anti-inflammatory agents that are at least equally effective, if not more so, and cause fewer side effects. In Asian countries, many species of mushrooms isolates are approved adjuvants for cancer or inflammation therapy. Investigation of the pathogenesis and treatment of IBD has been of increasing interest over the last decade.

We have recently demonstrated attenuation of experimental colitis in mice following treatment with orally administered glucans from the edible mushroom *P. pulmonarius* [48]. Our findings indicate that high doses of both hot-water soluble (HWS) or mycelium extract (ME) are effective at treating acute DSS-induced colitis and result in improved macroscopic and histological damage scores, decreased inflammation and lower levels of the pro-inflammatory cytokine IL-1β and TNF-α. Both treatments, when provided orally, were effective in preventing or curing colitis symptoms. Pleuran is a β-glucan isolated from *P. ostreatus*. When pleuran was locally administered, with or without concomitant parenteral pretreatment, it was effective in reducing colonic damage induced by acetic acid, but only when administered prophylactically by intraperitoneal route, but not when orally administerd [59]. Barley or yeast β-glucan, similarly to fungal β-glucan also show an immunostimulatory effect mediated by the activation of neutrophils, macrophages, monocytes and natural killer (NK) cells through specific receptors CR3 (CD11b/CD18), and through the β-glucan receptor. This effect is associated with stimulated production of cytokines, such as tumor necrosis factor alpha (TNF-α) and interleukin 1 (IL-1), resulting in increased immunological surveillance [59].

Choi et al., [55] have recently demonstrated that a water extract isolated from the mushroom *Inonotus obliquus* (IO) induces anti-inflammatory effects in colitis induced by dextran sodium sulfate in mice. IO has been demonstrated to induce anti-cancer [60–62] and anti-inflammatory activities [63]. Kim et al. [64] reported that the anti-inflammatory of IO is probably due to the inhibition of inducible NO synthase and cyclooxygenase-2 expression via the downregulation of NF-κB binding activity.

18.5 Effect on Irritable Bowel Syndrome

Irritable bowel syndrome (IBS) is a gastrointestinal disorder characterized by chronic abdominal pain associated with altered bowel habits. The prevalence of IBS is very high; IBS is one of the most common gastrointestinal disorders estimated to affect 7–15 % of the general population in the USA and other developed countries [65] thus, involves elevated health-care costs [66]. This situation calls for novel additional treatment protocols of this condition that could be beneficial; a case in point is edible mushroom water soluble polysaccharides. Asano et al. [67] examined the effect of β-glucan isolated from *A. pullulans* GM-NH-1A1 on fecal pellet output and visceral pain response in animal models of IBS [mice subjected to restraint stress and provided drugs that stimulate intestinal motility (bethanecol and 5-HT), cause diarrhea (castor oil) or cause constipation (loperamide and clonidine)]. Oral administration of the β-glucan suppressed the restraint stress- or drug-induced fecal pellet output. β-Glucan also suppressed the visceral pain response to colorectal distension. These results suggest that β-glucan could be therapeutically effective for the treatment of IBS [67]. In humans a recent study [68] demonstrated that a mixture of beta-glucan, inositol and digestive enzymes (Biointol) significantly improved some (but not all) of the IBS-related symptoms such as bloating, flatulence and abdominal pain in patients affected by IBS. Additional studies are needed to expand this promising area.

18.6 Hypolipidemic and Hypoglycemic Effects of Fungal Glucans

Diabetes is a group of metabolic disorders which includes alterations in carbohydrate, fat, and protein metabolisms associated with absolute or relative deficiencies in insulin secretion and insulin action. Diabetes associated dyslipidemia is a major risk factor for cardiovascular disease. There is an increasing demand for natural products with hypolipidemic and anti-diabetic activity, due to the side effects associated with the use of insulin and oral hypoglycemic agents. The search for substances present in edible mushrooms with hypolipidemic and/or hypoglycemic effects is advantageous, particularly in countries with a persistent incidence of hypercholesterolemia and/or hyperlipidemia and cardiovascular disease such as in the Western countries. Glucans harvested from various edible mushrooms have proven to exert significant hypolipidemic effects [69], hypocholesterolemic [70], hypoglycemic [71] and antidiabetic effects [72–75]. The polysaccharide isolated from *P. nameko* possesses significant anti-inflammatory activity in different models of inflammation and was recently demonstrated to be hypolipidemic [76]. β-glucans prepared from *A. blazei Murill* and oligosaccharides derived from the hydrolysis of β-glucans show anti-hyperglycemic, anti-hypertriglyceridemic, anti-hypercholesterolemic, and anti-arteriosclerotic activity indicating anti-diabetic activity as a whole in diabetic rats [77]. As far as we are aware, no human studies have been conducted with mushroom derived β-glucans [78]; nonetheless, a recent human trial using oat-derived β-glucan [79], demonstrated that a single daily ingestion of 3.5 g β-glucan, for 8 weeks did not change the lipid profile and hemoglobin A1c in type 2 diabetic subjects. In order to test the effect of fungal β-glucans, well designed studies are required in order to assess putative effects of fungal-derived β-glucans in hyperlipidemia, hyperglycemia and/or hypercholesterolemia.

18.7 Molecular Mechanisms Lying Underneath the Health Benefits of Edible Mushrooms

18.7.1 Mechanism of Action

Although mushroom 1-3-β-D-glucans (e.g. Lentinan from *Lentinula edodes*, D-fraction from *G. frondosa*, or Schyzophyllan from *Schizophyllum commune*) demonstrated promising anticancer effects in animal studies and limited clinical trials through their immunomodulatory activities, their therapeutic oral administration is limited due to the poor water solubility [80]. On the other hand, other water soluble mushroom polysaccharides (α- and β-glucans) also demonstrated significant biological activities. However, the molecular mechanism(s) responsible for their possible therapeutic effects is not clear. Interestingly, glucans demonstrated their potency to activate or to suppress specific immune responses and directly affect epithelial cells. The mechanisms involved in the glucan activities can be associated with their interactions with specific receptors which mediates intracellular signaling.

18.7.2 Immunomodulatory Activities

The original study with water-soluble 1-3-β-D-glucan isolated from the edible mushroom *Auricularia auricula-judae* demonstrated inhibition of Sarcoma 180 solid tumor in mice more than 30 years ago [81]. Recently, a variety of structurally related water soluble glucans and polysaccharides were isolated from different edible mushrooms. Water soluble 1-3-α,1-3-β1-6-β-d-glucan and 1-6-β-D-glucan from *P. florida* and 1-6-β-D-glucan from *P. florida* and 1-6-β-D-glucan from *Termitomyces robustus* induced NO production in peritoneal macrophages and stimulated proliferation of murine splenocytes and thymocytes, respectively [6, 82, 83]. In addition, a water soluble 1-6-α-D-glucan from *Sarcodon aspratus* induced proliferation of murine splenocytes [84]. 1-3-β,1-3-β,1-6-β-D-glucan and polysaccharide containing 1-6-β-Glu-1-6-α-Gal-1-3-β-Glu-1-4-β D-glucan backbone structure from *L. squarrosulus* also induced NO production in peritoneal macrophages and stimulated proliferation of murine splenocytes and thymocytes [85, 86]. A water soluble polysaccharide, branched-fuco-1-4-β-galacto-1-4-β,1-6-β,1-6-β-D-glucan from *C. indica* also induced NO production in macrophage cell line RAW264.7 and stimulated proliferation of murine splenocytes, thymocytes and bone marrow cells [87]. Water soluble polysaccharide rich extracts, containing β-D-glucan or 1-4-α-D-glucan, from *A. bisporus* induced production of TNF-α and NO in the bone marrow-derived macrophages [88]. In addition, other water soluble rich extracts, containing 1-6-α,1-4-α-D-glucan, 1-6-β-D-glucan and mannogalactan, from *A. bisporus* and *A. brasiliensis* induced mRNA levels of IL-1β, TNF-α, and COX-2 in THP-1 monocytes, respectively [89].

It is well established that β-glucans activate immune response through their interaction with membrane receptors found on the immune cells, including macrophages, neutrophils, monocytes, natural killer cells and dendritic cells [90]. These receptors: dectin-1, CR3 (complement receptor 3), LacCer (lactosylceramide), and TLR (toll-like receptor) mediate adaptive immunity. In addition, β-glucan interaction with dectin-1, CR3, LacCer and scavenger receptor induces innante immunity of neutrophils [91]. Mechanistically, binding of β-glucan to dectin-1: (1) activates Syk kinase which through CARD9 induces canonical NF-κB pathway, (2) activates Syk and induces ERK and MAP kinase pathways independently of CARD9, or (3) activates Raf1 kinase independently of Syk1 [90]. In addition, Dectin-1 interacts with MyD88 coupled TLR receptors (TLR-2,-4,-5,-7,-9) leading to the activation of NF-κB and production of a variety of cytokines [91]. The prominent ligands of dectin-1 are water insoluble 1-3-β-D-glucans, consisting of backbone chain of at least seven glucose subunits with minimally one glucose side chain (1-6-B) forming a helical structure [91]. Indeed, the helical structure is typical of 1-3-β-D-glucan among other glucans [92]. Therefore, dectin-1 differentially recognizes glucans based on structural factors such as polymer length and side-chain branching [91].

Since the above described water soluble mushroom glucans activities do not correspond to the typical water insoluble ligand for dectin-1, it is possible that other mechanisms are responsible for their immunomodulatory activities. Interestingly, water soluble 1-3-branched 1-6-β-D-glucan from *G. frondosa* (MD-fraction) induced TNF-α production and increased proliferation without excessive inflammation in resident macrophages [93]. However, MD-fraction activated ERK and MAPK resulting in the production of granulocyte macrophage colony-stimulating factor (GM-CSF) which subsequently activated dectin-1/Syk signaling [93]. In addition, increased production of TNF-α and NO by β-D-glucan or 1-4-α-D-glucan was independent of dectin-1 [88]. Although the binding of soluble β-glucans to dectin-1 and the biological activities of these β-glucans remain controversial [93], recent studies clearly demonstrate that structurally different mushroom α- or β-glucans hold significant immunomodulatory activities. Nevertheless, the mechanism of their interaction with dectin-1and TLR receptors remains to be elucidated.

18.7.3 Anti-inflammatory Activities

The anti-inflammatory activities (inhibition of DSS-induced colitis in mice) were originally demonstrated with an insoluble 1-3-β-D-glucan (Pleuran) from *P. ostreatus* in a rat model of acute colitis [54]. Moreover, we have recently demonstrated that water soluble α- and β-glucans from *P. pulmonarius* also suppressed colon inflammation as shown by the reversion of the shortening of colon length and inhibition of expression of TNF-α and IL-1β in colonic mucosa in the dextran sodium sulfate (DSS)-induced colitis and DSS/azoxymethane induced colon carcinogenesis in mice, respectively [19, 48]. These glucans also inhibited TNF-α-dependent activation of NF-κB in human intestinal cells in vitro [46]. In addition, soluble 1-6-branched 1-3-β-D-glucan from *P. pulmonarius* inhibited leukocyte migration to injured tissues and expression of IL-1β in a rodent model of inflammatory pain [94, 95]. We have also demonstrated that a water soluble extract from *P. ostreatus* (OMC), containing α- and β-glucans, suppressed LPS-induced dependent activation of TNF-α, IL-6 and IL-12, as well as COX-2 expression and production of prostaglandin E2 (PGE2) and iNOS expression and production of NO in macrophages [96]. These effects were mediated by the inhibition of LPS-dependent activation of NF-κB and AP-1 [96]. OMC also markedly suppressed LPS-induced production of TNF-α in mice and Concanavalin A induced proliferation and secretion of IFN-γ, IL-2 and IL-6 in mouse splenocytes ex vivo [96], further confirming anti-inflammatory activities of water soluble α- and β-glucans from the Pleurotus mushrooms.

In addition, a water soluble 1-3-β,1-4-β-D-glucan from edible *Collybia dryophila* markedly suppressed LPS- and IFN-γ-dependent expression of iNOS and NO production in macrophages [97]. Although 1-4-α-D-glucan from *A. bisporus* induced production of TNF-α and NO in macrophages in vitro [88], the same 1-4-β-D-glucan decreased LPS-dependent production of TNF-α in human peripheral mononuclear blood cells ex vivo [98].

Inflammatory response induced by bacterial endotoxin LPS is mediated by the TLR4 receptor. The formation of TLR4 signaling complex, which includes an extracellular protein MD-2, co-receptor CD14, LPS, and TLR4, results in the tyrosine phosphorylation of the intracellular Toll-IL-1 resistance (TIR) domain of TLR4 receptor [99]. TLR4 signaling is mediated through two separate pathways. The "MyD88-dependent"—MydD88-pathway, which engages the adapter proteins MyD88 and TIR domain containing adapter protein (TIRAP), and the "MyD88-independent" – TRIF-pathway, which uses the adapters TIR domain containing adapter-inducing interferon-β (TRIF) and TRIF-related adapter molecule (TRAM), finally resulting in the activation of a variety of kinases, leading to the activation of transcription factors NF-κB, AP-1 or IRF-3 and expression of pro-inflammatory molecules and cytokines. Therefore, the anti-inflammatory properties of different the soluble α- and β-glucans can be the results of the competition between LPS and glucans for the TLR4 binding (Fig. 18.2). Although extracts containing 1-6-α, 1-4-α-D-glucan, 1-6-β-D-glucan and mannogalactan from *A. bisporus* (ABS) and *A. brasiliensis* (ABL) induced expression of IL-1β and TNF-α; ABL containing 74.8 % of α- and β-glucans markedly suppressed LPS-dependent expression of IL-1β and TNF-α suggesting that these polysaccharides inhibit LPS-TLR binding [89].

18.7.4 Effects on Epithelial Cells

Although the majority of anti-cancer effect of soluble mushroom polysaccharides can be associated with the immunomodulatory activities described above, other studies demonstrated direct effect on cancer cells. We have previously isolated α-glucans from *P. ostreatus* which suppressed proliferation and induced apoptosis in

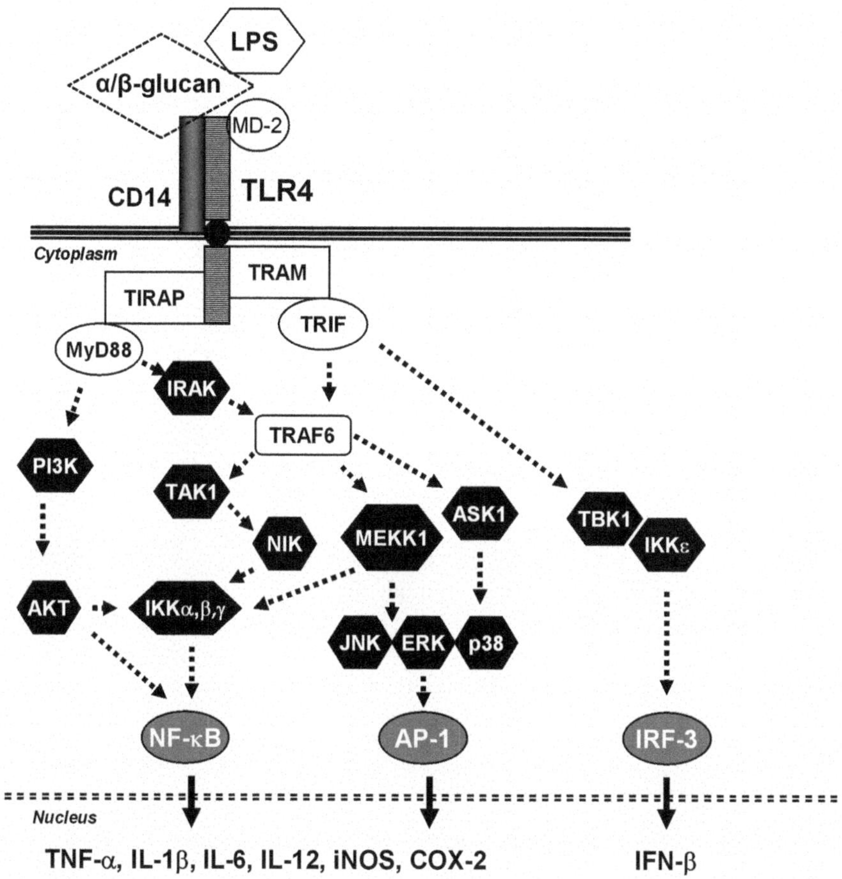

Fig. 18.2 Schematic representation of the TLR4 signaling

human cancer intestinal epithelial cells HT-29 through the up-regulation of expression of propapototic Bax [11]. Moreover, water soluble α- and β-glucans from *P. pulmonarius* induced apoptosis of HT-29 and HCT-116 cells, up-regulated pro-apoptotic Bax and suppressed survival Bcl-2 proteins increasing Bax/Bcl-2 ratio which is associated with the induction of apoptosis [46]. These glucans also inhibited TNF-α-dependent nuclear accumulation of p65 of NF-κB in intestinal epithelial cells [46]. Since NF-κB controls expression of genes that are involved in cell proliferation (e.g. cyclin D1) and apoptosis (e.g. Bcl-2), among others, NF-κB seems to be a suitable target for soluble α- or β-glucans. Indeed, we have previously shown that an extract, containing α- and β-glucans, from *P. ostreatus* suppressed expression of cyclin D1 in colonic tissue in a colitis-related colon carcinogenesis model in vivo [56]. In addition, water soluble β-D-glucan or 1-4-α-D-glucan, from *A. bisporus*, decreased IL-1β- and IFNγ-induced NF-κB transactivation in a reporter human epithelial colorectal adenocarcinoma cells Caco-2 [88].

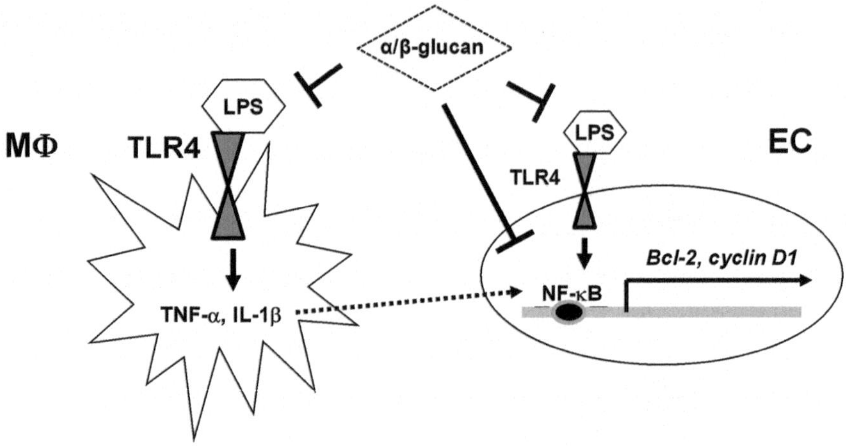

Fig. 18.3 Effects of water soluble glucans on the TLR4 signaling in macrophages and epithelial cells

A water soluble 1-3-β-1-4-β-D-glucan from edible *Poria cocos* suppressed proliferation and induced apoptosis through the down-regulation of expression of cyclin D1 and an increase of Bax/Bcl-2 ratio in epithelial breast cancer cells MCF-7, respectively [100]. Another water soluble hyperbranched β-D-glucans from edible *P. tuber-regium* or uncharacterized polysaccharides from *P. ostreatus* inhibited growth of human epithelial hepatocellular carcinoma cells HepG2 and prostate cancer cells PC-3 [101, 102]. Interestingly, water soluble polysaccharide, branched-fuco-1-4-β-galacto-1-4-β,1-6-β,1-6-β-D-glucan from *C. indica*, which activated immune cells, specifically suppressed proliferation and induced apoptosis of human cervical cancer HeLa cells but did not affect normal cells [87].

Mechanistically, glucans can inhibit interaction of TLR4 with LPS in macrophages which will results in the suppression of expression of TNF-α and IL-1β induced NF-κB-dependent expression of Bcl-2 and cyclin D1 in epithelial cells (Fig. 18.3). Alternatively, glucans can also inhibit interaction of TLR4 with LPS in epithelial cells resulting in the suppression of NF-κB activation and expression of Bcl-2 and cyclin D1. Since TLR4/NF-κB signaling was detected in intestinal epithelial cells and implicated in colon cancer [103], and TLR4 is overexpressed in breast and prostate cancer cells [104, 105], the inhibition of TLR4 signaling in epithelial cells by water soluble mushroom glucans is therapeutically promising. Finally, glucans can inhibit NF-κB in epithelial cells by another yet to be identified mechanism.

18.8 Conclusions and Future Perspectives

Since ancient times mushrooms have been claimed to exert antiviral, anti-inflammatory, hypotensive, hypocholesterolemic, hypoglycemic effects [27]. Mushroom polysaccharides have been associated to prevention of cancer, diabetes, hyperlipidemia, arteriosclerosis, hepatitis, and other chronic diseases [5]. Biologically active water soluble polysaccharides harvested from edible mushroom are widely distributed among the medicinal mushrooms. The bioactive polysaccharides isolated from mushroom fruiting-bodies, submerged cultured fungal biomass, or liquid culture fermentation broths are either water-soluble α- and β-D-glucans, β-D-glucans with heterosaccharide chains of xylose, mannose, galactose, or uronic acid, or β-D-glucan-protein complexes, β-(1,3)-D-glucans with β-(1,6) branches in other words, proteoglycans [5–7]. Polysaccharides differ in their primary structure (type of basic sugar), type of linkage (α, β, etc.), degree of branching, and molecular weight, among other parameters [8].

We aimed in this article to review the methodology to isolate the bioactive polysaccharides from mushrooms, to highlight some of the health-promoting activities of these polysaccharides and give an insight of the putative mechanism of action. We believe that molecular studies based on genomic information combined with advanced biochemistry and fermentation studies can lead to higher polysaccharides yields, quality and medicinal effectiveness. A focus should be directed towards polysaccharides produced by the mushrooms that currently are under investigation and show the highest and most practical potential to be integrated into medicine. Sources for additional, and may be more potent, medicinal mushrooms can be expanded through research in yet unexplored regions. De Silva et al. [2] proposed to explore tropical countries for effective mushrooms and to assay their bioactive metabolites for cure and prevention of chronic diseases. A recent study found high levels of β-glucans in wild mushrooms in Thailand [106]. To validate the know how accumulated during centuries in traditional medicine and in recent years in animal and in vitro experiments, state-of-the-art clinical studies should be conducted aiming at prevention and cure of specific disorders.

References

1. Zhang YY, Li S, Wang XH, Zhang LN, Cheung PCK (2011) Advances in lentinan: isolation, structure, chain conformation and bioactivities. Food Hydrocoll 25:196–206
2. De Silva DD, Rapior S, Fons F, Bahkali AH, Hyde KD (2012) Medicinal mushrooms in supportive cancer therapies: an approach to anti-cancer effects and putative mechanisms of action. Fungal Divers 55:1–35
3. Wasser SP (2002) Medicinal mushrooms as a source of antitumor and immunomodulating polysaccharides. Appl Microbiol Biotechnol 60:258–274
4. Wasser SP (2011) Current findings, future trends, and unsolved problems in studies of medicinal mushrooms. Appl Microbiol Biotechnol 89:1323–1332
5. Chang ST, Wasser SP (2012) The role of culinary-medicinal mushrooms on human welfare with a pyramid model for human health. Int J Med Mushrooms 14:95–134
6. Bhanja SK, Nandan CK, Mandal S, Bhunia B, Maiti TK, Mondal S et al (2012) Isolation and characterization of the immunostimulating beta-glucans of an edible mushroom *Termitomyces robustus* var. Carbohydr Res 357:83–89
7. Palacios I, Garcia-Lafuente A, Guillamon E , Villares A (2012) Novel isolation of water-soluble polysaccharides from the fruiting bodies of *Pleurotus ostreatus* mushrooms. Carbohydr Res 358: 72–77

8. Wang CR, Ng TB, Li L, Fang JC, Jiang Y, Wen TY et al (2011) Isolation of a polysaccharide with antiproliferative, hypoglycemic, antioxidant and HIV-1 reverse transcriptase inhibitory activities from the fruiting bodies of the abalone mushroom *Pleurotus abalonus*. J Pharm Pharmacol 63:825–832
9. Rop O, Mlcek J, Jurikova T (2009) Beta-glucans in higher fungi and their health effects. Nutr Rev 67:624–631
10. Stachowiak B, Regula J (2012) Health-promoting potential of edible macromycetes under special consideration of polysaccharides: a review. Eur Food Res Technol 234:369–380
11. Lavi I, Friesem D, Geresh S, Hadar Y, Schwartz B (2006) An aqueous polysaccharide extract from the edible mushroom *Pleurotus ostreatus* induces anti-proliferative and pro-apoptotic effects on HT-29 colon cancer cells. Cancer Lett 244:61–70
12. Chandra LC, Smith BJ, Clarke SL, Marlow D, D'Offay JM, Kuvibidila SR (2011) Differential effects of shiitake- and white button mushroom-supplemented diets on hepatic steatosis in C57BL/6 mice. Food Chem Toxicol 49:3074–3080
13. Sanchez C (2010) Cultivation of *Pleurotus ostreatus* and other edible mushrooms. Appl Microbiol Biot 85:1321–1337
14. Nikitina VE, Tsivileva OM, Pankratov AN, Bychkov NA (2007) Lentinula edodes biotechnology - From lentinan to lectins. Food Technol Biotech 45:230–237
15. Cohen R, Persky L, Hadar Y (2002) Biotechnological applications and potential of wood-degrading mushrooms of the genus *Pleurotus*. Appl Microbiol Biot 58:582–594
16. Gregori A, Svagelj M, Pohleven J (2007) Cultivation techniques and medicinal properties of *Pleurotus* spp. Food Technol Biotech 45:238–249
17. Salame TM, Knop D, Tal D, Levinson D, Yarden O, Hadar Y (2012) Predominance of a versatile-peroxidase-encoding gene, mnp4, as demonstrated by gene replacement via a gene targeting system for *Pleurotus ostreatus*. Appl Environ Microb 78:5341–5352
18. Khan MA, Tania M (2012) Nutritional and medicinal importance of pleurotus mushrooms: an overview. Food Rev Int 28:313–329
19. Lavi I, Levinson D, Peri I, Tekoah Y, Hadar Y, Schwartz B (2010) Chemical characterization, antiproliferative and antiadhesive properties of polysaccharides extracted from *Pleurotus pulmonarius* mycelium and fruiting bodies. Appl Microbiol Biotechnol 85:1977–1990
20. Barreto SM, Lopez MV, Levin L (2008) Effect of culture parameters on the production of the edible mushroom *Grifola frondosa* (maitake) in tropical weathers. World J Microb Biotechnol 24:1361–1366
21. Shen Q, Royse DJ (2001) Effects of nutrient supplements on biological efficiency, quality and crop cycle time of maitake (*Grifola frondosa*). Appl Microbiol Biotechnol 57:74–78
22. Svagelj M, Berovic M, Boh B, Menard A, Simcic S, Wraber B (2008) Solid-state cultivation of Grifola frondosa (Dicks: Fr) SF Gray biomass and immunostimulatory effects of fungal intra- and extracellular beta-polysaccharides. New Biotechnol 25:150–156
23. Montoya S, Orrego CE, Levin L (2012) Growth, fruiting and lignocellulolytic enzyme production by the edible mushroom *Grifola frondosa* (maitake). World J Microb Biotechnol 28:1533–1541
24. Chen GT, Ma XM, Liu ST, Liao YI, Zhao GQ (2012) Isolation, purification and antioxidant activities of polysaccharides from *Grifola frondosa*. Carbohydrate Polymers 89:61–66
25. Zhou XW, Su KQ, Zhang YM (2012) Applied modern biotechnology for cultivation of Ganoderma and development of their products. Appl Microbiol Biotechnol 93:941–963
26. Shi L, Fang X, Li MJ, Mu DS, Ren A, Tan Q et al (2012) Development of a simple and efficient transformation system for the basidiomycetous medicinal fungus *Ganoderma lucidum*. World J Microb Biotechnol 28:283–291
27. Jong SC, Birmingham JM (1992) Medicinal benefits of the mushroom Ganoderma. Adv Appl Microbiol 37:101–134
28. Laroche C, Michaud P (2007) New developments and prospective applications for beta (1,3) glucans. Recent Pat Biotechnol 1:59–73

29. Chen S, Oh SR, Phung S, Hur G, Ye JJ, Kwok SL et al (2006) Anti-aromatase activity of phytochemicals in white button mushrooms (*Agaricus bisporus*). Cancer Res 66:12026–12034
30. Jeong SC, Koyyalamudi SR, Jeong YT, Song CH, Pang G (2012) Macrophage immunomodulating and antitumor activities of polysaccharides isolated from *Agaricus bisporus* white button mushrooms. J Med Food 15:58–65
31. Sliva D (2004) Cellular and physiological effects of *Ganoderma lucidum* (Reishi). Mini Rev Med Chem 4:873–879
32. Thyagarajan A, Zhu J, Sliva D (2007) Combined effect of green tea and *Ganoderma lucidum* on invasive behavior of breast cancer cells. Int J Oncol 30:963–969
33. Vetvicka V, Thornton BP, Ross GD (1996) Soluble beta-glucan polysaccharide binding to the lectin site of neutrophil or natural killer cell complement receptor type 3 (CD11b/CD18) generates a primed state of the receptor capable of mediating cytotoxicity of iC3b-opsonized target cells. J Clin Invest 98:50–61
34. Rout D, Mondal S, Chakraborty I, Islam SS (2008) The structure and conformation of a water-insoluble (1 → 3)- (1 → 6)-beta-d-glucan from the fruiting bodies of *Pleurotus florida*. Carbohydr Res 343:982–987
35. Lavi I, Friesem D, Geresh S, Hadar Y, Schwartz B (2006) An aqueous polysaccharide extract from the edible mushroom *Pleurotus ostreatus* induces anti-proliferative and pro-apoptotic effects on HT-29 colon cancer cells. Cancer Lett 244:61–70
36. Akramiene D, Kondrotas A, Didziapetriene J, Kevelaitis E (2007) Effects of beta-glucans on the immune system. Medicina Kaunas 43:597–606
37. Deng G, Lin H, Seidman A, Fornier M, D'Andrea G, Wesa K et al (2009) A phase I/II trial of a polysaccharide extract from *Grifola frondosa* (Maitake mushroom) in breast cancer patients: immunological effects. J Cancer Res Clin Oncol 135:1215–1221
38. Shah SK, Walker PA, Moore-Olufemi SD, Sundaresan A, Kulkarni AD, Andrassy RJ (2011) An evidence-based review of a *Lentinula edodes* mushroom extract as complementary therapy in the surgical oncology patient. JPEN J Parenter Enteral Nutr 35:449–458
39. Sullivan R, Smith JE, Rowan NJ (2006) Medicinal mushrooms and cancer therapy: translating a traditional practice into Western medicine. Perspect Biol Med 49:159–170
40. Sze DM, Chan GC (2009) Supplements for immune enhancement in hematologic malignancies. Hematol Am Soc Hematol Educ Program 19:313–319
41. Nanba H (1995) Activity of maitake D-fraction to inhibit carcinogenesis and metastasis. Ann N Y Acad Sci 768:243–245
42. Hishida I, Nanba H, Kuroda H (1988) Antitumor activity exhibited by orally administered extract from fruit body of *Grifola frondosa* (maitake). Chem Pharm Bull (Tokyo) 36:1819–1827
43. Chihara G (1992) Recent progress in immunopharmacology and therapeutic effects of polysaccharides. Dev Biol Stand 77:191–197
44. Hetland G, Johnson E, Lyberg T, Bernardshaw S, Tryggestad AM, Grinde B (2008) Effects of the medicinal mushroom *Agaricus blazei Murill* on immunity, infection and cancer. Scand J Immunol 68:363–370
45. Rop O, Mlcek J, Jurikova T (2009) Beta-glucans in higher fungi and their health effects. Nutr Rev 67:624–631
46. Lavi I, Nimri L, Levinson D, Peri I, Hadar Y, Schwartz B (2012) Glucans from the edible mushroom *Pleurotus pulmonarius* inhibit colitis-associated colon carcinogenesis in mice. J Gastroenterol 47:504–518
47. Rice PJ, Adams EL, Ozment-Skelton T, Gonzalez AJ, Goldman MP, Lockhart BE et al (2005) Oral delivery and gastrointestinal absorption of soluble glucans stimulate increased resistance to infectious challenge. J Pharmacol Exp Ther 314:1079–1086
48. Lavi I, Levinson D, Peri I, Nimri L, Hadar Y, Schwartz B (2010) Orally administered glucans from the edible mushroom *Pleurotus pulmonarius* reduce acute inflammation in dextran sulfate sodium-induced experimental colitis. Br J Nutr 103:393–402

49. Battle J, Ha T, Li C, Della Beffa V, Rice P, Kalbfleisch J et al (1998) Ligand binding to the (1 → 3)-beta-D-glucan receptor stimulates NFkappaB activation, but not apoptosis in U937 cells. Biochem Biophys Res Commun 249: 499–504
50. Chen W, Zhao Z, Chen SF, Li YQ (2008) Optimization for the production of exopolysaccharide from *Fomes fomentarius* in submerged culture and its antitumor effect in vitro. Bioresour Technol 99:3187–3194
51. Xie JT, Wang CZ, Wicks S, Yin JJ, Kong J, Li J et al (2006) *Ganoderma lucidum* extract inhibits proliferation of SW 480 human colorectal cancer cells. Exp Oncol 28:25–29
52. Fujimiya Y, Suzuki Y, Oshiman K, Kobori H, Moriguchi K, Nakashima H et al (1998) Selective tumoricidal effect of soluble proteoglucan extracted from the basidiomycete, *Agaricus blazei Murill*, mediated via natural killer cell activation and apoptosis. Cancer Immunol Immunother 46:147–159
53. Kimura Y, Kido T, Takaku T, Sumiyoshi M, Baba K (2004) Isolation of an anti-angiogenic substance from *Agaricus blazei Murill*: its antitumor and antimetastatic actions. Cancer Sci 95:758–764
54. Bobek P, Nosalova V, Cerna S (2001) Effect of pleuran (beta-glucan from Pleurotus ostreatus) in diet or drinking fluid on colitis in rats. Nahrung 45:360–363
55. Choi SY, Hur SJ, An CS, Jeon YH, Jeoung YJ, Bak JP et al (2010) Anti-inflammatory effects of *Inonotus obliquus* in colitis induced by dextran sodium sulfate. J Biomed Biotechnol 2010:943516
56. Jedinak A, Dudhgaonkar S, Jiang J, Sandusky G, Sliva D (2010) *Pleurotus ostreatus* inhibits colitis-related colon carcinogenesis in mice. Int J Mol Med 26:643–650
57. Hendrickson BA, Gokhale R, Cho JH (2002) Clinical aspects and pathophysiology of inflammatory bowel disease. Clin Microbiol Rev 15:79–94
58. Loftus EV Jr, Sandborn WJ (2002) Epidemiology of inflammatory bowel disease. Gastroenterol Clin North Am 31:1–20
59. Nosal'ova V, Bobek P, Cerna S, Galbavy S, Stvrtina S (2001) Effects of pleuran (beta-glucan isolated from *Pleurotus ostreatus*) on experimental colitis in rats. Physiol Res 50:575–581
60. Kim YO, Park HW, Kim JH, Lee JY, Moon SH, Shin CS (2006) Anti-cancer effect and structural characterization of endo-polysaccharide from cultivated mycelia of *Inonotus obliquus*. Life Sci 79:72–80
61. Lemieszek MK, Langner E, Kaczor J, Kandefer-Szerszen M, Sanecka B, Mazurkiewicz W et al (2011) Anticancer effects of fraction isolated from fruiting bodies of Chaga medicinal mushroom, *Inonotus obliquus* (Pers.:Fr.) Pilat (Aphyllophoromycetideae): in vitro studies. Int J Med Mushrooms 13: 131–143
62. Zheng W, Zhao Y, Zheng X, Liu Y, Pan S, Dai Y et al (2011) Production of antioxidant and antitumor metabolites by submerged cultures of *Inonotus obliquus* cocultured with *Phellinus punctatus*. Appl Microbiol Biotechnol 89:157–167
63. Ko SK, Jin M, Pyo MY (2011) Inonotus obliquus extracts suppress antigen-specific IgE production through the modulation of Th1/Th2 cytokines in ovalbumin-sensitized mice. J Ethnopharmacol 137:1077–1082
64. Kim HG, Yoon DH, Kim CH, Shrestha B, Chang WC, Lim SY et al (2007) Ethanol extract of Inonotus obliquus inhibits lipopolysaccharide-induced inflammation in RAW 264.7 macrophage cells. J Med Food 10:80–89
65. Quigley EM, Abdel-Hamid H, Barbara G, Bhatia SJ, Boeckxstaens G, De Giorgio R et al (2012) A global perspective on irritable bowel syndrome: a consensus statement of the world gastroenterology organisation summit task force on irritable bowel syndrome. J Clin Gastroenterol 46:356–366
66. Agarwal N, Spiegel BM (2011) The effect of irritable bowel syndrome on health-related quality of life and health care expenditures. Gastroenterol Clin North Am 40:11–19
67. Asano T, Tanaka K, Suemasu S, Ishihara T, Tahara K, Suzuki T et al (2012) Effects of beta-(1,3–1,6)-D-glucan on irritable bowel syndrome-related colonic hypersensitivity. Biochem Biophys Res Commun 420:444–449

68. Ciacci C, Franceschi F, Purchiaroni F, Capone P, Buccelletti F, Iacomini P et al (2011) Effect of beta-glucan, inositol and digestive enzymes in GI symptoms of patients with IBS. Eur Rev Med Pharmacol Sci 15:637–643
69. Bobek P, Ozdin L, Kuniak L (1994) Mechanism of hypocholesterolemic effect of oyster mushroom (*Pleurotus ostreatus*) in rats: reduction of cholesterol absorption and increase of plasma cholesterol removal. Z Ernahrungswiss 33:44–50
70. Sugiyama K, Kawagishi H, Tanaka A, Saeki S, Yoshida S, Sakamoto H et al (1992) Isolation of plasma cholesterol-lowering components from ningyotake (*Polyporus confluens*) mushroom. J Nutr Sci Vitaminol (Tokyo) 38:335–342
71. Oh TW, Kim YA, Jang WJ, Byeon JI, Ryu CH, Kim JO et al (2010) Semipurified fractions from the submerged-culture broth of *Agaricus blazei Murill* reduce blood glucose levels in streptozotocin-induced diabetic rats. J Agric Food Chem 58:4113–4119
72. Chadwick WA, Roux S, van de Venter M, Louw J, Oelofsen W (2007) Anti-diabetic effects of *Sutherlandia frutescens* in Wistar rats fed a diabetogenic diet. J Ethnopharmacol 109:121–127
73. Hong L, Xun M, Wutong W (2007) Anti-diabetic effect of an alpha-glucan from fruit body of maitake (*Grifola frondosa*) on KK-Ay mice. J Pharm Pharmacol 59:575–582
74. Kim YW, Kim KH, Choi HJ, Lee DS (2005) Anti-diabetic activity of beta-glucans and their enzymatically hydrolyzed oligosaccharides from *Agaricus blazei*. Biotechnol Lett 27:483–487
75. Manohar V, Talpur NA, Echard BW, Lieberman S, Preuss HG (2002) Effects of a water-soluble extract of maitake mushroom on circulating glucose/insulin concentrations in KK mice. Diabetes Obes Metab 4:43–48
76. Li H, Zhang M, Ma G (2010) Hypolipidemic effect of the polysaccharide from Pholiota nameko. Nutrition 26:556–562
77. Bobek P, Galbavy S (1999) Hypocholesterolemic and antiatherogenic effect of oyster mushroom (*Pleurotus ostreatus*) in rabbits. Nahrung 43:339–342
78. Chen J, Raymond K (2008) Beta-glucans in the treatment of diabetes and associated cardiovascular risks. Vasc Health Risk Manag 4:1265–1272
79. Cugnet-Anceau C, Nazare JA, Biorklund M, Le Coquil E, Sassolas A, Sothier M et al (2010) A controlled study of consumption of beta-glucan-enriched soups for 2 months by type 2 diabetic free-living subjects. Br J Nutr 103:422–428
80. Chan GCF, Chan WK, Sze DMY et al (2009) The effects of beta-glucan on human immune and cancer cells. J Hematol Oncol 2: 25–35
81. Misaki A, Kakuta M, Sasaki T, Tanaka M, Miyaji H (1981) studies on interrelation of structure and anti-tumor effects of polysaccharides—anti-tumor action of periodate-modified, branched (1-]3)-Beta-D-Glucan of Auricularia-Auricula-Judae, and other polysaccharides containing (1-]3)-lycosidic linkages. Carbohydr Res 92:115–129
82. Dey B, Bhunia SK, Maity KK, Patra S, Mandal S, Maiti S et al (2012) Glucans of *Pleurotus florida* blue variant: Isolation, purification, characterization and immunological studies. Int J Biol Macromol 50:591–597
83. Roy SK, Das D, Mondal S, Maiti D, Bhunia B, Maiti TK et al (2009) Structural studies of an immunoenhancing water-soluble glucan isolated from hot water extract of an edible mushroom, *Pleurotus florida*, cultivar Assam Florida. Carbohydr Res 344:2596–2601
84. Han XQ, Chai XY, Jia YM, Han CX, Tu PF (2010) Structure elucidation and immunological activity of a novel polysaccharide from the fruit bodies of an edible mushroom, *Sarcodon aspratus* (Berk.) S. Ito. Int J Biol Macromol 47:420–424
85. Bhunia SK, Dey B, Maity KK, Patra S, Mandal S, Maiti S, et al (2010) Structural characterization of an immunoenhancing heteroglycan isolated from an aqueous extract of an edible mushroom, *Lentinus squarrosulus* (Mont.) Singer. Carbohydr Res 345: 2542–2549
86. Bhunia SK, Dey B, Maity KK, Patra S, Mandal S, Maiti S, et al (2011) Isolation and characterization of an immunoenhancing glucan from alkaline extract of an edible mushroom, *Lentinus squarrosulus* (Mont.) Singer. Carbohydr Res 346: 2039–2044

87. Mandal EK, Maity K, Maity S, Gantait SK, Maiti S, Maiti TK, et al (2011) Structural characterization of an immunoenhancing cytotoxic heteroglycan isolated from an edible mushroom *Calocybe indica* var. APK2. Carbohydr Res 346: 2237–2243
88. Volman JJ, Helsper JPFG, Wei S, Baars JJP, van Griensven LJLD, Sonnenberg ASM et al (2010) Effects of mushroom-derived beta-glucan-rich polysaccharide extracts on nitric oxide production by bone marrow-derived macrophages and nuclear factor-kappa B transactivation in Caco-2 reporter cells: Can effects be explained by structure? Mol Nutr Food Res 54:268–276
89. Smiderle FR, Ruthes AC, van Arkel J, Chanput W, Iacomini M, Wichers HJ, Van Griensven LJ (2011) Polysaccharides from *Agaricus bisporus* and *Agaricus brasiliensis* show similarities in their structures and their immunomodulatory effects on human monocytic THP-1 cells. Bmc Complem Altern M 11:58
90. Reid DM, Gow NAR, Brown GD (2009) Pattern recognition: recent insights from Dectin-1. Curr Opin Immunol 21:30–37
91. Adams EL, Rice PJ, Graves B, Ensley HE, Yu H, Brown GD et al (2008) Differential high-affinity interaction of dectin-1 with natural or synthetic glucans is dependent upon primary structure and is influenced by polymer chain length and side-chain branching. J Pharmacol Exp Ther 325:115–123
92. Zhang M, Cui SW, Cheung PCK, Wang Q (2006) Polysaccharides from mushrooms: a review on their isolation process, structural characteristics and antitumor activity. Trends Food Sci Tech 18:4–19
93. Masuda Y, Togo T, Mizuno S, Konishi M, Nanba H (2012) Soluble beta-glucan from *Grifola frondosa* induces proliferation and Dectin-1/Syk signaling in resident macrophages via the GM-CSF autocrine pathway. J Leukoc Biol 91:547–556
94. Smiderle FR, Olsen LM, Carbonero ER, Baggio CH, Freitas CS, Marcon R et al (2008) Anti-inflammatory and analgesic properties in a rodent model of a (1 → 3) (1 → 6)-linked beta-glucan isolated from *Pleurotus pulmonarius*. Eur J Pharmacol 597:86–91
95. Baggio CH, Freitas CS, Martins DF, Mazzardo L, Smiderle FR, Sassaki GL et al (2010) Antinociceptive effects of (1 → 3) (1 → 6)-Linked beta-glucan isolated from *Pleurotus pulmonarius* in models of acute and neuropathic pain in mice: evidence for a role for glutamatergic receptors and cytokine pathways. J Pain 11:965–971
96. Jedinak A, Dudhgaonkar S, Wu QL, Simon J, Sliva D (2011) Anti-inflammatory activity of edible oyster mushroom is mediated through the inhibition of NF-kappa B and AP-1 signaling. Nutr J 10:52
97. Pacheco-Sanchez M, Boutin Y, Angers P, Gosselin A, Tweddell RJ (2007) Inhibitory effect of CDP, a polysaccharide extracted from the mushroom Collybia dryophila, on nitric oxide synthase expression and nitric oxide production in macrophages. Eur J Pharmacol 555:61–66
98. Volman JJ, Mensink RP, van Griensven LJ, Plat J (2010) Effects of alpha-glucans from *Agaricus bisporus* on ex vivo cytokine production by LPS and PHA-stimulated PBMCs; a placebo-controlled study in slightly hypercholesterolemic subjects. Eur J Clin Nutr 64:720–726
99. Medvedev AE, Piao W, Shoenfelt J, Rhee SH, Chen H, Basu S et al (2007) Role of TLR4 tyrosine phosphorylation in signal transduction and endotoxin tolerance. J Biol Chem 282:16042–16053
100. Zhang M, Chiu LC, Cheung PC, Ooi VE (2006) Growth-inhibitory effects of a beta-glucan from the mycelium of *Poria cocos* on human breast carcinoma MCF-7 cells: cell-cycle arrest and apoptosis induction. Oncol Rep 15:637–643
101. Gu YH, Sivam G (2006) Cytotoxic effect of oyster mushroom *Pleurotus ostreatus* on human androgen-independent prostate cancer PC-3 cells. J Med Food 9:196–204
102. Tao Y, Zhang L, Cheung PC (2006) Physicochemical properties and antitumor activities of water-soluble native and sulfated hyperbranched mushroom polysaccharides. Carbohydr Res 341:2261–2269
103. Liu L, Li YH, Niu YB, Sun Y, Guo ZJ, Li Q et al (2010) An apple oligogalactan prevents against inflammation and carcinogenesis by targeting LPS/TLR4/NF-kappaB pathway in a mouse model of colitis-associated colon cancer. Carcinogenesis 31:1822–1832

104. Hua D, Liu MY, Cheng ZD, Qin XJ, Zhang HM, Chen Y et al (2009) Small interfering RNA-directed targeting of Toll-like receptor 4 inhibits human prostate cancer cell invasion, survival, and tumorigenicity. Mol Immunol 46:2876–2884
105. Yang H, Zhou H, Feng P, Zhou X, Wen H, Xie X et al (2010) Reduced expression of Toll-like receptor 4 inhibits human breast cancer cells proliferation and inflammatory cytokines secretion. J Exp Clin Cancer Res 29:92
106. Boonyanuphap J, Hansawasdi C (2011) Spatial distribution of Beta glucan containing wild mushroom communities in subtropical dry forest Thailand. Fungal Divers 46:29–42

Author Biography

Betty Schwartz PhD in Clinical Biochemistry at Ben-Gurion University of the Negev, Beer-Sheva, Israel in 1990. From 1990 to 1992 she was a postdoctoral fellow at the University of San-Francisco, California and from 1992–1996 she was a Lecturer in Clinical Biochemistry Faculty of Health Sciences, School of Medicine, Ben-Gurion University of the Negev, Beer-Sheva, Israel. From 1996-to now she was appointed as Lecturer and now as Associate Professor at the Institute of Biochemistry, Food Science and Nutrition, the Hebrew University of Jerusalem, Israel. From 2005–2011 she served as Head of the School of Nutritional Sciences. From 2005 up to now she is the Academic Co-ordinator of the International M.Sc. program of the Hebrew University of Jerusalem and from 2010 up to now she is the Academic Co-ordinator of the Nutrigenomics Center. Between 2003–2004 she was a visiting scientist at the Department of Cancer Biology, M. D. Anderson Cancer Center, University of Texas, Houston, TX, USA. Betty's research interests are focused on the role of nutrients for their anticarcinogenic and anti-inflammatory activities especially in the intestinal tissue.

Yitzhak Hadar received B.Sc.(1975) and M.Sc.(1977) in Plant Protection and Ph.D. in 1980 from the Hebrew University. Post doc at Cornell University studying biological control. Joined the Hebrew University, Faculty of Agriculture in 1981 as a lecturer and promoted to Professor in 1995. He was a visiting Scientist at Repligen—Sandoz Research Corporation, Lexington, MA, USA (1988-9) and Honorary Lecturer at the Department of Biological Sciences, The University Waikato, Hamilton, New Zealand (1996–1997). Served as Dean of the Faculty of Agriculture, Food and Environment (2001–2005), Head of the Department of Microbiology and Plant Pathology (2006–2010), President of the Israeli Society for Microbiology (2006–2008) and currently, Head of the Center for Pre-Academic Studies. Hadar's research interest is in environmental microbiology and fungal biology.

Daniel Sliva studied Food and Biochemical Technology at the University of Chemical Technology and Biochemistry at the Charles University, Prague, Czech Republic. He completed PhD in Biology and Genetics, at the Charles University in Prague, Czech Republic in 1993. From 1993 to 1996 he worked at the Department of Medical Nutrition, Karolinska Institute, Stockholm, Sweden, as a postdoctoral researcher focused on the growth hormone signal transduction. From 1996 to 1999 he worked at the Walter Oncology Center, Indiana University School of Medicine, Indianapolis, USA, as a postdoctoral fellow focused on the interleukin-9 receptor signalling. He is currently Senior Investigator at the Methodist Research Institute, Indiana University Health, and adjunct Associate Professor at the Department of Medicine, Indiana University Simon Cancer Center, Indiana University School of Medicine, Indianapolis, Indiana, USA. His research interests are focused on the molecular mechanisms responsible for the invasive/metastatic behavior of cancer and the use of the natural/dietary compounds for the prevention and treatment of cancer.

Chapter 19
Essential Oil Constituents: Biodiversity and Their Applicability for Cancer Therapy

Daniel P. Bezerra, Emmanoel V. Costa and Paulo Cesar L. Nogueira

Abstract Essential oils are odoriferous substances traditionally used in the perfumery, food, and pharmaceutical industries. The most volatile fraction (constituting 90–95 % of total oil) comprises, in most cases, complex mixtures that may contain hundreds of compounds, which are composed mainly of terpenoids. Each of these constituents contributes for the biological effects of these essential oils. In this chapter, a total of 20 essential oil constituents, which presented positive results on cytotoxic drug screening, were selected; among them, ascaridole, α-bisabolol, (*E*)-caryophyllene, β-elemene, β-eudesmol, D-limonene, terpinen-4-ol, and thymol have been extensively studied with promissory results. Herein, we highlighted the recent advances in the knowledge of the chemical and anticancer properties of these compounds, establishing new goals for future research.

19.1 Introduction

Plant secondary metabolites have amazing structural diversity and biological activities, including anticancer proprieties [1, 2]. Among these, essential oils are odoriferous substances traditionally used in the perfumery, food, and pharmaceutical industries. Essential oils may be found in different plant parts generally in flowers (e.g. *Acacia* spp., *Dianthus caryophyllus*, *Jasminum* spp., *Lavandula* spp., *Rosa* spp., *Rosmarinus officinalis*, *Syzygium aromaticum*, etc.), fruits (*Citrus* spp., *Juniperus communis,* etc.), leaves (e.g. *Cymbopogon* spp., *Mentha* spp., *Ocimum* spp., etc.), leaves and stems (e.g. *Aloysia citriodora*, *Cinnamomum* spp., *Pelargonium* spp., *Pogostemon cablin*, etc.). Moreover, bark (e.g. *Betula*

D. P. Bezerra (✉)
Department of Physiology, Federal University of Sergipe, São Cristóvão, Sergipe, Brazil
e-mail: danielpbezerra@gmail.com

E. V. Costa · P. C. L. Nogueira
Department of Chemistry, Federal University of Sergipe, São Cristóvão, Sergipe, Brazil
e-mail: emmanoelvc@gmail.com

E. F. Fang and T. B. Ng (eds.), *Antitumor Potential and Other Emerging Medicinal Properties of Natural Compounds*, DOI: 10.1007/978-94-007-6214-5_19,

pubescens, Cinnamomum cassia), rhizomes (e.g. *Acorus calamus, Curcuma longa, Zingiber officinale*, etc.), roots (e.g. *Angelica archangelica, Vetiveria zizanioides*, etc.), seeds (e.g. *Coriandrum sativum, Foeniculum vulgare, Myristica* spp., etc.), and wood (e.g. *Aniba rosaeodora, Eremanthus erythropappus, Pinus sylvestris, Santalum album*, etc.) are also natural sources of commercially important essential oils [3, 4].

The most volatile fraction (constituting 90–95 % of total oil) comprises, in most cases, complex mixtures that may contain hundreds of compounds which are composed mainly of terpenoids (mono-, sesqui-, and even diterpenes). Besides, it may contain benzenoids, phenylpropanoids, aliphatic aldehydes, alcohols, and esters. Terpenoids (mono- and sesquitepenes) are the primary constituents of the essential oils of many types of herbs; many of them are commercially important, and are widely used as flavoring agents, perfumes, insecticides, antimicrobial agents, and raw material for important chemicals [5].

In this chapter, a total of 20 essential oil constituents, which presented positive results on cytotoxic drug screening, were selected (Fig. 19.1). In addition, we highlighted the recent advances in the knowledge of the chemical and anticancer properties of these compounds, establishing new goals for future research.

19.2 Chemical and Botanical Data

A rigid scheme for classifying secondary metabolites is not applicable due to their immense structural diversity; however, three main classes are often used: terpenoids and steroids; fatty-acid derivatives and polyketides; and alkaloids. Terpenoids constitute the largest and one of the most diverse classes of secondary metabolites and they are classified according to the number of containing five-carbon units coupled through biosynthetic pathways [6, 7]. In volatile fraction, we found terpenoids that are classified as hemiterpenes (C_5), monoterpenes (C_{10}), sesquiterpenes (C_{15}), and even some diterpenes (C_{20}).

Despite their diversity, all plant terpenoids derive from the common building units isopentenyl diphosphate (IPP) and its isomer dimethylallyl diphosphate (DMAPP) that are synthesized via two parallel pathways known as mevalonate (MVA) and methylerythritol 4-phosphate (MEP) [6]. In general, MVA pathway leads to the synthesis of some sesquiterpenes and triterpenes (sterols) in the cytoplasm, while the MEP pathway is responsible for the synthesis of monoterpenes, diterpenes, tetraterpenes (carotenoids) and polyterpenes [7, 8].

In the most volatile fraction of essential oils, terpenoids are generally unsaturated compounds which easily decomposable by light, heat or oxygen to produce undesirable compounds. So, the steps of isolation, concentration and purification in methods for extraction of essential oil become critical. The most commonly used technique is the so-called traditional methods [9], i.e., those based on mechanical pressing (e.g. citrus oils) and by hydro- or steam-distillation. Steam distillation is the worldwide procedure for extraction of essential oils from plant

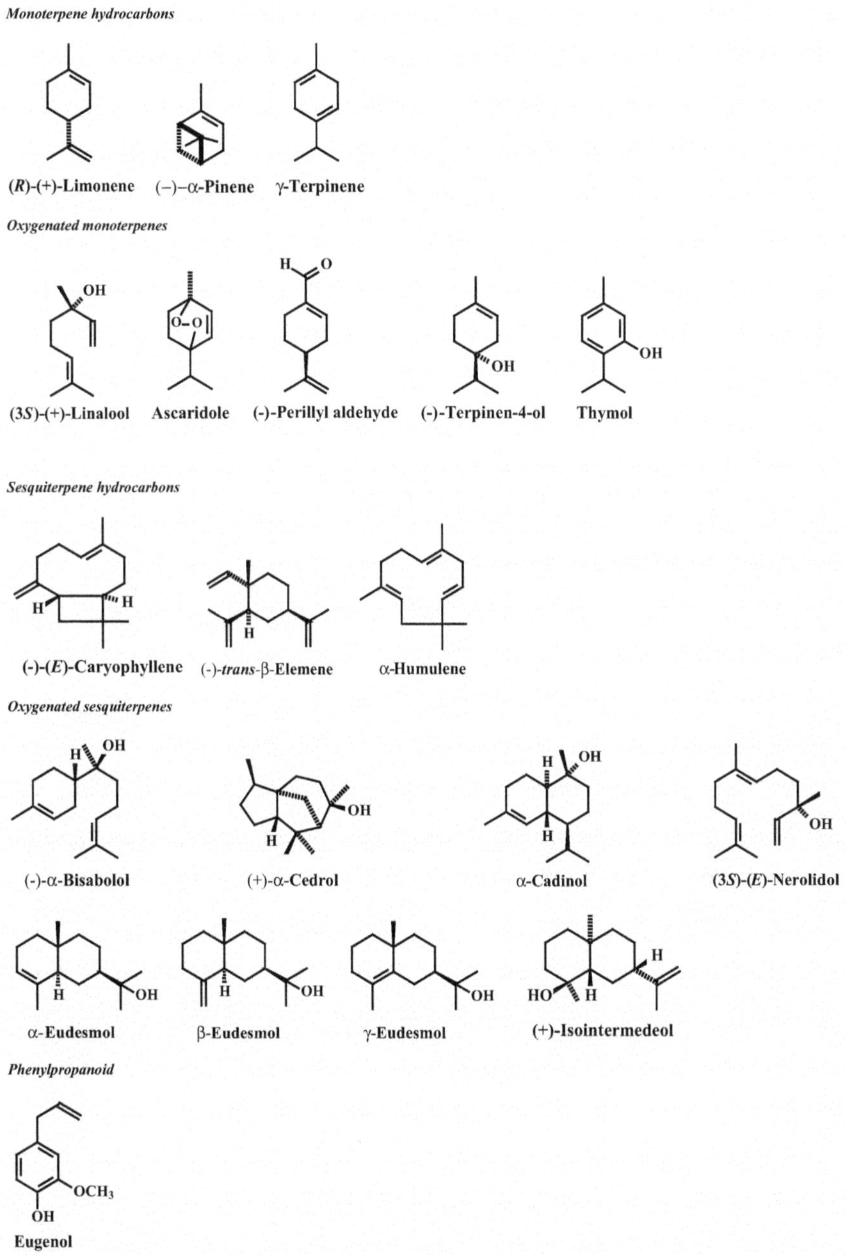

Fig. 19.1 Chemical structures of selected essential oil components

material, which is usually made by Clevenger-type apparatus in laboratory scale (hydro-distillation). The main drawback of this technique is related to the decomposition of labile compounds and the possibility of formation of non-natural compounds (artifacts). Therefore, it is desirable to employ mild conditions to avoid oxidation, thermal deterioration or other chemical changes. Modern methods of sampling using small volume or no organic solvent have emerged in recent years. Techniques such as supercritical fluid extraction with CO_2 (SFE-CO_2), microwave assisted extraction (MAE) and pressure liquid extraction (PLE) have gained in interest as a green approach in volatile fraction extractions which can be used in small and large scale. These techniques have advantages and limitations, which have been recently reviewed [10–12].

Gas chromatograph (GC) coupled with a mass spectrometer (MS) has been the main analytical technique for the chemical characterization of essential oils [10] and together forms a powerful tool for high quality quantitative and qualitative analysis. However, GC/MS analysis of this complex mixture often is a time-consuming task due to limitations in techniques and instrumentation. Advances in this field have led to increased sensitivity, reproducibility and shorter times for analysis of volatile components. For example, fast GC often yields faster analysis times than conventional GC while maintaining resolution and allowing more samples to be analyzed per shift. Nevertheless, the use of fast GC for essential oils analysis is still exploring [13, 14].

For identification of essential oil constituents, computerized matching of acquired mass spectrum with those stored in the mass spectral libraries (NIST and Wiley) and with that of an authentic standard, comparison of retention indices [15] determined at least on two columns with distinct polarities together are currently the most widely used for structural identification [7]. Quadrupole mass spectra obtained by electron ionization (EI) at 70 eV has been preferred to the ion-trap-derived mass spectra due to resulting in fragmentation characteristics of each compound [16], especially for the identification of unknown compounds. Tandem mass spectrometry (MS–MS), Fourier transform infrared spectroscopy (FT-IR) and time-of-flight mass spectrometry (TOF–MS) are alternative techniques for detection employed to identify closely related isomers or overlapping compounds. The latter has been the detector of choice for GC $\times$ GC analysis [10, 17].

In addition, enantiomer separation and determination of enantiomeric ratio or enantiomeric excess (*e.e.*) remains with increasing interest mainly due to possible difference in enantiomer biological properties [18]. Two-dimensional (2D) chromatographic approach has been required for chiral recognition of components in complex samples. The most used techniques is heart-cutting multidimensional GC (MDGC) and comprehensive 2D GC (GC $\times$ GC). Advantages and limits of both techniques are well-known, although recently mass spectrometry as a second dimension in detection has gained further interest because of the role it can play in speeding up enantioselective GC analysis (es-GC) [18, 19]. Derivatized cyclodextrins (CDs) based columns are the most popular chiral stationary phase for es-GC [20] and has been successfully applied in the 2D GC analysis of chiral components from essential oils [21].

Recently there has been growing interest in terpenoids compounds due to the biological activity shown by some of them, especially as anticancer agents. In addition, terpenoid enantiomers are particularly useful chiral building blocks for the chemical and biotechnological syntheses [22]. Thus, essential oils are a rich source of highly value compounds and their commercialization has increasing because they are also used widely in prevent and therapy several diseases [23].

Despite the importance of the issue, to our knowledge, little attention was paid on the investigation of chiral components of essential oils and their anticancer activity [24, 25].

It is widely known that many substances produced by living organisms are chiral and can occur alone (only one enantiomer) or can be present in varying amounts or even as racemic mixtures, exerting the same or distinct activities. For example, (4*S*)-(+)-carvone smell like caraway and (4*R*)-(−)-carvone has a distinct sweet spearmint odor with both enantiomers presenting different effects on the central nervous system [26]. Moreover, the isomers of the monoterpene alcohol (3*R*)-(−)- and (3*S*)-(+)-linalool have distinctive odors [27], but show the same antifungal activity [28].

The biosynthetic pathways of the essential oil components can produce one form only of an optically active chemical. Sometimes one form of an enantiomer may be produced in larger amounts, but often the relative abundances of both enantiomeric forms are very specific to the species and geographical origin of an essential oil [29].

Except for α-humulene, γ-terpinene, eugenol, and thymol exists chirality (stereogenic centers) in the other molecules mentioned in Fig. 19.1. Moreover, enantiomeric distribution of constituents from essential oils is still scarce, especially for sesquiterpenes. Nevertheless, some papers have been published mainly on the enantiomeric variation of monoterpenes such as (±)-limonene, (±)-linalool, (±)-α-pinene, and (±)-terpinen-4-ol [27, 28, 30, 31].

The enantiomers of the **linalool are found in variable** distribution from different plant species [21, 28]. In general, the optically active (3*S*)-(+)- and (3*R*)-(−)-linalool were isolated from lavender oil and coriander oil respectively, but both forms can be found in variable proportion in the wood and leaves essential oils of Brazilian rosewood (*Aniba rosaeodora* Ducke) [32]. It is noteworthy that the chiral stability may be influenced by the developmental stage of plant material, pH value and sampling techniques. In comparison, increasing amounts of the (3*S*)-(+) linalool was detected in oils produced by hydro-distillation of longer than 1 h duration [33]. In the same way, ascaridole, a monoterpene endoperoxide found as major component (up to 92 % of total oil) of essential oil from *Chenopodium ambrosioides* from various origins, is a heat-sensitive compound which rearranges to isoascaridole [34].

The enantiomeric composition of the monoterpene limonene is different for the various plant parts and changes during the development of the umbels as well as antimicrobial efficacy of the individual enantiomers and the racemic mixture showed variation [35]. For example, (4*R*)-(+)-limonene, the major component orange and other citrus peel oils, is one of the most investigated monoterpenes regarding the prevention of chemically induced tumors [36]. From a biosynthetic point of view, (4*S*)-(−)-limonene then serves as a precursor to other oxygenated monocyclic monoterpenes such as (−)-perillyl alcohol and (−)-perillaldehyde which are minor components in many aromatic plants including *Perilla frutescens* (Lamiaceae) and also have been linked to anticancer activity [23, 24, 36].

The sesquiterpene α-(−)-bisabolol was first isolated from the blossoms of *Matricaria chamomilla* (Asteraceae) which may exist in three others possible stereoisomers: (+)-α-bisabolol and (+)- and (−)-*epi*-α-bisabolol. Besides, chamomile, another source of α-(−)-bisabolol is sage (*Salvia runcinata*) that contain up to 90 % and candeia (*Eremanthus erythropappus*) which may contain up to 85 % from the wood [37].

(−)-*trans*-β-elemene is a sesquiterpene that has attracted attention due to recent developments on their use in the antitumor therapy of many kinds of cancer. A comprehensive review that includes its natural occurrence, biogenesis, anticancer activity, and synthesis and chemical characterization was recently published [38].

Caryophyllene is probably the most widely distributed sesquiterpene in nature. The pure form of the (−)-(*E*)-caryophyllene is frequently present in the essential oil from *Humulus lupulus*, *Piper nigrum*, *Syzygium aromaticum* leaf, and *Copaifera officinalis*, etc. However, (+)-(*E*)-caryophyllene is rare and has been found in essential oils of liverworts [39].

Phenylpropanoid volatile compounds are found in essential oil composition of many plant species presenting significant biological activities. One such compound is eugenol which has been used by humans since antiquity for food preservation and flavoring and for medicinal purpose. Essential oil from clove (*Syzigium aromaticum*) is the main natural source of eugenol (up to 90 % of total oil) [40].

19.3 Anticancer Proprieties

As already mentioned, we selected a total of 20 compounds that presented positive results on cytotoxic drug screening; among them, ascaridole, α-bisabolol, (*E*)-caryophyllene, β-elemene, β-eudesmol, D-limonene, terpinen-4-ol, and thymol have been extensively studied with promissory results. Therefore, a special attention was given to these compounds. Moreover, the relevant compounds were summarized and included in a condensed form in Table 19.1.

Ascaridole exhibits cytotoxic activity against leukemia, melanoma, brain, and colon cancer cell lines, as well as, multiple drug resistance cancer cell lines [41, 42]. It also presented in vivo antitumor in a sarcoma murine model with no toxic side effects [42]. More recently, cell cycle and DNA damage analyses revealed a remarkable NER (nucleotide excision repair) specificity of ascaridol. Ascaridole decreased the G_1 phase in three cells lines, but it caused G_2/M phase arrest only in NER-deficient cells. It also induced an increase in the subG_1 peak which was considerably higher in NER-deficient cells than in proficient cells. Moreover, DNA damage induction was substantially higher in NER-deficient cells. In addition, ascaridole led to a dose-dependent increase in intracellular levels of reactive oxygen species at cytotoxic concentrations, but only NER-deficient cells showed a strongly induced amount of 8-oxodG sites [43].

Table 19.1 Summary of the anticancer proprieties of essential oil constituents

No.	Compounds	Anticancer proprieties	Range of IC_{50} values (μg/ml)	References
1	Ascaridole	In vitro and in vivo antitumor activities against several tumor cell lines G_2/M phase arrest and DNA-damage in NER-deficient cells Increase in intracellular levels of reactive oxygen species	6.3–18.4	[41–43]
2	α-Bisabolol	Selective cytotoxic effect against several tumor cell lines Apoptotic activity by liberation of cytochrome c and via Fas receptor	22.8–39.9	[44–46]
3	α-Cadinol	In vitro cytotoxic activity against several tumor cell lines	0.78–13.05	[82]
4	(*E*)-Caryophyllene	Selective cytotoxic effect against several tumor cell lines Induction of apoptosis by caspase-3 catalytic activity	20.1–21.8	[47–49]
5	α-Cedrol	In vitro cytotoxic activity against several tumor cell lines	41–44	[84]
6	β-Elemene	In vitro and in vivo antitumor activity against several tumor cells G_2/M phase arrest and apoptosis by reduction of Bcl-2 protein expression Enhances of taxanes- and cisplatin-induced cytotoxicity	45.3–56.3	[50–59]
7	α-Eudesmol	In vitro cytotoxic activity against several tumor cell lines	5.1–19.4	[64]
8	β-Eudesmol	In vitro and in vivo antitumor activity against several tumor cells Induces apoptosis by mitochondrial apoptotic pathway via JNK signaling Inhibits angiogenesis by blocking extracellular regulated protein kinases (ERK) MAPK signaling	20–25.1	[62–65]
9	γ-Eudesmol	In vitro cytotoxic activity against several tumor cell lines	0.01–20.6	[64, 85]

(continued)

Table 19.1 (continued)

No.	Compounds	Anticancer proprieties	Range of IC_{50} values (µg/ml)	References
10	Eugenol	In vitro and in vivo melanoma growth inhibition through inhibition of E2F1 transcriptional activity Apoptosis induction by translocation of phospho-ser 15-p53 into mitochondria	~0.08	[86, 87]
11	α-Humulene	In vitro cytotoxic activity against breast tumor cell line Induction of the decrease in cellular GSH content and increase in ROS production	~ 14.9	[87]
12	Isointermedeol	In vitro cytotoxic activity against tumor cell line Induction of apoptosis by both intrinsic and extrinsic pathways	~20	[88]
13	D-limonene	Inhibition of the growth of gastric cancer cell through apoptotic pathways Anti-angiogenic activity by down-regulation of VEGF	34.1–681.2	[66–71]
14	Linalool	In vitro cytotoxic activity against tumor cell line	~23	[83]
15	(*E*)-Nerolidol	In vitro cytotoxic activity against tumor cell lines	5.8–6.4	[89]
16	α-Pinene	In vitro cytotoxic activity against tumor cell line	~186.0	[90]
17	Perillaldehyde	In vitro cytotoxic activity against tumor cell lines	37.6–751.1	[91]
18	γ-Terpinene	In vitro cytotoxic activity against tumor cell line	~ 156.9	[90]
19	Terpinen-4-ol	Selective cytotoxic effect against several tumor cell lines Cell-cycle arrest and cell death through p53-dependent apoptosis pathways In vivo antitumor activity against xenograft tumor	54.84–189.7	[72–74]
20	Thymol	Induction of caspase-dependent and -independent apoptosis and necrosis Induces a $[Ca^{2+}]_i$ rise by inducing phospholipase C- and protein kinase C-dependent Ca^{2+} release from the endoplasmic reticulum and Ca^{2+} entry via non store-operated Ca^{2+} channels	~ 60.1	[79–81]

α-Bisabolol showed cytotoxic effect on several human and rat malignant cell lines. The action of α-bisabolol seems to be selective as its effect in normal mouse astroglial cells was not cytotoxic. It also exhibited apoptotic activity by induction of the liberation of cytochrome c and via Fas receptor [44, 45]. In addition, Gomes-Carneiro et al. [46] showed that α-bisabolol is non-mutagenic in the *Salmonella* microsomal test, and it can even neutralize the effect of various mutagenic substances. Moreover, α-bisabolol also showed an antigenotoxic effect against the hydrogen peroxide effect [45].

(*E*)-Caryophyllene has been reported to have cytotoxic activity over a wide range of tumor cell lines, but not against normal cells [47–49]. In addition, it caused an induction of apoptosis accompanied by DNA ladder and caspase-3 catalytic activity in tumor cell lines [49].

β-Elemene exhibits in vitro and in vivo antitumor activity on human and murine tumor cells. Many studies showed that the cell proliferation inhibited by β-elemene is correlated to G_2/M phase arrest and induction of apoptotic cell death by reduction of Bcl-2 protein expression. β-Elemene also enhances caspase-3 activity, and inhibits protein expression of eukaryotic initiation factors eIFs (4E, 4G), basic fibroblast growth factor (bFGF), and vascular endothelial growth factor (VEGF) [50–57].

Furthermore, β-elemene markedly enhanced taxanes or cisplatin-induced cytotoxicity [57–59]. The combination treatments induced increased cytochrome c release from mitochondria, significant caspase-8 and -3 cleavage, and down-regulation of Bcl-2 and Bcl-XL expression. The suppression of specific 'survival' gene expression appears to be the key action leading to the synergistic effect of combination treatments with β-elemene and taxanes [58, 59]. In vivo, the growth of laryngeal cancer cell-transplanted tumors in nude mice was inhibited by intraperitoneal injection of elemene. Compared with control groups, elemene significantly inhibited the protein expression of eIFs (4E and 4G), bFGF, and VEGF and decreased the microvessel density (MVD) [57]. Moreover, some clinical trials indicated that the possible side effects of β-elemene given intravenously include slight fever (usually lower than 38 °C), gastro-intestinal reactions, allergic reactions, local pain, and phlebitis. No bone marrow, liver, cardiac, or renal toxicities were found to be related to clinical treatment with β-elemene [50, 60, 61].

β-Eudesmol produced inhibitory effect on the growth of various tumor cells lines, but it had no effect on the proliferation of the rat aortic smooth muscle cells and astrocytes [62–64]. In addition, β-eudesmol induced apoptosis accompanied by cleavage of caspase-3, caspase-9, and poly (ADP-ribose) polymerase; downregulation of Bcl-2 expression; release of cytochrome c from mitochondria; and decrease in mitochondrial membrane potential (MMP). Activation of c-Jun N-terminal kinases (JNK) mitogen-activated protein kinases was observed in β-eudesmol-treated cells, and the inhibitor of JNK blocked the β-eudesmol-induced apoptosis, downregulation of Bcl-2, and the loss of MMP, suggesting that β-eudesmol induces apoptosis by mitochondrial apoptotic pathway, which is controlled through JNK signaling [63].

β-Eudesmol also inhibited angiogenesis by blocking extracellular regulated protein kinases (ERK) MAPK signaling [62]. Moreover, it inhibited the growth of mouse H22 and S180 tumor and the formation of new blood vessels in tumor tissues in vivo [65].

D-limonene is a known anticarcinogenic compound that it was proved to have antitumor activity [66–70]. In recent studies, D-limonene showed able to inhibit the growth of human gastric cancer cell in vitro through a mechanism of inducing the apoptosis of tumor cells [70].

In xenograft model, D-limonene alone or combined with 5-FU decreased drastically, the metastasis to liver, peritoneum and the occurrence of ascites were inhibited significantly compared with the control group. In addition, a notably decreased expression of MVD and VEGF in D-limonene and combined group were observed, suggesting the anti-angiogenic mechanism of D-limonene via down-regulation of VEGF [71].

Many studies have indicated that terpinen-4-ol exerts cytotoxic effects against several tumor cell lines without affecting normal cells. It is also able to induce cell-cycle arrest and cell death through apoptosis or necrosis pathways [72, 73]. In addition, Wu et al. [74] demonstrated that caspase-dependent mitochondrial dysfunction is the mechanism of terpinen-4-ol-induced apoptosis. Downregulation of Bcl-2, XIAP and survivin suggests that terpinen-4-ol increases the susceptibility of cancer cells to apoptosis induction. Notably, the ability of terpinen-4-ol to induce apoptosis in tumor cells was p53-dependent. Furthermore, the growth of s.c. xenograft tumors was remarkably inhibited by intratumoral injection of terpinen-4-ol, indicating that the agent also has potential for clinical anticancer activity.

The cytoprotective and antimutagenic effects of thymol has been extensively reported [75–78]. Anyway, some studies also have indicated the antitumor activity of thymol [79–81]. Thymol caused activation of caspase-9, -8 and -3 and concomitant PARP cleavage and it induced disruption of mitochondrial membrane potential, which is associated with caspase-dependent apoptosis. The disruption of mitochondrial membrane and activation of apoptosis appears to be dependent on reactive oxygen species. The translocation of AIF from mitochondria to cytosol and then to nucleus indicates thymols' ability to induce apoptosis through caspase independent pathway as well [79]. In addition, Hsu et al. [80] demonstrated that thymol induces a [Ca^{2+}] concentration rise by inducing phospholipase C- and protein kinase C-dependent Ca^{2+} release from the endoplasmic reticulum and Ca^{2+} entry via non store-operated Ca^{2+} channels. Thymol induced cell death that may involve apoptosis.

19.4 Conclusions and Perspectives

A great amount of essential oil constituents with anticancer potential are found. The most of them presented cytotoxic activity only high range of IC_{50} values; therefore, they present weak clinical potential use. On the other hand, some of them have been extensively studied with promissory results. In short, 20 compounds were identified for their activities in the experimental models used for cytotoxic drug screening; among them, ascaridole, α-bisabolol, (*E*)-caryophyllene, β-elemene, β-eudesmol, D-limonene, terpinen-4-ol, and thymol have been shown promissory results. Anyway, further investigations are necessary to validate these compounds as novel clinically useful cancer chemotherapeutic agents.

References

1. Kinghorn AD, de Blanco EJC, Chai HB, Orjala J, Farnsworth NR, Soejarto DD, Oberlies NH, Wani MC, Kroll DJ, Pearce CJ, Swanson SM, Kramer RA, Rose WC, Fairchild CR, Vite GD, Emanuel S, Jarjoura D, Cope FO (2009) Discovery of anticancer agents of diverse natural origin. Pure Appl Chem 81:1051–1063
2. Adorjan B, Buchbauer G (2010) Biological properties of essential oils: an update review. Flavour Fragr J 25:407–426
3. Handa SS (2008) An overview of extraction techniques for medicinal and aromatic plants. In: Handa SS, Khanuja SPS, Longo G, Rakesh DD (eds) Extraction technologies for medicinal and aromatic plants. ICS-UNIDO, Trieste, pp 21–52
4. Bizzo HR, Hovell AMC, Rezende CM (2009) Brazilian essential oils: general view, developments and perspectives. Quim Nova 32:588–594
5. Bakkali F, Averbeck S, Averbeck D, Idaomar M (2008) Biological effects of essential oils: a review. Food Chem Toxicol 46:446–475
6. Croteau R, Kutchan TM, Lewis NG (2000) Natural products (secondary metabolites). In: Buchanan B, Gruissem W, Jones R (eds) Biochemistry & molecular biology of plants. American Society of Plant Physiologists, Rockville, pp. 1250–1348
7. Tholl D, Boland W, Hansel A, Loreto F, Röse USR, Schnitzler JP (2006) Practical approaches to plant volatile analysis. Plant J 45:540–560
8. Takita MA, Berger IJ, Basílio-Palmieri AC, Borges KM, de Souza JM, Targon MLNP (2007) Terpene production in the peel of sweet orange fruits. Genet Mol Biol 30:841–847
9. Luque de Castro MD, Jiménez-Carmona MM, Fernández-Pérez V (1999) Towards more rational techniques for the isolation of valuable essential oils from plants. Trends Anal Chem 18:708–709
10. Marriot PJ, Shellie R, Cornwell C (2001) Gas chromatographic technologies for the analysis of essential oils. J Chromatogr A 936:1–22
11. Ormenõ E, Goldstein A, Niinemets Ü (2011) Extracting and trapping biogenic volatile organic compounds stored in plant species. Trends Analyt Chem 30:978–989
12. Sovilj MN, Nikolovski BG, Spasojevic MD (2011) Critical review of supercritical fluid extraction of selected spice plant materials. Maced J Chem Chem Eng 30:197–220
13. Matisová E, Dömötörová M (2003) Fast gas chromatography and its use in trace analysis. J Chromatogr A 1000:199–221
14. Rubiolo P, Liberto E, Sgorbini B, Russo R, Veuthey JL, Bicchi C (2008) Fast-GC-conventional quadrupole mass spectrometry in essential oil analysis. J Sep Sci 31:282–290
15. Van Den Dool H, Kratz PD (1963) A generalization of the retention index system including linear temperature programmed gas–liquid partition chromatography. J Chromatogr 11:463–471
16. Adams RP (2007) Identification of essential oil components by gas chromatography/mass spectrometry, 4th edn. Allured, Carol Stream
17. Baharum SN, Banawan H, Ghani MA, Mustapha WAW, Noor NM (2010) Analysis of the chemical composition of the essential oil of *Polygonium minus* Huds. using two-dimensional gas chromatography-time-of-flight mass spectrometry (GC-TOF-MS). Molecules 15:7006–7015

18. Bicchi C, Cagliero C, Rubiolo P (2011) New trends in the analysis of the volatile fraction of matrices of vegetal origin: a short overview. A review. Flavour Fragr J 26:321–325
19. Tranchida PQ, Sciarrone D, Dugo P, Modello L (2012) Heart-cutting multidimensional gas chromatography: a review of recent evolution, applications, and future prospects. Anal Chim Acta 716:66–75
20. Schurig V (2002) Chiral separations using gas chromatography. Trends Anal Chem 21:647–661
21. König WA, Hochmuth DH (2004) Enantioselective gas chromatography in flavor and fragrance analysis: strategies for the identification of known and unknown plant volatiles. J Chromatogr Sci 42:423–439
22. De Carvalho CCCR, da Fonseca MMR (2006) Biotransformation of terpenes. Biotechnol Adv 24:134–142
23. Edris AE (2007) Pharmaceutical and therapeutic potentials of essential oils and their individual volatile constituents: a review. Phytother Res 21:308–323
24. Elson CE, Yu SG (1994) The chemoprevention of cancer by mevalonate-derived constituents of fruits and vegetables. J Nutr 124:607–614
25. Bicas JL, Neri-Numa IA, Ruiz ALTG, De Carvalho JE, Pastore GM (2011) Evaluation of the antioxidant and antiproliferative potential of bioflavors. Food Chem Toxicol 49:1610–1615
26. Souza DP, Nóbrega FFF, Almeida RN (2007) Influence of the chirality of (*R*)-(−)- and (*S*)-(+)-carvone in the central nervous system: a comparative study. Chirality 19:264–268
27. Chanotiya CS, Yadav A (2009) Enantiomeric composition of (3R)-(−)- and (3S)-(+)-linalool in various essential oils of Indian origin by enantioselective capillary gas chromatography-flame ionization and mass spectrometry detection methods. Nat Prod Commun 4:563–566
28. Özek T, Tabanca N, Demirci F, Wedge DE, Baser KHC (2010) Enantiomeric distribution of some linalool containing essential oils and their biological activities. Rec Nat Prod 4:180–192
29. Dickschat JS (2011) Isoprenoids in three-dimensional space: the stereochemistry of terpene biosynthesis. Nat Prod Rep 28:1917–1936
30. Del Castillo MLR, Caja MM, Blanch GP, Herraiz M (2003) Chiral evaluation of aroma-active compounds in real complex samples. J Food Sci 68:770–774
31. Bonaccorsi I, Sciarrone D, Cotroneo A, Mondello L, Dugo P, Dugo G (2011) Enantiomeric distribution of key volatile components in *Citrus* essential oils. Braz J Pharmacogn 21:841–849
32. Zellner BD, Presti ML, Barata LES, Dugo P, Dugo G, Mondello L (2006) Evaluation of leaf-derived extracts as an environmentally sustainable source of essential oils by using gas chromatography-mass spectrometry and enantioselective gas chromatography-olfactometry. Anal Chem 78:883–890
33. Kreis P (1992) Chiral compounds of essential oils. Part XII. Authenticity control of rose oils using enantioselective multidimensional gas chromatography. Flavour Fragr J 7(4):199–203
34. Cavalli J-F, Tomi F, Bernardini A-F, Casanova J (2004) Combined analysis of the essential oil of *Chenopodium ambrosioides* by GC, GC-MS and ^{13}C-NMR spectroscopy: quantitative determination of ascaridole, a heat-sensitive compound. Phytochem Anal 15(5):275–279
35. Van Vuuren SF, Viljoen AM (2007) Antimicrobial activity of limonene enantiomers and 1,8-cineole alone and in combination. Flavour Fragr J 22:540–544
36. Miller JA, Thompson PA, Hakim IA, Chow HHS, Thomson CA (2011) D-Limonene: a bioactive food component from citrus and evidence for a potential role in breast cancer prevention and treatment. Oncol Rev 5:31–42
37. Kamatou GPP, Viljoen AM (2010) A review of the application and pharmacological properties of α-bisabolol and α-bisabolol-rich oils. J Am Oil Chem Soc 87:1–7
38. Adio AM (2009) (−)-*trans*-β-elemene and related compounds: occurrence, synthesis, and anticancer activity. Tetrahedron 65:5145–5159
39. Fricke C, Rieck A, Hardt IH, König WA, Muhle H (1995) Identification of (+)-β-caryophyllene in essential oils of liverworts by enantioselective gas chromatography. Phytochemistry 39:1119–1121

40. Kamatou GP, Vermaak I, Viljoen AM (2012) Eugenol-from the remote Maluku islands to the international market place: a review of a remarkable and versatile molecule. Molecules 17:6953–6981
41. Efferth T, Olbrich A, Sauerbrey A, Ross DD, Gebhart E, Neugebauer M (2002) Activity of ascaridol from the anthelmintic herb *Chenopodium anthelminticum* L. against sensitive and multidrug-resistant tumor cells. Anticancer Res 22:4221–4224
42. Bezerra DP, Marinho Filho JD, Alves AP, Pessoa C, Moraes MO, Pessoa OD, Torres MC, Silveira ER, Viana FA, Costa-Lotufo LV (2009) Antitumor activity of the essential oil from the leaves of *Croton regelianus* and its component ascaridole. Chem Biodivers 6:1224–1231
43. Abbasi R, Efferth T, Kuhmann C, Opatz T, Hao X, Popanda O, Schmezer P (2012) The endoperoxide ascaridol shows strong differential cytotoxicity in nucleotide excision repair-deficient cells. Toxicol Appl Pharmacol 259:302–310
44. Cavalieri E, Mariotto S, Fabrizi C, de Patri AC, Gottardo R, Leone S, Berra LV, Lauro GM, Ciampa AR, Suzuki H (2004) Alphabisabolol, a nontoxic natural compound, strongly induces apoptosis in glioma cells. Biochem Biophys Res Commun 315:589–594
45. Anter J, Romero-Jiménez M, Fernández-Bedmar Z, Villatoro-Pulido M, Analla M, Alonso-Moraga A, Muñoz-Serrano A (2011) Antigenotoxicity, cytotoxicity, and apoptosis induction by apigenin, bisabolol, and protocatechuic acid. J Med Food 14:276–283
46. Gomes-Carneiro MR, Dias DM, De-Oliveira AC, Paumgartten FJ (2005) Evaluation of mutagenic and antimutagenic activities of alpha-bisabolol in the Salmonella/microsome assay. Mutat Res 585:105–112
47. Kubo I, Chaudhuri SK, Kubo Y, Sanchez Y, Ogura T, Saito T, Ishikawa H, Haraguchi H (1996) Cytotoxic and antioxidative sesquiterpenoids from *Heterotheca inuloides*. Planta Med 62:427–430
48. Ashour ML, El-Readi M, Youns M, Mulyaningsih S, Sporer F, Efferth T, Wink M (2009) Chemical composition and biological activity of the essential oil obtained from *Bupleurum marginatum* (Apiaceae). J Pharm Pharmacol 61:1079–1087
49. Amiel E, Ofir R, Dudai N, Soloway E, Rabinsky T, Rachmilevitch S (2012) β-Caryophyllene, a compound isolated from the biblical balm of gilead (*Commiphora gileadensis*), is a selective apoptosis inducer for tumor cell lines. Evid Based Complement Alternat Med 2012:872394
50. Yang H, Wang X, Yu L (1996) The antitumor activity of elemene is associated with apoptosis. Zhonghua Zhong Liu Za Zhi 18:169–172
51. Zou L, Liu W, Yu L (2001) beta-elemene induces apoptosis of K562 leukemia cells. Zhonghua Zhong Liu Za Zhi 23:196–198
52. Zhou HY, Shen JK, Hou JS, Qiu YM, Luo QZ (2003) Experimental study on apoptosis induced by elemene in glioma cells. Ai Zheng 22:959–963
53. Xu YH, Dong B, Luo QZ, Zhou HY, Jia YC, Yang YF, Wang YZ (2005) Influence of elemene on the expression of Bcl-2 family genes in rat C6 glioma cells. Zhonghua Yi Xue Za Zhi 85:1700–1703
54. Wang G, Li X, Huang F, Zhao J, Ding H, Cunningham C, Coad JE, Flynn DC, Reed E, Li QQ (2005) Antitumor effect of beta-elemene in non-small-cell lung cancer cells is mediated via induction of cell cycle arrest and apoptotic cell death. Cell Mol Life Sci 62:881–893
55. Wang XS, Yang W, Tao SJ, Li K, Li M, Dong JH, Wang MW (2006) Effect of delta-elemene on Hela cell lines by apoptosis induction. Yakugaku Zasshi 126:979–990
56. Zhou HY, Hou JS, Wang Y (2006) Dose-and time-dependence of elemene in the induction of apoptosis in two glioma cell lines. Zhonghua Zhong Liu Za Zhi 28:270–271
57. Tao L, Zhou L, Zheng L, Yao M (2006) Elemene displays anti-cancer ability on laryngeal cancer cells *in vitro* and *in vivo*. Cancer Chemother Pharmacol 58:24–34
58. Li X, Wang G, Zhao J, Ding H, Cunningham C, Chen F, Flynn DC, Reed E, Li QQ (2005) Antiproliferative effect of betaelemene in chemoresistant ovarian carcinoma cells is mediated through arrest of the cell cycle at the G_2-M phase. Cell Mol Life Sci 62:894–904
59. Zhao J, Li QQ, Zou B, Wang G, Li X, Kim JE, Cuff CF, Huang L, Reed E, Gardner K (2007) *In vitro* combination characterization of the new anticancer plant drug beta-elemene with taxanes against human lung carcinoma. Int J Oncol 31:241–252

60. Wang J, Zhang H, Sun Y (1996) Phase III clinical trial of elemenum emulsion in the management of malignant pleural and peritoneal effusions. Zhonghua Zhong Liu Za Zhi 18:464–467
61. Tan P, Zhong W, Cai W (2000) Clinical study on treatment of 40 cases of malignant brain tumor by elemene emulsion injection. Zhongguo Zhong Xi Yi Jie He Za Zhi 20:645–648
62. Tsuneki H, Ma EL, Kobayashi S, Sekizaki N, Maekawa K, Sasaoka T, Wang MW, Kimura I (2005) Antiangiogenic activity of beta-eudesmol in vitro and in vivo. Eur J Pharmacol 512:105–115
63. Li Y, Li T, Miao C, Li J, Xiao W, Ma E (2012) β-Eudesmol induces JNK-dependent apoptosis through the mitochondrial pathway in HL60 cells. Phytother Res 14. doi:10.1002/ptr.4727
64. Britto ACS, Oliveira ACA, Henriques RM, Cardoso GMB, Bomfim DS, Carvalho AA, Moraes MO, Pessoa C, Pinheiro MLB, Costa EV, Bezerra DP (2012) *In vitro* and in vivo antitumor effects of the essential oil from the leaves of *Guatteria friesiana*. Planta Med 78:409–414
65. Ma EL, Li YC, Tsuneki H, Xiao JF, Xia MY, Wang MW, Kimura I (2008) beta-Eudesmol suppresses tumour growth through inhibition of tumour neovascularisation and tumour cell proliferation. J Asian Nat Prod Res 10:159–167
66. Uedo N, Tatsuta M, Iishi H, Baba M, Sakai N, Yano H, Otani T (1999) Inhibition by D-limonene of gastric carcinogenesis induced by N-methyl-N′-nitro-N-nitrosoguanidine in Wistar rats. Cancer Lett 137:131–136
67. Broitman SA, Wilkinson J 4th, Cerda S, Branch SK (1996) Effects of monoterpenes and mevinolin on murine colon tumor CT-26 in vitro and its hepatic "metastases" in vivo. Adv Exp Med Biol 401:111–130
68. Kaji I, Tatsuta M, Iishi H, Baba M, Inoue A, Kasugai H (2001) Inhibition by D-limonene of experimental hepatocarcinogenesis in Sprague-Dawley rats does not involve p21(ras) plasma membrane association. Int J Cancer 93:441–444
69. Stratton SP, Dorr RT, Alberts DS (2000) The state-of-the-art in chemoprevention of skin cancer. Eur J Cancer 36:1292–1297
70. Lu XG, Feng BA, Zhan LB, Yu ZH (2003) D-limonene induces apoptosis of gastric cancer cells. Zhonghua Zhongliu Zazhi 25:325–327
71. Lu XG, Zhan LB, Feng BA, Qu MY, Yu LH, Xie JH (2004) Inhibition of growth and metastasis of human gastric cancer implanted in nude mice by D-limonene. World J Gastroenterol 10:2140–2144
72. Greay SJ, Ireland DJ, Kissick HT, Levy A, Beilharz MW, Riley TV, Carson CF (2010) Induction of necrosis and cell cycle arrest in murine cancer cell lines by *Melaleuca alternifolia* (tea tree) oil and terpinen-4-ol. Cancer Chemother Pharmacol 65:877–888
73. Calcabrini A, Stringaro A, Toccacieli L, Meschini S, Marra M, Colone M, Salvatore G, Mondello F, Arancia G, Molinari A (2004) Terpinen-4-ol, the main component of *Melaleuca alternifolia* (tea tree) oil inhibits the in vitro growth of human melanoma cells. J Invest Dermatol 122:349–360
74. Wu CS, Chen YJ, Chen JJ, Shieh JJ, Huang CH, Lin PS, Chang GC, Chang JT, Lin CC (2012) Terpinen-4-ol induces apoptosis in human nonsmall cell lung cancer in vitro and in vivo. Evid Based Complement Alternat Med 2012:818261
75. Archana PR, Nageshwar Rao B, Ballal M, Satish Rao BS (2009) Thymol, a naturally occurring monocyclic dietary phenolic compound protects Chinese hamster lung fibroblasts from radiation-induced cytotoxicity. Mutat Res 680:70–77
76. Mezzoug N, Elhadri A, Dallouh A, Amkiss S, Skali NS, Abrini J, Zhiri A, Baudoux D, Diallo B, El Jaziri M, Idaomar M (2007) Investigation of the mutagenic and antimutagenic effects of *Origanum compactum* essential oil and some of its constituents. Mutat Res 629:100–110
77. Horvathova E, Turcaniova V, Slamenova D (2007) Comparative study of DNA damaging and DNA-protective effects of selected components of essential plant oils in human leukemic cells K562. Neoplasma 54:478–483
78. Mahmud H, Mauro D, Foller M, Lang F (2009) Inhibitory effect of thymol on suicidal erythrocyte death. Cell Physiol Biochem 24:407–414

79. Deb DD, Parimala G, Saravana Devi S, Chakraborty T (2011) Effect of thymol on peripheral blood mononuclear cell PBMC and acute promyelotic cancer cell line HL-60. Chem Biol Interact 193:97–106
80. Hsu SS, Lin KL, Chou CT, Chiang AJ, Liang WZ, Chang HT, Tsai JY, Liao WC, Huang FD, Huang JK, Chen IS, Liu SI, Kuo CC, Jan CR (2011) Effect of thymol on Ca^{2+} homeostasis and viability in human glioblastoma cells. Eur J Pharmacol 670:85–91
81. Chang HT, Hsu SS, Chou CT, Cheng JS, Wang JL, Lin KL, Fang YC, Chen WC, Chien JM, Lu T, Pan CC, Cheng HH, Huang JK, Kuo CC, Chai KL, Jan CR (2011) Effect of thymol on Ca^{2+} homeostasis and viability in MG63 human osteosarcoma cells. Pharmacology 88:201–212
82. He K, Zeng L, Shi G, Zhao GX, Kozlowski JF, McLaughlin JL (1997) Bioactive compounds from *Taiwania cryptomerioides*. J Nat Prod 60:38–40
83. Loizzo MR, Tundis R, Menichini F, Saab AM, Statti GA, Menichini F (2008) Antiproliferative effects of essential oils and their major constituents in human renal adenocarcinoma and amelanotic melanoma cells. Cell Prolif 41:1002–1012
84. Hsieh TJ, Chang FR, Chia YC, Chen CY, Chiu HF, Wu YC (2001) Cytotoxic constituents of the fruits of *Cananga odorata*. J Nat Prod 64:616–619
85. Ghosh R, Nadiminty N, Fitzpatrick JE, Alworth WL, Slaga TJ, Kumar AP (2005) Eugenol causes melanoma growth suppression through inhibition of E2F1 transcriptional activity. J Biol Chem 280:5812–5819
86. Park BS, Song YS, Yee SB, Lee BG, Seo SY, Park YC, Kim JM, Kim HM, Yoo YH (2005) Phospho-ser 15-p53 translocates into mitochondria and interacts with Bcl-2 and Bcl-xL in eugenol-induced apoptosis. Apoptosis 10:193–200
87. Legault J, Dahl W, Debiton E, Pichette A, Madelmont JC (2003) Antitumor activity of balsam fir oil: production of reactive oxygen species induced by alpha-humulene as possible mechanism of action. Planta Med 69:402–407
88. Kumar A, Malik F, Bhushan S, Sethi VK, Shahi AK, Kaur J, Taneja SC, Qazi GN, Singh J (2008) An essential oil and its major constituent isointermedeol induce apoptosis by increased expression of mitochondrial cytochrome c and apical death receptors in human leukaemia HL-60 cells. Chem Biol Interact 171:332–347
89. Sylvestre M, Pichette A, Lavoie S, Longtin A, Legault J (2007) Composition and cytotoxic activity of the leaf essential oil of *Comptonia peregrina* (L.) Coulter. Phytother Res 21:536–540
90. Döll-Boscardin PM, Sartoratto A, Sales Maia BH, Padilha de Paula J, Nakashima T, Farago PV, Kanunfre CC (2012) *In vitro* cytotoxic potential of essential oils of *Eucalyptus benthamii* and its related terpenes on tumor cell lines. Evid Based Complement Alternat Med 2012:342652
91. Sonboli A, Esmaeili MA, Gholipour A, Kanani MR (2010) Composition, cytotoxicity and antioxidant activity of the essential oil *of Dracocephalum surmandinum* from Iran. Nat Prod Commun 5:341–344

Author Biography

Daniel Pereira Bezerra studied Pharmacy at the Federal University of Ceará, Brazil, and completed his PhD at the same University, in 2008. In 2009, he was appointed as Adjunct Professor in Pharmacology at the Federal University of Sergipe, Brazil. His research interests have focused on the preclinical pharmacological and toxicological studies of the synthetic/natural products with anticancer potential.

Emmanoel Vilaça Costa studied Chemistry at the Federal University of Amazonas, Brazil, and completed his PhD at the Federal University of Paraná, Brazil in 2009. From 2009 to 2010 he worked at the Federal University of Sergipe, Brazil, as a post-doctoral researcher working with isolation and characterization of natural products from Brazilian plants, especially from the families Annonaceae and Verbenaceae, focused in the search of new and safer bioactive compounds. Since 2010 he receives from CNPq a post-doctoral fellowship to work with natural products with insecticidal properties at the same University as a researcher of the Program for Regional Scientific Development (DCR/CNPq). His research interests have focused in chemistry of natural products from Brazilian plants, chemotaxonomy and systematic, essential oils, nuclear magnetic resonance (NMR), and chromatography.

Paulo Cesar de Lima Nogueira studied Industrial Chemistry at the Federal University of Ceará, Brazil, and completed his PhD at the State University of Campinas, Brazil, in 2002. From 2003 to 2006 he worked at the Federal University of Sergipe, Brazil, as a researcher of the Program for Regional Scientific Development (DCR/CNPq) in organic chemistry. In 2006 he was appointed as Adjunct Professor in Organic Chemistry. His research interests have focused on the use of renewable natural resources such as the isolation and identification of bioactive secondary metabolites, including essential oils particularly from medicinal and aromatic plants belonging to the Apocynaceae, Euphorbiaceae and Verbenaceae families.

Part VI
Medicinal Compounds from the Garden: A Bird's Eye View of One Mushroom, One Fruit, One Insect, and Others

Chapter 20
Recent Research on Pharmacological Activities of the Medicinal Fungus *Cordyceps sinensis*

Tzi Bun Ng, Jack Ho Wong and Evandro Fei Fang

Abstract The *Cordyceps sinensis* (CS) fungus is an herbal medicinal product in China with a history of over two millenniums. It exhibits a spectacular array of pharmacological activities encompassing anticancer, antidiabetic, anti-adipogenic, analgesic, anti-inflammatory, neuroprotective, anti-aging, eryptosis inducing activities; ameliorating effect on pulmonary fibrosis, hypoxic pulmonary hypertension, and hypertension; protective action against viral myocarditis, stimulatory effect on corticosterone and androgen production, and enhancing effect on exercise performance in healthy older subjects. In this chapter we try to sketch out the recent progress on the pharmacological studies of CS, identify its current research hurdles, and recommend directions for future research.

20.1 Introduction

Cordyceps sinensis (CS) is a Chinese endemic fungus which mainly locates on the Tibetan plateau. It is a parasite on larvae (caterpillars) of two moth genera named *Hepialus* and *Thitarodes*. After an overwintering growth, the fungus multiplies to a perithecial stroma which links to the caterpillar below ground [1]. In traditional Chinese medicine, either the stroma, or the stroma with the caterpillar is used as CS product. The medicinal function of this prized medicine was documented early in Wu Yiluo's Ben Cao Cong Xin (New Compilation of Materia Medica) in 1757 of Qing dynasty, and expanded in the latest decade or so by the disclosure of a diversity of exploitable activities [2]. The pharmacological activities of a number

T. B. Ng (✉) · J. H. Wong · E. F. Fang
School of Biomedical Sciences, The Chinese University of Hong Kong, Shatin, Hong Kong
e-mail: b021770@mailserv.cuhk.edu.hk

J. H. Wong
e-mail: jack1993@yahoo.com

E. F. Fang
Laboratory of Molecular Gerontology, National Institute on Aging, National Institutes of Health, Baltimore, MD 21224, USA

E. F. Fang and T. B. Ng (eds.), *Antitumor Potential and Other Emerging Medicinal Properties of Natural Compounds*, DOI: 10.1007/978-94-007-6214-5_20,

of *Cordyceps* species including *Cordyceps militaris* are available in the literature. This chapter aims at reviewing information pertaining to the activities of CS published after the account given in [2].

20.2 Medicinal Activities of CS

20.2.1 Anticancer Activity

CS contains a large number of medicinal compounds with both a mixture and different isolated compounds showing anticancer activities. For example, Wu et al. sequentially extracted the cultured mycelium of a CS fungus with petroleum ether, ethyl acetate, ethanol and hot water to acquire bioactive metabolites [3]. All solvent extracts except the hot water extract expressed anti-proliferative activity toward B16 mouse melanoma, HL-60 human premyelocytic leukemia, HepG2 human hepatocellular carcinoma and MCF-7 breast cancer cells. The ethyl acetate extract was the most potent, with IC_{50} of 12 μg/ml on B16 cells and 5 μg/ml on MCF-7 cells. In contrast, it had much lower cytotoxicity against normal mouse bone marrow cells. The ethyl acetate extract contained carbohydrates, adenosine, ergosterol and a trace amount of cordycepin, of which ergosterol and related compounds were identified as a major class of active constituents contributing to the in vitro cytotoxicity. The ethyl acetate extract brought about 60 % reduction of melanoma size over 27 days in C57BL/6 mice [3]. Moreover, CS ameliorated the suppressive effects of a reputable antitumor compound Taxol on bone marrow function, as indexed by the enhanced expression of core binding factor-a and bone morphogenetic protein with concurrent suppression of osteoclast differentiation factor ligand [4].

Small isolated components in CS also exhibit excellent antitumor potential. A 7.7-kDa neutral mannoglucan 1 showed weak cytotoxicity activity against SPC-I ($IC_{50} = 63$ μg/mL) cancer cell line, but was lacking in cytotoxicity against BCAP37 cancer lines [5]. The exopolysaccharide of CS inhibited tumor growth in lungs and livers in mice and reduced the c-Myc, c-Fos, and VEGF levels in these organs [6]. It also exerted an effect in H22 tumor-bearing mice by regulation of immune-modulatory factors [7]. The polysaccharide fraction of CS enhanced triptolide-induced apoptosis in HL-60 cells and inhibited the expression of NF-κB and caspases −3, −6, −7, and −9 [8]. Mixed compounds in CS mycelium included [5 alpha, 8 alpha-epidioxy-22E-ergosta-6, 22-dien-3beta-ol], [5 alpha, 8 alpha-epidioxy-22E-ergosta-6,9(11), 22-trien-3beta-ol], [5 alpha, 6 alpha-epoxy-5 alpha-ergosta-7,22-dien-3beta-ol] with a peroxide ring or an epoxide ring exhibited cytotoxic activity with IC_{50} values of 7.3–7.8 μg/ml. The induction of apoptosis by these compounds depends on the activation of caspases-3/7 [9]. It should be noted that there were some reports showing the apoptosis-inhibiting activity of crude CS extract or its purified compounds, please refer to the review for further information [1]. Furthermore, the expression of vascular endothelial growth factor and basic fibroblast growth factor in H157 (non-small cell lung cancer) cells and the cell viability were diminished in cells treated with a combination of cisplatin and CS polysaccharide. Thus the polysaccharide represents a potential adjuvant chemotherapeutic agent in the treatment of non-small cell lung cancer [10]. In addition, CS polysaccharide (EPS) induced dendritic cell sarcoma (DCS) cell line to develop mature characteristics with a mechanism probably related to suppression of the Janus kinase 2 (p-JAK2)/signal transducer and activator of transcription 3 (p-STAT3) signal pathway and stimulation of the NF-κB signal pathway [11]. Interestingly, it also exhibited a protective action against DNA damage of human skin cells after UV B exposure [12].

Cordycepin (also named 3′-deoxyadenosine) is a compound extracted from CS and *C. militaris*. It exhibits anti-viral, anti-bacterial, immune-stimulatory, and anti-tumor effects. The apoptotic effect of cordycepin on OEC-M1, a human oral squamous cancer cell line, has been studied. The cells became rounded up and the plasma membrane exhibited blebbing and phosphatidylserine flipping on cell membrane and the cell viability declined. The percentage of cells in G1 phase dropped whereas there was an increase of cells in G2/M and subG1 phases [13]. It also inhibited cell growth in human leukemia cells by inducing apoptosis which was accompanied by production of reactive oxygen species, mitochondrial dysfunction, activation of caspases, and cleavage of PARP [14]. Recently, Jeong et al. reported that cordycepin inhibited cell motility and invasiveness of LNCaP cells associated with the tightening of tight junctions [15]. The levels of claudin proteins, principal constituents of tight junctions that have a pivotal role in regulation and selectivity of paracellular transport. Furthermore, cordycepin down-regulated the expression and activity of matrix metalloproteinase (MMP)-2 and MMP-9, while it augmented levels of tissue inhibitor of metalloproteinase (TIMP)-1 and TIMP-2. These actions were associated with inactivation of the phosphoinositide 3-kinase (PI3K)/Akt pathway in LNCaP cells. The observations indicate that cordycepin suppresses the migration and invasion of LNCaP cells by inhibiting the activity of TJs and MMPs, probably by reduction of Akt activation [15].

The medicinal activities of CS are also demonstrable in in vivo and clinical studies. Intraperitoneal administration of a water extract of CS reduced the number of metastatic surface nodules of B16-F0 cells in the liver and prolonged the survival of C57BL/6Cr mice. Hepatocyte growth factor-accelerated invasion of B16-F0 cells in vitro was also suppressed [16]. On the other hand, treatment of patients with hepatocellular carcinoma, following a regimen of four agents prepared from natural products including CS, brought about prolonged survival in some of the patients [12].

20.2.2 Anti-Diabetic Activity

Many natural crude extracts and isolated compounds are shown to have promising anti-diabetic activity, such as bitter melon and others [17]. The fermented mycelia and broth as well as fruiting bodies of CS manifested anti-hyperglycemic activity in diabetic rats [15]. Shi et al. in 2009 found that the onset of type 1 diabetes in

NOD mice was associated with an imbalance of $CD4^+CD25^+FoxP3^+$ regulatory T (Treg) cells and IL-17 producing Th17 cells. A reduced incidence of diabetes due to an increase in the ratio of Treg cells to Th17 in the spleen and pancreatic lymph nodes was effected after oral administration of CS [18]. A vey recent study showed that CS extract elevated HDL/LDL ratios in KK/HIJ diabetic mice fed a high-fat diet at 4 weeks and decreased body weight gain at 8 weeks. Although the extract did not correct hyperglycemia or insulin resistance, it exerted a protective action on pancreatic beta cells against the toxic effects of streptozotocin. The extract attenuated the accumulation of mesangial matrix and collagen deposition [19]. In corroboration of this study, Zahraa et al. found that CS elevated serum insulin level, β cell function, HDL-cholesterol level, total antioxidant capacity, and pancreatic content of reduced glutathione. CS and taurine enhanced, both in the presence and absence of insulin, glucose uptake by diaphragms of normal and diabetic rats [20].

CS may also have synergistic activity with components of other origins. For example, cordyceps and taurine, either alone or in conjunction, were less efficacious than glibenclamide in lowering the blood glucose level although they displayed more potent antioxidant activity and were better able to reduce insulin resistance [20]. What is more, in rats with diabetic nephropathy, CS and *Tipterygium wilfordii* polyglycosidium mitigated proteinuria, and protected and repaired damaged podocytes by promoting expression of nephrin and podocin. This recipe could improve the efficacy and reduce adverse reactions to the latter [21].

20.2.3 Anti-Adipogenic Activity

The anti-diabetic function of CS may be associated with its anti-adipogenic and anti-hyperlipidemic activities. The anti-adipogenic action of cordycepin is brought about by intervention in the mTORC1-CCAAT/enhancer-binding protein-PPARγ pathway. CS prevented adipogenesis and lipid accumulation [22]. Daily administration of cordycepin to male Syrian golden hamsters fed a high-fat diet reduced serum levels of triglycerides, total cholesterol, low-density lipoprotein cholesterol, and relative retroperitoneal fat. CS prevented hyperlipidemia via activation of phospho-AMP-activated protein kinase. It also improved insulin sensitivity [23]. In conclusion, it has a good potential as a therapeutic agent for treatment of obesity and obesity-related diseases.

20.2.4 Anti-Aging Activity

In view of the increased life expectancy, it is speculated that there will be over 1.2 billion people aged above 60 worldwide in 2025, and natural compounds serve as a good source to the exploration of anti-aging drugs [24]. CS has been shown to improve exercise performance in a double-bind, placebo-controlled trial on 20 healthy old adults (50–75 years) [25]. The mechanisms may include immune-regulatory activity (details provided in Sect. 20.5) and the reduction of different disease risks commonly in old adults, such as thrombosis and osteoporosis. An example is found in a 31-kDa serine protease named CSP, with fibrinolytic activity and a free cysteine residue near the active site. It hydrolyzed fibrinogen, fibrin and casein with a high efficiency, and serum albumins of bovine and human origins to a lesser extent. CSP was a plasmin-like protease, but not a plasminogen activator. It may be a potential therapeutic agent for thrombosis [26]. Secondly, CS has reportedly anti-osteoporosis activity in ovariectomized osteopenic rats by reduction of serum alkaline phosphatase activity, tartarate-resistant acid phosphatase activity, C-terminal crosslinked telopeptides of collagen type I level, IFN-γ level, and elevation of osteocalcin and estradiol levels [27]. The same research group further revealed that CS up-regulated estradiol synthesis and prevented osteoporosis in ovariectomized rats. It decreased markers of bone turnover and increased the osteocalcin levels in HLS rats. Micro-CT analysis from the L4 vertebra revealed that CS prevented the decrease of the bone volume fraction, connectivity density, trabecula number and thickness, and improved trabecula separation and structure model index in HLS rats [28]. Metabolic studies conducted on CS consolidated the findings. The extract of CS increased the activity of the age-related enzymes superoxide dismutase, glutathione peroxidase and catalase and lowered the level of lipid peroxidation and monoamine oxidase activity in aging mice. The extract shortened escape latency, prolonged step-down latency and decreased the number of errors in these mice in water maze and step-down type avoidance tests. The study demonstrated that the extract augmented brain function and anti-oxidative enzyme activity in mice with D-galactose-induced senescence and increased sexual function in castrated rats [29].

Some pre-aging diseases, such as Cockayne syndrome and xeroderma pigmentosum, are extremely sensitive to sunlight exposure and irradiation, and the syndrome then becomes severely deteriorated [30, 31].The radio-protective activity of CS may also be linked with its anti-aging activity. Liu et al. reported that a hot-water extract of CS prolonged the median time to death in mice from 13 to 20 days following 8 Gy total-body irradiation, and from 9 to 18 days following 10 Gy total-body irradiation by the scavenging of free radical species [32]. But no in-depth mechanistic studies were reported.

Other activities of CS connected to its anti-aging function include the induction of steroidogenesis. An interesting report by Chen found that CS increased steroidogenesis in Leydig cells and improved male reproductive function by elevation of steroidogenic acute regulatory protein mRNA expression and the mitochondria electrochemical gradient [33].

20.2.5 Analgesic, Neuroprotective, Anti-Inflammatory, and Anti-Oxidant Activities

Cordymin is a 10906-Da peptide in CS which manifested analgesic activity. It inhibited acetic acid-induced abdominal constrictions in mice. In the hot-plate test, it delayed the reaction time to heat stimuli. It demonstrated marked activity against neurolysin [34, 35]. CS also manifested neuroprotective activity. Using the right middle cerebral artery occlusion model, it was demonstrated in rats that CS extract improved the outcome as reflected by neurobehavioral function after cerebral ischemia and reperfusion. Supplementation with CS extract augmented the defense mechanism against cerebral ischemia by up-regulating antioxidant activity [36].

Evidence for close association of the neuroprotective activity of CS with its anti-oxidant activity has been further consolidated. For example, CS mycelium inhibited brain ischemia-reperfusion-induced up-regulation of NF-κB activation and brain production of IL-1β, TNF-α, adhesion molecule, inducible NO synthase, and cyclooxygenase [37]. A mycelial extract of CS named C.S.3 reduced airway inflammation in ovalbumin-induced allergic mice. It attenuated extracellular signal-regulated kinase 1/2 signaling pathway and hence inhibited the activity of NF-κB in lung cells and cultured airway smooth muscle cells. Thus C.S.3 may furnish a treatment for asthma [38]. The anti-inflammatory activity of CS is at least contributed by the following isolated components. For example, [1-(5-hydroxymethyl-2-furyl)-β-carboline] suppressed superoxide anion production and elastase release with IC_{50} values of 0.45 ± 0.15 and 1.68 ± 0.32 μM, respectively [39]. Another polysaccharide also displayed antioxidant activity and scavenged the $ABTS^{\cdot +}$ free radical in a concentration-dependent fashion [40]. In addition, CS also contains an 8.1-kDa glucomannogalactan (named CPS1) with potent antioxidant and hydroxyl radical scavenging activities [41].

The inhibitory activity of CS on receptor activator of NF-κB ligand (RANKL)-induced osteoclast differentiation is closely linked to its anti-inflammatory activity. A study performed by Mizuha et al. unveiled that the water extract of CS dose-dependently inhibited the receptor activator of RANKL-induced osteoclast differentiation as disclosed by tartrate-resistant acid phosphatase staining [42]. Murine bone marrow cells and monocyte/macrophage cell line RAW264.7 were applied for investigation and the dose of CS used in the experiment had no detectable toxicity on the cells. The mRNA expression of osteoclast-related genes (calcitonin receptor, cathepsin K, matrix metalloprotease 9 and nuclear factor of activated T cells c1) was suppressed. The activation of NF-κB was inhibited by preventing IκB-α phosphorylation [42].

20.2.6 Immuno-Modulatory Activity

Though there are some overlaps among the antitumor, anti-inflammatory and immune-modulatory activities, here we would like to place a special focus on the immune-modulatory activity of CS. This activity of CS first gained international attention when it was reported that several Chinese runners who broke world records in 1993 used CS as part of their training program [1]. Recent studies have made significant progress on the underling mechanisms. Firstly, CS could up-regulate IL-10, IL-1β, IL-6, IL-8, and TNF-α in cultured peripheral blood mononuclear cells, and down-regulate phytohemagglutinin-induced production of IL-2, IL-4, IL-5, IFN-γ and IL-12 [43]. Secondly, CS increased the proportion of $CD4^+$ and $CD8^+$ T cells and decreased IL-4 and IL-10 in mesenteric lymph node lymphocytes in C57Bl/6 N mice [44]. Thirdly, in LPS-activated RAW264.7 murine macrophage cells, CS blocked NF-κB activation and the production of NF-κB as well as some pro-inflammatory cytokines (IL-1β, IL-6, and TNF-α) [45].

However, there are some conflicting reports on the mechanisms of its immune-regulatory activity. For instance, by using RAW264.7 macrophages, Chen et al. found that the polysaccharide fraction of MC stimulated macrophage phagocytosis and nitric oxide production by activating the IκB/NF-κB pathway [46]. Similarly, in RA synovial fibroblasts, another bioactive compound cordycepin, that existed in both CS and *C. militaris*, significantly inhibited IL-1β-induced p38/JNK and AP-1 activation, but had no detectable inhibition on ERK pathway and NF-κB activation [47]. Moreover, CS increased phagocytosis in U937 cells by the up-regulation of IFN-γ, IL-12, p35 and p40, and TNF-α. However the last one was reported to be inhibited in Ref. [45]. Reasons for the discrepancy may comprise the lack of a well-defined mechanism to authenticate the sample, the application of different CS components (crude extract, or a purified component), and the employment of different cell models.

20.2.7 Hepatoprotecitve, Lung Protective, Heart Protective, and Renoprotective Activities

Work on the dimethylnitrosamine-induced liver fibrosis rat model showed that CS relieved liver fibrosis by increasing the degradation of collagens [48]. Xu et al. investigated the combinational effect of CS with glucocorticoid on the protection of bleomycin-induced damage in a rat model of experimental pulmonary fibrosis. It was shown that the treatment alleviated the expression of pulmonary fibrosis, CTGF, hydroxyproline, and transforming growth factor-beta 1 [49]. In addition, CS may find an application in treating hypoxic pulmonary hypertension. A report revealed that CS inhibited hypoxia-induced proliferation of rat pulmonary artery smooth muscle cells by down-regulation of PCNA, c-fos, c-jun, and a cell cycle arrest [50]. Additionally, the alcoholic extract of CS had a protective action against Coxsackie virus B3-induced murine myocarditis by up-regulation of IFN-γ and $CD3^+$, $CD8^+$ T lymphocytes [51].

CS has been reported to protect the kidneys from toxicity by inhibition of both caspase-9 and caspase-8-associated apoptotic pathways. Recently, Zhu et al. described that CS extract elicited tolerance to ischemia–reperfusion injury in NRK-52E rat renal tubular epithelial cells and spontaneously hypertensive rats. The mechanisms of this protection are associated with (a) reduction of anti-apoptotic proteins of Toll-like receptor TLR-4 and Bax and induction of Bcl-2, all tested at mRNA level [52]; (b) the reduction of intercellular adhesion molecule-1 and vascular cell adhesion molecule-1 [53]; (c) suppression of caspase-3 activation by the apoptosis inducer angiotensin II [54]; and (d) up-regulation of the expression of hypoxia inducible factor-1α and down-regulation of neutrophil gelatinase-associated lipocalin in kidney tissues [55]. There were also clinical trials on the renoprotective activity of CS. Ding et al. found that CS alleviated the side effects produced as a consequence of long-term treatment with cyclosporine A [56]. Clinical data also showed that CS would be a good alternative medicine in improving the renal function of patients with chronicallograft nephropathy [57].

20.2.8 Eryptosis Inducing Activity

Suicidal death of erythrocytes (termed as eryptosis) is an apoptosis-like process and induced by erythrocyte injury after different stressors. The damaged cell is characterized by membrane blebbing, cell shrinkage, activation of proteases, and phosphatidylserine exposure at the out membrane leaflet [58]. As mentioned above, cordycepin is a major component in CS. The eryptosis-inducing activity of cordycepin was first reported by the group of Prof. Kong who found that this phenomenon was produced via a calcium-dependent pathway in the absence of mitochondria and caspase-3 activation [59].

20.3 Conclusion and Future Perspectives

During the preparation of this chapter, we noted that there are conflicting results regarding the pharmacological activities and significant challenges on the research on CS. For example, the discrepancies in results from apoptosis studies (pro-apoptosis or anti-apoptotic activity), and the effect on different cytokines were noticed in this work and elsewhere [1]. Such problems are not intractable and may be solved by a focus on the following points. First, the application of some state-of-the-art techniques, such as AFLP-based DNA fingerprinting, for strain authentication is necessary. There are many *Cordyceps* species with similar morphological features, and also the products in different forms (such as a comprehensive one or in the form of powder) challenging the traditional identification methods. Second, the application of a pure bioactive compound for pharmacological investigation will help to make the results compatible. Third, in view of the existence of synergistic effects of some natural compounds, it is common and unavoidable to use a crude extract for both experimental and clinical use. A recommendation is to establish a consensus strain (or some strains) as standard sample for experimental studies worldwide. Fourth, high-throughput genomics, such as cDNA microarray, is an effective way to investigate the effect of CS on the genomic spectrum [60].

However, undoubtedly CS displays a multitude of pharmacological activities some of which may have therapeutic value. This accounts for the existence of a voluminous literature on the medicinal fungus and the popular use of it as a tonic.

The active principles responsible for some of the pharmacological activities have been structurally elucidated. They comprise cordycepin, other small molecules, polysaccharides, and proteins. *C. militaris* has also been extensively investigated and in general demonstrates analogous activities. Research on CS, *C. militaris* and related species will continue to bear many fruits beneficial to mankind in the years to come.

References

1. Buenz EJ, Bauer BA, Osmundson TW, Motley TJ (2005) The traditional Chinese medicine *Cordyceps sinensis* and its effects on apoptotic homeostasis. J Ethnopharmacol 96:19–29
2. Ng TB, Wang HX (2005) Pharmacological actions of Cordyceps, a prized folk medicine. J Pharm Pharmacol 57:1509–1519
3. Wu JY, Zhang QX, Leung PH (2007) Inhibitory effects of ethyl acetate extract of *Cordyceps sinensis* mycelium on various cancer cells in culture and B16 melanoma in C57BL/6 mice. Phytomedicine: Int J Phytotherapy Phytopharmacol 14:43–49
4. Liu WC, Chuang WL, Tsai ML, Hong JH, McBride WH, Chiang CS (2008) *Cordyceps sinensis* health supplement enhances recovery from taxol-induced leukopenia. Exp Biol Med 233:447–455
5. Wu Y, Hu N, Pan Y, Zhou L, Zhou X (2007) Isolation and characterization of a mannoglucan from edible *Cordyceps sinensis* mycelium. Carbohydr Res 342:870–875
6. Yang J, Zhang W, Shi P, Chen J, Han X, Wang Y (2005) Effects of exopolysaccharide fraction (EPSF) from a cultivated *Cordyceps sinensis* fungus on c-Myc, c-Fos, and VEGF expression in B16 melanoma-bearing mice. Pathol Res Pract 201:745–750
7. Zhang W, Li J, Qiu S, Chen J, Zheng Y (2008) Effects of the exopolysaccharide fraction (EPSF) from a cultivated *Cordyceps sinensis* on immunocytes of H22 tumor bearing mice. Fitoterapia 79:168–173
8. Shen YD, Shao XT, Ni YD, Xu H, Tong XM (2009) *Cordyceps sinensis* polysaccharide enhances apoptosis of HL-60 cells induced by triptolide. Zhejiang Da Xue Xue Bao. Yi Xue Ban. J Zhejiang Univ (Med Sci) 38:158–162
9. Matsuda H, Akaki J, Nakamura S, Okazaki Y, Kojima H, Tamesada M et al (2009) Apoptosis-inducing effects of sterols from the dried powder of cultured mycelium of *Cordyceps sinensis*. Chem Pharm Bull 57:411–414
10. Ji NF, Yao LS, Li Y, He W, Yi KS, Huang M (2011) Polysaccharide of *Cordyceps sinensis* enhances cisplatin cytotoxicity in non-small cell lung cancer H157 cell line. Integr Cancer Ther 10:359–367
11. Song D, He Z, Wang C, Yuan F, Dong P, Zhang W (2012) Regulation of the exopolysaccharide from an anamorph of *Cordyceps sinensis* on dendritic cell sarcoma (DCS) cell line. Eur J Nutr(in press)
12. Wong WC, Wu JY, Benzie IF (2011) Photoprotective potential of Cordyceps polysaccharides against ultraviolet B radiation-induced DNA damage to human skin cells. Br J Dermatol 164:980–986
13. Wu WC, Hsiao JR, Lian YY, Lin CY, Huang BM (2007) The apoptotic effect of cordycepin on human OEC-M1 oral cancer cell line. Cancer Chemother Pharmacol 60:103–111
14. Jeong JW, Jin CY, Park C, Hong SH, Kim GY, Jeong YK et al (2011) Induction of apoptosis by cordycepin via reactive oxygen species generation in human leukemia cells. Toxicol In Vitro (International journal published in association with BIBRA) 25:817–824
15. Jeong JW, Jin CY, Park C, Han MH, Kim GY, Moon SK et al (2012) Inhibition of migration and invasion of LNCaP human prostate carcinoma cells by cordycepin through inactivation of Akt. Int J Oncol 40:1697–1704

16. Kubo E, Yoshikawa N, Kunitomo M, Kagota S, Shinozuka K, Nakamura K (2010) Inhibitory effect of *Cordyceps sinensis* on experimental hepatic metastasis of melanoma by suppressing tumor cell invasion. Anticancer Res 30:3429–3433
17. Fang EF, Ng TB (2011) Bitter gourd (*Momordica charantia*) is a cornucopia of health: a review of its credited antidiabetic, anti-HIV, and antitumor properties. Curr Mol Med 11:417–436
18. Lo HC, Hsu TH, Tu ST, Lin KC (2006) Anti-hyperglycemic activity of natural and fermented *Cordyceps sinensis* in rats with diabetes induced by nicotinamide and streptozotocin. Am J Chin Med 34:819–832
19. Kan WC, Wang HY, Chien CC, Li SL, Chen YC, Chang LH et al (2012) Effects of extract from solid-state fermented *Cordyceps sinensis* on type 2 diabetes mellitus. Evidence-based Complement Altern Med: eCAM 2012:743107
20. El Zahraa ZEAF, Mahmoud MF, El Maraghy NN, Ahmed AF (2012) Effect of *Cordyceps sinensis* and taurine either alone or in combination on streptozotocin induced diabetes. Food Chem Toxicol (An international journal published for the British Industrial Biological Research Association) 50:1159–1165
21. Hao L, Pan MS, Zheng Y (2012) Effects of *Cordyceps sinensis* and *Tripterygium wilfordii* Polyglycosidium on the podocytes in rats with diabetic nephropathy. Zhongguo Zhong xi yi jie he za zhi Zhongguo Zhongxiyi jiehe zazhi = Chin J Integr Tradit Western Med/Zhongguo Zhong xi yi jie he xue hui, Zhongguo Zhong yi yan jiu yuan zhu ban 32:261–265
22. Takahashi S, Tamai M, Nakajima S, Kato H, Johno H, Nakamura T et al (2012) Blockade of adipocyte differentiation by cordycepin. Br J Pharmacol 167:561–575
23. Guo P, Kai Q, Gao J, Lian ZQ, Wu CM, Wu CA et al (2010) Cordycepin prevents hyperlipidemia in hamsters fed a high-fat diet via activation of AMP-activated protein kinase. J Pharmacol Sci 113:395–403
24. Fang EF, Scheibye-Knudsen M, Bohr VA, Ng TB (2012) The anti-aging efficacy of natural compounds. Med Aromat Plants 1:1–2
25. Chen S, Li Z, Krochmal R, Abrazado M, Kim W, Cooper CB (2010) Effect of Cs-4 (*Cordyceps sinensis*) on exercise performance in healthy older subjects: a double-blind, placebo-controlled trial. J Altern Complement Med 16:585–590
26. Li HP, Hu Z, Yuan JL, Fan HD, Chen W, Wang SJ et al (2007) A novel extracellular protease with fibrinolytic activity from the culture supernatant of *Cordyceps sinensis*: purification and characterization. Phytotherapy Res: PTR 21:1234–1241
27. Qi W, Wang PJ, Guo WJ, Yan YB, Zhang Y, Lei W (2011) The mechanism of *Cordyceps sinensis* and strontium in prevention of osteoporosis in rats. Biol Trace Elem Res 143:302–309
28. Qi W, Yan YB, Lei W, Wu ZX, Zhang Y, Liu D et al (2012) Prevention of disuse osteoporosis in rats by *Cordyceps sinensis* extract. Osteoporosis Int (A journal established as result of cooperation between the European Foundation for Osteoporosis and the National Osteoporosis Foundation of the USA) 23:2347–2357
29. Ji DB, Ye J, Li CL, Wang YH, Zhao J, Cai SQ (2009) Antiaging effect of *Cordyceps sinensis* extract. Phytotherapy Res: PTR 23:116–122
30. Scheibye-Knudsen M, Ramamoorthy M, Sykora P, Maynard S, Lin PC, Minor RK et al (2012) Cockayne syndrome group B protein prevents the accumulation of damaged mitochondria by promoting mitochondrial autophagy. J Exp Med 209:855–869
31. DiGiovanna JJ, Kraemer KH (2012) Shining a light on xeroderma pigmentosum. J Invest Dermatol 132:785–796
32. Liu WC, Wang SC, Tsai ML, Chen MC, Wang YC, Hong JH et al (2006) Protection against radiation-induced bone marrow and intestinal injuries by *Cordyceps sinensis*, a Chinese herbal medicine. Radiat Res 166:900–907
33. Chen YC, Huang BM (2010) Regulatory mechanisms of *Cordyceps sinensis* on steroidogenesis in MA-10 mouse Leydig tumor cells. Biosci Biotechnol Biochem 74:1855–1859
34. Wang J, Liu YM, Cao W, Yao KW, Liu ZQ, Guo JY (2012) Anti-inflammation and antioxidant effect of Cordymin, a peptide purified from the medicinal mushroom *Cordyceps sinensis*, in middle cerebral artery occlusion-induced focal cerebral ischemia in rats. Metab Brain Dis 27:159–165

35. Qian GM, Pan GF, Guo JY (2012) Anti-inflammatory and antinociceptive effects of cordymin, a peptide purified from the medicinal mushroom *Cordyceps sinensis*. Nat Prod Res 26(24):2358–2362
36. Liu Z, Li P, Zhao D, Tang H, Guo J (2010) Protective effect of extract of *Cordyceps sinensis* in middle cerebral artery occlusion-induced focal cerebral ischemia in rats. Behav Brain Functions: BBF 6:61
37. Liu Z, Li P, Zhao D, Tang H, Guo J (2011) Anti-inflammation effects of *Cordyceps sinensis* mycelium in focal cerebral ischemic injury rats. Inflammation 34:639–644
38. Chiou YL, Lin CY (2012) The extract of *Cordyceps sinensis* inhibited airway inflammation by blocking NF-kappaB activity. Inflammation 35:985–993
39. Yang ML, Kuo PC, Hwang TL, Wu TS (2011) Anti-inflammatory principles from *Cordyceps sinensis*. J Nat Prod 74:1996–2000
40. Yu HM, Wang BS, Huang SC, Duh PD (2006) Comparison of protective effects between cultured *Cordyceps militaris* and natural *Cordyceps sinensis* against oxidative damage. J Agric Food Chem 54:3132–3138
41. Wang Y, Wang M, Ling Y, Fan W, Wang Y, Yin H (2009) Structural determination and antioxidant activity of a polysaccharide from the fruiting bodies of cultured *Cordyceps sinensis*. Am J Chin Med 37:977–989
42. Mizuha Y, Yamamoto H, Sato T, Tsuji M, Masuda M, Uchida M et al (2007) Water extract of *Cordyceps sinensis* (WECS) inhibits the RANKL-induced osteoclast differentiation. BioFactors 30:105–116
43. Zhou X, Luo L, Dressel W, Shadier G, Krumbiegel D, Schmidtke P et al (2008) Cordycepin is an immunoregulatory active ingredient of *Cordyceps sinensis*. Am J Chin Med 36:967–980
44. Park DK, Choi WS, Park PJ, Kim EK, Jeong YJ, Choi SY et al (2008) Immunoglobulin and cytokine production from mesenteric lymph node lymphocytes is regulated by extracts of *Cordyceps sinensis* in C57Bl/6N mice. J Med Food 11:784–788
45. Shin S, Lee S, Kwon J, Moon S, Lee S, Lee CK et al (2009) Cordycepin suppresses expression of diabetes regulating genes by inhibition of lipopolysaccharide-induced inflammation in macrophages. Immune Netw 9:98–105
46. Chen W, Zhang W, Shen W, Wang K (2010) Effects of the acid polysaccharide fraction isolated from a cultivated *Cordyceps sinensis* on macrophages in vitro. Cell Immunol 262:69–74
47. Noh EM, Kim JS, Hur H, Park BH, Song EK, Han MK et al (2009) Cordycepin inhibits IL-1beta-induced MMP-1 and MMP-3 expression in rheumatoid arthritis synovial fibroblasts. Rheumatology 48:45–48
48. Li FH, Liu P, Xiong WG, Xu GF (2006) Effects of *Cordyceps sinensis* on dimethylnitrosamine-induced liver fibrosis in rats. Zhong Xi Yi Jie He Xue Bao/J Chin Integr Med 4:514–517
49. Xu H, Li S, Lin Y, Liu R, Gu Y, Liao D (2011) Effectiveness of cultured *Cordyceps sinensis* combined with glucocorticosteroid on pulmonary fibrosis induced by bleomycin in rats. Zhongguo Zhong Yao Za Zhi = Zhongguo Zhongyao Zazhi = China J Chin Mater Med 36:2265–2270
50. Gao BA, Yang J, Huang J, Cui XJ, Chen SX, Den HY et al (2010) *Cordyceps sinensis* extract suppresses hypoxia-induced proliferation of rat pulmonary artery smooth muscle cells. Saudi Med J 31:974–979
51. Li F, Gao XY, Rao BF, Liu L, Dong B, Cui LQ (2006) Effects of *Cordyceps sinensis* alcohol extractive on serum interferon-gamma level and splenic T lymphocyte subset in mice with viral myocarditis. Xi Bao Yu Fen Zi Mian Yi Xue Za Zhi = Chin J Cell Mol Immun 22:321–323
52. Zhou Q, Hu S (2010) Effect of *Cordyceps sinensis* extractant on apoptosis and expression of Toll-like receptor 4 mRNA in the ischemia-reperfusion injured NRK-52E cells. Zhong Nan Da Xue Xue Bao. Yi Xue Ban = J Central South Univ (Med Sci) 35:77–84
53. Wu R, Zhou Q, Lin S, Ao X, Chen X, Yang J (2010) Effect of Cordceps Sinensis on the expression of ICAM-1 and VCAM-1 in the kidney of spontaneously hypertensive rats. Zhong Nan Da Xue Xue Bao. Yi Xue Ban = J Central South Univ (Med Sci) 35:152–158

54. Tu S, Zhou Q, Tang R, Tang T, Hu S, Ao X (2012) Proapoptotic effect of angiotensin II on renal tubular epithelial cells and protective effect of *Cordyceps sinensis*. Zhong Nan Da Xue Xue Bao. Yi Xue Ban = J Central South Univ (Med Sci) 37:67–72
55. Yu H, Zhou Q, Huang R, Yuan M, Ao X, Yang J (2012) Effect of *Cordyceps sinensis* on the expression of HIF-1alpha and NGAL in rats with renal ischemia-reperfusion injury. Zhong Nan Da Xue Xue Bao. Yi Xue Ban = J Central South Univ (Med Sci) 37:57–66
56. Ding C, Tian P, Jia L, Li Y, Ding X, Xiang H et al (2009) The synergistic effects of *C. Sinensis* with CsA in preventing allograft rejection. Front Biosci (a journal and virtual library) 14:3864–3871
57. Zhang Z, Wang X, Zhang Y, Ye G (2011) Effect of *Cordyceps sinensis* on renal function of patients with chronic allograft nephropathy. Urol Int 86:298–301
58. Lang F, Qadri SM (2012) Mechanisms and significance of eryptosis, the suicidal death of erythrocytes. Blood Purif 33:125–130
59. Lui JC, Wong JW, Suen YK, Kwok TT, Fung KP, Kong SK (2007) Cordycepin induced eryptosis in mouse erythrocytes through a Ca2+-dependent pathway without caspase-3 activation. Arch Toxicol 81:859–865
60. Li CY, Chiang CS, Cheng WC, Wang SC, Cheng HT, Chen CR et al (2012) Gene expression profiling of dendritic cells in different physiological stages under *Cordyceps sinensis* treatment. PLoS One 7:e40824

Chapter 21
The Bitter Fruit with Sweet Health Benefits: A Comprehensive Synopsis of Recent Research Progress on Medicinal Properties of *Momordica Charantia*

Evandro Fei Fang and Tzi Bun Ng

Abstract Though eliciting a somewhat 'bitter' taste to our palate, *Momordica charantia* (MC) is 'sweet' for our health. As a popular vegetable indigenous to tropical countries, MC is frequently encountered in the fields of both culinary usage and folklore medicine, especially in Asia. The advance of scientific techniques helps build a platform for investigating the molecular basis of the pharmaceutical functions of medicinal herbs including MC. Over the past few years, a number of pure bioactive components with potential medicinal applications have been procured as a consequence of *in vitro* and *in vivo* investigations. In this review, some representative and promising medicinal constituents of MC have been categorized and their emerging medicinal effects such as antitumor, anti-HIV, and anti-diabetic activities are accompanied by the pertinent molecular mechanisms. In addition, the wide range of antimicrobial and pesticidal activities will further validate the medicinal value of MC. Further large-scale clinical trials for those components are needed to fully ascertain their medicinal effects and potential drawbacks, and a new horizon in the treatment of tumor, AIDS, and diabetes may then be opened.

21.1 Introduction

Advantages of herbal formulas such as good therapeutic effects on some devastating diseases in a holistic approach and nominal side reactions have put them in the spotlight in medicinal research [1–4]. Though an insufficiency of high-quality

E. F. Fang · T. B. Ng (✉)
School of Life Sciences, Faculty of Science, The Chinese University of Hong Kong, Shatin, Hong Kong
e-mail: b021770@mailserv.cuhk.edu.hk

E. F. Fang
Laboratory of Molecular Gerontology, National Institute on Aging, National Institutes of Health, Baltimore, MD, USA

E. F. Fang and T. B. Ng (eds.), *Antitumor Potential and Other Emerging Medicinal Properties of Natural Compounds*, DOI: 10.1007/978-94-007-6214-5_21,

scientific evidence on the medicinal efficacy of Traditional Chinese medicine (TCM) has retarded its market application in the past, the status has been ameliorated by the cutting-edge scientific progress in genetics and cell and molecular biology. *Momordica charantia* (hereafter referred to as MC) is a good example of medicinal plant. The common name of MC is bitter melon, and also alternatively called bitter gourd, balsam pear, and karela (karolla). MC is a vegetable indigenous to tropical areas, including Asia, Africa, and South America. In Asia, MC is used in the culinary field and folklore medicine [1, 5]. Its dried fruit is sold as a health supplement.

In recent years, a number of investigations have targeted on the multifarious medicinal activities of MC: it has been reported with antitumor, anti-HIV, anti-diabetic, anti-obesity, and pesticidal activities, among others. These activities are contributed by different components ranging from crude extract, fruit/leaf juice, to pure proteins/compounds [1, 5]. In this review, besides crude extract, some representative and promising medicinal components in MC are categorized, and their emerging medicinal effects encompassing antitumor, anti-HIV, and anti-diabetic activities are summarized, accompanied by a discussion of potential molecular mechanisms. The components comprise lectins, ribosome inactivating proteins, ribonucleases, protease inhibitors, saponins, and others. A synopsis of their general properties is shown in Table 21.1.

21.2 Medicinal Effects of Crude Extract

21.2.1 Antitumor Activity of Crude Extract

Crude extracts from MC (mostly seeds, fruit, or leaves) are always prepared by extraction with water, ethanol, or methanol. Medicinal effects of MC, especially those on the treatments of tumor, diabetes, and pests and viruses are well documented.

Although there are insufficient data from clinical trials, *in vitro* investigations unveiled that MC crude extract exhibited anti-proliferative activity toward several types of cancer cells, including human breast cancer cells [4], prostate cancer cells [2], colon cancer cells [6], and skin cancer cells [7]. Firstly, Ray et al. found that the anti-proliferative activity of MC extract (young fruit) on breast cancer cells was attributed to arrest of cells in G2/M phase of the cell cycle and induction of apoptosis which appeared as up-regulation of p53, p21, and pChk1/2, and down-regulation of cyclin B1, and cyclin D1 [8]. This activity may be at least partially associated with the proteins MAP30 and RNase MC2 in MC [5, 9]. Secondly, non-toxic concentrations of leaf extracts significantly inhibited the migration and invasion of a rat prostate cancer cell line (PLS10) *in vitro* [10]. This was concomitant with an early study by Claflin and coworker in 1978, which found that crude extract of MC (ripe fruit) also manifested anti-prostate cancer activity which

Table 21.1 A summary of MC compounds with potential medicinal applications

Name[a]	Source	MW (kDa) Structure	Classification	Medicinal activities	Ref.
MCL	Seeds	130 Tetrameric Pre-crystal structure	Lectin Type II RIP[b]	Antitumor, insulinomimetic, antilipolytic, lipogenic, immunoadjuvant	[1, 33]
A cytostatic factor	n.d.[c]	~40 n.d.[c] No crystal structure	n.d.[c]	Antitumor Anti-viral	[88]
MAP30	Seeds	30 Monomeric Crystal structure	Type I RIP	Antitumor, anti-viral, antimicrobial	[3, 43]
α-MMC	Seeds	30 Monomeric Crystal structure	Type I RIP	Antitumor, anti-HIV, abortifacient	[44]
β-MMC	Seeds	29 Monomeric Crystal structure	Type I RIP	Antitumor, anti-HIV, abortifacient, insulinomimetic	[50]
γ-MMC	Seeds	11.5 Monomeric No crystal structure	Type I RIP	N-glycosidase	[53]
MOM-I[d] MOM-II[d]	Seeds	30 Monomeric Crystal structure	Type I RIP	Antitumor, N-glycosidase	[54, 55]
RNase MC1	Seeds	21.2 Monomeric Crystal structure	RNase T2	RNase	[60–64]
Polypeptide-P	Fruit, seeds	18.5 Monomeric No crystal structure	n.d.[c]	Anti-diabetic	[86]

(continued)

Table 21.1 continued

Name[a]	Source	MW (kDa) Structure	Classification	Medicinal activities	Ref.
Napin-like protein	Seeds	~12 Dimeric No crystal structure	2S albumin	Anti-fungal	[85]
Napin-like RIP	Seeds	9.7 Monomeric No crystal structure	Type I RIP	N-glycosidase	[91]
Serpins	Seeds	~9 ~3.5 Monomeric Crystal structure	Potato-type Squash-type	Protease-inhibitory, antitumor, anti-pest	[1, 69]
Compounds	Fruits, seeds, vines, roots	<0.5 – Chemical structure	Saponin Flavonoids Etc.	Anti-diabetic, anti-HIV, antitumor	[76–81]

[a]*Abbreviations MCL M. charantia* lectin; *MAP30* a 30-kDa *M. charantia* anti-HIV protein; *α-MMC* α-momorcharin; *β-MMC* β-momorcharin; γ-MMC, γ-momorcharin; *MOM-I* momordin-I; *MOM-II* momordin-II; RNase *MC1* the first ribonuclease from *M. charantia*; *Serpins* serine protease inhibitors. [b]*RIP* ribosome inactivating protein. [c]*n.d.* not determined. [d]Despite the fact that MOM-I and MOM-II are separate entities, some investigators do not discriminate between them in view of high sequence homology

was contributed some part by guanylate cyclase inhibitor [2], α-momorcharin, and β-momorcharin [11]. Most recently, Ru et al. unveiled the potential molecular mechanism of this activity which was associated with the impairment of prostate cancer cell cycle progression and retardation of prostatic xenograft growth in TRAMP (transgenic adenocarcinoma of mouse prostate) mice [12]. Thirdly, Yasui and coworkers investigated the strong anti-proliferative and apoptosis-inducing effects of MC seed oil on colon cancer Caco-2 cells. This might be contributed by its components including 9cis, 11trans, and 13trans-conjugated linolenic acids in MC oil through up-regulation of GADD45, p53 and PPARγ [6]. MC leaf extracts also manifested antitumor activity. The leaf extracts demonstrated a chemopreventive potential on dimethyl benz(a)nthracene-induced skin tumorigenesis, melanoma and cytogenicity [7].

21.2.2 Anti-Diabetic Activity of Crude Extract

Diabetes mellitus, characterized by hyperglycaemia, is a metabolic disorder caused by an abnormality in insulin secretion and/or insulin action. It has become a major global public health concern. It is estimated that about one-third of diabetic patients resort to complementary and alternative medicine, and there is a long history of using MC for the treatment of diabetes in China, India, South America, and East Africa [13, 14]. MC is a good source to acquire natural anti-diabetic drugs as evidenced by its various activities including: (a) enhancement of insulin sensitivity, (b) regulation of adipokine profiles, (c), potent hypoglycemic, hypolipidemic actions and (d) enhancement of immune system. For the details of the anti-diabetic mechanism, please see our recent review [1]. Though a cohort *in vitro* bench top studies and *in vivo* investigations on rodents, high-quality (large-scaled, randomized double-blind, and placebo-controlled) clinical trials are only sporadic, thus limiting the application of MC to diabetic patients. Such clinical investigations should deserve a high priority in future studies on MC.

21.2.3 Anti-Microbial and Anti-Pest Properties of Crude Extract

The crude extract of MC also has anti-microbial activity toward *Staphylococcus aureus*, and measles, and anti-pest properties toward leishmania and some lepidopteran pests, among others. Phytochemical screening study showed that there are a list of anti-microbial components in BG, such as tannins, phenols, flavonoids, and alkaloids [15].

In both *in vitro* and *in vivo* experiments, the aqueous extract of MC fruits expressed potent anti-leishmania activity which was mediated through inhibition of parasite superoxide dismutase (SOD), one of the key enzymes of the oxidative burst. This activity is (partially) contributed by a compound, 4-(o-carboethoxyphenyl) butanol, named momordicatin [16]. Furthermore, both the fruit and seeds of MC manifest anti-microbial activity on a variety of microorganisms, including *Bacillus cereus*, *Bacillus subtilis*, *Pseudomonas aeruginosa*, *Escherichia coli*, *Candida albicans*, *Staphylococcus aureus* (methicilin-resistant), *Enterococcus faecalis*, *Aspergillus niger*, *Klebsiella pneumoniae*, and *Salmonella typhimurium* [17]. The highest activity was toward *E. coli* [15]. The seed oil of MC contains components which were effective on *Staphylococcus aureus* [17], and the malarial fever mosquito *Anopheles stephensi* [18]. The main constituents in the oil are trans-nerolidol, apiole, cis-dihydrocarveol and germacrene D, which are active components that need further purification [17]. The larvicidal efficacy of MC is also noteworthy. MC seeds retarded growth and development of two lepidopteran pests, *Helicoverpa armigera* and *Spodoptera litura*, with some protease inhibitors as active components [19]. The methanol extract of the leaves of MC demonstrated potent anti-pest activity against the larvae of mosquitos (*Aedes aegypti L.* and *Culex quinquefasciatus*) which can serve as an eco-friendly approach for the control of dengue fever vector (*Aedes aegypti*), and the lymphatic filariasis vector (*Culex quinquefasciatus*) [20–22]. The ethanolic leaf extract of MC display significant antifeedant and antioviposition activities against *Liriomyza sativae* adults [23].

21.2.4 Antioxidant Activity of MC

MC may serve as a good source of antioxidants for use in the prevention and treatment of illnesses characterized with oxidative stress. Antioxidants are a number of components which in relation to oxidizable substrates, significantly inhibit or delay oxidative processes, while often being oxidized themselves [24]. MC evinced a higher antioxidant activity compared with some other medicinal and food plants [25, 26]. MC fruits displayed radical-scavenging and cytoprotective effects against oxidative stress by multiple causes, including hydrogen peroxide (H_2O_2), hypoxanthine-xanthine oxidase (HX-XO), immobilization stress, and toxic chemicals [27–29].

The *in vitro* study on fibroblasts displayed that fruit extract of MC offered good and consistent cytoprotection against the oxidants of H_2O_2 and HX-XO [27]. MC extract provided a protection against lipid peroxidation induced by immobilization stress in albino rats, and the potential mechanism was associated with an augmentation of antioxidants like reduced glutathione and catalase [29]. Recently, a study uncovered the antioxidant potential of MC fruit extract in ammonium chloride-induced hyperammonemic rats exposed to a high oxidative stress environment [28]. The antioxidant activity of MC may be at least contributed by its phenolic compounds (such as catechin, gallic acid, gentisic acid, chlorogenic acid, and epicatechin), and some peroxidases.

During recent years, there has been increasing interest in the medicinal applications of antioxidants [24]. Depending on (a) the etiological bases of some chronic degenerative diseases characterized with oxidative stress, such as Alzheimer's disease, Parkinson's disease, atherosclerosis, precocious aging, and the above described diabetes, and (b) the enrichment of antioxidants in MC, the effects of using MC as an alternative for the treatment of such severe diseases are warranted for future investigations.

21.2.5 Other Bioactive Activities of MC Crude Extract

In addition to the abovementioned medicinal activities, there may be a wider application of the crude extract of MC. Ono and coworkers found that bitter melon pulp extract induced hepatocyte growth factor production by activation of MAPKs and proliferation of human dermal fibroblasts [30]. Both seem to be contributed by a 14-kDa protein [30]. MC has also been used for the treatment of measles [31]. The methanolic extract of MC fruit exhibited anti-ulcer activity. A study by Alam, et al., indicated that the methanolic extract of MC fruit increased healing of gastric ulcer and prevented its further development in rats [32].

21.3 Lectins/Hemagglutinins in MC

Over a century ago, researchers started to demonstrate the presence of proteins with erythrocyte-agglutinating and specific carbohydrate-binding activities which were given the nomenclature of lectins/hemagglutinins. In addition to being useful tools for the investigation of carbohydrates on cell surfaces, and for the isolation and characterization of glycoproteins, lectins, especially those of plant origins, have been reported with medicinal applications relying on their antitumor, anti-HIV, nitric oxide-inducing activities [33]. For the story of marine lectins, please refer to Chap. 4 by Prof. Koji Muramoto and coworkers for more information.

To date, three lectins have been isolated from the seeds of MC including momordin agglutinin (originally named 'momordin', we renamed it momordin agglutinin here to distinguish it from the ribosome inactivating proteins (RIPs) referred to as momordins in part 21.4.3 below), momordica agglutinin, and *Momordica charantia* lectin (MCL) [34, 35]. Momordin agglutinin is a 23.7-kDa toxic polypeptide purified from the seeds of MC. Its lethality (LD_{50}) was determined to be 5 mg/kg body weight by intraperitoneal injection into Swiss white mice. It also manifested antitumor activity on Ehrlich ascites tumor cells by the inhibition of protein biosynthesis. Momordin agglutinin exhibited slight/no effect on RNA and DNA biosynthesis [34]. Compared with momordin agglutinin, momordica agglutinin is a non-toxic polypeptide of 31.8 kDa. It agglutinated human O type erythrocytes and Ehrlich ascites tumor cells which may be associated with its galactose (or its derivatives)-binding characteristic [34].

In addition, there exists a galactose-binding, 130-kDa tetrameric lectin, MCL, which seems exclusively expressed in the seeds, but not in the roots, shoots, leaves, and cotyledons [35, 36]. It manifests the highest hemagglutinating activity toward human type-O(Rh^+) erythrocytes followed by A, B, and O^h_m types, and blood from rabbits, rats and pigs [36, 37]. Intriguingly, besides hemagglutinating activity, MCL also exhibits N-glycosidase activity which subsumes it into type II RIPs. Progress on its sequence and structural studies provides an answer to its dual activities [35, 36]. The four chains of MCL are linked through a disulfide bridge between two identical copies of a two-chain unit. Based on the total sequence (GenBank: BAH05018), a predicted 3-D structure of MCL is shown in Fig. 21.1–(1). MCL consists of two identical A chains (each possessing 263 amino acids and characterized by hemagglutinating activity) and two identical B chains (each possessing 261 amino acids and characterized by N-glycosidase activity) [35, 36]. Tryptophan and tyrosine residues are crucial to the hemagglutinating activity.

In comparison with the large amount of work on purification and structure of MCL, only sporadic investigations on its medicinal activity were carried out. For example, Huang et al. indicated that MCL was a T cell-independent B cell activator and a polyclonal immunoglobulin (Ig) inducer [38]. MCL could selectively bind to B cells and stimulated the B cell subset of splenocytes via membrane Ig in the B cell surface, as well as in a dose- and time-dependent manner. Moreover, after 96-h co-culture, MCL triggered splenocytes to produce a large amount of non-specific IgM. These data unveiled the immunomodulatory effect of MCL [38]. MCL was demonstrated to inhibit protein synthesis in Yoshida AH-130 ascites-tumor cells [37]. Recently, we found that MCL manifested antitumor activity toward human nasopharyngeal carcinoma cells *in vitro* and *in vivo* [39]. The underlying mechanisms were associated with the attack of G1-phase cell cycle, regulation of MAPKs phosphorylation, promoted nitric oxide production, and the induction of both caspase-8 and caspase-9 regulated caspase cascades [39].

21.4 Medicinal Activities of Ribosome Inactivating Proteins (RIPs)

RIPs are a group of proteins which inactivate ribosomes by cleaving the adenine base from ribose at the glycosidic bond at position A-4324 in the highly conserved α-sarcin/ricin loop of 28S rRNA [40]. RIPs are generally classified into 3 types depending on the number of subunits [41]. Type I RIPs are single polypeptides with an enzymatically active (N-glycosidase) chain while type II RIPs consist of an N-glycosidase chain and another chain with hemagglutinating activity. Recently, a maize RIP was reported and subsumed as an atypical type I or type III RIP for its special two-stage structural changes [41]. RIPs account for a large part of the active medicinal ingredients in MC. To date, approximately ten RIPs in MC have been purified and their biochemical and functional activities were elucidated. They comprise a number of type I RIPs (MAP30, α-momorcharin, β-momorcharin,

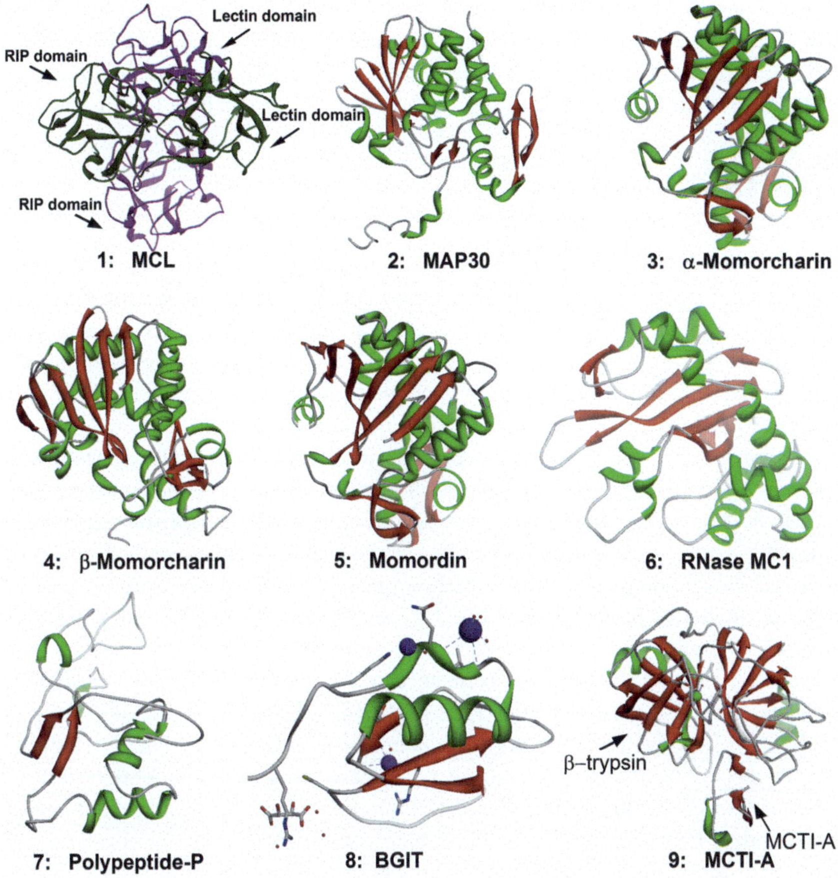

Fig. 21.1 Stereoscopic views of representative proteins from bitter gourd. Data of crystal structures are as follows: (1) MCL (GenBank: BAH05018), (2) MPA30 (PDB ID: 1D8V_A), (3) α-Momorcharin (PDB ID: 1AHA) [92], (4) β-Momorcharin (PDB ID: 1CF5) [93], (5) Momordin (PDB ID: 1MOM) [54], (6) RNase MC1 (PDB ID: 1BK7) [64], (7) Polypeptide-P (GenBank: ADO14327.1), (8) BGIT (PDB ID: 1VBW) [69], MCTI-A (PDB ID: 1F2S) [71]. Only the ribbon structures of MCL and polypeptide-P are predicted using the on-line Phyre server with Chimera 1.5 for visualization as used before [87, 94, 95]. For MCL, it is composed of two identical copies, colored magenta and forest green respectively, and each copy contains a RIP domain and a lectin domain. The helix and strand are shown in *green* and *red* colors, respectively

γ-momorcharin, momordin-I, and momordin-II), some of which may be the same or similar, and a type II RIP MCL, as described in the previous section.

21.4.1 MAP30

MAP30, the abbreviated name of 'a 30-kDa anti-HIV protein from *Momordica charantia* seeds', was first reported by Prof. Sylvia Lee-Huang et al. in 1990 [42]. This group of investigators and others carried out in-depth studies on MAP30 ranging from total amino acid sequence, crystal structure (Fig. 21.1–(2)), to anti-HIV, anti-herpes simplex virus, anti-hepatitis B virus (HBV), and antitumor activities [1, 3, 43]. Please go to Chap. 8 for the interesting story on MAP30 written by it first discoverer.

21.4.2 Momorcharins

Alpha-momorcharin and β-momorcharin are two type I RIPs from MC [1]. They both have a molecular weight of about 30kDa, but can be separated by ion exchange chromatography on a Mono S column based on the lesser binding ability of α-momorcharin [44]. The N-glycosidase activity of both proteins is optimal at pH 7, and β-momorcharin is more unstable to Mn^{2+} ions than α-momorcharin [45]. Both proteins exhibit antitumor potential.

As shown in Fig. 21.1–(3), alpha momorcharin (α-MMC) is a glycoprotein with a sugar chain attached to Asn-227 and its crystal structure has been elucidated [46]. Besides N-glycosidase and abortifacient activities, α-MMC also displayed tumoricidal effects toward a wide range of cancer cells which augmented its potential medicinal value. Some early *in vitro* studies reported that it inhibited proliferation of choriocarcinoma cells, melanoma cells [47], P388 (mouse monocyte-macrophage) cells, J774 (Balb/c macrophage) cells, JAR (human placental choriocarcinoma) cells and sarcoma S180 cells. It also enhanced the tumoricidal effect of mouse macrophages on mouse mastocytomal (P815) cells [48]. Recently, some derivatives of α-MMC-PGE (polyethylene glycol) with untarnished antitumor activity but reduced side effects have been synthesized using a bio-engineering technique [49].

As with α-MMC, the structure of beta momorcharin (β-MMC) has been extensively studied (Fig. 21.1–(4)). The molecular mass of β-MMC is 29 kDa with 249 amino acids. It is also a glycoprotein, with a branched hexasaccharide chain linked to Asn51 [50]. The crystal structure reveals that its transient folding/unfolding intermediate state adopts a specific conformation near the active site [50, 51]. The active site responsible for its N-glycosidase activity is dependent on four residues, including Tyr70, Tyr109, Glu158, and Arg161, whereas an adjacent Trp190 residue covers up the lid for the active site [50, 51]. However, there are few studies on the antitumor activity of β-MMC. Our data indicate it inhibits growth of human squamous carcinoma of the tongue and larynx [52]. A recent report uncovered the antitumor activity of MCP30 (a combination of α-MMC and β-MMC), on human prostate cancer [11]. The investigators observed that MCP30 selectively induced apoptosis of prostatic intraepithelial neoplasia and prostate cancer, inhibited the activity of histone deacetylase-1 (HDAC-1), and promoted histone-3 and histone-4 acetylation, but was devoid of an effect on normal prostate cells [11]. The clinical potential of MCP30 seems attractive and deserves further investigation.

Pu et al. have purified an 11.5 kDa RIP from the seeds of MC named γ-momorcharin [53]. The N-glycosidase activity of γ-momorcharin is targeted at adenosine position 4324, which is located in a highly conserved loop of 28S rat liver rRNA. Information pertaining to its medicinal activity is not studied.

21.4.3 Momordin-I and Momordin-II

There are another two type I RIPs with a molecular mass around 30 kDa and designated as momordin-I/a (MOM-I) and momordin-II/b (MOM-II), respectively. Despite the fact that MOM-I and MOM-II are distinct proteins, some investigators do not discriminate between them due to their pronounced sequence similarity. A crystal structure of MOM-I is presented in Fig. 21.1–(5) [54]. Besides N-glycosidase activity based on residues His140, Tyr165, and Lys231, MOM-I also displayed antitumor activity. For example, a monoclonal anti-Thy 1.1 antibody (OX7) linked momordin demonstrated selective antitumor activity toward the Thy 1.1-expressing mouse lymphoma cell line AKR-A [55]. The anti-CD5-momordin conjugate might be of value in the treatment of CD5-positive leukemia and lymphoma [56].

21.5 Ribonucleases

Ribonucleases (RNases) are a type of nucleases that catalyze the degradation of RNA into smaller components, and some manifest a wide spectrum of medicinal applications on cancer and AIDS [57]. Antitumor RNases are a group of small (10–28 kDa) proteins existing as members of both RNase A and T1 super-families. They bind to the negatively charged cell membrane, enter cells by endocytosis, and are translocated to the cytosol where they evade mammalian protein ribonuclease inhibitor and degrade RNA [58]. Examples of RNases that hold good prospects of being developed into therapeutic tools on tumors are α-sarcin, binase, Ginseng RNase, and onconase which we have previously reviewed [57].

Knowledge about RNases in MC is confined to RNase MC1 and RNase MC2. An effective procedure for the concurrent purification of α-, β-momorcharins, and RNase MC1 has been reported [44]. RNase MC1 is a 21-kDa protein belonging to the RNase T2 family [59–64]. It is poly(U)-specific: it specifically recognizes the uracil base and cleaves the phosphodiester bond of NpU ('N' stands for any RNA nucleotide) [44, 63]. The structure of RNase MC1 has been thoroughly studied and shown in Fig. 21.1–(6). There are 190 amino acids. Tyr101, Phe102, Ala105, Pro125, Gly127, Gly144, Leu162, Val165, and Phe190 regulate its stability [63]. Asn71 and Leu73 determine its substrate specificity by changing the size and shape of its uracil binding site [60, 61]. Four catalytic residues including His46, Glu105, and His109 are responsible for its RNase activity [64]. To date, there are no medicinal investigations on RNase MC1. Because RNases are a new class of antitumor components which manifest an entirely different mechanisms of action than the drugs currently used in the clinic [58, 65], future *in vitro* and *in vivo* studies on RNase MC1 to explore its clinical potential are worth undertaking.

Recently, we purified the second RNase in MC seeds, named RNase MC2 [5, 66]. It is a 14-kDa protein with potent RNA-cleavage activity toward baker's yeast tRNA, and tumor cell rRNA. Further biomedical studies found that it demonstrated toxicity against both human breast and liver cancer cells. In MCF-7 cells, RNase MC2 induced various MAPKs (p38, JNK and ERK) and Akt, activated caspase-8, caspase-9, caspase-7, increased the production of Bak and cleaved PARP, which in turn contributed to the apoptotic response [5]. Furthermore, the antitumor activity of RNase MC2 in human liver cancer Hep G2 cells was also effective as confirmed by both *in vitro* and *in vivo* experimental models [66]. The function was also ascribed to the induction of both caspase-9 and caspase-8 regulated pathways of apoptosis.

21.6 Protease Inhibitors

The proteasome is a multicatalytic protease complex that mediates the controlled degradation of intracellular proteins, such as the key components of pathways involved in cancer cell growth and immune cell signaling [67]. Furthermore, the ubiquitin–proteasome system has become a promising novel molecular target in cancer due to its critical role in cellular protein degradation, its interaction with the cell cycle and regulation of apoptosis and its unique mechanism of action [68]. To date, a number of structurally and mechanistically distinct protease inhibitors have reached the stage of clinical development in oncology [67]. Protease inhibitors can be classified into five classes: serine-, cysteine-, metallo-, aspartic, and threonine protease inhibitors. Among them, serine protease inhibitors (serpins) are further subdivided into: Kunitz type, Bowman-Birk, potato type I, Kazal, and squash type inhibitors. There are other classifications available, described by Professor Krzysztof Rolka in Chap. 12 of this book.

Currently, two types of serpins in MC are known. They include BGIT (bitter gourd inhibitor against acidic amino acid-specific proteinase of *Streptomyces griseus*), BGIA (bitter gourd trypsin inhibitor), MCI-3, and MCI-1 (*Momordica charantia* trypsin inhibitor-3 and -1), which belong to potato type 1 types, as well as MCEIs (*Momordica charantia* elastase inhibitor) (four isoforms) and MCTIs (*Momordica charantia* trypsin inhibitors) (five isoforms), pertaining to squash types [1]. On the one hand, BGTI, BGIA, MCI-3, and MCI-1 are peptides with a molecular mass in the vicinity of 8-kDa that manifest a striking resemblance in sequence [1]. The crystal structure of BGIT is shown in Fig. 21.1–(8) [69]. On the other hand, MCEIs and MCTIs are smaller peptides with a molecular mass of about 3.5–4 kDa and exhibit a high degree of sequence homology to each other [70]. Figure 21.1–(9) shows the crystal structure of the complex formed between bovine β-trypsin and MCTI-A, a squash trypsin inhibitor of MCTIs [71].

Several 4-kDa bitter gourd protease inhibitors (BGPIs), which elicit pesticide-like activity against *Helicoverpa armigera* and *Spodoptera litura*, have been purified from MC [19]. At least four BGPIs with inhibitory activity against both trypsin and *Helicoverpa armigera* gut proteinases were detected. Two (BGPI-1 and-2) are slow-migrating proteins with major activity and the others (BGPI-3 and -4) had minor activity. Because no sequence data of BGPIs are available, there is some identity overlap between BGPIs and MCEIs based on the identity in molecular mass and trypsin inhibitory activity. Further sequence work is warranted.

Considering the compelling evidence on the efficacy of protease inhibitors with possible medicinal applications to therapy of AIDS and tumor [72–74], and the fact that there are no antitumor and anti-HIV investigations on protease inhibitors in MC, future research in this field is recommended.

21.7 Saponins

Saponins, found in many plants, are glycosides that consist of a steroid, steroid alkaloid or triterpenoid aglycone and one or more sugar chains covalently linked by glycosidic binding to the aglycone moiety. Advantages of saponins include small molecular size, potent antitumor, anti-HIV, and anti-diabetic activities which place saponins in the limelight of research on MC [75].

MC is a rich source of triterpenes including two classes of saponins known as cucurbitane and oleanane-type triterpenoids [76]. Hitherto over fifty triterpenes have been isolated from MC fruits [77–79], seeds, stems/vines [80, 81], leaves [81], and roots [82]. Eighteen triterpenes have been purified from MC fruits. They all manifested inhibitory effects on Epstein-Barr virus early antigen (EBV-EA) induction by 12-O-tetradecanoylphorbol-13-acetate (TPA) in Raji cells. In addition, compounds 1 and 2 demonstrated significant inhibitory effects in both 7,12-dimethylbenz[a] anthracene (DMBA)- and peroxynitrite (ONOO-; PN)-induced mouse skin carcinogenesis tests [77]. Five octanorcucurbitane triterpenes, including octanorcucurbitacins A-D and kuguacin M, have been isolated from the stems of MC, and compound 3 inhibited tert-butyl hydroperoxide (t-BHP)-induced hepatotoxicity against Hep G2 cells [80]. Fourteen cucurbitane triterpenoids, named kuguacins F–S, which exhibited weak anti-HIV-1 activities *in vitro*, were isolated from vines and leaves of MC [81]. Eight cucurbitacins were isolated from the roots of MC. Among them, compounds 3 and 5 showed moderate anti-HIV-1 activity with EC_{50} values of 8.45 and 25.62 microg/ml, respectively, and exerted minimal cytotoxicity against C8166 cells which are human leukemia lymphocytes, and are particularly susceptible to human HIV infection ($IC_{50} > 200$ microg/ml) [82].

In addition to antitumor and anti-HIV activities, triterpenes in MC leaves also play a role as oviposition deterrents to a polyphagic leafminer (*Liriomyza trifolii*). The methanol extract of MC leaves strongly deterred the females from laying eggs on a substrate which had been treated at a concentration of 1 g leaf equivalent extract/ml. The activity was contributed by momordicine IV and momordicine II. There was no synergistic inhibitory effect on oviposition when the two compounds were combined in a ratio mimicking their natural abundance [83].

Furthermore, quantitative and mechanistic investigations of the anti-diabetic activity of MC at the molecular level revealed the important role of triterpenes. Using 3T3-L1 cells as an *in vitro* model, Popovich et al. found that MC triterpenoids extract reduced preadipocyte viability due to a G2/M arrest in the cell cycle, and prevented lipid accumulation [76]. The anti-diabetic activity of MC triterpenoids was associated with activation of the AMP-activated protein kinase (AMPK) pathway [75]. In another study, four MC triterpenoids, momordicosides Q, R, S, and T, and stereochemistry-established karaviloside XI, exhibited marked anti-diabetic and anti-obesity activities in both *in vitro* and *in vivo* experiments. Potential mechanisms include: (a) a stimulation of GLUT4 translocation to the cell membrane by activation of the AMPK pathway which mediates glucose uptake and fatty acid oxidation; (b) an enhancement of fatty acid oxidation and glucose disposal in both insulin-sensitive and insulin-resistant mice [75]. These findings disclose a novel anti-diabetic therapeutic potential.

21.8 Other Proteins

Besides the above categories of proteins/components, there are occasional reports of other bioactive components that deserve mention. A napin-like protein was purified from MC and later cloned [84, 85]. Both the native and recombinant forms exhibited strong antifungal activity against *Trichoderma hamatum* [84].

Polypeptide-P is a hypoglycemic peptide reported by Khanna et al. in the fruits and seeds of MC. The natural form contains 166 amino acid residues [86]. Recently its complete sequence was published (GenBank: ADO14327.1, 172 amino acids) and its 3-D structure was predicted as shown in Fig. 21.1–(7) [87]. It evinced a very potent hypoglycemic activity in both rodents (gerbils, langurs) and juvenile- (type 1) and maturity-onset (type 2) diabetic patients [86]. There is no further research on polypeptide-P and an in-depth exploration on its structure and mechanism investigation is needed.

In addition, an early study showed that there was a 40-kDa protein with cytostatic, cytotoxic, and anti-viral activities from MC. This protein manifested cytostatic activity toward BHK-21 cells in a dose-dependent manner, and inhibited both RNA and protein synthesis in these cells. Its complete sequence is not known [88].

21.9 Conclusion and Perspectives

The present review summarizes the bioactive components in MC with a focus on their biochemical characteristics, structures, and medicinal activities, including anti-diabetic, antitumor, anti-HIV, antimicrobial and pesticidal activities. It is noteworthy to mention that there may be a big market for MC, for it is nutritious and inexpensive, and has very little or no side effects based on long-time consumption of MC, and emerging beneficial data from scientific investigations [89, 90]. Ingestion of MC appears to be beneficial, not only for the treatment of specific illnesses, but also for preventing illnesses and promoting health (Table 21.1).

MC is one of a few plants that is commonly used in meals and medical treatments that has been scientifically studied. In spite of a voluminous literature on

MC, there are still arenas that invite investigations. Those areas may include: (a) a call for better-designed clinical trials to provide conclusive evidence for the applicability of MC in diabetes; (b) further research on the anti-obesity activity of MC; (c) studies on the medicinal activities and potential applications of other components of MC, such as MCL and the large number of serpins; (d) the isolation and characterization of other bioactive components from MC (Fig. 21.1).

All in all, MC is a readily available and inexpensive substance with a wide spectrum of medicinal activities and negligible drawbacks. The medicinal potentials of MC typified as anti-diabetic, anti-obesity, antitumor, and anti-HIV activities conform to the prevailing principles of healthy eating and represent a new, promising approach to broaden the therapeutic windows of such illnesses.

Acknowledgments We thank Dr. Jennifer Illuzzi at NIH for her invaluable feedback and critically reading the manuscript. This research was supported in part by the Intramural Research Program of the NIH, National Institute on Aging, USA.

References

1. Fang EF, Ng TB (2011) Bitter gourd (*Momordica charantia*) is a cornucopia of health: a review of its credited antidiabetic, anti-HIV, and antitumor properties. Curr Mol Med 11:417–436
2. Claflin AJ, Vesely DL, Hudson JL, Bagwell CB, Lehotay DC, Lo TM et al (1978) Inhibition of growth and guanylate cyclase activity of an undifferentiated prostate adenocarcinoma by an extract of the balsam pear (*Momordica charantia* abbreviata). Proc Natl Acad Sci U S A 75:989–993
3. Wang YX, Neamati N, Jacob J, Palmer I, Stahl SJ, Kaufman JD et al (1999) Solution structure of anti-HIV-1 and anti-tumor protein MAP30: structural insights into its multiple functions. Cell 99:433–442
4. Ray RB, Raychoudhuri A, Steele R, Nerurkar P (2010) Bitter melon (*Momordica charantia*) extract inhibits breast cancer cell proliferation by modulating cell cycle regulatory genes and promotes apoptosis. Cancer Res 70:1925–1931
5. Fang EF, Zhang CZ, Fong WP, Ng TB (2012) RNase MC2: a new *Momordica charantia* ribonuclease that induces apoptosis in breast cancer cells associated with activation of MAPKs and induction of caspase pathways. Apoptosis Int J Program Cell Death 17:377–387
6. Yasui Y, Hosokawa M, Sahara T, Suzuki R, Ohgiya S, Kohno H et al (2005) Bitter gourd seed fatty acid rich in 9c,11t,13t-conjugated linolenic acid induces apoptosis and up-regulates the GADD45, p53 and PPARgamma in human colon cancer Caco-2 cells. Prostaglandins Leukot Essent Fatty Acids 73:113–119
7. Agrawal RC, Beohar T (2010) Chemopreventive and anticarcinogenic effects of *Momordica charantia* extract. Asian Pac J Cancer Prev APJCP 11:371–375
8. Ray RB, Raychoudhuri A, Steele R, Nerurkar P Bitter melon (*Momordica charantia*) extract inhibits breast cancer cell proliferation by modulating cell cycle regulatory genes and promotes apoptosis. Cancer Res 70:1925–1931
9. Lee-Huang S, Huang PL, Sun Y, Chen HC, Kung HF, Murphy WJ (2000) Inhibition of MDA-MB-231 human breast tumor xenografts and HER2 expression by anti-tumor agents GAP31 and MAP30. Anticancer Res 20:653–659
10. Pitchakarn P, Ogawa K, Suzuki S, Takahashi S, Asamoto M, Chewonarin T et al (2010) *Momordica charantia* leaf extract suppresses rat prostate cancer progression in vitro and in vivo. Cancer Sci 101:2234–2240
11. Xiong SD, Yu K, Liu XH, Yin LH, Kirschenbaum A, Yao S et al (2009) Ribosome-inactivating proteins isolated from dietary bitter melon induce apoptosis and inhibit histone

deacetylase-1 selectively in premalignant and malignant prostate cancer cells. Int J Cancer 125:774–782
12. Ru P, Steele R, Nerurkar PV, Phillips N, Ray RB (2011) Bitter melon extract impairs prostate cancer cell-cycle progression and delays prostatic intraepithelial neoplasia in TRAMP model. Cancer Prev Res 4:2122–2130
13. Leung L, Birtwhistle R, Kotecha J, Hannah S, Cuthbertson S (2009) Anti-diabetic and hypoglycaemic effects of *Momordica charantia* (bitter melon): a mini review. Br J Nutr 102:1703–1708
14. Shih CC, Lin CH, Lin WL (2008) Effects of *Momordica charantia* on insulin resistance and visceral obesity in mice on high-fat diet. Diabetes Res Clin Pract 81:134–143
15. Mahomoodally MF, Gurib-Fakim A, Subratty AH Screening for alternative antibiotics: an investigation into the antimicrobial activities of medicinal food plants of Mauritius. J Food Sci 75:M173–177
16. Gupta S, Raychaudhuri B, Banerjee S, Das B, Mukhopadhaya S, Datta SC Momordicatin purified from fruits of *Momordica charantia* is effective to act as a potent antileishmania agent. Parasitol Int 59:192–197
17. Braca A, Siciliano T, D'Arrigo M, Germano MP (2008) Chemical composition and antimicrobial activity of *Momordica charantia* seed essential oil. Fitoterapia 79:123–125
18. Rajkumar S, Jebanesan A (2007) Repellent activity of selected plant essential oils against the malarial fever mosquito *Anopheles stephensi*. Trop Biomed 24:71–75
19. Telang M, Srinivasan A, Patankar A, Harsulkar A, Joshi V, Damle A et al (2003) Bitter gourd proteinase inhibitors: potential growth inhibitors of *Helicoverpa armigera* and *Spodoptera litura*. Phytochemistry 63:643–652
20. Rahuman AA, Venkatesan P (2008) Larvicidal efficacy of five cucurbitaceous plant leaf extracts against mosquito species. Parasitol Res 103:133–139
21. Maurya P, Sharma P, Mohan L, Batabyal L, Srivastava CN (2009) Evaluation of larvicidal nature of fleshy fruit wall of *Momordica charantia* Linn. (family: cucurbitaceae) in the management of mosquitoes. Parasitol Res 105:1653–1659
22. Batabyal L, Sharma P, Mohan L, Maurya P, Srivastava CN (2009) Relative toxicity of neem fruit, bitter gourd, and castor seed extracts against the larvae of filaria vector, *Culex quinquefasciatus* (Say). Parasitol Res 105:1205–1210
23. Ling B, Xiang YL, Wang GC, Chen SH, Zhang MX (2009) Antifeedant and antioviposition activities of *Momordica charantia* leaf ethanol extract against *Liriomyza sativae*. Ying Yong Sheng Tai Xue Bao 20:836–842
24. Vaya J, Aviram M (2001) Nutritional antioxidants: mechanisms of action, analyses if activities and medical applications. Curr Med Chem Immunol Endocr Metab Agents 1:99–117
25. Santos AK, Costa JG, Menezes IR, Cansancao IF, Santos KK, Matias EF et al. Antioxidant activity of five Brazilian plants used as traditional medicines and food in Brazil. Pharmacogn Mag 6:335–338
26. Scartezzini P, Speroni E (2000) Review on some plants of Indian traditional medicine with antioxidant activity. J Ethnopharmacol 71:23–43
27. Kumar R, Balaji S, Sripriya R, Nithya N, Uma TS, Sehgal PK In vitro evaluation of antioxidants of fruit extract of *Momordica charantia* L. on fibroblasts and keratinocytes. J Agric Food Chem 58:1518–1522
28. Thenmozhi AJ, Subramanian P (2010) Antioxidant potential of *Momordica charantia* in Ammonium Chloride-induced Hyperammonemic Rats. Evid Based Complement Alternat Med [Epub ahead of print]
29. Chaturvedi P (2009) Bitter melon protects against lipid peroxidation caused by immobilization stress in albino rats. Int J Vitam Nutr Res 79:48–56
30. Ono T, Tsuji T, Sakai M, Yukizaki C, Ino H, Akagi I et al (2009) Induction of hepatocyte growth factor production in human dermal fibroblasts and their proliferation by the extract of bitter melon pulp. Cytokine 46:119–126
31. Sonibare MA, Moody JO, Adesanya EO (2009) Use of medicinal plants for the treatment of measles in Nigeria. J Ethnopharmacol 122:268–272

32. Alam S, Asad M, Asdaq SM, Prasad VS (2009) Antiulcer activity of methanolic extract of *Momordica charantia* L. in rats. J Ethnopharmacol 123:464–469
33. Fang EF, Zhang CZ, Ng TB, Wong JH, Pan WL, Ye XJ et al (2012) *Momordica charantia* lectin, a type II ribosome inactivating protein, exhibits antitumor activity toward human nasopharyngeal carcinoma cells in vitro and in vivo. Cancer Prev Res 5:109–121
34. Lin JY, Hou MJ, Chen YC (1978) Isolation of toxic and non-toxic lectins from the bitter pear melon *Momordica charantia* Linn. Toxicon 16:653–660
35. Chandran T, Sharma A, Vijayan M (2010) Crystallization and preliminary X-ray studies of a galactose-specific lectin from the seeds of bitter gourd (*Momordica charantia*). Acta Crystallogr Sect F Struct Biol Cryst Commun 66:1037–1040
36. Tanaka H, Toyama J, Akashi R (2009) Molecular characterization of a galactose-binding lectin from *Momordica charantia* seeds and its expression in tobacco cells. Asian J Plant Sci 8:544–550
37. Barbieri L, Zamboni M, Lorenzoni E, Montanaro L, Sperti S, Stirpe F (1980) Inhibition of protein synthesis in vitro by proteins from the seeds of *Momordica charantia* (bitter pear melon). Biochem J 186:443–452
38. Huang L, Adachi T, Shimizu Y, Goto Y, Toyama J, Tanaka H et al (2008) Characterization of lectin isolated from *Momordica charantia* seed as a B cell activator. Immunol Lett 121:148–156
39. Fang EF, Zhang CZY, Ng TB, Wong JH, Pan WL, Ye XJ et al (2012) *Momordica charantia* Lectin, a Type II ribosome inactivating Protein, Exhibits antitumor activity toward human nasopharyngeal carcinoma cells in vitro and in vivo. Cancer Prev Res 5:109–121
40. Wang YX, Neamati N, Jacob J, Palmer I, Stahl SJ, Kaufman JD et al (1999) Solution structure of anti-HIV-1 and anti-tumor protein MAP30: structural insights into its multiple functions. Cell 99:433–442
41. Yang Y, Mak AN, Shaw PC, Sze KH Solution structure of an active mutant of maize ribosome-inactivating protein (MOD) and its interaction with the ribosomal stalk protein P2. J Mol Biol 395:897–907
42. Lee-Huang S, Huang PL, Nara PL, Chen HC, Kung HF, Huang P et al (1990) MAP 30: a new inhibitor of HIV-1 infection and replication. FEBS Lett 272:12–18
43. Fang EF, Zhang CZ, Wong JH, Shen JY, Li CH, Ng TB (2012) The MAP30 protein from bitter gourd (*Momordica charantia*) seeds promotes apoptosis in liver cancer cells in vitro and in vivo. Cancer Lett
44. Fong WP, Mock WY, Ng TB (2000) Intrinsic ribonuclease activities in ribonuclease and ribosome-inactivating proteins from the seeds of bitter gourd. Int J Biochem Cell Biol 32:571–577
45. Fong WP, Poon YT, Wong TM, Mock JW, Ng TB, Wong RN et al (1996) A highly efficient procedure for purifying the ribosome-inactivating proteins alpha- and beta-momorcharins from *Momordica charantia* seeds, N-terminal sequence comparison and establishment of their N-glycosidase activity. Life Sci 59:901–909
46. Zhu G, Huang Q, Qian M, Tang Y (2001) Crystal structure of alpha-momorcharin in 80 % acetonitrile–water mixture. Biochim Biophys Acta 1548:152–158
47. Tsao SW, Ng TB, Yeung HW (1990) Toxicities of trichosanthin and alpha-momorcharin, abortifacient proteins from Chinese medicinal plants, on cultured tumor cell lines. Toxicon 28:1183–1192
48. Ng TB, Liu WK, Sze SF, Yeung HW (1994) Action of alpha-momorcharin, a ribosome inactivating protein, on cultured tumor cell lines. Gen Pharmacol 25:75–77
49. Bian X, Shen F, Chen Y, Wang B, Deng M, Meng Y PEGylation of alpha-momorcharin: synthesis and characterization of novel anti-tumor conjugates with therapeutic potential. Biotechnol Lett 32:883–890
50. Yuan YR, He YN, Xiong JP, Xia ZX (1999) Three-dimensional structure of beta-momorcharin at 2.55 A resolution. Acta Crystallogr D Biol Crystallogr 55:1144–1151
51. Fukunaga Y, Nishimoto E, Yamashita K, Otosu T, Yamashita S (2007) The partially unfolded state of beta-momorcharin characterized with steady-state and time-resolved fluorescence studies. J Biochem 141:9–18

52. Chan WY, Ng TB, Yeung HW (1992) Beta-Momorcharin, a plant glycoprotein, inhibits synthesis of macromolecules in embryos, splenocytes and tumor cells. Int J Biochem 24:1039–1046
53. Pu Z, Lu BY, Liu WY, Jin SW (1996) Characterization of the enzymatic mechanism of gamma-momorcharin, a novel ribosome-inactivating protein with lower molecular weight of 11,500 purified from the seeds of bitter gourd (*Momordica charantia*). Biochem Biophys Res Commun 229:287–294
54. Husain J, Tickle IJ, Wood SP (1994) Crystal structure of momordin, a type I ribosome inactivating protein from the seeds of *Momordica charantia*. FEBS Lett 342:154–158
55. Stirpe F, Wawrzynczak EJ, Brown AN, Knyba RE, Watson GJ, Barbieri L et al (1988) Selective cytotoxic activity of immunotoxins composed of a monoclonal anti-Thy 1.1 antibody and the ribosome-inactivating proteins bryodin and momordin. Br J Cancer 58:558–561
56. Porro G, Bolognesi A, Caretto P, Gromo G, Lento P, Mistza G et al (1993) In vitro and in vivo properties of an anti-CD5-momordin immunotoxin on normal and neoplastic T lymphocytes. Cancer Immunol Immunother 36:346–350
57. Fang EF, Ng TB (1815) Ribonucleases of different origins with a wide spectrum of medicinal applications. Biochim Biophys Acta 2011:65–74
58. Ardelt W, Ardelt B, Darzynkiewicz Z (2009) Ribonucleases as potential modalities in anticancer therapy. Eur J Pharmacol 625:181–189
59. Ide H, Kimura M, Arai M, Funatsu G (1991) The complete amino acid sequence of ribonuclease from the seeds of bitter gourd (*Momordica charantia*). FEBS Lett 289:126
60. Numata T, Suzuki A, Kakuta Y, Kimura K, Yao M, Tanaka I et al (2003) Crystal structures of the ribonuclease MC1 mutants N71T and N71S in complex with 5'-GMP: structural basis for alterations in substrate specificity. Biochemistry 42:5270–5278
61. Suzuki A, Yao M, Tanaka I, Numata T, Kikukawa S, Yamasaki N et al (2000) Crystal structures of the ribonuclease MC1 from bitter gourd seeds, complexed with 2'-UMP or 3'-UMP, reveal structural basis for uridine specificity. Biochem Biophys Res Commun 275:572–576
62. Numata T, Suzuki A, Yao M, Tanaka I, Kimura M (2001) Amino acid residues in ribonuclease MC1 from bitter gourd seeds which are essential for uridine specificity. Biochemistry 40:524–530
63. Kimura K, Numata T, Kakuta Y, Kimura M (2004) Amino acids conserved at the C-terminal half of the ribonuclease T2 family contribute to protein stability of the enzymes. Biosci Biotechnol Biochem 68:1748–1757
64. Nakagawa A, Tanaka I, Sakai R, Nakashima T, Funatsu G, Kimura M (1999) Crystal structure of a ribonuclease from the seeds of bitter gourd (*Momordica charantia*) at 1.75 A resolution. Biochim Biophys Acta 1433:253–260
65. Fang EF, Ng TB (1815) Ribonucleases of different origins with a wide spectrum of medicinal applications. Biochim Biophys Acta-Rev Cancer 2011:65–74
66. Fang EF, Zhang CZ, Zhang L, Fong WP, Ng TB (2012) In vitro and in vivo anticarcinogenic effects of RNase MC2, a ribonuclease isolated from dietary bitter gourd, toward human liver cancer cells. Int J Biochem Cell Biol 44:1351–1360
67. Bennett MK, Kirk CJ (2008) Development of proteasome inhibitors in oncology and autoimmune diseases. Curr Opin Drug Discov Devel 11:616–625
68. Sterz J, von Metzler I, Hahne JC, Lamottke B, Rademacher J, Heider U et al (2008) The potential of proteasome inhibitors in cancer therapy. Expert Opin Investig Drugs 17:879–895
69. Kondo H, Nakanose J, Nakagawa A, Tanaka I, Kimura M, Funatsu G (1997) Crystallization and preliminary X-ray studies of two serine proteinase inhibitors, BGIA and BGIT, from the seeds of bitter gourd. J Struct Biol 120:204–206
70. Hamato N, Koshiba T, Pham TN, Tatsumi Y, Nakamura D, Takano R et al (1995) Trypsin and elastase inhibitors from bitter gourd (*Momordica charantia* LINN.) seeds: purification, amino acid sequences, and inhibitory activities of four new inhibitors. J Biochem 117:432–437

71. Zhu Y, Huang Q, Qian M, Jia Y, Tang Y (1999) Crystal structure of the complex formed between bovine beta-trypsin and MCTI-A, a trypsin inhibitor of squash family, at 1.8-A resolution. J Protein Chem 18:505–509
72. Fang EF, Wong JH, Ng TB Thermostable Kunitz trypsin inhibitor with cytokine inducing, antitumor and HIV-1 reverse transcriptase inhibitory activities from Korean large black soybeans. J Biosci Bioeng 109:211–217
73. Losso JN (2008) The biochemical and functional food properties of the bowman-birk inhibitor. Crit Rev Food Sci Nutr 48:94–118
74. Drag M, Salvesen GS Emerging principles in protease-based drug discovery. Nat Rev Drug Discov 9:690–701
75. Tan MJ, Ye JM, Turner N, Hohnen-Behrens C, Ke CQ, Tang CP et al (2008) Antidiabetic activities of triterpenoids isolated from bitter melon associated with activation of the AMPK pathway. Chem Biol 15:263–273
76. Popovich DG, Li L, Zhang W Bitter melon (*Momordica charantia*) triterpenoid extract reduces preadipocyte viability, lipid accumulation and adiponectin expression in 3T3-L1 cells. Food Chem Toxicol 48:1619–1626
77. Akihisa T, Higo N, Tokuda H, Ukiya M, Akazawa H, Tochigi Y et al (2007) Cucurbitane-type triterpenoids from the fruits of *Momordica charantia* and their cancer chemopreventive effects. J Nat Prod 70:1233–1239
78. Li QY, Chen HB, Liu ZM, Wang B, Zhao YY (2007) Cucurbitane triterpenoids from *Momordica charantia*. Magn Reson Chem 45:451–456
79. Nhiem NX, Kiem PV, Minh CV, Ban NK, Cuong NX, Tung NH et al Alpha-Glucosidase inhibition properties of cucurbitane-type triterpene glycosides from the fruits of *Momordica charantia*. Chem Pharm Bull (Tokyo) 58:720–724
80. Chang CI, Chen CR, Liao YW, Shih WL, Cheng HL, Tzeng CY et al Octanorcucurbitane triterpenoids protect against tert-butyl hydroperoxide-induced hepatotoxicity from the stems of *Momordica charantia*. Chem Pharm Bull (Tokyo) 58:225–229
81. Chen JC, Liu WQ, Lu L, Qiu MH, Zheng YT, Yang LM, Kuguacins F-S et al (2009) Cucurbitane triterpenoids from *Momordica charantia*. Phytochemistry 70:133–140
82. Chen J, Tian R, Qiu M, Lu L, Zheng Y, Zhang Z (2008) Trinorcucurbitane and cucurbitane triterpenoids from the roots of *Momordica charantia*. Phytochemistry 69:1043–1048
83. Mekuria DB, Kashiwagi T, Tebayashi S, Kim CS (2006) Cucurbitane glucosides from *Momordica charantia* leaves as oviposition deterrents to the leafminer. Liriomyza trifolii. Z Naturforsch C 61:81–86
84. Vashishta A, Sahu T, Sharma A, Choudhary SK, Dixit A (2006) In vitro refolded napin-like protein of *Momordica charantia* expressed in *Escherichia coli* displays properties of native napin. Biochim Biophys Acta 1764:847–855
85. Neumann GM, Condron R, Polya GM (1996) Purification and sequencing of napin-like protein small and large chains from *Momordica charantia* and *Ricinus communis* seeds and determination of sites phosphorylated by plant Ca(2 +)-dependent protein kinase. Biochim Biophys Acta 1298:223–240
86. Khanna P, Jain SC, Panagariya A, Dixit VP (1981) Hypoglycemic activity of polypeptide-p from a plant source. J Nat Prod 44:648–655
87. Kelley LA, Sternberg MJ (2009) Protein structure prediction on the Web: a case study using the Phyre server. Nat Protoc 4:363–371
88. Takemoto DJ, Jilka C, Rockenbach S, Hughes JV (1983) Purification and characterization of a cytostatic factor with anti-viral activity from the bitter melon. Prep Biochem 13:371–393
89. Nerurkar P, Ray RB Bitter melon: antagonist to cancer. Pharm Res 27:1049–1053
90. Grover JK, Yadav SP (2004) Pharmacological actions and potential uses of *Momordica charantia*: a review. J Ethnopharmacol 93:123–132
91. Parkash A, Ng TB, Tso WW (2002) Purification and characterization of charantin, a napin-like ribosome-inactivating peptide from bitter gourd (*Momordica charantia*) seeds. J Pept Res 59:197–202

92. Ren J, Wang Y, Dong Y, Stuart DI (1994) The N-glycosidase mechanism of ribosome-inactivating proteins implied by crystal structures of alpha-momorcharin. Structure 2:7–16
93. Xiong JP, Xia ZX, Zhang L, Ye GJ, Jin SW, Wang Y (1994) Crystallization and preliminary crystallographic study of beta-momorcharin. J Mol Biol 238:284–285
94. Fang EF, Ng TB (2011) Ribonucleases of different origins with a wide spectrum of medicinal applications. Biochim Biophys Acta-Rev Cancer 1815:65–74
95. Pettersen EF, Goddard TD, Huang CC, Couch GS, Greenblatt DM, Meng EC et al (2004) UCSF Chimera–a visualization system for exploratory research and analysis. J Comput Chem 25:1605–1612

Chapter 22
Bioactive Constituents of the Silk Worm *Bombyx mori*

Lixin Xia, Tzi Bun Ng, Evandro Fei Fang and Jack Ho Wong

Abstract Silkworm is economically important and also has a large number of activities encompassing anticancer, antibacterial, antibacterial, antiviral, antihypertensive, angiotensin-converting enzyme activity, hepatoprotective, superoxide dismutase enzyme activity, free radical scavenging and tyrosinase inhibitory, antidiabetic, antityrosinase, cognition enhancing, alternative complement pathway-inhibiting, carotenoid binding, chymotrypsin inhibiting, and chitinase activities. These activities may be beneficial to human health.

22.1 Introduction

Silkworm, the caterpillar of the domesticated silkmoth *Bombyx mori*, produces silk and hence has great economical value [1]. It does not occur naturally in the wild and depends on humans for reproduction. Silkworms have been bred for the production of raw silk in China for thousands of years from where it spread to Korea and Japan, and later to India and the West [2]. The silkworm was domesticated from the wild silkmoth *Bombyx mandarina*. The domesticated *B. mori* and the wild *B. mandarina* can still breed and sometimes produce hybrids [3]. Silkworm is a model organism for investigations on Lepidopterans and arthropods such as studies on pheromones, genetics and genomics [4, 5]. Research on

L. Xia (✉)
Institute of Allergy and Immunology School of Medicine, Shenzhen University, Shenzhen, China
e-mail: xialixin@126.com

T. B. Ng · E. F. Fang · J. H. Wong
School of Biomedical Sciences, The Chinese University of Hong Kong, Shatin, Hong Kong

E. F. Fang
Laboratory of Molecular Gerontology, National Institute on Aging, National Institutes of Health, Baltimore, MD 21224, USA

E. F. Fang and T. B. Ng (eds.), *Antitumor Potential and Other Emerging Medicinal Properties of Natural Compounds*, DOI: 10.1007/978-94-007-6214-5_22,

the genome also raises the possibility of genetically engineering silkworms to produce proteins, including pharmacological drugs, in the place of silk proteins [4]. Reports on the various pharmacological activities of the silkworm have been accumulating. It is timely to review these findings.

22.2 Bioactivities of Silkworm

22.2.1 Anticancer Activity

Selenium, a chemical element with symbol Se and atomic number 34, has a vital role in prevention of cancer. When selenium-rich amino acids extracted from Ziyang silkworm pupas were added to cultured human hepatoma cells SMMC-7721, cell viability declined, alterations in cell morphology and cell cycle were observed, and apoptosis was induced. This inhibitory effect on SMMC-7721 cells could not be reproduced by using normal amino acids. The selenium-rich amino acids increased the production of intracellular reactive oxygen species [5].

22.2.2 Antibacterial Activity

A large number of proteins/peptides have been demonstrated to be involved in the immune response of insects and protect them from bacterial infecton. Examples are cecropins, hemolins, attacins, and lysozymes [6]. These components can be found in silkworm. Two cDNA clones encoding the antibacterial protein cecropin B were isolated from a fat body cDNA library of the silkworm. The deduced amino acid sequences of silkworm cecropin B displayed closer resemblance to Lepidopteran than Dipteran cecropins [7]. Discovered in 1922, lysozymes are a type of bacteriolytic enzmes that hydrolyze β-1,4,-glycosidic linkages between *N*-acetylglucosamine and *N*-acetylumramic acid of the peptidoglycan layer in the bacterial cell wall [8]. For silkworm lysozyme, it was first isolated in 1973 and serves as an important inducible antibacterial protein in the insect defense system [9]. Subsequently, an interesting work carried out by Yu et al. compared the biochemical and antibacterial activities of silkworm lysozyme, and the lysozymes from another two lepidopteran larvae, *Galleria mellonella* and *Agrius convolvuli* [8]. The results unveiled that all three lysozymes had strong activities against Gram-positive bacteria and lower anti-Gram-negative bacterial activites. On the other hand, lysozyme-like proteins (LLPs) demonstrate partial similarity with lysozymes. However, the absence of catalytic amino acids crucial to muramidase activity in LLPs distinguishes them functionally from classical lysozymes. Two LLPs, one from *B. mori* (BLLP1) expressed in a recombinant system, displayed inhibitory action against a broad spectrum of bacteria. BLLP1 is bacteriostatic rather than bactericidal against *E. coli* and *Micrococcus luteus*. Its antibacterial mechanism relies on peptidoglycan binding in contrast to peptidoglycan hydrolysis or membrane permeabilization induced by lysozymes and most other antimicrobial peptides [10].

Peptides are short polymers of amino acid monomers and normally contain less than 50 amino acids. They serve as prized candidates for medicinal applications [11]. There are many antibacterial peptides produced by the silkworm. For example, a 6-kDa lysozyme-like hemolymph protein, with trypsin-labile and heat-labile antibacterial activity induced by injection of live *Escherichia coli* into silkworm larvae, have similar activity against *E. coli* and *Micrococcus luteus* [12]. Hara et al. had isolated three structurally related and novel antibacterial peptides from the haemolymph of the silkworm (*Bombyx mori*), immunized with *Escherichia coli* [13]. These peptides were 32 amino acids long and characteristically rich in proline residues. A unique threonine residue in each peptide was O-glycosylated and the modification seemed to be important for expression of antibacterial activity. The primary structure and antibacterial characteristic of the novel peptides resemble those of abaecin (41 % identity in amino acid sequence), an antibacterial peptide of the honeybee, although abaecin is not O-glycosylated [13]. Furthermore, moricin, a highly basic 42-amino acid peptide from silkworm hemolymph inducible by bacterial injection, inhibits several Gram-negative and -positive bacteria. It is more potent against Gram-positive bacteria than cecropin B1. It targets the bacterial cytoplasmic membrane. The N-terminal portion of the peptide with a predicted alpha-helix causes an increase in the membrane permeability [14]. Two foreign gene expression systems in Escherichia coli were adopted to procure a sizeable amount of moricin to facilitate further studies [15]. In addition, Li et al. isolated a 35-residue cationic, amphipathic α-helical peptide, named CM4 (ABP-CM4) from silkworm, and this peptide exhibited broad-spectrum antimicrobial activity [16]. Interestinly, recombinant CM4 also showed antimicrobial activity against *E. coli*, *Penicillium chrysogenum*, *Aspergillus niger* and *Gibberella saubinetii* [16].

22.2.3 Antiviral Activity

Both the crude extract and purified components in silkworm have been reported with promising antiviral activity. A study in 1997 showed that silkworm faecal extract display edantiviral activity on enveloped viruses, but not on a non-enveloped virus. It suppressed the synthesis of a virus- specific gene of HVJ Sendai virus, herpes simplex virus type, and HIV-1 without altering viral adsorption and entry into the host cell. The active principle was a chlorophyll-like substance with a molecular mass of about 530 Da. Its antiviral activity was light irradiation- and temperature-dependent [17]. Later in 2007, the effect of silkworm extract on hepatititis B and C viruses were investigated. This extract contained natural iminosugars, such as 1-deoxynojirimycin (1-DNJ). It exhibited an over thousand fold higher antiviral effect against bovine viral diarrhea virus compared to 1-deoxynojirimycin. Glycoprotein processing of bovine viral diarrhea virus envelope proteins became impaired following treatment. Glycosylation of woodchuck hepatitis virus envelope proteins was affected largely by treatment with the silkworm extract. The mechanism of action may entail production of defective particles incapable of starting the next cycle of infection as shown by inhibition of GB virus-B in vitro. The iminosugars in the silkworm extract may synergize in their antiviral effects on the inhibition of intact maturation of hepatitis virus particles and may serve to complement conventional treatments [18].

The above-mentioned antiviral activity may to some extent be attributed to silk fibroins and NADH-oxidoreductase-like proteins. For example, two differentially sulfated silk fibroins, SclFib30 and SclFib31, exhibited anti-HIV-1 activity in vitro due to interference with virus particle adsorption to $CD4^+$ cells, and prevention of virus binding to the cells. Cell-to-cell infection-induced syncytium formation upon coculture of MOLT-4 and MOLT-4/HIV-IIIB cells in the presence of the sulfated silk fibroins was absent, suggesting that they would interfere with gp120 and prevent the formation of gp120/CD4 complex. Silk is used in biomaterials such as surgical sutures and it is believed to be a safe material for humans. In accordance with low anticoagulant activity and high anti-HIV-1 activity against both X4 HIV-1 and R5 HIV-1 strains, sulfated silk fibroins have potential as antiviral material such as that used for a vaginal anti-HIV formulation [19]. Another example is a 26.5-kDa soluble NADH-oxidoreductase-like protein. It is overexpressed in the intestinal juice of disease-resistant multivoltine races. It possesses anti-BmNPV activity which attenuates the ability of the viral particles to infect BmN cells in vitro [20].

22.2.4 Antihypertensive Activity

Angiotensin I converting enzyme (ACE) is an enzyme that catalyzes the cleavage of the inactive decapeptide angiotensin I to the active octapeptide angiotensin II, and this process is associated with an elevation of blood pressure [21]. Though a study on a group of Caucasians indicated that the ACE gene is not associated with genetic predisposition to high blood pressure [22], in recent years ACE inhibitors are used as antihypertensive drugs [21]. The screening of natural ACE inhibitors is a hot research field aimed at ascertaining ways to treat cardiovascular diseases. There are some ACE inhibitors in silkworms. For instance, the hydrolyzates of silkworm pupal protein produced as a result of the catalytic actions of neutrase, pepsin, Asperqiius usamii acidic protease, flavourzyme, alcalase, and trypsin inhibited angiotensin I-converting enzyme. The molecular weight of the active principles ranged from below 500 to about 1000 Da [23]. A promising purified candidate is a 238.2-Da peptide which showed antihypertensive activity in both in vitro and in vivo experiments [24]. Zhou et al. found that this peptide exhibited ACE inhibiting activity and also blood pressure lowering activity in spontaneously hypertensive rats after chronic oral administration [25]. In addition to peptide(s), antihypertensive proteins may also exist in silkworm. A clone encoding a putative angiotensin-converting enzyme-related gene from the silkworm wing disc cDNA library has been isolated by Quan et al. [26]. The predicted open reading frame encoded a protein composed of 648 amino acids with nearly 50 % identity to angiotensin-converting enzyme from the fruit fly *Drosophila melanogaster*.

22.2.5 Hepatoprotective Activity

Raghavendra et al. purified a 35-kDa protein from silk worm and investigated its protective activity against carbon tetrachloride-induced hepatotoxicity [27]. The protein exerted in rats a protective effect against carbon tetrachloride-induced oxidative stress and liver damage as evidenced by a reduction of activities of enzymes in blood such as alkaline phosphatase, aspartate aminotransferase, alanine aminotransferase, and total bilirubin. The protein also displayed 2,2-diphenylpicrylhydrazyl radical and superoxide anion scavenging activity. However, the protein had no detectable antiviral effect on camelpox and goatpox viruses at its maximum non-toxic concentration.

22.2.6 Antidiabetic Activity

Type 2 diabetes is an inflammatory disease with the mechanisms linked to the adipose tissue of obese subjects. It is a serious health problem worldwide, and novel therapeutic drugs are urgently needed [28]. The silkworm cocoon shell of *B. mori* is composed of silk fibroin fiber enveloped by a sericin layer containing sericin and non-sericin components. The non-sericin components include carbohydrate, salt, wax, flavonoids and derivatives. Sericin and non-sericin compounds were obtained from the sericin layer of five types of cocoon shell by degumming in water followed by extraction and separation in ethanol. These ethanol extracts contain flavonoids and free amino acids with 2,2-diphenyl-1-picrylhydrazyl free radical scavenging and tyrosinase inhibiting activities, much more potent than the corresponding activities of the purified sericin proteins. The extracts but not sericins strongly inhibited α-glucosidase. The activities of the non-sericin component were much more potent than that of sericin alone. The non-sericin component was also effective in vivo as shown by a reduction of streptozocin-induced diabetes in mice. Hence ethanol extract of the sericin layer of cocoon shells furnishes a stock which, together with sericin protein, has potential functional food, biotechnological and medical applications [29]. It is noteworthy that two anti-diabetic compounds have been isolated by Lee et al. The tripeptides Gly-Glu-Tyr (MW = 367 Da) and Gly-Tyr-Gly (MW = 295 Da) manifested α-glucosidase inhibitory activities with an IC_{50} of 2.7 and 1.5 mg/mL, respectively [30].

22.2.7 Cognition Enhancing Activity

Brain factor-7 (BF-7) is a natural compound in silkworm, and a series of studies on this natural component has been carried out by a team of Korean researchers [31]. There are solid data showing the learning and memory-enhancing activities of this component. Examples include the positive results from cell culture studies, experiments using rats, and investigations on humans (involving 46 school children) [31]. The mechanisms of the effect are shown to be associated with the augmented production of the neurotransmitter acetylcholine, reduction of reactive oxidative species, and attenuation of apoptosis [31]. Thus BF-7 should be an excellent cognition-enhancing drug.

22.2.8 Antioxidant Activity

Natural compounds with antioxidant activity have been linked with antitumor, antidiabetic, immunomodulatory, cognition-enhancing, and life-prolonging functions, and are promising candidates for new drug discovery [31, 32]. Silkworm has been reported with free radical scavenging and tyrosinase inhibitory activities, as well as superoxide dismutase activity. For example, Manosroi et al. conducted an experiment on five Thai native silkworms (Keawsakol, Nangnoi, Somrong, Nangleung, and Noneruesee) and extracted bioactive components (such as oils and sericin) and tested their antioxidant activity [33]. The oils were extracted by a hot process using Soxhlet apparatus and a cold process using petroleum ether, while sericin was extracted by basic hydrolysis and autoclaving. Sericin from the five Thai native silkworms displayed free radical scavenging activity which was lower than those of vitamin C, vitamin E, and BHT (used as standard reagents for comparison) by about 20–100 fold, but the activity of all oils was higher than that of the standard linoleic acid by 11-22 fold. Oil extracted from Noneruesee by the cold process gave the highest DPPH scavenging activity. All sericin samples showed tyrosinase inhibiting activity with IC_{50} values in the range of 1.2–18.76 mg/mL, but only oils from Noneruesee extracted using the hot process, and Nangleung, Somrong, and Noneruesee extracted using the cold process, demonstrated this activity. Oil extracted using the hot process and sericin by basic hydrolysis from Noneruesee yieded the highest antityrosinase activity, but lower than that of the standards vitamin C and kojic acid by 20–49 and 3–8 times, respectively [33].

As a ubiquitous metalloenzyme in aerobic organisms, superoxide dismutase (SOD) is an enzyme that catalyzes the elimination of superoxide anions from living organisms and especially protects the organisms from the side effects of ROS. Silkworm superoxide dismutase is 60 % identical in sequence to its *Drosophila melanogaster* counterpart. Its activity was preserved following incubation at pH 4–11 for 24 h and also after incubation for 30 min at pH 7 and temperatures under 40 degree Centigrade [34]. The above-mentioned antitumor and antidiabetic activities of silkworm can be at least partially explained by its antioxidant activity as compiled here.

22.2.9 Immuno-Regulatory Activity

In vertebrates, the complement system (CS, containing around 30 proteins) consists of the classical, alternative, and lectin activation pathways, and plays a key role as an effector in immune reactions [35]. The alternative complement pathway (ACP) is an innate component of the immune system with the activity to defend external infections. Two isoforms of a protein (34- and 32-kDa) with the activity to inhibit ACP were isolated from silkworms, but no in-depth investigation has-been undertaken. It is worthwhile to investigate functional as well as molecular properties of these proteins.

22.2.10 Arginase Kinase as Silkworm Allergen

Different parts of silkworm such as silk, urine and dander contain allergens which can trigger or aggravate allergic diseases including allergic asthma, hypersensitivity pneumonia, allergic rhinitis and atopic dermatitis [36]. Serum samples collected from ten Chinese patients showing a positive skin prick test reaction to silkworm crude extract reacted with a 42-kDa protein in a Western blot using silkworm crude extract as the antigen. Preincubation of these serum samples with recombinant silkworm arginine kinase abolished the reactivity of the patients' samples to this 42-kDa band. All patient sera also exhibited positive reactivity to silkworm crude extract in ELISA. Silkworm arginine kinase demonstrated cross-reactivity with the recombinant cockroach enzyme. Thus silkworm arginine kinase is an allergen which cross reacts with cockroach arginine kinase [37]. Further studies in this field will help people in contact with silkworm and its associated products to combat the silkworm allergen.

22.2.11 Other Activities

There are few reports on other activities in silkworm, such as chymotrypsin-inhibitory activity, chitinase activity, anti-tyrosinase activity, and carotenoid-binding activity. A 7.5-kDa thermostable and pH-stable Kunitz-type chymotrypsin inhibitor has been isolated from silkworm hemolymph [38]. In view of the fact that other protease inhibitors have been reported with antitumor and anti-HIV-1 reverse transcriptase activity, whether these activities exist in this 7.5-kDa protein remains to be elucidated [39]. Furthermore, a 88-kDa silk worm chitinase, with a K_m of 22.3 μmol/L. and the optimal temperature at 45 °C and optimum pH at pH 6.0, was inhibited by Cu^{2+} ions but activated by high concentrations of Mn^{2+} ions and sodium dodecyl sulfate. However, no antifungal activity of this chitinase has been reported [40]. In addition, silk sericin extracted using urea exhibited the highest antityrosinase activity, whereas alkali-degraded silk sericin did not inhibit mushroom tyrosinase. Pigments, mainly flavonoids and carotenoids from silk cocoons, increased tyrosinase inhibition of silk sericin [41].

22.3 Conclusions and Future Perspectives

The economic importance of silkworms is indisputable. Silkworm pupas are rich in proteins and are used in Chinese cuisine and traditional Chinese medicine. Silkworm cell lines are also employed in scientific research to express proteins. The aforementioned biological activities of silkworms may be exploited to the welfare of mankind. It remains to be seen whether regular consumption of silkworm pupas for people without allergy to silkworm would promote health. More research on silkworms should be invited and encouraged to unveil further useful activities.

Acknowledgments The award of a research grant (81273275) to Lixin Xia from the National Science Foundation of China is gratefully acknowledged.

References

1. Omenetto FG, Kaplan DL (2010) New opportunities for an ancient material. Science 329:528–531
2. Arunkumar KP, Metta M, Nagaraju J (2006) Molecular phylogeny of silkmoths reveals the origin of domesticated silkmoth, *Bombyx mori* from Chinese *Bombyx mandarina* and paternal inheritance of *Antheraea proylei* mitochondrial DNA. Mol Phylogenet Evol 40:419–427
3. Ohnishi A, Kaji M, Hashimoto K, Matsumoto S (2011) Screening for the genes involved in bombykol biosynthesis: identification and functional characterization of *Bombyx mori* Acyl carrier protein. Front Endocrinol 2:92
4. Goldsmith MR, Shimada T, Abe H (2005) The genetics and genomics of the silkworm, *Bombyx mori*. Annu Rev Entomol 50:71–100
5. Hu D, Liu Q, Cui H, Wang H, Han D, Xu H (2005) Effects of amino acids from selenium-rich silkworm pupas on human hepatoma cells. Life Sci 77:2098–2110
6. Lockey TD, Ourth DD (1996) Purification and characterization of lysozyme from hemolymph of *Heliothis virescens* larvae. Biochem Biophys Res Commun 220:502–508
7. Taniai K, Kato Y, Hirochika H, Yamakawa M (1992) Isolation and nucleotide sequence of cecropin B cDNA clones from the silkworm, *Bombyx mori*. Biochim et Biophys Acta 1132:203–206
8. Yu KH, Kim KN, Lee JH, Lee HS, Kim SH, Cho KY et al (2002) Comparative study on characteristics of lysozymes from the hemolymph of three lepidopteran larvae, *Galleria mellonella, Bombyx mori, Agrius convolvuli*. Dev Comp Immun 26:707–713
9. Powning RF, Davidson WJ (1973) Studies on insect bacteriolytic enzymes. I. Lysozyme in haemolymph of *Galleria mellonella* and *Bombyx mori*. Comp Biochem Physiol B: Comp Biochem 45:669–686
10. Gandhe AS, Janardhan G, Nagaraju J (2007) Immune upregulation of novel antibacterial proteins from silkmoths (Lepidoptera) that resemble lysozymes but lack muramidase activity. Insect Biochem Mol Biol 37:655–666

11. Ng TB (2004) Antifungal proteins and peptides of leguminous and non-leguminous origins. Peptides 25:1215–1222
12. Abraham EG, Nagaraju J, Salunke D, Gupta HM, Datta RK (1995) Purification and partial characterization of an induced antibacterial protein in the silkworm, *Bombyx mori*. J Invertebr Pathol 65:17–24
13. Hara S, Yamakawa M (1995) A novel antibacterial peptide family isolated from the silkworm, *Bombyx mori*. Biochem J 310(Pt 2):651–656
14. Hara S, Yamakawa M (1995) Moricin, a novel type of antibacterial peptide isolated from the silkworm, *Bombyx mori*. The Journal of biological chemistry 270:29923–29927
15. Hara S, Yamakawa M (1996) Production in Escherichia of moricin, a novel type antibacterial peptide from the silkworm, *Bombyx mori*. Biochem Biophys Res Commun 220:664–669
16. Li BC, Zhang SQ, Dan WB, Chen YQ, Cao P (2007) Expression in *Escherichia coli* and purification of bioactive antibacterial peptide ABP-CM4 from the Chinese silk worm, *Bombyx mori*. Biotechnol Lett 29:1031–1036
17. Hiraki A, Yukawa M, Kim J, Ueda S (1997) Antiviral substance from silkworm faeces: purification and its chemical characterization. Biol Pharm Bull 20:547–555
18. Jacob JR, Mansfield K, You JE, Tennant BC, Kim YH (2007) Natural iminosugar derivatives of 1-deoxynojirimycin inhibit glycosylation of hepatitis viral envelope proteins. J Microbiol 45:431–440
19. Gotoh K, Izumi H, Kanamoto T, Tamada Y, Nakashima H (2000) Sulfated fibroin, a novel sulfated peptide derived from silk, inhibits human immunodeficiency virus replication in vitro. Biosci Biotechnol Biochem 64:1664–1670
20. Selot R, Kumar V, Shukla S, Chandrakuntal K, Brahmaraju M, Dandin SB et al (2007) Identification of a soluble NADPH oxidoreductase (BmNOX) with antiviral activities in the gut juice of *Bombyx mori*. Biosci Biotechnol Biochem 71:200–205
21. Masuyer G, Schwager SL, Sturrock ED, Isaac RE, Acharya KR (2012) Molecular recognition and regulation of human angiotensin-I converting enzyme (ACE) activity by natural inhibitory peptides. Sci Rep 2:717
22. Harrap SB, Davidson HR, Connor JM, Soubrier F, Corvol P, Fraser R et al (1993) The angiotensin I converting enzyme gene and predisposition to high blood pressure. Hypertension 21:455–460
23. Wang W, Shen S, Chen Q, Tanga B, He G, Ruan H et al (2008) Hydrolyzates of silkworm pupae (*Bombyx mori*) protein is a new source of angiotensin I-converting enzyme inhibitory peptides (ACEIP). Curr Pharm Biotechnol 9:307–314
24. Ni L, Tao GJ, Dai J, Wang Z, Xu SY (2001) Separation, purification and identification of angiotensin converting enzyme inhibitory silk fibroin peptide. Se Pu = Chin J Chromatogr/Zhongguo Hua Xue Hui 19:222–225
25. Zhou F, Xue Z, Wang J (2010) Antihypertensive effects of silk fibroin hydrolysate by alcalase and purification of an ACE inhibitory dipeptide. J Agric Food Chem 58:6735–6740
26. Quan GX, Mita K, Okano K, Shimada T, Ugajin N, Xia Z et al (2001) Isolation and expression of the ecdysteroid-inducible angiotensin-converting enzyme-related gene in wing discs of *Bombyx mori*. Insect Biochem Mol Biol 31:97–103
27. Raghavendra R, Neelagund S, Kuluvar G, Bhanuprakash V, Revanaiah Y (2010) Protective effect of partially purified 35 kDa protein from silk worm (*Bombyx mori*) fecal matter against carbon tetrachloride induced hepatotoxicity and in vitro anti-viral properties. Pharm Biol 48:1426–1431
28. Leiherer A, Mundlein A, Drexel H (2012) Phytochemicals and their impact on adipose tissue inflammation and diabetes. Vasc Pharmacol 58(1–2):3–20
29. Wang HY, Wang YJ, Zhou LX, Zhu L, Zhang YQ (2012) Isolation and bioactivities of a non-sericin component from cocoon shell silk sericin of the silkworm *Bombyx mori*. Food Function 3:150–158
30. Lee HJ, Lee HS, Choi JW, Ra KS, Kim JM, Suh HJ (2011) Novel tripeptides with alpha-glucosidase inhibitory activity isolated from silk cocoon hydrolysate. J Agric Food Chem 59:11522–11525

31. Kim K, Park S, Yoo HK, Lee JY, Jung HY, Kim DH et al (2009) Brain factor-7 extracted from *Bombyx mori* enhances cognition and attention in normal children. J Med Food 12:643–648
32. Fang EF, Ng TB (2011) Bitter gourd (*Momordica charantia*) is a cornucopia of health: a review of its credited antidiabetic, anti-HIV, and antitumor properties. Curr Mol Med 11:417–436
33. Manosroi A, Boonpisuttinant K, Winitchai S, Manosroi W, Manosroi J (2010) Free radical scavenging and tyrosinase inhibition activity of oils and sericin extracted from Thai native silkworms (*Bombyx mori*). Pharm Biol 48:855–860
34. Yamamoto K, Zhang P, Banno Y, Fujii H, Miake F, Kashige N et al (2005) Superoxide dismutase from the silkworm, *Bombyx mori*: sequence, distribution, and overexpression. Biosci Biotechnol Biochem 69:507–514
35. Stephan AH, Barres BA, Stevens B (2012) The complement system: an unexpected role in synaptic pruning during development and disease. Annu Rev Neurosci 35:369–389
36. Ji KM, Zhan ZK, Chen JJ, Liu ZG (2008) Anaphylactic shock caused by silkworm pupa consumption in China. Allergy 63:1407–1408
37. Liu Z, Xia L, Wu Y, Xia Q, Chen J, Roux KH (2009) Identification and characterization of an arginine kinase as a major allergen from silkworm (*Bombyx mori*) larvae. Int Arch Allergy Immunol 150:8–14
38. Zhao P, Xia Q, Li J, Fujii H, Banno Y, Xiang Z (2007) Purification, characterization and cloning of a chymotrypsin inhibitor (CI-9) from the hemolymph of the silkworm, *Bombyx mori*. Protein J 26:349–357
39. Fang EF, Hassanien AA, Wong JH, Bah CS, Soliman SS, Ng TB (2011) Isolation of a new trypsin inhibitor from the Faba bean (Vicia faba cv. Giza 843) with potential medicinal applications. Protein Pept Lett 18:64–72
40. Liu M, Zhang H, Hu X, Wei Q (2010) Purification and characterization of a chitinase from *Bombyx mori*. Sheng Wu Gong Cheng Xue Bao = Chin J Biotechnol 26:404–409
41. Aramwit P, Damrongsakkul S, Kanokpanont S, Srichana T (2010) Properties and antityrosinase activity of sericin from various extraction methods. Biotechnol Appl Biochem 55:91–98

Author Biography

Dr. Lixin Xia is Associate Professor at the Medical School of Shenzhen University, China. He received his PhD degree from the Chinese University of Hong Kong and postdoctoral training from the Department of Biological Science, Florida State University, USA. He has conducted research on cashew allergen, silkworm allergen and other proteins including lectins, antifungal proteins and ribonucleases.

Chapter 23
Proteins with Anticancer and Antimicrobial Activities from Mammals, Submammalian Vertebrates and Invertebrates

Tzi Bun Ng, Evandro Fei Fang and Jack Ho Wong

Abstract The intent of this chapter is to review proteins with anticancer and antimicrobial activities from various animal species. The proteins comprise venom proteins from snakes, scorpions, and bees; peptides from frogs and toads; bovine pancreas and frogs; and mammalian proteins and peptides including lactoferrin, cathelicidins, tumor necrosis factor, interferons, interleukins, and hormonal peptides. To avoid duplication of efforts, the medicinal applications of mammalian lectins and ribonucleases fall in the realm of other chapters in this book. The current progress and future work on animal constituents with medicinal activity depicted in this chapter would help to shed light on their clinical potentials or applications.

23.1 Introduction

A countless number of people are tortured by debilitating malignant diseases and eventually succumb. Some chemotherapeutic agents elicit untoward side effects. Thus a search for new therapeutics is warranted. The following account discloses the constellation of proteins and peptides from diverse species in the animal kingdom with the potential of combating cancer. Existing drugs for antimicrobial therapy suffer from the drawback of declining efficacy owing to emerging drug

T. B. Ng (✉) · E. F. Fang · J. H. Wong
School of Biomedical Sciences, The Chinese University of Hong Kong, Shatin, Hong Kong
e-mail: b021770@mailserv.cuhk.edu.hk

J. H. Wong
e-mail: jack1993@yahoo.com

E. F. Fang
Laboratory of Molecular Gerontology, National Institute on Aging, National Institutes of Health, Baltimore, MD 21224, USA
e-mail: fange@mail.nih.gov

E. F. Fang and T. B. Ng (eds.), *Antitumor Potential and Other Emerging Medicinal Properties of Natural Compounds*, DOI: 10.1007/978-94-007-6214-5_23,

resistance in pathogens. This necessitates a search for new drugs. Plants constitute a predominant source of drugs and complementary/alternative medicine as demonstrated in this book and elsewhere. In addition, various constituents derived from animal tissues also serve as important medicinal ingredients. For example, a recent list of traditional Chinese medicinal materials is composed of 11,146 herbs, 80 minerals, and 1,581 zoological materials [1]. In this book, two chapters on animal and human lectins written by Prof. Hans-Joachim Gabius, and the third one contributed by Prof. Koji Muramoto have compiled some promising mammalian and marine lectins which are prospective drug candidates. The narrative in this chapter introduces animal proteins with the potential of circumventing the abovementioned health problems including cancer and infectious diseases.

23.2 Medicinal Proteins from Mammals, Submammalian Vertebrates and Invertebrates

The first line of human defence against different microorganisms is the innate immune system with Toll-like receptors as initiators and antimicrobial peptides as key effectors [2]. A large number of antimicrobial peptides/proteins are located in granules of human polymorphonuclear neutrophilic leukocytes, such as bactericidal/permeability increasing protein, azurocidin (CAP37), cathelicidins, and defensins among others [3]. They manifest direct cytotoxicity on cancer cells or activate adaptive immunity and are differentially regulated in cancers such as oral squamous cell carcinomas. Dysregulation of these antimicrobial peptides may lead to cancer development and some immune-deficiency diseases [4]. Here, we focus on cathelicidin, human defensins, and other mammalian bioactive proteins with regard to their bioactivities and mechanisms of action.

23.2.1 Cathelicidins

Cathelicidins are a class of endotoxin-binding and wide-spectrum antimicrobial preproproteins which share a conserved 12 kDa N-terminal sequence [3, 5, 6]. Until now, hCAP-18 is the only human cathelicidin found, and the active peptide (designated as LL-37) resides in the 37 amino acids at the C-terminal liberated by extracellular cleavage with proteinase 3 [3]. Besides manifesting potent antimicrobial activity against a wide range of gram-negative and gram-positive bacteria, LL-37 also exhibited promising gut-protective, anti-fungal, anti-HIV, and antitumor activities.

Cathelicidin-encoding *Lactococcus lactis* manifested protective effects on murine ulcerative colitis. Ulcerative colitis induced in mice by oral consumption of 3 % dextran sulfate sodium for seven days was attenuated by intrarectal administration of murine cathelicidin-related antimicrobial peptide (mCRAMP) produced and secreted from transformed *Lactococcus lactis*. The symptoms were mitigated, and crypt integrity and mucus content were maintained. The number of cells undergoing apoptosis, the activity of myeloperoxidase, the level of malondialdehyde, and fecal microbiota in animals with colitis underwent a decline. In contrast to mCRAMP-encoding *L. lactis*, effective doses of the reference drug sulfasalazine alleviated only the symptoms but failed to lessen mucosal damage in the colon [7].

We recently reported that LL37 and its fragments LL13-37 and LL17-32 were approximately equipotent in hindering growth in the yeast *Candida albicans* [8]. Following exposure to 0.5 and 5 μM LL13-37, the hyphae assumed an increasingly slender appearance, with some abnormal budding and cell death. Only the yeast form and no hyphal form was discernible following incubation in the presence of 50 μM LL13-37. LL13-37 at 5 μM concentration modified membrane permeability of both the yeast and hyphal forms of *C. albicans* as witnessed by localization of the nuclear stain SYTOX Green in both forms. Mycelia exposed to LL13-37 took up SYTOX Green but not MitoTracker deep red, revealing the adverse effect of LL13-37 on the mitochondria. Bimane-labeled LL13-37 gained entry into some of the hyphae, indicative of the impairment of hyphal membrane permeability Reactive oxygen species was present in the yeast form of *C. albicans* cells exposed to LL13-37 but not in the untreated cells. Thus the changes in membrane permeability brought about by LL13-37 might not be the only factor accounting for cell death. It might ensue in uptake of LL13-37, which might have some intracellular targets [8].

In addition to anti-fungal activity, it has also been found that LL-37 and its fragments exhibited mild anti-HIV activity [9]. An in vitro experiment showed that LL37, LL13-37 and LL17-32 inhibited HIV-1 reverse transcriptase with an IC_{50} value of 15, 7, and 70 μM, respectively. However, only 20–30 % inhibition of HIV-1 protease was brought about by the three peptides at 100 μM concentration, through a mechanism which involved protein–protein interaction as disclosed by surface plasmon resonance. Nuclear translocation of HIV-1 integrase labeled with green fluorescent protein, however, was not inhibited. Cytotoxicity toward human peripheral blood mononuclear cells was not apparent [9].

Though the effect of LL-37 on tumor growth is complex, it expressed significant antitumor activity in some circumstances. For example, combination treatment with CpG oligodeoxynucleotides and LL-37 improved antitumor effectiveness, lengthened survival in mice bearing ovarian tumor, and triggered proliferation and activation of peritoneal natural killer cells [10]. In two different xenograft tumor mouse models (B16.F10 and RMA-S), tumor grew more rapidly in cathelicidin-knockout mice ($Camp^{-/-}$) than in wild type controls. Natural killer cells derived from $Camp^{-/-}$ versus wild type mice demonstrated attenuated cytotoxic activity toward tumor. These findings could not be totally accounted for by perforin deficiency in freshly isolated $Camp^{-/-}$ natural killer cells, because the deficiency could be partially made up for by IL-2, whereas cytotoxic activity remained impaired in IL-2-activated $Camp^{-/-}$ natural killer cells. Thus, cathelicidin is important in the antitumor function of natural killer cells [11].

23.2.2 Human Defensins

Mammalian defensins are small cysteine-rich cationic peptides with β-pleated sheets and three intramolecular disulphide bonds [2]. In humans, defensins are categorized into α-, β-, and θ-defensins. Among them, α-defensins comprise human α-defensin-1 (or human neutrophil peptide 1/HNP1 since it is mainly expressed by neutrophils), -2 (HNP2), -3 (HNP3), -4 (HNP4), -5 (HD5), and -6 (HD6). There are 28 human β-defensins and around six of them (HBD-1, 2,…-6) are expressed mainly by epithelial cells, whereas three θ-defensins named RTD-1 (rhesus θ-defensin-1), RTD-2, and RTD-3 have been isolated from rhesus macaques [2, 12]. Some cationic antimicrobial peptides demonstrate promising clinical activities, such as wide-spectrum antimicrobial activity, and cytotoxic effects against a broad range of human cancer cells, including those exhibiting multi-drug-resistance [13].

In 2002, the group headed by Prof. David Ho reported that three human α-defensins (1, 2, and 3) isolated from supernatant fluids of CD8 T lymphocyte cultures manifested anti-HIV-1 activity. These defensins are peptides with a molecular mass of 3371.9, 3442.5, and 3486.5 daltons, respectively, and all belong to the same group on the basis of specific antibody recognition and amino acid sequence [14]. Though 2 years later, they retracted their claim due to the observation that α-defensins could not account for the CAF (CD8 antiviral factor) activity in experimental systems that do not employ allogeneic feeders, the anti-HIV-1 activity of these defensins that they described holds true [15].

Two human peptides, α-defensins HD5 and HD6, are generated by human Paneth cells, and the lack of these defensins is the etiology of a chronic inflammatory bowel disease, Crohn's disease [16, 17]. For example, HD5 prevents infection of non-enveloped viruses, such as human adenoviruses and polymaviruses. Arginine residues play an important role in their antiviral activity against human adenoviruses and polyomaviruses. Arg-28 is crucial for both human adenoviruses and polyomaviruses, whereas Arg-9 is critical for human adenoviruses. Hydrophobicity at residue 29 is a dominant determinant of anti-adenoviral activity. Hence HD5-mediated neutralization of human adenoviruses entails binding to the viral capsid through interactions brought about partly by certain key arginine residues, hydrophobicity at residue 29, and multimerization of HD5, which enhances initial viral binding to the cell but inhibits subsequent viral uncoating and genome delivery to the nucleus [16]. For HD6, though it has little bactericidal activity, a recent report in Science showed that it promoted mucosal innate immunity through self-associated peptide nanonets [17]. It manifests a protective action against invasion of enteric bacteria in vitro and in vivo. After stochastic binding to bacterial surface proteins, it forms fibrils and nanonets to envelop and entangle bacteria [17].

Two human β-defensins, HBD2 and HBD3, show substantial sequence identity and structural similarity. Despite their similar activities against *E. coli*, HBD3 kills *Staphylococcus aureus* with a potency several folds higher than that of HBD2. One of the HBD2/HBD3-chimeric molecules (chimera C3) kills both *E. coli* and

S. aureus with a potency even higher than HBD2/HBD3 [18]. On the other hand, θ-defensins have inhibitory activity against HIV-1, HSV-2, and influenza virus [2]. For more details concerning the medicinal activities of these human defensins, please refer to a review by Klotman et al. [2].

23.2.3 Other Mammalian Defensins

In 2000, Johnstone and colleagues showed that short mammalian-derived extended-helical cationic peptides and insect-derived alpha-helical peptides lysed cancer cells with a potency of 2–50 times higher than that against non-malignant cells. They manifested similar cytotoxic activity against sensitive and multidrug-resistant cells and were capable of potentiating the effect of doxorubicin against multidrug-resistant tumor cells in vitro [19]. Another study further confirmed the tumor specificity of other mammalian peptides. Four enantiomeric 9-mer peptides designated as d-peptide A, B, C and D were synthesized on the basis of two 43-mer beetle defensins. The d-9-mer peptides displayed bacterial membranolytic activity similar to the parent peptides. d-Peptide B demonstrated the highest selective cytotoxicity against P3-X63-Ag8.653 mouse myeloma cells, and disrupted the cell membrane. On the contrary, normal leukocytes remained intact. The d-9-mer peptides exerted negative charge-dependent selective cytotoxicity targeting cancer cell membrane phosphatidylserine. A combination of d-peptide B with dexamethasone synergistically inhibited the growth of mouse myeloma [20]. A recent report unveiled that a β-defensin-1-like antimicrobial peptide isolated from tree shrew (*Tupaia belangeri*) serum inhibited a variety of Gram-negative and -positive bacteria and fungi but exhibited very little hemolytic activity [21].

23.2.4 Lactoferrin

Named as a 'biological drug', lactoferrin is an 80-kDa bilobed glycoprotein located in milk and other bodily fluids, and in the secretory granules of neutrophils [22]. It manifests antiviral, anti-bacterial, anti-fungal activities, and recently data pertaining to its anticancer activity in various cancer cell lines, animal models and clinical trials have become available [22–25]. For example, lactoferrin acts by induction of apoptosis, inhibition of angiogenesis, and modulation of carcinogen-metabolizing enzymes and perhaps plays the role of an iron scavenger. Supplementing cows with selenium increases the content of selenoproteins in milk, which on isolation inhibited colon tumorigenesis in the rat [25]. Xu et al. showed that bovine lactoferrin brought about apoptosis of SGC-7910 human stomach cancer cells by inhibiting Akt activation and modulating its downstream protein phosphorylation [26].

The recombinant form of human lactoferrin also exhibits the same antitumor activity as the natural form. Varadhachary et al. showed that recombinant human lactoferrin suppressed the growth of squamous cell carcinoma tumors in T cell-immunocompromised nu/nu mice and in syngeneic, immunocompetent BALB/c mice. In 2008, Jonasch et al. conducted a phase 2 trial of talactoferrin (recombinant lactoferrin) on previously treated patients with metastatic renal cell carcinoma. The results demonstrated that talactoferrin was devoid of toxicity and this unveiled prospects of clinical application of talactoferrin [27].

In addition to antitumor activity, there is also progress on the anti-viral and anti-bacterial activities of both lactoferrin and its peptide fragments. The C-lobe of bovine lactoferrin is responsible for inhibition of hemagglutination and cell infection of influenza virus. A number of major virus subtypes, comprising H1N1 and H3N2, were inhibited. The C-lobe bound strongly to the HA_2 region of viral hemagglutinin, which is the highly conserved region containing the fusion peptide. By employing molecular docking, three C-lobe fragments were identified which inhibited viral hemagglutination and infection at fentomolar concentrations [28]. The peptide L10 (WFRKQLKW) with minimal hemolytic activity, designed by modifying the first eight N-terminal residues of bovine lactoferrin, demonstrated inhibitory activity toward clinical isolates of extended-spectrum beta-lactamase producing Gram-negative bacteria and Candida strains. The peptide targeted the membrane of Gram-negative bacteria as evidenced by interaction with lipopolysaccharide and lipid A. It also induced membrane permeabilization of Candida cells. The peptide produced its anti-inflammatory effects by inhibiting cyclooxygenase-2 [29]. Furthermore, lactoferricin and lactoferrampin are two peptides in the N-terminal lobe of bovine lactoferrin with inhibitory activity against *Candida albicans* and a variety of Gram-positive and Gram-negative bacteria. A heterodimer composed of lactoferrampin linked to a lactoferricin fragment at their C-termini through the two amino groups of a single Lys residue, designated as LFchimera, manifested more potent activity than lactoferricin, lactoferrampin, and an equimolar mixture of the two peptides [30].

23.2.5 Cytokines: Interferons, Tumor Necrosis Factor and Interleukins

Cytokines are secreted or membrane-bound proteins that are implicated in the regulation of the growth, differentiation and activation of immune cells, and participate in different aspects of the functioning of the human immune system [31]. In response to a diverse range of cellular stresses, cytokines are released by the human body to stimulate a host response (such as the innate and adaptive immune systems) to fight pathogen invasion, thwart tumor development and progression, among others [31, 32]. Defects in cytokine signaling pathways or imbalance of the production of cytokines contribute to the etiology of different immunodeficiency diseases, and cancer pathogenesis [31, 33]. Representative cytokines are type-I (such as IL-2, IL-3, IL-4, IL-5, IL-6, IL-7, IL-9, IL-11, IL-15, IL-21, TSLP, GM-CSF, etc.), and type-II cytokines (IL-10, IL-20, IF-22, and IFN-α, INF-β, and INF-γ, etc.) [33].

Interferons (INFs) from fibroblasts, leukocytes and other cell types are referred to as type-1 INFs whereas immune interferon is named type-2 INF. Type-1 INFs comprise INF-alpha and INF-beta and their anti-proliferative activity won the approval of US Food and Drug Administration for treatment of AIDs related Kaposi sarcoma, chronic myelogenous leukemia (CML), malignant melanoma (MM), and hairy cell leukemia [34, 35]. For referring to the historical perspectives of INFs and their anti-proliferative properties, a recent review is recommended [35].

Tumor necrosis factor-α (TNF-α) is a 51-kDa homotrimer, first isolated from serum samples of animals challenged with cell wall lipopolysaccharide (LPS) of gram-negative bacteria. It inhibited the growth of cultured human melanoma cells [36] and caused certain tumors to undergo necrosis when administered to tumor-bearing animals. Owing to the painstaking effort of many investigators on the protein for about four decades, TNF-α is now applied in the therapy of some cancers, such as locally advanced soft tissue sarcomas, metastatic melanomas, and other irresectable tumors [37]. Future investigations on TNF-α will hopefully lead to clinical applications.

Interleukins (ILs) are a family of cytokines that were first discovered in leukocytes and until now over 40 ILs and IL-related cytokines have been elucidated. A recent review by Brocker et al. described the nomenclature and classification of ILs [38]. They have been confirmed to participate in complex immune-modulatory activities in human bodies. Though many ILs contain antitumor activities in different in vitro and in vivo settings, LCs are also participate in cancer pathogenesis. It is recommended that future researches should not only focus on the discovery of cytokine-inducing compounds with antitumor activity, but also focus on the understanding of cytokines-tumor-cell interactions [31], and a combination of the two may help to open a window for improving cancer immunotherapy.

23.2.6 Peptide Hormones of the Hypothalamus and Gastrointestinal Tract

Some of the peptides produced by mammals serve an endocrine function. Though some peptide hormones are closed linked with tumor progression and so can serve as tumor biomarkers [39], other peptide hormones manifest antitumor activity [40]. There are some examples. First, Abarelix, a gonadotropin releasing hormone antagonist, has been indicated for prostate cancer therapy [41]. Second, growth hormone releasing hormone (GHRH) receptor antagonists inhibit growth and metastasis of a large number of experimental tumors expressing the pituitary GHRH receptor (pGHRH-R). A positive correlation exists between the endocrine potency and the antiproliferative efficacy of GHRH antagonists in tumors strongly expressing pGHRH-R [42]. Third, the substance P analog (D-Arg(1), D

Trp (s,7,9), Leu (11)) substance P suppressed bombesin-induced mitogenic transduction mediated by both G(q) and G (12) in Swiss 3T3 cells. Other substance P analogs are also broad-spectrum neuropeptide antagonists and potentially useful as anticancer agents [43]. Fourth, bombesin analogs with alpha-aminoisobutyric acid replacing Ala 9 or Gly 11 inhibited proliferation in human colon cancer (HT29, PTC, SW620) and pancreatic (MiaPaca-1) cells. They inhibited tumor growth in athymic nude mice bearing xenografts of colon cancer [44]. Furthermore, Jaggi et al. demonstrated that a combination of synthetic analogs of bombesin, substance P, somatostatin, and vasoactive intestinal peptide exerted antiproliferative activity against human cancer cells of the colon, duodenum and pancreas. The combination also exhibited an anticancer effect on xenograft models of colon and duodenal cancer with no acute or chronic toxicity [45]. In addition, a report showed that endothelin B receptor antagonists inhibited proliferation and triggered apoptosis in glioma cells [46].

23.2.7 Venom Proteins

In a traditional view, venoms are toxic fluids that inflict sudden death or paralysis in the host/prey. Based on the modern classification of this type of toxins, they have been defined as secretions delivered through a wound that interferes with normal physiological processes to facilitate feeding or defense by the venom-producing animal [47]. Recent studies unveil the medicinal applications of different venoms and their active components [48].

Snakes such as southern copperhead (*Agkistrodon contortrix*), India viper *(Dabia russelli russselli*), Indian black scorpion (*Heterometrus bengalensis*), and Taiwan cobra (*Naja atra atra*) are the main source of venoms. A homodimeric disintegrin from venom of the southern copperhead, known as contortrostatin, inhibited adhesion of human epithelial cancer cells of the ovary to several extracellular matrix proteins and suppressed cancer cell invasion through an artificial basement membrane. It also prevented dissemination of ovarian cancer in nude mice and recruitment of new blood vessels to tumor cells at secondary sites [49]. A 7.2-kDa toxin from venom of India viper inhibited proliferation of human leukemia U937 and K562 cells by induction of apoptosis [50]. The 72-kDa protein designated as bengaline from venom of Indian black scorpion inhibited proliferation of human leukemic cells U937 (histiocytic lymphoma) and K562 (chronic myelogenous leukemia) also through induction of apoptosis [51].

A polypeptide with 60 amino acid residues designated as Cardiotoxin III (CTX III) was isolated from the venom of Taiwan cobra. The antitumor activity of CTX-III against different tumor cell lines has been investigated. It exerted anti-proliferative activity toward K562 cells by induction of apoptosis [52]. The cytotoxicity of CTX-III towards human colorectal cancer (Colo205) cells was attributed to induction of mitochondrial- and caspase-dependent apoptosis [53]. It inhibited HL-60 human leukaemia cells by activation of c-jun N-terminal kinase-dependent apoptosis [54]. The antitumor activity of CTX-III was also extended to two breast cancer cell lines, including MDA-MB-231 breast cancer cells through blocking of the EGF/EGFR-mediated signaling pathway [55], and MCF-7 cells by activation of the caspase-9-regulated apoptotic pathway and inhibition of the NF-κB pathway [56]. Based on the results, it seems that CTX-III could induce apoptosis in different tumor cells. However, the following additional information would make the picture more complete: (a) whether CTX-III exerts any toxicity on normal cells (b) whether the cytotoxicity of CTX-III is also attributed to the induction of necrosis and/or autophagy, and (c) whether CTX-III displays an antitumor action in vivo.

23.2.8 Amphibian Peptides

Amphibians are a good source for drug discovery. In addition to being a source of medicinal ribonucleases (such as onconase) [57], some types of frogs and toads also produce bioactive peptides. Examples are the Australian Green and Golden Bell frog (*Litoria aurea*) and the Southern Bell Frog (*L. raniformis*).The former species produces 17 and the latter 16 aurein peptides, 10 of which are common to both species. Aurein peptides with 13 amino acid residues display anticancer as well as antibiotic activities [58, 59]. Furthermore, the gastric tissue of the Asian toad (*Bufo bufo gangarizars*) produces peptides known as buforins with anticancer and antimicrobial activities,. Buforin II has 22 amino acid residues and is derived from buforin I which has 39. The two peptides exhibit sequence identity with the N-terminal region of histone H2A which directly interacts with nuclei acids [60]. Pentadactylin is a peptide with anticancer and antimicrobial activities isolated from skin secretion of the frog *Leptodactylus labyrinthicus.* Treatment of mouse melanoma (B16F10) cells with pentadactylin brought about a dose-dependent reduction of cell viability. The cells underwent morphological changes characteristic of apoptosis [61].

23.3 Conclusion and Future Perspectives

The foregoing account discloses that a variety of animal proteins with different molecular masses and structures are capable of inhibiting proliferation of cancer cells and cancer growth. These proteins range from insect lectins, toxins in insect, scorpion and snake venoms, frog lectins, ribonucleases, and antimicrobial peptides; mammalian cytokines, hormones, hormone antagonists, antimicrobial peptides, and milk proteins. Among these proteins and peptides, the mammalian cytokines, hormones, and hormone antagonists, and frog onconase, have been used clinically with some success. Peptide hormone analogs have been employed to target peptide hormone receptors and tumor blood vessels for cancer therapy. Cytotoxic peptide hormone analogs containing an anticancer compound such as doxorubicin can be used. The use of a combination of peptides has also been reported.

Some of the aforementioned proteins are also endowed with antimicrobial activity and represent candidates for development into new agents for combatting emerging pathogen resistance to existing antimicrobial drugs. With continued research, hopefully novel anticancer and antimicrobial therapeutic agents and technology will be developed to save the lives of cancer patients and immune-compromised patients highly susceptible to infections.

Acknowledgments The award of research grants (Research Funds for the Control of Infectious Diseases project number 10090812, 12110672, and 12110282) from Food and Health Bureau, Government of Hong Kong, is gratefully acknowledged.

References

1. Sheng-Ji P (2001) Ethnobotanical approaches of traditional medicine studies: some experiences from Asia. Pharm Biol 39(Suppl 1):74–79
2. Klotman ME, Chang TL (2006) Defensins in innate antiviral immunity. Nat Rev Immunol 6:447–456
3. Sorensen OE, Follin P, Johnsen AH, Calafat J, Tjabringa GS, Hiemstra PS et al (2001) Human cathelicidin, hCAP-18, is processed to the antimicrobial peptide LL-37 by extracellular cleavage with proteinase 3. Blood 97:3951–3959
4. Meyer JE, Harder J (2007) Antimicrobial peptides in oral cancer. Curr Pharm Des 13:3119–3130
5. Coffelt SB, Marini FC, Watson K, Zwezdaryk KJ, Dembinski JL, LaMarca HL et al (2009) The pro-inflammatory peptide LL-37 promotes ovarian tumor progression through recruitment of multipotent mesenchymal stromal cells. Proc Natl Acad Sci U S A 106:3806–3811
6. Zanetti M, Gennaro R, Romeo D (1995) Cathelicidins: a novel protein family with a common proregion and a variable C-terminal antimicrobial domain. FEBS Lett 374:1–5
7. Wong CC, Zhang L, Li ZJ, Wu WK, Ren SX, Chen YC et al (2012) Protective effects of cathelicidin-encoding Lactococcus lactis in murine ulcerative colitis. J Gastroenterol Hepatol 27:1205–1212
8. Wong JH, Ng TB, Legowska A, Rolka K, Hui M, Cho CH (2011) Antifungal action of human cathelicidin fragment (LL13-37) on *Candida albicans*. Peptides 32:1996–2002
9. Wong JH, Legowska A, Rolka K, Ng TB, Hui M, Cho CH et al (2011) Effects of cathelicidin and its fragments on three key enzymes of HIV-1. Peptides 32:1117–1122
10. Chuang CM, Monie A, Wu A, Mao CP, Hung CF (2009) Treatment with LL-37 peptide enhances antitumor effects induced by CpG oligodeoxynucleotides against ovarian cancer. Hum Gene Ther 20:303–313
11. Buchau AS, Morizane S, Trowbridge J, Schauber J, Kotol P, Bui JD et al (2010) The host defense peptide cathelicidin is required for NK cell-mediated suppression of tumor growth. J Immunol 184:369–378
12. Tran D, Tran PA, Tang YQ, Yuan J, Cole T, Selsted ME (2002) Homodimeric theta-defensins from rhesus macaque leukocytes: isolation, synthesis, antimicrobial activities, and bacterial binding properties of the cyclic peptides. J Biol Chem 277:3079–3084
13. Mader JS, Hoskin DW (2006) Cationic antimicrobial peptides as novel cytotoxic agents for cancer treatment. Expert Opin Investig Drugs 15:933–946
14. Zhang L, Yu W, He T, Yu J, Caffrey RE, Dalmasso EA et al (2002) Contribution of human alpha-defensin 1, 2, and 3 to the anti-HIV-1 activity of CD8 antiviral factor. Science 298:995–1000

15. Zhang L, Lopez P, He T, Yu W, Ho DD (2004) Retraction of an interpretation. Science 303:467
16. Gounder AP, Wiens ME, Wilson SS, Lu W, Smith JG (2012) Critical determinants of human alpha-defensin 5 activity against non-enveloped viruses. J Biol Chem 287:24554–24562
17. Chu H, Pazgier M, Jung G, Nuccio SP, Castillo PA, de Jong MF et al (2012) Human alpha-defensin 6 promotes mucosal innate immunity through self-assembled peptide nanonets. Science 337:477–481
18. Spudy B, Sonnichsen FD, Waetzig GH, Grotzinger J, Jung S (2012) Identification of structural traits that increase the antimicrobial activity of a chimeric peptide of human beta-defensins 2 and 3. Biochem Biophys Res Commun 427:207–211
19. Johnstone SA, Gelmon K, Mayer LD, Hancock RE, Bally MB (2000) In vitro characterization of the anticancer activity of membrane-active cationic peptides. I. Peptide-mediated cytotoxicity and peptide-enhanced cytotoxic activity of doxorubicin against wild-type and p-glycoprotein over-expressing tumor cell lines. Anticancer Drug Des 15:151–160
20. Iwasaki T, Ishibashi J, Tanaka H, Sato M, Asaoka A, Taylor D et al (2009) Selective cancer cell cytotoxicity of enantiomeric 9-mer peptides derived from beetle defensins depends on negatively charged phosphatidylserine on the cell surface. Peptides 30:660–668
21. Yang X, Dong L, Lv L, Yan X, Lai R, Liu R (2012) A beta-defensin 1-like antimicrobial peptide from the tree shrew. *Tupaia belangeri*. Gene 509:258–262
22. Vogel HJ (2012) Lactoferrin, a bird's eye view. Biochemistry and cell biology. Biochimie et biologie cellulaire 90:233–244
23. Mohan KV, Gunasekaran P, Varalakshmi E, Hara Y, Nagini S (2007) In vitro evaluation of the anticancer effect of lactoferrin and tea polyphenol combination on oral carcinoma cells. Cell Biol Int 31:599–608
24. Varadhachary A, Wolf JS, Petrak K, O'Malley BW Jr, Spadaro M, Curcio C et al (2004) Oral lactoferrin inhibits growth of established tumors and potentiates conventional chemotherapy. Int J Cancer. Journal International du Cancer 111:398–403
25. Parodi PW (2007) A role for milk proteins and their peptides in cancer prevention. Curr Pharm Des 13:813–828
26. Xu XX, Jiang HR, Li HB, Zhang TN, Zhou Q, Liu N (2010) Apoptosis of stomach cancer cell SGC-7901 and regulation of Akt signaling way induced by bovine lactoferrin. J Dairy Sci 93:2344–2350
27. Jonasch E, Stadler WM, Bukowski RM, Hayes TG, Varadhachary A, Malik R et al (2008) Phase 2 trial of talactoferrin in previously treated patients with metastatic renal cell carcinoma. Cancer 113:72–77
28. Ammendolia MG, Agamennone M, Pietrantoni A, Lannutti F, Siciliano RA, De Giulio B et al (2012) Bovine lactoferrin-derived peptides as novel broad-spectrum inhibitors of influenza virus. Pathog Glob Health 106:12–19
29. Mishra B, Leishangthem GD, Gill K, Singh AK, Das S, Singh K, Xess I, Dinda A, Kapil A, Patro IK, Dey S (2013) A novel antimicrobial peptide derived from modified N-terminal domain of bovine lactoferrin: design, synthesis, activity against multidrug-resistant bacteria and Candida. Biochim Biophys Acta 1828:677–686
30. Haney EF, Nazmi K, Bolscher JG, Vogel HJ (2012) Influence of specific amino acid side-chains on the antimicrobial activity and structure of bovine lactoferrampin. Biochemistry and cell biology (Biochimie et Biologie Cellulaire) 90:362–377
31. Dranoff G (2004) Cytokines in cancer pathogenesis and cancer therapy. Nat Rev Cancer 4:11–22
32. Fang EF, Wong JH, Ng TB (2010) Thermostable Kunitz trypsin inhibitor with cytokine inducing, antitumor and HIV-1 reverse transcriptase inhibitory activities from Korean large black soybeans. J Biosci Bioeng 109:211–217
33. Leonard WJ (2001) Cytokines and immunodeficiency diseases. Nat Rev Immunol 1:200–208
34. Safai B, Bason M, Friedman-Birnbaum R, Nisce L (1990) Interferon in the treatment of AIDS-associated Kaposi's sarcoma: the American experience. J Invest Dermatol 95:166S–169S

35. Bekisz J, Baron S, Balinsky C, Morrow A, Zoon KC (2010) Anti-proliferative properties of Type I and Type II interferon. Pharmaceuticals 3:994–1015
36. Helson L, Green S, Carswell E, Old LJ (1975) Effect of tumour necrosis factor on cultured human melanoma cells. Nature 258:731–732
37. van Horssen R, Ten Hagen TL, Eggermont AM (2006) TNF-alpha in cancer treatment: molecular insights, antitumor effects, and clinical utility. Oncologist 11:397–408
38. Brocker C, Thompson D, Matsumoto A, Nebert DW, Vasiliou V (2010) Evolutionary divergence and functions of the human interleukin (IL) gene family. Human Genomics 5:30–55
39. Mezo G, Manea M (2010) Receptor-mediated tumor targeting based on peptide hormones. Expert Opin Drug Deliv 7:79–96
40. Saba SR, Vesely DL (2006) Cardiac natriuretic peptides: hormones with anticancer effects that localize to nucleus, cytoplasm, endothelium, and fibroblasts of human cancers. Histol Histopathol 21:775–783
41. Debruyne F, Bhat G, Garnick MB (2006) Abarelix for injectable suspension: first-in-class gonadotropin-releasing hormone antagonist for prostate cancer. Future Oncology 2:677–696
42. Kovacs M, Schally AV, Hohla F, Rick FG, Pozsgai E, Szalontay L et al (2010) A correlation of endocrine and anticancer effects of some antagonists of GHRH. Peptides 31:1839–1846
43. Sinnett-Smith J, Santiskulvong C, Duque J, Rozengurt E (2000) [D-Arg (1), D-Trp (5,7,9), Leu (11)]Substance P inhibits bombesin-induced mitogenic signal transduction mediated by both G(q) and G(12) in Swiss 3T3cells. J Biol Chem 275:30644–30652
44. Prasad S, Mathur A, Gupta N, Jaggi M, Singh AT, Rajendran P et al (2007) Bombesin analogs containing alpha-amino-isobutyric acid with potent anticancer activity. J Pept Sci (an official Publication of the European Peptide Society) 13:54–62
45. Jaggi M, Prasad S, Singh AT, Praveen R, Dutt S, Mathur A et al (2008) Anticancer activity of a peptide combination in gastrointestinal cancers targeting multiple neuropeptide receptors. Invest New Drugs 26:489–504
46. Paolillo M, Russo MA, Curti D, Lanni C, Schinelli S (2010) Endothelin B receptor antagonists block proliferation and induce apoptosis in glioma cells. Pharmacol Res (the official journal of Italian Pharmacological Society) 61:306–315
47. Asgari S, Rivers DB (2011) Venom proteins from endoparasitoid wasps and their role in host-parasite interactions. Annu Rev Entomol 56:313–335
48. Gomes A, Bhattacharjee P, Mishra R, Biswas AK, Dasgupta SC, Giri B (2010) Anticancer potential of animal venoms and toxins. Indian J Exp Biol 48:93–103
49. Markland FS, Shieh K, Zhou Q, Golubkov V, Sherwin RP, Richters V et al (2001) A novel snake venom disintegrin that inhibits human ovarian cancer dissemination and angiogenesis in an orthotopic nude mouse model. Haemostasis 31:183–191
50. Gomes A, Choudhury SR, Saha A, Mishra R, Giri B, Biswas AK et al (2007) A heat stable protein toxin (drCT-I) from the Indian Viper (*Daboia russelli russelli*) venom having antiproliferative, cytotoxic and apoptotic activities. Toxicon (official journal of International Society on Toxinology) 49:46–56
51. Das Gupta S, Debnath A, Saha A, Giri B, Tripathi G, Vedasiromoni JR et al (2007) Indian black scorpion (*Heterometrus bengalensis* Koch) venom induced antiproliferative and apoptogenic activity against human leukemic cell lines U937 and K562. Leuk Res 31:817–825
52. Yang SH, Lu MC, Chien CM, Tsai CH, Lu YJ, Hour TC et al (2005) Induction of apoptosis in human leukemia K562 cells by cardiotoxin III. Life Sci 76:2513–2522
53. Tsai CH, Yang SH, Chien CM, Lu MC, Lo CS, Lin YH et al (2006) Mechanisms of cardiotoxin III-induced apoptosis in human colorectal cancer colo205 cells. Clin Exp Pharmacol Physiol 33:177–182
54. Chien CM, Yang SH, Yang CC, Chang LS, Lin SR (2008) Cardiotoxin III induces c-jun N-terminal kinase-dependent apoptosis in HL-60 human leukaemia cells. Cell Biochem Funct 26:111–118
55. Tsai PC, Hsieh CY, Chiu CC, Wang CK, Chang LS, Lin SR (2012) Cardiotoxin III suppresses MDA-MB-231 cell metastasis through the inhibition of EGF/EGFR-mediated

signaling pathway. Toxicon (official journal of International Society on Toxinology) 60:734–743
56. Chiu CC, Lin KL, Chien CM, Chang LS, Lin SR (2009) Effects of cardiotoxin III on NF-kappaB function, proliferation, and apoptosis in human breast MCF-7 cancer cells. Oncol Res 17:311–321
57. Fang EF, Ng TB (1815) Ribonucleases of different origins with a wide spectrum of medicinal applications. Biochim Biophys Acta 2011:65–74
58. Rozek T, Bowie JH, Wallace JC, Tyler MJ (2000) The antibiotic and anticancer active aurein peptides from the Australian Bell Frogs *Litoria aurea* and *Litoria raniformis*. Part 2. Sequence determination using electrospray mass spectrometry. Rapid Commun Mass Spectrom 14:2002–2011
59. Rozek T, Wegener KL, Bowie JH, Olver IN, Carver JA, Wallace JC et al (2000) The antibiotic and anticancer active aurein peptides from the Australian Bell Frogs *Litoria aurea* and *Litoria raniformis* the solution structure of aurein 1.2. Eur J Biochem/FEBS 267:5330–5341
60. Cho JH, Sung BH, Kim SC (2009) Buforins: histone H2A-derived antimicrobial peptides from toad stomach. Biochim Biophys Acta 1788:1564–1569
61. Liberio MS, Joanitti GA, Azevedo RB, Cilli EM, Zanotta LC, Nascimento AC et al (2011) Anti-proliferative and cytotoxic activity of pentadactylin isolated from *Leptodactylus labyrinthicus* on melanoma cells. Amino Acids 40:51–59

Chapter 24
Defense Proteins with Antiproliferative and Antimicrobial Activities from Fungi and Bacteria

Jack Ho Wong, Tzi Bun Ng, Evandro Fei Fang and He-Xiang Wang

Abstract Defensins are a group of proteins characterized by a conservedcysteine-stabilized alpha-helix and beta-sheet structural motif. They are widely distributed among vertebrates and invertebrates. In addition to the presence of defensins in mammals and submammalian vertebrates enumerated in Chap. 23, fungi and bacteria also produce a variety of defensins with antiproliferative activity toward tumor cells and anticancer activity in tumor bearing mice. The fungal proteins include ribonucleases, antifungal proteins, ubiquitin-like peptides, ribosome inactivating proteins, antibacterial proteins, hemolysins, hemagglutinins/lectins, laccases, and proteases. The bacterial proteins comprise ribonucleases and antifungal proteins. Some of the aforementioned proteins also display antimicrobial activities toward pathogenic microbes.

24.1 Introduction

Some fungal products, like the cholesterol-lowering hydroxymethylglutaryl Co A reductase inhibitor lovastatin and the antibiotic penicillin, are from the fungi *Aspergillus terreus* and *Penicillium* species. The intent of the present article is to

J. H. Wong (✉) · T. B. Ng · E. F. Fang
School of Biomedical Sciences, Faculty of Medicine, The Chinese University of Hong Kong, Shatin New Territories, Hong Kong
e-mail: jack1993@yahoo.com

T. B. Ng
e-mail: b021770@mailserv.cuhk.edu.hk

E. F. Fang
Laboratory of Molecular Gerontology, National Institute on Aging, National Institutes of Health, Baltimore, MD, USA
e-mail: fange@mail.nih.gov

H.-X. Wang
State Key Laboratory for Agrobiotechnology and Department of Microbiology, China Agricultural University, Beijing 100193, China
e-mail: hxwang@cau.edu.cn

E. F. Fang and T. B. Ng (eds.), *Antitumor Potential and Other Emerging Medicinal Properties of Natural Compounds*, DOI: 10.1007/978-94-007-6214-5_24,

review fungal and bacterial proteins capable of inhibiting cancer cell proliferation and cancer growth and microbial proliferation. These proteins may find application in the treatment of cancer and infectious diseases.

24.2 Fungal Antibacterial Defensins

24.2.1 Plectasin

Animals and plants produce cysteine-rich peptides with antimicrobial activity called defensins. Defensins are characterized by a conserved cysteine-stabilized alpha-helix and beta-sheet (CSalphabeta) structural motif. Zhu et al. reported the computational identification of six families of fungal defensin-like peptides in which three known types of defensin comprising antibacterial ancient invertebrate-type defensins, antibacterial classical insect-type defensins, and antifungal plant/insect-type defensins can be clearly assigned [1].

Plectasin is the first fungal defensin (4.4-kDa) with therapeutic potential isolated from the saprophytic ascomycete *Pseudoplectania nigrella*. Its primary, secondary and tertiary structures are similar to those of the defensins of arthropods including dragonflies, scorpions, spiders, and mollusks including mussels [2]. Plectasin was comparable in efficacy to vancomycin and penicillin for treating pneumonia induced by *S. pneumoniae* in mice. It demonstrated very low toxicity in mice [2]. Recombinant plectasin also exhibited compatible antimicrobial activity in its native form as other commonly used antibiotics [3, 4]. In the investigation of Jing et al. [4], the coding sequence of plectasin was optimized, cloned into pET32a$^+$ vector and then expressed as a thioredoxin fusion protein in *Escherichia coli*. The soluble fusion protein in the supernatant of the cell lysate was purified using Ni^{2+}-chelating affinity chromatography, cleaved by Factor Xa protease to liberate mature plectasin, and then rechromatographed. At a concentration of 2.56 mg/ml, plectasin manifested activity *against Staphylococcus aureus*, *S. epidermidis*, *S. suis*, and *S. pneumoniae*, similar to those of 0.32 mg/ml vancomycin, 0.64 mg/ml penicillin, 0.32 mg/ml vancomycin and 0.16 mg/ml vancomycin, respectively. Its anti-*S. aureus* activity was resistant to pepsin and papain, heat stable as well as pH-stable [3]. Plectasin bound the bacterial cell-wall precursor Lipid II leading to the formation of an equimolar stoichiometric complex. Amino acid residues in plectasin important to complex formation were identified with nuclear magnetic resonance spectroscopy and computational modeling [5]. The heme sensing response regulator HssR in *S. aureus*, but not its homolog RR23 from *Listeria monocytogenes,* confers susceptibility to plectasin and the plectasin-like defensin eurocin. This explains why *S. aureus* was much more susceptibe than *L. monocytogenes* to plectasin [6]. Synthetic L-plectasin showed antimicrobial activity, while synthetic D-plectasin was devoid of such activity. Racemic crystallography was employed to determine the X-ray structure of plectasin [7].

Additionally, recombinant plectasin and its variant NZ2114 were potent against *Streptococcus pneumonia* and *Staphylococcus aureus*, including drug-resistant strains. By using an experimental meningitis model, plectasin penetration into the cerebrospinal fluid (CSF) of infected and uninfected rabbits and the bactericidal activities in CSF of the plectasin variant NZ2114 and ceftriaxone against a penicillin-resistant *S. pneumoniae* strain were studied. NZ2114 showed a significantly greater extent of CSF penetration through inflamed than through noninflamed rabbit meninges. NZ2114 brought about a significantly higher reduction in CSF concentrations of a penicillin-resistant *S. pneumoniae* strain than ceftriaxone. A significantly larger number of animals possessed sterile CSF at 5 and 10 h after administration of NZ2114 than following ceftriaxone treatment [8]. In an investigation, Xiong and colleagues compared the effectiveness of NZ2114 with those of vancomycin and daptomycin, in an experimental rabbit model of infective endocarditis induced by a methicillin-resistant *Staphylococcus aureus* (MRSA) strain (ATCC 33591) [9]. All three doses of NZ2114 employed (5, 10, and 20 mg/kg body weight, twice daily intravenous administration for 3 days) significantly lowered MRSA densities in cardiac vegetations, spleen, and kidneys compared to untreated controls, except in splenic MRSA counts in the group receiving the 5 mg/kg dose. The efficacy of NZ2114 was clearly dose-dependent in all target tissues. At 20 mg/kg, the efficacy of NZ2114 resembled that of daptomycin but was higher than that of vancomycin. Bacterial counts in cardiac vegetations, kidneys, and spleen rose in post -therapy relapse in vancomycin- and daptomycin-treated animals but not in NZ2114 (10- and 20-mg/kg) treated groups [9].

In addition to a wide-range of anti-microbial activities, the application of plectasin also has some other advantages. Plectasin showed no cytotoxicity to A549 cells, normal human bronchial epithelial cells, or lung fibroblasts, and it did not induce IL-8 transcription or production in A549 cells [10]. On the other hand, it was demonstrated using an agar diffusion bioassay that serum samples from plectasin-immunized animals did not reduce the efficacy of the peptide. Hyperimmune sera from animals immunized with plectasin in Freund's incomplete adjuvant lowered the effectiveness of plectasin at the lowest concentration tested. Experiments adopting the mouse peritonitis model revealed that plectasin was highly effective in treating *S. pneumoniae* infections, both in naive animals and in animals having anti-drug antibodies. No animals developed hypersensitivity or injection site reactions toward plectasin. Its half-life was the same in mice with and without antibodies [11].

24.2.2 Mushroom Antifungal Proteins

We have purified a number of bioactive proteins from both edible and toxic mushrooms, such as agrocybin from *Agrocybe cylindracea* [12], and cordymin from the medicinal fungus *Cordyceps militaris*. They share some similarities in their antifungal and anti-HIV reverse transcriptase (RT) activities. For example, the 9-kDa agrocybin exhibited anti-HIV RT activity at an IC_{50} of 60 μM, while the IC_{50} for cordymin (10906 Da) was 55 μM [13]. Furthermore, cordymin exhibited antiproliferative activity toward MCF-7 breast cancer cells but was inactive toward HT-29 colon cancer cells [13].

24.2.3 Mushroom Ribosome Inactivating Proteins

Ribosome inactivating proteins (RIPs) are toxic RNA N-glycosidases that cleave an adenine-ribose glycosidic bond at adenine (position 4324) with the conserved ricin/alpha-sarcin loop in eukaryotic 28S ribosomal RNA. This type of proteins has attracted considerable attention due to their diverse and potent defensive activities, and inter alia, their antitumor and anti-HIV activities [14]. In part III of this book, there are three chapters that are focused on the bioactive RIPs mainly from plants. Here, we extend the origins of RIPS to mushrooms, such as hypsin, calcaelin, and marmorin. Hypsin, an RIP from *Hypsizygus marmoreus,* exhibited antiproliferative activity against mouse leukemia cells and human leukemia and hepatoma cells [15]. Fresh fruiting bodies of the mushroom *Calvatia caelata*, commonly known as the mosaic puffball, produced a 39-kDa protein, designated as calcaelin, with two subunits possessing a molecular mass of 19 kDa and 20 kDa, respectively. It inhibited translation in rabbit reticulocyte lysate with an IC_{50} value of 4 nM. It exhibited an antimitogenic activity toward mouse splenocytes, and curtailed the viability of breast cancer cells [16]. The RIP marmorin from *H. marmoreus* exerted more potent antiproliferative activity toward hepatoma ($IC_{50} = 0.15$ μM) and breast cancer cells ($IC_{50} = 5$ μM) than RIPs from a number of plants and the mushroom *Flammulina velutipes* [17]. Hypsin and marmorin suppressed cell-free translation with an IC_{50} of 7 nM and 0.7 nM, and reduced HIV-1 RT activity with an IC50 of 8 μM and 0.7 nM, respectively. Unlike marmorin which was without antifungal activity, hypsin hindered mycelial growth in *Botrytis cinerea, Fusarium oxysporum, Mycospharella arachidicola,* and *Physalospora piricola*, with an IC_{50} of 0.06, 14.2, 2.7, and 2.5 μM, respectively.

24.2.4 Mushroom Ribonucleases (RNases)

Pleurotus sajor RNase reduced viability of L1210 leukemia cells and HepG2 hepatoma cells with an IC_{50} of 0.1 μM and 0.22 μM, respectively [18]. RNase of the brown oyster mushroom exerted an antiproliferative activity toward leukemia cells [19]. *Hypsizigus marmoreus* RNase inhibited (3H-methyl)-thymidine uptake with an IC50 of 60 μM [20]. *Russula delica* RNase inhibited proliferation of hepatoma HepG2 cells and MCF-7 breast cancer cells with an IC_{50} of 8.6 μM and 7.2 μM, respectively [21]. *Lyophyllum shimeiji* RNase inhibited the proliferation of hepatoma and breast cancer cells with an IC_{50} of 10 μM and 6.2 μM, respectively [22]. *Pleurotus djamor* RNase inhibited the same cells with an IC_{50} of

3.9 μM and 3.4 μM [23]. *Lactarius flavidulus* RNase suppressed proliferation of HepG2 cells and L1210 cells with an IC_{50} of 3.19 μM and 6.52 μM, respectively. It also inhibited the activity of HIV-1 RT with an IC_{50} of 2.55 μM [24].

P. sajor-caju, *P. ostreatus*, *H. marmoreus*, *L. flavidulus* and *R. delica* RNase had a molecular mass of 12 kDa, 9 kDa, 18 kDa 14.6 kDa and 14 kDa, respectively. The RNases except that of *H. marmoreus* were similar in chromatographic behavior in that they did not bind to DEAE-cellulose, but bound to CM-cellulose/SP-Sepharose, Affi-gel blue gel/Red Sepharose, Q-Sepharose and Heparin-Sepharose. *H. marmoreus* did not bind to CM-cellulose and Affi-gel blue gel. They showed a variety of polyhomoribonucleotide specificities, being poly U > poly A > poly C > poly G >>> Poly U, Poly C > poly A >>> poly U = poly G, and poly C > poly G > poly A > poly U, respectively. The optimum pH values were 5.5, 8, 5, 5.5 and 5, respectively. The optimum temperatures were 60, 50–70, 70, 70,and 60 °C, respectively*sajor-caju* RNase but not the other abovementioned RNases inhibited mycelial growth in *Fusarium oxysporum* and *M. arachidicola* with an IC_{50} of 95 μM and 73 μM. It suppressed (3H-methyl)-thymidine uptake by murine splenocytes and cell-free translation with an IC_{50} of 65 nM and 158 nM, respectively. *P. sajor-caju* RNase inhibited *Pseudomonas aeruginosa* and *Staphylococcus aureus* and exhibited some DNase activity toward DNA from calf thymus and herring sperm. The ribonucleolytic activity of the RNases from oyster mushroom and *P. sajor-caju* was inhibited by Cu^{2+}, Ca^{2+}, Mg^{2+}, Zn^{2+} and Fe^{2+} ions [18–24].

24.2.5 Mushroom Ubiquitin-Like Peptides

The 9.5-kDa *Agrocybe cylindracea* ubiquitin-like peptide (ULP) exerted antiproliferative activity toward leukemia cells (IC_{50} = 10 μM) and hepatoma cells (IC_{50} = 100 μM) [25] while the 8-kDa ULP from the mosaic puffball mushroom *Calvatia caelata* suppressed the viability of breast cancer cells with an IC_{50} of 100 nM [26]. The ULPs were unadsorbed on DEAE-cellulose but adsorbed on Affi-gel blue gel and Mono S. *C. caelata* ULP exerted anti-mitogenic activity toward mouse splenocytes with an IC50 of 100 nM and inhibited translation in the cell-free rabbit reticulocyte lysate system. *A. cylindracea* ULP stimulated nitric oxide production by murine peritoneal macrophages with the potency similar to that of lipopolysaccharide. This action may contribute to its antiproliferative activity.

24.2.6 Mushroom Hemolysins

Eryngeolysin from *Pleurotus eryngii* with sequence homology to hemolysins from *P. ostreatus* and *A. cylindracea* exerted cytotoxicity toward leukemia L1210 cells. It exhibited antibacterial activity toward *Bacillus* sp. and exhibited anti-mitogenic activity toward mouse splenocytes but did not manifest antifungal activity. Polyethylene glycol 10000 protected its hemolytic activity but did not affect its antiproliferative or anti-mitogenic activity [27].

24.2.7 Mushroom Lectins and Hemagglutinins

The lectins TML-1 and TML-2 from *Tricholoma mongolicum* exhibited antiproliferative activity against mouse monocyte-macrophage PU 5-1.8 cells and mouse mastocytoma p815 cells [28]. *Pleurotus eous* lectin inhibitied proliferation of MCF-7, K562 and HepG2 cells with the same IC_{50} of 125 nM and that of SK-N-NC cells with an IC_{50} of 3 μM [29]. *Polyporus adusta* lectin exhibited antiproliferative activity toward a variety of tumor cells [30]. *Ganoderma capense* lectin inhibited proliferation of leukemia and hepatoma cells [31]. *Schizophyllum commune* lectin exhibited potent antiproliferative activity toward tumor cells [32]. *Armillaria luteo-virens* lectin inhibited proliferation of MBL2 cells, Hela cells and L1210 cells with an IC_{50} of 2.5, 5, and 10 μM, respectively [33]. *Cordyceps militaris* hemagglutinin exhibited some antiproliferative activity toward hepatoma cells [34]. *Clitocybe nebularis* lectin exerted an antiproliferative action on human leukemic T cells and the effect was probably mediated by binding to carbohydrate receptors on the cells [35]. *Pholiota adiposa* lectin inhibited proliferation of hepatoma and breast cancer cells with an IC_{50} of 2.1 μM and 3.2 μM [36]. *Russula delica* lectin did so with an IC_{50} of 0.88 μM and 0.52 μM [37]. *Russula lepida* lectin, when injected intraperitoneally at a dose of 5 mg/kg body weight/day for 20 days, caused 68% decrease in S-18- tumor weight. The IC50 values of the lectin on proliferation of hepatoma and breast cancer cells were 1.6 μM and 0.9 μM, respectively [38]. *Hericium erinaceum* lectin suppressed proliferation of hepatoma and breast cancer cells with an IC50 of 56.1 μM and 76.5 μM, respectively [39].

The dimeric 37-kDa *Tricholoma mongolicum* lectins were lactose-binding and relatively thermostable. They were both nonglycoproteins and possess hydroxyproline residues, although they had different proline and tyrosine contents [28]. The 16-kDa *P. eous* lectin possessed a neutral sugar content of 9% [29]. *Polyporus adusta* lectin was dimeric and possessed a molecular mass of 24 kDa. It exerted a mitogenic action toward mouse spleen cells. It was acid and alkali-labile, and stable up to 70 °C. A number of divalent metal chlorides did not affect the hemagglutinating activity of the lectin [30]. *G. capense* lectin had a molecular mass of 18 kDa. Its mitogenic activity toward mouse splenocytes was more potent than Con A. it bound galactose and galactosamine specifically. It was thermostable and unaffected by a number of monvalent divalent chlorides. It lacked inhibitory activity toward mycelial growth and HIV-1 RT [31]. *S. commune* lectin, a 64-kDa dimeric lactose-binding protein with an N-terminal sequence resembling a partial sequence of the cell division protein from *Gleobacter violaceus*, exhibited mitogenic activity toward mouse splenocytes and inhibited HIV-1 [32]. The 29.4-kDa dimeric *A. luteo-viren* lectin was acid- and alkali-labile and yet relatively thermostable (up to 70 °C). Its hemagglutinating activity was inhibited by inulin but not by simple sugars. It manifested potent mitogenic activity toward mouse splenocytes [33].

C. militaris hemagglutinin exhibited a molecular mass of 28 kDa. Its hemagglutinating activity could not be inhibited by simple sugars or heparin and was relatively thermostable (up to 60 °C) and pH-stable (within the pH range 2-13). Chemical modification of tryptophan and tyrosine residues did not affect its hemagglutinating activity. It inhibited HIV-1 RT with an IC_{50} of 50 μM. However, there was no mitogenic activity toward splenocytes, nitric oxide inducing activity toward macrophages, or inhibitory activity toward mycelial growth [34]. *Clitocybe nebularis* lectin, a 31.8-kDa homodimeric ricin b-like lectin, bound galactose containing sugars and lactose. The lectin exerted antiproliferative activity specific to human leukemic T cells. Its antiproliferative effect was a consequence of binding to carbohydrate receptors on human leukemic T cells [35]. *Pholiota adiposa* lectin, a 32-kDa dimeric protein, was acid-and alkali-labile and thermolabile. The lectin showed antiproliferative activity toward hepatoma Hep G2 cells and breast cancer MCF7 cells with an IC_{50} of 2.1 μM and approximately 3.2 μM, respectively. It exhibited HIV-1 RT inhibitory activity with an IC50 of 1.9 μM [36]. The 60-kDa dimeric lectin from *R. delica* inhibited HIV-1 RT with an IC_{50} of 0.26 μM [21]. *Russula lepida* lectin is a relatively thermostable 32-kDa protein with HIV-1 RT inhibitory activity. Its hemagglutinating activity was reduced by Mn^{2+}, Ca^{2+},and Hg^{2+} Ions but potentiated by Cu^{2+} ions [38].

Hericium erinaceum lectin is a 52-kDa protein with relatively high pH stability (pH 1.9-12.1) and thermostability (up to 70 °C). It was inhibited by Cu^{2+}, Hg^{2+} and Fe^{3+} ions. It inhibited HIV-1 RT with an IC50 of 31.7 μM and elicited a mitogenic response from murine splenocytes [39]. *Paxillus involutus* lectin is a homotetrameric 28 kDa protein with an N-terminal amino acid sequence CTCAVFLNNTTVKS, which showed a low level of similarity to mushroom lectin sequences reported previously. The lectin was inhibited by inulin and o-nitrophenyl-β-D-galactopyranoside. It manifested antiphytovirus activity towards tobacco mosaic virus with 70.61% inhibition at 200 μg/mL. This lectin demonstrated antiproliferative activity against A-549 lung cancer cells and HCT-8 human colon cancer cells [40]. *Lactarius flavidulus* lectin is a dimeric 29.8-kDa protein whose hemagglutinating activity was inhibited by a variety of simple sugars, such as lactose, p-nitrophenyl α-D-glucopyranoside, p-nitrophenyl β-D-glucopyranoside and inositol, and by the polysaccharide inulin. It suppressed the proliferation of hepatoma (HepG2) and leukemic (L1210) cells and inhibited the activity of HIV-1 RT with an IC_{50} of 8.90 μM, 6.81 μM and 5.68 μM, respectively [41]. The wild ascomycete mushroom *Xylaria hypoxylon* produced a 28.8-kDa dimeric lectin with an N-terminal sequence analogous to a part of *Aspergillus oryzae* genome and only slight similarity to *Flammulina velutipes* immunomodulatory protein. The hemagglutinating activity of the lectin was inhibited by inulin and xylose. The lectin exhibited strong antiproliferative activity toward tumor cell lines, and exerted a potent anti-mitogenic action on murine splenocytes [42].

24.2.8 Mushroom Peptide with HIV-1 RT Inhibitory Activity

A 4.5-kDa peptide with an N-terminal amino acid sequence KREHGQHCEHIV-1 purified from a hot water extract of *Russula paludosa* exhibited potent HIV-1RT inhibitory activity with an IC_{50} of 11 μM [43].

24.2.9 Mushroom Laccases

Hu et al. (2010) demonstrated that *Agrocybe cylindracea* laccase inhibited proliferation of hepatoma HepG2 cells and breast cancer cells with an IC_{50} of 5.6 μM and 6.5 μM, respectively. A 68-kDa laccase from the mycorrhizal fungus *Agaricus placomyces* exhibited the N-terminal amino acid sequence DVIGPQAQVTLANQD with only slight homology to sequences of other fungal laccases. It potently inhibited proliferation of Hep G2 hepatoma cells and MCF 7 human breast cancer cells and inhibited human HIV-1 RT activity with an IC_{50} of 1.7, 1.8, and 1.25 μM [44]. A laccase, with HIV-1 RT inhibitory activity (IC_{50} = 12.7 μM) and antiproliferative activity against HepG2 cells (IC_{50} = 5.6 μM) and MCF7 cells (IC_{50} = 6.5 μM), was purified from fresh fruiting bodies of the edible white common *Agrocybe cylindracea* mushroom [45]. A 66-kDa laccase from the edible mushroom *Pleurotus cornucopiae* inhibited proliferation of murine leukemia cell line L1210 and human hepatoma cell line HepG2, and reduced the activity of HIV-1 RT with an IC_{50} of 22 μM9 [46]. The IC_{50} values of the HIV-1 RT inhibitory activity of other mushroom laccases are as follows: 0.06 μM for *Hericium coralloides* laccase [47], 9.5 μM for *Hericium erinaceum* laccase [48], 2.2 μM for *Tricholoma giganteum* laccase [49], 2.2 μM for *Pleurotus eryngii laccase* [50], 1.2 μM for *Ganoderma lucidum* laccase [51], and 7.5 μM for *Lentinus edodes* laccase [52].

Protective vaccines and effective drugs against hepatitis C virus (HCV) are unavailable. Sustained virological response to INF/ribavirin only has about 50% efficiency. Incubation of peripheral blood cells and hepatoma HepG2 cells with a 58-kDa laccase from oyster mushroom (*Pleurotus ostreatus*) followed by HCV infection did not prevent HCV from entering the cells. Direct interaction between HCV and the laccase at 2.0 and 2.5 mg/ml completely prevented viral entry after one week of incubation. The laccase inhibited HCV replication at 1.25 and 1.5 mg/ml after the first treatment for four days and at 0.75, 1.0, 1.25 and 1.5 mg/ml after the second treatment for another four days [53].

24.2.10 Mushroom Proteases

A 15-kDa protease from fresh fruiting bodies of the wild mushroom *Amanita farinosa* inhibited cancer cell proliferation with an IC_{50} of 25 μM [54]. *Cordyceps sobolifera* produces a protease with HIV-1 RT inhibitory activity [55].Proteases with fibrinolytic activity have been reported from *Cordyceps militaris* [56], *Pleurotus ostreatus* [57], *Ganoderma lucidum* [58], and *Schizophyllum commune* [59].

24.2.11 Mushroom Antibacterial Protein

A 44-kDa dimeric antibacterial protein from the wild mushroom *Clitocybe sinopica* manifested antibacterial activity against *Agrobacterium rhizogenes, A. tumefaciens, A. vitis, Xanthomonas malvacearum* and *X. oryzae* with a minimum inhibitory concentration mostly below 0.6 μM [60].

24.3 Bacterial Defensins

24.3.1 Bacterial RNases

Binase, a 12.2-kDa RNase from *Bacillus intermedius,* exerted antiproliferative and apoptosis-inducing activities toward human myelogenic erythroleukemia K562 cells, and human lung carcinoma A549 cells, but was ineffective toward human peripheral blood mononuclear cells [61, 62]. Binase caused transformed myeloid cells to undergo apoptosis, and attacked artificially expressing activated C-kit myeloid progenitor FDC cells and chronic myelogenous leukemia K562 cells. However, leukocytes of healthy donors and normal myeloid progenitor cells were unaffected [63]. Barnase from *B. amyloliquefaciens* closely resembles binase in amino acid sequence and is highly stable [61, 62]. Two barnase molecules were fused serially to the single chain variable fragment (scFv) of humanized 4D5 antibody which is directed against the extracellular domain of human epidermal growth receptor 2 (HER2). HER2 was overexpressed in a variety of human cancers. The 4D5 antibody facilitated the delivery of barnase to HER2/neu-pisitive cells and the penetration of barnase into target cells [64]. The immunotoxin induced apoptosis in HER2-overexpressing SKBR-3 and BT-474 human cancer cells in vitro. It suppressed human breast cancer xenografts in nude mice [65].

24.3.2 Bacterial Antimicrobial Proteins

Baciamin, a50-kDa antifungal protein from the culture broth of the bacterium *Bacillus amyloliquefaciens,* permeabilized fungal membrane but not rabbit erythrocyte membrane. Its antifungal activity was characterized by fair trypsin-, pH- and heat stability. It inhibited HIV-1 RT as well as the proliferation of breast cancer, colon cancer and hepatoma cells. It upregulated production of nitric oxide by murine macrophages [66]. A monomeric 31-kDa antifungal protein produced by *Bacillus licheniformis* strain BS-3, designated as F2 protein, hindered the growth of *Rhizoctonia solan* (EC_{50} = 35.82 μg/mL), *Aspergillus niger*, and *Magnaporthe oryzae*. It was capable of hydrolyzing casein, but not o-nitrophenyl-N-acetylglucosaminide, xylan, CM-cellulose, and colloidal chitin [67]. The lantibiotic nisin is a post-translationally modified antimicrobial peptide produced by some *Lactococcus lactis* strains, and utilized as a safe and natural preservative in the food industry. The nisin precursor was expressed as a His6-tagged peptide in *Escherichia coli* and purified with a nickel affinity column [68]. Laterosporulin, a novel 5.6-kDa bacteriocin (bacterial antimicrobial peptide) effective against a wide range of Gram-positive and Gram-negative bacteria, was purified from supernatant of the culture of *Brevibacillus sp.* strain GI-9 grown under optimal conditions [69].

BCP6 (ca. 1.1 kDa)1 exhibiting the N-terminal amino acid sequence A-I-N-X-D-A-A-Y-L, has been purified from *Bacillus* sp. CS61 freshly isolated from the traditional fermented food kimchi.It demonstated activity against multidrug-resistant bacteria comprising methicillin-resistant *Staphylococcus aureus,* vancomycin-resistant *S. aureus* and vancomycin-resistant enterococci with minimal inhibitory concentrations in the range 0.625-20 μg/mL [70]. DC-specific ICAM-3-grabbing nonintegrin (DC-SIGN, CD209) is a cell-surface adhesion factor which facilitates entry of viruses belonging to several virus families. A virtually complete suppression of JUNV infection was achieved after exposure of 3T3 cells stably expressing DC-SIGN to surface-layer (S-layer) protein of *Lactobacillus acidophilus* ATCC 4365. On the other hand, inhibition was indiscernible in 3T3 wild type cells or in 3T3 cells expressing liver/lymph node-specific ICAM-3 grabbing nonintegrin (L-SIGN or DC-SIGNR or CD209L). Inhibition by S-layer protein was detectable only in the initial stages of viral infection. The S-layer protein is probably crucial for the exclusion of pathogens in this probiotic strain and may be useful as an antiviral agent [71]. Tostadin, a thermostable antibacterial nonapeptide purified from the fermentation medium of *Brevibacillus brevis* XDH, which is a wide -spectrum antagonistic bacterium isolated from the soil of Mountain Tai in China, exerted potent inhibitory effects on the growth of *Staphylococcus aureus* and *Escherichia coli* in vitro [72].

24.4 Conclusion and Future Perspectives

Polysaccharopeptide from the medicinal mushroom *Coriolus versicolor* has been used popularly in the orient such as China and Japan as complementary and alternative medicine for cancers and has been reported to reduce the untoward side effects of chemotherapy and radiation therapy [73]. A polysaccharide with antiproliferative, hypoglycemic, antioxidant and HIV-1 reverse transcriptase inhibitory activities has been reported from the fruiting bodies of the edible abalone mushroom *Pleurotus abalonus* [74].It is known some lectins can pass through the gastrointestinal tract without loss of biological activity. Hence it is likely that some of the aforementioned mushroom lectins behave similarly and their antiproliferative/antitumor activity may persist after oral consumption. Consumption of mushrooms, which contain lectins, laccases, proteases, antifungal proteins, RIPs, RNases, polysaccharopeptides and polyasccharides with antiproliferative/anticancer activity may reduce the incidence of cancer and infectious diseases [74]. In addition, these mushroom proteins have the potential of development into new drugs with less side effects or drugs for combination therapy. Bacteria also produce a number of proteins and peptides with potentially exploitable activities for treatment of cancer and infections. It is hoped that continued research endeavors will disclose novel fungal and bacterial proteins/peptides with potent activity and new mechanisms of action against malignant and infectious diseases.

Acknowledgments We thank National Science Foundation of China for award of a research grant (number 81201270).

References

1. Zhu S (2008) Discovery of six families of fungal defensin-like peptides provides insights into origin and evolution of the CSalphabeta defensins. Mol Immunol 45:828–838
2. Mygind PH, Fischer RL, Schnorr KM, Hansen MT, Sonksen CP, Ludvigsen S et al (2005) Plectasin is a peptide antibiotic with therapeutic potential from a saprophytic fungus. Nature 437:975–980
3. Zhang J, Yang Y, Teng D, Tian Z, Wang S, Wang J (2011) Expression of plectasin in Pichia pastoris and its characterization as a new antimicrobial peptide against Staphyloccocus and Streptococcus. Protein Expr Purif 78:189–196
4. Jing XL, Luo XG, Tian WJ, Lv LH, Jiang Y, Wang N et al (2010) High-level expression of the antimicrobial peptide plectasin in Escherichia coli. Curr Microbiol 61:197–202
5. Schneider T, Kruse T, Wimmer R, Wiedemann I, Sass V, Pag U et al (2010) Plectasin, a fungal defensin, targets the bacterial cell wall precursor Lipid II. Science 328:1168–1172
6. Thomsen LE, Gottlieb CT, Gottschalk S, Wodskou TT, Kristensen HH, Gram L et al (2010) The heme sensing response regulator HssR in Staphylococcus aureus but not the homologous RR23 in Listeria monocytogenes modulates susceptibility to the antimicrobial peptide plectasin. BMC Microbiol 10:307
7. Mandal K, Pentelute BL, Tereshko V, Thammavongsa V, Schneewind O, Kossiakoff AA et al (2009) Racemic crystallography of synthetic protein enantiomers used to determine the X-ray structure of plectasin by direct methods. Protein Sci: Publ Protein Soc 18:1146–1154
8. Ostergaard C, Sandvang D, Frimodt-Moller N, Kristensen HH (2009) High cerebrospinal fluid (CSF) penetration and potent bactericidal activity in CSF of NZ2114, a novel plectasin variant, during experimental pneumococcal meningitis. Antimicrob Agents Chemother 53:1581–1585
9. Xiong YQ, Hady WA, Deslandes A, Rey A, Fraisse L, Kristensen HH et al (2011) Efficacy of NZ2114, a novel plectasin-derived cationic antimicrobial peptide antibiotic, in experimental endocarditis due to methicillin-resistant Staphylococcus aureus. Antimicrob Agents Chemother 55:5325–5330
10. Hara S, Mukae H, Sakamoto N, Ishimoto H, Amenomori M, Fujita H et al (2008) Plectasin has antibacterial activity and no affect on cell viability or IL-8 production. Biochem Biophys Res Commun 374:709–713
11. Brinch KS, Sandberg A, Baudoux P, Van Bambeke F, Tulkens PM, Frimodt-Moller N et al (2009) Plectasin shows intracellular activity against Staphylococcus aureus in human THP-1 monocytes and in a mouse peritonitis model. Antimicrob Agents Chemother 53:4801–4808

12. Ngai PH, Zhao Z, Ng TB (2005) Agrocybin, an antifungal peptide from the edible mushroom Agrocybe cylindracea. Peptides 26:191–196
13. Wong JH, Ng TB, Wang H, Sze SC, Zhang KY, Li Q et al (2011) Cordymin, an antifungal peptide from the medicinal fungus Cordyceps militaris. Phytomedicine : Int J Phytother Phytopharmacol 18:387–392
14. Fang EF, Ng TB, Shaw PC, Wong RN (2011) Recent progress in medicinal investigations on trichosanthin and other ribosome inactivating proteins from the plant genus Trichosanthes. Curr Med Chem 18:4410–4417
15. Lam SK, Ng TB (2001) Hypsin, a novel thermostable ribosome-inactivating protein with antifungal and antiproliferative activities from fruiting bodies of the edible mushroom Hypsizigus marmoreus. Biochem Biophys Res Commun 285:1071–1075
16. Ng TB, Lam YW, Wang H (2003) Calcaelin, a new protein with translation-inhibiting, antiproliferative and antimitogenic activities from the mosaic puffball mushroom Calvatia caelata. Planta Med 69:212–217
17. Wong JH, Wang HX, Ng TB (2008) Marmorin, a new ribosome inactivating protein with antiproliferative and HIV-1 reverse transcriptase inhibitory activities from the mushroom Hypsizigus marmoreus. Appl Microbiol Biotechnol 81:669–674
18. Ngai PH, Ng TB (2004) A ribonuclease with antimicrobial, antimitogenic and antiproliferative activities from the edible mushroom Pleurotus sajor-caju. Peptides 25:11–17
19. Xia L, Chu KT, Ng TB (2005) A low-molecular mass ribonuclease from the brown oyster mushroom. J Pept Res : Off J Am Pept Soc 66:1–8
20. Guan GP, Wang HX, Ng TB (2007) A novel ribonuclease with antiproliferative activity from fresh fruiting bodies of the edible mushroom Hypsizigus marmoreus. Biochim Biophys Acta 1770:1593–1597
21. Zhao S, Zhao Y, Li S, Zhang G, Wang H, Ng TB (2010) An antiproliferative ribonuclease from fruiting bodies of the wild mushroom Russula delica. J Microbiol Biotechnol 20:693–699
22. Zhang RY, Zhang GQ, Hu DD, Wang HX, Ng TB (2010) A novel ribonuclease with antiproliferative activity from fresh fruiting bodies of the edible mushroom Lyophyllum shimeiji. Biochem Genet 48:658–668
23. Wu X, Zheng S, Cui L, Wang H, Ng TB (2010) Isolation and characterization of a novel ribonuclease from the pink oyster mushroom Pleurotus djamor. J Gen Appl Microbiol 56:231–239
24. Wu Y, Wang H, Ng T (2012) Purification and characterization of a novel RNase with antiproliferative activity from the mushroom Lactarius flavidulus. J Antibiot 65:67–72
25. Ngai PH, Wang HX, Ng TB (2003) Purification and characterization of a ubiquitin-like peptide with macrophage stimulating, antiproliferative and ribonuclease activities from the mushroom Agrocybe cylindracea. Peptides 24:639–645
26. Lam YW, Ng TB, Wang HX (2001) Antiproliferative and antimitogenic activities in a peptide from puffball mushroom Calvatia caelata. Biochem Biophys Res Commun 289:744–749
27. Ngai PH, Ng TB (2006) A hemolysin from the mushroom Pleurotus eryngii. Appl Microbiol Biotechnol 72:1185–1191
28. Wang HX, Ng TB, Liu WK, Ooi VE, Chang ST (1995) Isolation and characterization of two distinct lectins with antiproliferative activity from the cultured mycelium of the edible mushroom Tricholoma mongolicum. Int J Pept Protein Res 46:508–513
29. Mahajan RG, Patil SI, Mohan DR, Shastry P (2002) Pleurotus Eous mushroom lectin (PEL) with mixed carbohydrate inhibition and antiproliferative activity on tumor cell lines. JBMBB : off J Fed Asian Ocean Biochem Mol Biol 6:341–345
30. Wang H, Ng TB, Liu Q (2003) A novel lectin from the wild mushroom Polyporus adusta. Biochem Biophys Res Commun 307:535–539
31. Ngai PH, Ng TB (2004) A mushroom (Ganoderma capense) lectin with spectacular thermostability, potent mitogenic activity on splenocytes, and antiproliferative activity toward tumor cells. Biochem Biophys Res Commun 314:988–993

32. Han CH, Liu QH, Ng TB, Wang HX (2005) A novel homodimeric lactose-binding lectin from the edible split gill medicinal mushroom Schizophyllum commune. Biochem Biophys Res Commun 336:252–257
33. Feng K, Liu QH, Ng TB, Liu HZ, Li JQ, Chen G et al (2006) Isolation and characterization of a novel lectin from the mushroom Armillaria luteo-virens. Biochem Biophys Res Commun 345:1573–1578
34. Wong JH, Wang H, Ng TB (2009) A haemagglutinin from the medicinal fungus Cordyceps militaris. Biosci Rep 29:321–327
35. Pohleven J, Obermajer N, Sabotic J, Anzlovar S, Sepcic K, Kos J et al (2009) Purification, characterization and cloning of a ricin B-like lectin from mushroom Clitocybe nebularis with antiproliferative activity against human leukemic T cells. Biochim Biophys Acta 1790:173–181
36. Zhang GQ, Sun J, Wang HX, Ng TB (2009) A novel lectin with antiproliferative activity from the medicinal mushroom Pholiota adiposa. Acta Biochim Pol 56:415–421
37. Zhao S, Zhao Y, Li S, Zhao J, Zhang G, Wang H et al (2010) A novel lectin with highly potent antiproliferative and HIV-1 reverse transcriptase inhibitory activities from the edible wild mushroom Russula delica. Glycoconj J 27:259–265
38. Zhang G, Sun J, Wang H, Ng TB (2010) First isolation and characterization of a novel lectin with potent antitumor activity from a Russula mushroom. Phytomedicine : Int J Phytother Phytopharmacol 17:775–781
39. Li Y, Zhang G, Ng TB, Wang H (2010) A novel lectin with antiproliferative and HIV-1 reverse transcriptase inhibitory activities from dried fruiting bodies of the monkey head mushroom Hericium erinaceum. J Biomed & Biotechnol 2010:716515
40. Wang SX, Zhang GQ, Zhao S, Xu F, Zhou Y, Geng XL, et al (2012) Purification and characterization of a novel lectin with antiphytovirus activities from the wild mushroom Paxillus involutus. Protein Pept Lett
41. Wu Y, Wang H, Ng TB (2011) Purification and characterization of a lectin with antiproliferative activity toward cancer cells from the dried fruit bodies of Lactarius flavidulus. Carbohydr Res 346:2576–2581
42. Liu Q, Wang H, Ng TB (2006) First report of a xylose-specific lectin with potent hemagglutinating, antiproliferative and anti-mitogenic activities from a wild ascomycete mushroom. Biochim Biophys Acta 1760:1914–1919
43. Wang J, Wang HX, Ng TB (2007) A peptide with HIV-1 reverse transcriptase inhibitory activity from the medicinal mushroom Russula paludosa. Peptides 28:560–565
44. Sun J, Chen QJ, Cao QQ, Wu YY, Xu LJ, Zhu MJ et al (2012) A Laccase with Antiproliferative and HIV-I Reverse Transcriptase Inhibitory Activities from the Mycorrhizal Fungus Agaricus placomyces. J Biomed & Biotechnol 2012:736472
45. Hu DD, Zhang RY, Zhang GQ, Wang HX, Ng TB (2011) A laccase with antiproliferative activity against tumor cells from an edible mushroom, white common Agrocybe cylindracea. Phytomedicine : Int J Phytother Phytopharmacol 18:374–379
46. Wong JH, Ng TB, Jiang Y, Liu F, Sze SC, Zhang KY (2010) Purification and characterization of a Laccase with inhibitory activity toward HIV-1 reverse transcriptase and tumor cells from an edible mushroom (Pleurotus cornucopiae). Protein Pept Lett 17:1040–1047
47. Zou YJ, Wang HX, Ng TB, Huang CY, Zhang JX (2012) Purification and characterization of a novel laccase from the edible mushroom Hericium coralloides. J Microbiol 50:72–78
48. Wang HX, Ng TB (2004) A new laccase from dried fruiting bodies of the monkey head mushroom Hericium erinaceum. Biochem Biophys Res Commun 322:17–21
49. Wang HX, Ng TB (2004) Purification of a novel low-molecular-mass laccase with HIV-1 reverse transcriptase inhibitory activity from the mushroom Tricholoma giganteum. Biochem Biophys Res Commun 315:450–454
50. Wang HX, Ng TB (2006) Purification of a laccase from fruiting bodies of the mushroom Pleurotus eryngii. Appl Microbiol Biotechnol 69:521–525

51. Wang HX, Ng TB (2006) A laccase from the medicinal mushroom Ganoderma lucidum. Appl Microbiol Biotechnol 72:508–513
52. Sun J, Wang H, Ng TB (2011) Isolation of a laccase with HIV-1 reverse transcriptase inhibitory activity from fresh fruiting bodies of the Lentinus edodes (Shiitake mushroom). Indian J Biochem Biophys 48:88–94
53. El-Fakharany EM, Haroun BM, Ng TB, Redwan ER (2010) Oyster mushroom laccase inhibits hepatitis C virus entry into peripheral blood cells and hepatoma cells. Protein Pept Lett 17:1031–1039
54. Sun J, Zhao Y, Chai H, Wang H, Ng TB (2011) A novel alkaline protease with antiproliferative activity from fresh fruiting bodies of the toxic wild mushroom Amanita farinosa. Acta Biochim Pol 58:567–572
55. Wang SX, Liu Y, Zhang GQ, Zhao S, Xu F, Geng XL et al (2012) Cordysobin, a novel alkaline serine protease with HIV-1 reverse transcriptase inhibitory activity from the medicinal mushroom Cordyceps sobolifera. J Biosci Bioeng 113:42–47
56. Kim JS, Sapkota K, Park SE, Choi BS, Kim S, Nguyen TH et al (2006) A fibrinolytic enzyme from the medicinal mushroom Cordyceps militaris. J Microbiol 44:622–631
57. Shen MH, Kim JS, Sapkota K, Park SE, Choi BS, Kim S et al (2007) Purification, characterization, and cloning of fibrinolytic metalloprotease from Pleurotus ostreatus mycelia. J Microbiol Biotechnol 17:1271–1283
58. Kumaran S, Palani P, Nishanthi R, Kaviyarasan V (2011) Studies on screening, isolation and purification of a fibrinolytic protease from an isolate (VK12) of Ganoderma lucidum and evaluation of its antithrombotic activity. Medical Mycol J 52:153–162
59. Lee SY, Kim JS, Kim JE, Sapkota K, Shen MH, Kim S et al (2005) Purification and characterization of fibrinolytic enzyme from cultured mycelia of Armillaria mellea. Protein Expr Purif 43:10–17
60. Zheng S, Liu Q, Zhang G, Wang H, Ng TB (2010) Purification and characterization of an antibacterial protein from dried fruiting bodies of the wild mushroom Clitocybe sinopica. Acta Biochim Pol 57:43–48
61. Zelenikhin PV, Cherepnev GV, Kern F, Ilinskaia ON (2006) Binase does not induce polyclonal T-cell response. Dokl Bio Sci : Proc Acad Sci USSR, Biol Sci Sect (translated from Russian) 407:195–197
62. Fang EF, Ng TB (1815) Ribonucleases of different origins with a wide spectrum of medicinal applications. Biochim Biophys Acta 2011:65–74
63. Makarov AA, Kolchinsky A, Ilinskaya ON (2008) Binase and other microbial RNases as potential anticancer agents. BioEssays: News Rev Mol, Cell Dev Biol 30:781–790
64. Glinka EM, Edelweiss EF, Sapozhnikov AM, Deyev SM (2006) A new vector for controllable expression of an anti-HER2/neu mini-antibody-barnase fusion protein in HEK 293T cells. Gene 366:97–103
65. Balandin TG, Edelweiss E, Andronova NV, Treshalina EM, Sapozhnikov AM, Deyev SM (2011) Antitumor activity and toxicity of anti-HER2 immunoRNase scFv 4D5-dibarnase in mice bearing human breast cancer xenografts. Invest New Drugs 29:22–32
66. Wong JH, Hao J, Cao Z, Qiao M, Xu H, Bai Y et al (2008) An antifungal protein from Bacillus amyloliquefaciens. J Appl Microbiol 105:1888–1898
67. Cui TB, Chai HY, Jiang LX (2012) Isolation and partial characterization of an antifungal protein produced by Bacillus licheniformis BS-3. Molecules 17:7336–7347
68. Karakas-Sen A, Narbad A (2012) Heterologous expression and purification of NisA, the precursor peptide of lantibiotic nisin from Lactococcus lactis. Acta Biol Hung 63:301–310
69. Singh PK (2012) Chittpurna, Ashish, Sharma V., Patil P.B., Korpole S., Identification, purification and characterization of laterosporulin, a novel bacteriocin produced by Brevibacillus sp. strain GI-9. PLoS ONE 7:e31498
70. Choi YH, Cho SS, Simkhada JR, Yoo JC (2012) A novel thermotolerant and acidotolerant peptide produced by a Bacillus strain newly isolated from a fermented food (kimchi) shows activity against multidrug-resistant bacteria. Int J Antimicrob Agents 40:80–83

71. Martinez MG (2012) Prado Acosta M., Candurra N.A., Ruzal S.M., S-layer proteins of Lactobacillus acidophilus inhibits JUNV infection. Biochem Biophys Res Commun 422:590–595
72. Song Z, Liu Q, Guo H, Ju R, Zhao Y, Li J et al (2012) Tostadin, a novel antibacterial peptide from an antagonistic microorganism Brevibacillus brevis XDH. Bioresour Technol 111:504–506
73. Ng TB (1998) A review of research on the protein-bound polysaccharide (polysaccharopeptide, PSP) from the mushroom Coriolus versicolor (Basidiomycetes: Polyporaceae). Gen Pharmacol 30:1–4
74. Wang CR, Ng TB, Li L, Fang JC, Jiang Y, Wen TY et al (2011) Isolation of a polysaccharide with antiproliferative, hypoglycemic, antioxidant and HIV-1 reverse transcriptase inhibitory activities from the fruiting bodies of the abalone mushroom Pleurotus abalonus. J Pharmacy Pharmacol 63:825–832

Author Biography

Dr. Jack Ho Wong is Research Associate at the Faculty of Medicine, Chinese University of Hong Kong, China. He received his PhD degree and postdoctoral training from the same university. He has published over 70 peer-reviewed papers on antifungal proteins and peptides of diverse origins including cathelicidins, defensins and thaumatin-like proteins, hemagglutinins/lectins, laccases and ribosome inactivating proteins.

Prof. Hexiang Wang is Professor at State Key Laboratory for Agrobiotechnology and Department of Microbiology, China Agricultural University, Beijing, China. He received his PhD degree and posdoctoral training from the Chinese University of Hong Kong. Over the past 2 decades in collaboration with Prof. TB Ng and other national and international researchers, his team has published over 250 papers in international journal, on a diversity of bioactive proteins and peptides of diverse origins with an emphasis on mushrooms and plants.

Part VII
Modernizing Investigations of Traditional Chinese Medicine at the Molecular Level

Chapter 25
The Role of the GP-TCM Research Association to Modernization and Globalization of Traditional Chinese Medicine

Rudolf Bauer, De-An Guo, Peter Hylands, Tai-Ping Fan and Qihe Xu

Traditional Chinese Medicine (TCM) is an ancient medical system using experience-based therapies such as acupuncture and herbal medicine, which has been practised for thousands of years in the diagnosis, prevention and treatment of diseases in Asia [1, 2]. Recently, TCM has been promoted worldwide, is undergoing continuous globalization [3], and plans to be integrated with Western medicine [4]. China has continuously increased its investment in R&D of TCM and has promoted international dialogues and collaborations, especially since the late 2000s [5]. In 2009, the European Union funded a coordination project under its 7th Framework Programme (FP7) entitled "*Good Practice in Traditional Chinese Medicine (GP-TCM) Research in the Post-genomic Era*", in order to coordinate EU-China dialogues and collaborations in TCM research [6]. During the FP7 GP-TCM project a large collaborative network of more than 200 scientists and clinicians from 112 institutions in 24 countries was established to work on the future perspectives of TCM research. The participating members were engaged for three and a half years in discussions on good practice issues related to various aspects of Chinese herbal medicine and acupuncture research. The project resulted

R. Bauer (✉)
TCM Research Center Graz, Institute of Pharmaceutical Sciences, University of Graz, Graz, Austria
e-mail: rudolf.bauer@uni-graz.at

D.-A. Guo
Shanghai Research Center for TCM Modernization, Shanghai Institute of Materia Medica, Chinese Academy of Sciences, Shanghai, China

P. Hylands
Institute of Pharmaceutical Science, King's College London, London, UK

T.-P. Fan
Department of Pharmacology, University of Cambridge, Cambridge, UK

Q. Xu
Department of Renal Medicine, King's College London, London, UK

E. F. Fang and T. B. Ng (eds.), *Antitumor Potential and Other Emerging Medicinal Properties of Natural Compounds*, DOI: 10.1007/978-94-007-6214-5_25,

in recommendations on best practice related to various aspects of Chinese herbal medicine and acupuncture research, leading to state-of-the-art reports and guidelines published in an open-access special issue of *Journal of Ethnopharmacology* (2012;140:455–643) [7].The Good Practice in TCM Research Association (GP-TCM RA) has been founded as a sustainable follow-up organization of the FP7 project. Dedicated to (TCM) research, it will disseminate, validate and further develop good practice guidelines through continued interregional, interdisciplinary and intersectoral collaborations (http://www.gp-tcm.org). A Board of Directors (BoD) has been elected for the management of the Association (Table 25.1), and an Executive Council (ExC) consisting of the President, President-Elect, Past President, Secretary-General, and Treasurer will exercise executive power of the Association and perform the daily business according to established bylaws (http://www.gp-tcm.org/about/bylaws/). GP-TCM RA was officially launched in Leiden, the Netherlands, on April 16, 2012, and will be registered as a not-for-profit organization/charity in the UK. Membership of the Association is open to scientists and practitioners who support the objectives and abide by the rules of the Association. The following Interest Groups will be established, based on the activities of the members:

- Quality Control;
- Pharmacology and Toxicology;
- Clinical Studies;
- Regulatory Aspects;
- Acupuncture—Moxibustion and Meridians.

Table 25.1 Founding board of directors of the GP-TCM RA

Prof. Dr. Rudolf Bauer, President; Institute of Pharmaceutical Sciences, Department of Pharmacognosy, Karl-Franzens-University Graz, Graz, Austria
Prof. Dr. De-an Guo, Vice President, President-Elect, Shanghai Research Centre for TCM Modernisation, Shanghai Institute of Materia Medica, Chinese Academy of Sciences, Shanghai, China
Dr. Tai-Ping Fan, Secretary-General, Department of Pharmacology, University of Cambridge, Cambridge, UK
Prof. Peter Hylands, Treasurer, Institute of Pharmaceutical Science, Department of Pharmacy, King's College London, London, UK
Dr. Qihe Xu, Vice President and Deputy Secretary-General, Department of Renal Medicine, King's College London, London, UK
Prof. Kelvin Chan, The University of Sydney and University of Western Sydney, Sydney, Australia
Prof. Pierre Duez, Laboratory of Pharmacognosy, Bromatology and Human Nutrition, Université Libre de Bruxelles, Brussels, Belgium
Prof. Aiping Lu, School of Chinese Medicine, Hong Kong Baptist University, Hong Kong, China
Prof. Nicola Robinson, Allied Health Sciences, Faculty of Health and Social Care, London South Bank University, London, UK
Prof. Monique Simmonds, Royal Botanic Gardens, Kew, Richmond, Surrey, UK
Prof. Rob Verpoorte, Institute of Biology, University of Leiden, Leiden, the Netherlands
Prof. Vivian Taam Wong, University of Hong Kong & Open University of Hong Kong, Hong Kong, China

The major objectives of GP-TCM RA have been defined as follows:

- Perpetuate the interactive network established by the FP7 GP-TCM consortium;
- Promote discussion and implementation of good practice in TCM research and development, including the use of sustainably sourced materials;
- Advocate high-quality evidence-based research and development on TCM as well as on its integration with conventional medicine;
- Organise and co-organise scientific meetings;
- Nurture young TCM researchers at different levels supporting an interdisciplinary approach, including BSc, MSc, PhD and post-doctoral programmes;
- Facilitate collaborations and sharing of resources, expertise and good practice among members, industry and regulatory agencies;
- Encourage collaborations with existing relevant societies, consortia and organisations;
- Strengthen interdisciplinary, interregional, and intersectoral collaborations in TCM research and development;
- Perpetuate good practice in publishing TCM research outcomes; and
- Disseminate scientific research outcomes and latest developments in regulatory sciences to stakeholders, industry, professional groups and the public.

The GP-TCM RA has identified the following fields in which there is a need for modernization.

25.1 TCM Resources and Quality Control

According to China's State Administration of TCM (SATCM), 11,146 herbal, 1,581 zoological, and 80 mineral materials have been used in TCM in China [8]. In the current version of the Chinese Pharmacopoeia (CP2010) 2,165 official monographs of TCM raw materials and formulae are listed [9]. Also the European Pharmacopoeia is now publishing monographs on Chinese herbs [10]. For the future development of TCM, it must be guaranteed that the raw materials are produced in a sustainable way. Cultivation under controlled conditions and according to good agricultural practice (GAP) should be achieved as often as possible [11, 12]. The principles of *daodi* should be respected [13]. The identity and pharmaceutical quality must be controlled according to the pharmacopoeia standards [14]. Sometimes DNA-based methods may be needed for unambiguous authentication [15]. Classical quality control measures such as those applied to single chemical drugs are not suited to ensure quality, efficacy and safety of TCM materials and products. In order to focus not only on single quality marker compounds, more holistic concepts need to be developed. New strategies, like metabolic fingerprinting [16, 17] or PhytomicsQC [18], which combines chemical fingerprinting, *in vivo* activity testing, and gene expression profiling, may be used in the future. Processing (*paozhi*) and post-harvest treatment are common features for Chinese herbal medicine [19]. In order to provide plant material of consistent quality, these processes need to be scientifically investigated and Standard Operating Procedures (SOPs) and specific endpoints need to be defined. The issue of authentication and quality control of the materials used in clinical, animal and *in vitro* studies has been recognized also as an area in research that needs urgent improvement. Funding bodies, journals, academia and commercial suppliers should collaborate to address this fundamental issue on the basis of a checklist and good practice guidelines which have been developed during the FP-7 GP-TCM project [20].

25.2 Efficacy and Effectiveness of Chinese Herbal Medicine

Although TCM has been primarily an experience-based type of medicine, for future acceptance and for integration with Western medicine it is essential to demonstrate efficacy in an objective and evidence-based manner. Randomized, controlled clinical trials are still the Gold Standard of evidence-based medicine and should be applied also in clinical TCM research [21, 22]. In contrast to the reductionist approach of Western medicine, TCM is based on a different theory with a personalized and holistic view to describe health and disease, e.g. the Yin-Yang theory emphasizing the balance of functional systems [23]. Therefore, clinical trials with TCM may have to follow a special strategy considering syndrome differentiation [24, 25]. In addition, it is also important to perform research on the principles and significance of TCM diagnosis by systemic biomarker analysis and an omics-based approach [26, 27].

25.3 Acupuncture

A number of large-scale studies on the adverse effects, the efficacy and effectiveness of acupuncture have recently been published [28–32]. A meta-analysis, that pooled the data of 29 clinical trials on chronic pain, concluded that acupuncture is statistically superior to sham acupuncture [33]. Nevertheless, debates about acupuncture point specificity, acupuncture techniques and placebo effects do continue, and further research is necessary. Within the GP-TCM project, a summary of the current scientific understanding of the acupuncture meridian system and has been published [34]. The paper suggests also future strategies and approaches in the application of omics techniques and a systems biology approach to acupuncture research. In a survey of practitioners of traditional acupuncture in the EU and China, differences in acupuncture practice, education and training, as well as research priorities have been identified, which will influence the methodologies of future international multi-centre clinical trials [35].

25.4 Mechanisms and Active Compounds

In order to understand the mechanisms of action of TCM, pharmacological *in vitro* and *in vivo* studies need to be performed considering systems biology and the latest omics technologies [36–38] and recommendations for animal studies [39]. For the assignment of the active principles, databases of known constituents and *in silico* studies can be used [40]. Activity-guided isolation is still another efficient approach to isolate unknown active compounds with lead structures. However, multicompound and multitarget effects need to be considered. Chinese medicine has been considered as a treasure trove for new lead structures, and many highly effective compounds, like artemisinin or camptothecin have been already isolated [41]. In China in 2008, the "Herbalome Project" has been launched to clarify the chemical composition, structure and function of Chinese herbs and TCM formulae, to establish a standard resource library, and to interpret the synergistic and complementary mechanisms of multiple components in TCM drugs on multiple targets [42, 43].

25.5 Safety and Adverse Effects

Traditional use is only an indication of acute safety but certainly not a proof of absence of mid- or long-term toxicities. Intoxications with aristolochic acid containing herbs [44, 45], toxicity of pyrrolizidine alkaloids [46], aconite poisoning [47], and herb-drug interactions [48], have become a major issue. Therefore research on adverse side effects [49] and establishment of an efficient pharmacovigilance system is highly needed [50, 51]. Also a pharmacogenomics-based approach for the prediction of toxicity may be used [52].

25.6 Legislation and Regulation of Chinese Herbal Medicine

In many countries, regulations for herbal medicines are in place [53]. However, these regulations are not always adequate and may need adjustment [54]. Also in this field, omics-based methodologies have been suggested in order to consider the complexity of herbal products [55]. While in United States botanicals are mostly considered as dietary supplements, Europe has established a sophisticated regulatory system for herbal medicinal products (HMPs) with monographs on quality, efficacy and safety (Fig. 25.1) [56]. The European Directorate for the Quality of Medicine and HealthCare (EDQM) in Strasbourg is elaborating monographs for the European Pharmacopoeia, which is legal in 36 European countries. These monographs define the quality of herbal drugs and preparations and provide validated quality control

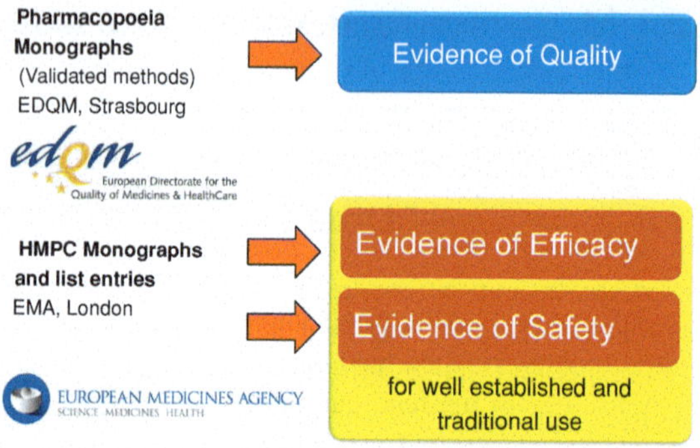

Fig. 25.1 Monographs used for providing evidence in European drug legislation

methods. They can be used to provide quality data in the application for marketing authorization of corresponding products. A TCM Working Party has been established and is currently elaborating monographs on Chinese herbs. So far 31 monographs of Chinese herbs have been adopted by the European Pharmacopoeia Commission. Drafts of 17 other monographs have been published with open access in PHARMEUROPA (http://pharmeuropa.edqm.eu/home/). Interested parties are encouraged to follow this process and to send in comments via the national authorities before the final adoption by the European Pharmacopoeia Commission.

The Committee on HMP (HMPC) at the European Medicines Agency (EMA) in London is responsible for monographs covering efficacy and safety issues, and is preparing a draft list of herbal substances, preparations and combinations (a "European positive list") and establishing Community monographs for well established and traditional HMP. These monographs can be used for the application for marketing authorization of products. Since no Chinese herbal drugs have been considered so far, it is suggested to establish a Sino-European Cooperative to draft such monographs.

25.7 Conclusion

The various challenges of TCM research can only be solved in an interdisciplinary network, using the most advanced methodologies of the post-genomic era. The expertise of the FP7 GP-TCM consortium and the ideas of new members will enable the Good Practice in TCM Research Association to develop and implement good practice in TCM research and to modernize TCM in an integrative manner.

References

1. Tang JL, Liu BY, Ma KW (2008) Traditional Chinese medicine. Lancet 372:1938–1940
2. Unschuld PU (1999) The past 1000 years of Chinese medicine. Lancet 354(4):SIV9
3. Scheid V (1999) The globalization of Chinese medicine. Lancet 354(4):SIV10

4. Cai J (1988) Integration of traditional Chinese medicine with western medicine—right or wrong? Soc Sci Med 27:521–529
5. Qiu J (2007) China plans to modernize traditional medicine. Nature 446:590–591
6. Uzuner H, Fan TP, Dias A, Guo DA, El-Nezami HS, Xu Q (2010) Establishing an EU-China consortium on traditional Chinese medicine research. Chin Med 5:42
7. Uzuner H, Bauer R, Fan TP, Guo DA, Dias A, El-Nezami H, Efferth T, Williamson EM, Heinrich M, Robinson N, Hylands PJ, Hendry BM, Cheng YC, Xu Q (2012) Traditional Chinese medicine research in the post-genomic era: good practice, priorities, challenges and opportunities. J Ethnopharmacol 140:458–468
8. Sheng-Ji P (2001) Ethnobotanical approaches of traditional medicine studies: some experiences from Asia. Pharm Biol 39(1):74–79
9. Qian ZZ, Dan Y, Liu YZ, Peng Y (2010) Pharmacopoeia of the People's Republic of China (2010 edition): a milestone in development of China's healthcare. Chin Herb Med 2(2):157–160
10. Bauer R, Franz G (2010) Modern European monographs for quality control of Chinese herbs. Planta Med 76(17):2004–2011
11. Zhang B, Peng Y, Zhang Z, Liu H, Qi Y, Liu S, Xiao P (2010) GAP production of TCM herbs in China. Planta Med 76(17):1948–1955
12. Heuberger H, Bauer R, Friedl F, Heubl G, Hummelsberger J, Nögel R, Seidenberger R, Torres-Londoño P (2010) Cultivation and breeding of Chinese medicinal plants in Germany. Planta Med 76(17):1956–1962
13. Zhao Z, Guo P, Brand E (2012) The formation of daodi medicinal materials. J Ethnopharmacol 140:476–481
14. Pferschy-Wenzig E-M, Bauer R (2009) Quality control of Chinese herbal drugs. In: Houghton P, Mukherjee PK (eds) Evaluation of herbal medicinal products: perspectives on quality, safety, and efficacy. Pharmaceutical Press, London, pp 1–502
15. Heubl G (2010) New aspects of DNA-based authentication of Chinese medicinal plants by molecular biological techniques. Planta Med 76(17):1963–1974
16. Liang YZ, Xie PS, Chan K (2010) Perspective of chemical fingerprinting of Chinese herbs. Planta Med 76(17):1997–2003
17. Sheridan H, Krenn L, Jiang R, Sutherland I, Ignatova S, Marmann A, Liang X, Sendker J (2012) The potential of metabolic fingerprinting as a tool for the modernisation of TCM preparations. J Ethnopharmacol 140:482–491
18. Tilton R, Paiva AA, Guan JQ, Marathe R, Jiang Z, van Eyndhoven W, Bjoraker J, Prusoff Z, Wang H, Liu SH, Cheng YC (2010) A comprehensive platform for quality control of botanical drugs (PhytomicsQC): a case study of Huangqin Tang (HQT) and PHY906. Chin Med 5:30
19. Zhao ZZ, Liang ZT, Chan K, Lu GH, Lee ELM, Chen HB, Li L (2010) A unique issue in the standardization of Chinese materia medica: processing. Planta Med 76:1975–1986
20. Chan K, Shaw D, Simmonds MS, Leon CJ, Xu Q, Lu A, Sutherland I, Ignatova S, Zhu YP, Verpoorte R, Williamson EM, Duez P (2012) Good practice in reviewing and publishing studies on herbal medicine, with special emphasis on traditional Chinese medicine and Chinese materia medica. J Ethnopharmacol 140:469–475
21. Wang J (2010) Evidence-based medicine in China. Lancet 375:532–533
22. Jiang M, Yang J, Zhang C, Liu B, Chan K, Cao H, Lu A (2010) Clinical studies with traditional Chinese medicine in the past decade and future research and development. Planta Med 76(17):2048–2064
23. Normile D (2003) Asian medicine. The new face of traditional Chinese medicine. Science 299:188–190
24. Flower A, Witt C, Liu JP, Ulrich-Merzenich G, Yu H, Lewith G (2012) Guidelines for randomised controlled trials investigating Chinese herbal medicine. J Ethnopharmacol 140:550–554
25. Jiang M, Lu C, Zhang C, Yang J, Tan Y, Lu A, Chan K (2012) Syndrome differentiation in modern research of traditional Chinese medicine. J Ethnopharmacol 140:634–642

26. van der Greef J (2011) Perspective: all systems go. Nature 480:S87
27. van der Greef J, Hankemeier T, McBurney RN (2006) Metabolomics-based systems biology and personalized medicine: moving towards n = 1 clinical trials? Pharmacogenomics 7:1087–1094
28. Melchart D, Weidenhammer W, Streng A, Reitmayr S, Hoppe A, Ernst E, Linde K (2004) Prospective investigation of adverse effects of acupuncture in 97733 patients. Arch Intern Med 164:104–105
29. Witt CM, Streng A, Weidenhammer W, Wagenpfeil S, Brinkhaus B, Willich SN, Melchart D (2007) The impact of patient expectations on outcomes in four randomized controlled trials of acupuncture in patients with chronic pain. Pain 128:264–271
30. Berman BM, Lao L, Langenberg P, Lee WL, Gilpin AM, Hochberg MC (2004) Effectiveness of acupuncture as adjunctive therapy in osteoarthritis of the knee: a randomized, controlled trial. Ann Intern Med 141:901–910
31. Haake M, Müller HH, Schade-Brittinger C, Basler HD, Schäfer H, Maier C, Endres HG, Trampisch HJ, Molsberger A (2007) German acupuncture trials (GERAC) for chronic low back pain: randomized, multicenter, blinded, parallel-group trial with 3 groups. Arch Intern Med 167:1892–1898
32. Linde K, Streng A, Jürgens S, Hoppe A, Brinkhaus B, Witt C, Wagenpfeil S, Pfaffenrath V, Hammes MG, Weidenhammer W, Willich SN, Melchart D (2005) Acupuncture for patients with migraine: a randomized controlled trial. JAMA 293:2118–2125
33. Vickers AJ, Rees RW, Zollman CE, McCarney R, Smith CM, Ellis N, Fisher P, Van Haselen R (2004) Acupuncture for chronic headache in primary care: large, pragmatic, randomised trial. BMJ 328:744
34. Jia J, Yu Y, Deng JH, Robinson N, Bovey M, Cui YH, Liu HR, Ding W, Wu HG, Wang XM (2012) A review of omics research in acupuncture: the relevance and future prospects for understanding the nature of meridians and acupoints. J Ethnopharmacol 140:594–603
35. Robinson N, Lorenc A, Ding W, Jia J, Bovey M, Wang XM (2012) Exploring practice characteristics and research priorities of practitioners of traditional acupuncture in China and the EU-A survey. J Ethnopharmacol 140:604–613
36. van der Greef J, van Wietmarschen H, Schroën J, Wang M, Hankemeier T, Xu G (2010) Systems biology-based diagnostic principles as pillars of the bridge between Chinese and Western medicine. Planta Med 76(17):2036–2047
37. Buriani A, Garcia-Bermejo ML, Bosisio E, Xu Q, Li H, Dong X, Simmonds MS, Carrara M, Tejedor N, Lucio-Cazaña J, Hylands PJ (2012) Omic techniques in systems biology approaches to traditional Chinese medicine research: present and future. J Ethnopharmacol 140:535–544
38. Zhang A, Sun H, Wang Z, Sun W, Wang P, Wang X (2010) Metabolomics: towards understanding traditional Chinese medicine. Planta Med 76(17):2026–2035
39. Tejedor Garcia N, Garcia Bermejo L, Fernandez Martinez AB, Olmos Centenera G, Kumari R, Xu Q, Cheng X, Watson S, Lucio Cazaña FJ (2012) MEDLINE-based assessment of animal studies on Chinese herbal medicine. J Ethnopharmacol 140:545–549
40. Barlow DJ, Buriani A, Ehrman T, Bosisio E, Eberini I, Hylands PJ (2012) In-silico studies in Chinese herbal medicines' research: Evaluation of in silico methodologies and phytochemical data sources, and a review of research to date. J Ethnopharmacol 140:526–534
41. Chen ST, Dou J, Temple R, Agarwal R, Wu KM, Walker S (2008) New therapies from old medicines. Nat Biotechnol 26:1077–1083
42. Biochemistry Stone R (2008) Lifting the veil on traditional Chinese medicine. Science 319:709–710
43. Zhang X, Liu Y, Guo Z, Feng J, Dong J, Fu Q, Wang C, Xue X, Xiao Y, Liang X (2012) The herbalome—an attempt to globalize Chinese herbal medicine. Anal Bioanal Chem 402:573–581
44. Shaw D (2010) Toxicological risks of Chinese herbs. Planta Med 17:2012–2018
45. Debelle FD et al (2008) Aristolochic acid nephropathy: a worldwide problem. Kidney Int 74:158–169

46. Fu PP, Xia QS, Lin G, Chou MW (2004) Pyrrolizidine alkaloids-genotoxicity, metabolism enzymes, metabolic activation, and mechanisms. Drug Metab Rev 36:1–55
47. Tai YT, But PP, Young K, Lau CP (1993) Adverse effects from traditional Chinese medicine. Lancet 341:892
48. Fugh-Berman A (2000) Herb-drug interactions. Lancet 355:134–138
49. Ouedraogo M, Baudoux T, Stévigny C, Nortier J, Colet JM, Efferth T, Qu F, Zhou J, Chan K, Shaw D, Pelkonen O, Duez P (2012) Review of current and "omics" methods for assessing the toxicity (genotoxicity, teratogenicity and nephrotoxicity) of herbal medicines. J Ethnopharmacol 140:492–512
50. Shaw D, Graeme L, Pierre D, Elizabeth W, Kelvin C (2012) Pharmacovigilance of herbal medicine. J Ethnopharmacol 140:513–518
51. Zhang L, Yan J, Liu X, Ye Z, Yang X, Meyboom R, Chan K, Shaw D, Duez P (2012) Pharmacovigilance practice and risk control of Traditional Chinese Medicine drugs in China: current status and future perspective. J Ethnopharmacol 140:519–525
52. Youns M, Hoheisel JD, Efferth T (2010) Toxicogenomics for the prediction of toxicity related to herbs from traditional Chinese medicine. Planta Med 76(17):2019–2025
53. Fan TP, Deal G, Koo HL, Rees D, Sun H, Chen S, Dou JH, Makarov VG, Pozharitskaya ON, Shikov AN, Kim YS, Huang YT, Chang YS, Jia W, Dias A, Wong VC, Chan K (2012) Future development of global regulations of Chinese herbal products. J Ethnopharmacol 140:568–586
54. Dobos GJ, Tan L, Cohen MH, McIntyre M, Bauer R, Li X, Bensoussan A (2005) Are national quality standards for traditional Chinese herbal medicine sufficient? Current governmental regulations for traditional Chinese herbal medicine in certain Western countries and China as the Eastern origin country. Complement Ther Med 13(3):183–190
55. Pelkonen O, Pasanen M, Lindon JC, Chan K, Zhao L, Deal G, Xu Q, Fan TP (2012) Omics and its potential impact on R&D and regulation of complex herbal products. J Ethnopharmacol 140:587–593
56. Vlietinck A, Pieters L, Apers S (2009) Legal requirements for the quality of herbal substances and herbal preparations for the manufacturing of herbal medicinal products in the European union. Planta Med 75(7):683–688

Author Biography

Dr. Rudolf Bauer studied pharmacy at the University of Munich; 1984 graduation as Ph.D. at the Institute of Pharmaceutical Biology, University of Munich; 1993–2002 Associate Professor at the Institute of Pharmaceutical Biology, University of Düsseldorf; since 2002 full professor of pharmacognosy, and since 2004 Head of the Institute of Pharmaceutical Sciences at University of Graz. He has long experience in natural product chemistry, analysis and the bioassay-guided isolation of constituents from medicinal plants. He has published more than 280 research papers. He has been president of the Society for Medicinal Plant and Natural Product Research (GA) and is founding president of The GP-TCM Research Association, as well as Editor of *Planta Medica* and Associate Editor in Chief of *Chinese Journal of Natural Medicines*; Prof Bauer has been active in the development methods for quality control of Chinese herbs for 21 years. He is member of the expert group 13A and of the TCM Working Party of the European Pharmacopoeia Commission and involved in the development of monographs for the European Pharmacopoeia; he is member of the TAM Advisory Board of the Austrian Minister of Health; together with Professor Litscher (Medical University of Graz) he is heading the TCM Research Center Graz.

Chapter 26
High-Throughput Screening in Traditional Chinese Medicine-Based Drug Discovery

Catherine C.K. Wong and Yung Hou Wong

Abstract Traditional Chinese medicine (TCM) has demonstrated its effectiveness in the treatment of ailments and diseases over a long history of application, but its wider acceptance has been hindered by a lack of scientific evidence for efficacy and safety as well as a barrier of languages and cultures. In the 1990s, biotech and pharmaceutical companies have started to take interest in traditional medicines in their efforts to replenish the drying pipelines, yet the results were not sufficiently impressive to warrant extensive attention. Recently, along with the rapid advancements in biology and new technologies, there is a resurgent interest in the potential of the ancient medicine. Given our proximity to the Chinese mainland, our institute has utilized TCM as a source of potential drug candidates as well as a probe for understanding complex signaling pathways. This chapter describes the application of a number of versatile high throughput techniques for identifying therapeutic agents from TCM preparations. The rationale and principles for each approach will be discussed and illustrated with examples.

26.1 Introduction

In recent years we have witnessed an unprecedented accumulation of knowledge on molecules and molecular processes that sustain the functions of cells. This together with significant advances in technological development has provided a rich knowledge base for understanding human disease physiology. However, the productivity performance of drug research has been disappointing over the past two decades. New strategies are required to revitalize the discovery and development process, and to improve the introduction of new medicinal agents to meet unmet medical needs.

C. C. K. Wong · Y. H. Wong
Biotechnology Research Institute and Division of Life Science, Hong Kong University of Science and Technology, Clear Water Bay, Kowloon, Hong Kong, China

Y. H. Wong (✉)
Division of Life Science, State Key Laboratory of Molecular Neuroscience, Hong Kong University of Science and Technology, Clear Water Bay, Kowloon, Hong Kong, China
e-mail: boyung@ust.hk

E. F. Fang and T. B. Ng (eds.), *Antitumor Potential and Other Emerging Medicinal Properties of Natural Compounds*, DOI: 10.1007/978-94-007-6214-5_26,

The emerging field of systems biology provides an answer by enabling the study of living systems from a holistic perspective. It represents an integration of concepts and ideas from a wide array of disciplines; and strives to gain a system-level understanding of biological systems by examining the structure and dynamics of cellular and organismal functions, rather than the characteristics of isolated parts of a cell or an organism [1]. This new framework of analysis opens up the possibility of studying the synergistic effects of complex mixtures in biological systems [2], and puts forward a new concept of "magic shrapnel" drug which acts on multiple molecular targets [3]. Traditional medicines, such as TCM and Indian Ayurveda, have been reinvestigated with this new approach and are becoming important sources for the discovery of novel therapeutic agents and for designing multi-targets drugs [4].

TCM has evolved over thousands of years with remarkable effectiveness in the treatment of various ailments and diseases. It has been reported that there are 12,870 kinds of TCM resources, including plants, animals and mineral substances [5]. The wealth of empirical experience and the vast variety of TCM materials increase the prospects for obtaining biologically active hits to specific diseases, representing a tremendous advantage to drug discovery and development. The best known TCM-derived drug around the world is probably the potent antimalarial artemisinin [6, 7]. Other classic examples of TCM successfully adapted by the Western pharmaceutical and nutraceutical industries include *Tripterygium wilfordii* which is being developed as a botanical drug for rheumatoid arthritis, and *Camellia sinensis* which is used as functional beverage and a component of dietary supplements [8]. A recent analysis using the artificial intelligence methodology indicates that ~60 % of the molecules in Chinese Medicine Compound Database possess drug-likeness, affirming that TCM is an excellent source of drug-like molecules [9]. As illustrated by the recent dedication of a *Nature* supplementary to the theme (Vol. 480 No. 7378 supp ppS81–S121), there is a growing interest worldwide in harnessing the therapeutic potential of TCM.

High-throughput screening (HTS) is a core function in the drug discovery process that enables the testing of large number of candidate molecules in a biological assay. The primary goal is to identify hits or leads, compounds that affect a particular biological target in the desired manner. It involves large, highly automated liquid handling and detection systems that allow thousands of compounds to be screened for biological activity against a test assay. This chapter provides an overview of the application of HTS in TCM-based drug discovery, from the selection of herbal materials to preclinical assessments (Fig. 26.1), with examples to illustrate the viability of this approach.

26.2 TCM Selection

The common approaches to develop new treatments from TCM focus on: (1) single molecules from the herbs, (2) standardized extracts from a single herb, and (3) herbal formula with definite composition. A broadly applicable methodology to select study

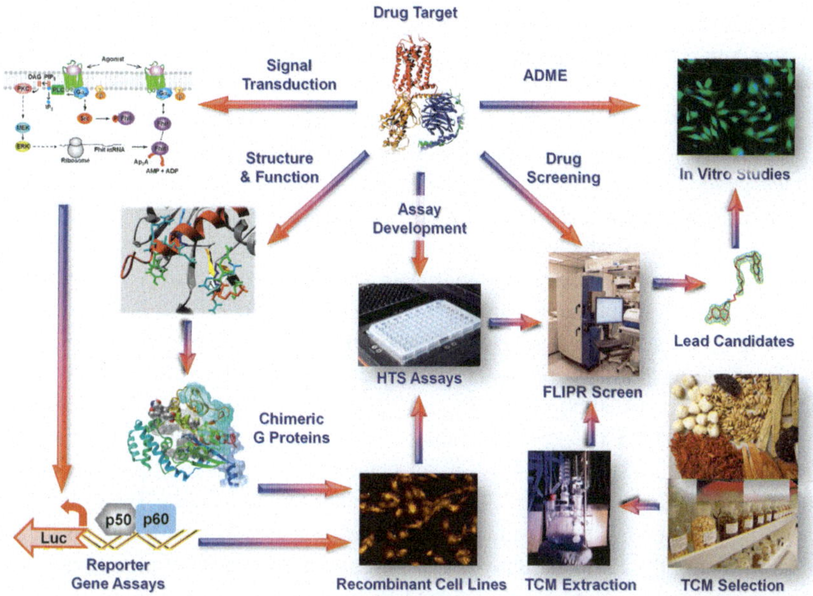

Fig. 26.1 An overview of TCM-based HTS process. The selection and matching of TCM and drug targets are based on extensive literature review. Selected TCM herbs are extracted by standardized procedures. Drug targets can be incorporated into HTS assays directly or engineered into recombinant cell lines expressing adapter proteins (e.g., chimeric G proteins). A thorough understanding of the structure and functions of the drug target, as well as the signal transduction pathways being regulated, can facilitate the development of assay tools (e.g., reporter genes). Hits identified in the HTS platforms will undergo secondary confirmation and lead optimization, before being subjected to *in vitro* ADME

samples is to exploit the rich medical literature which provides details on the empirical indications and usage of the TCM products. The effectiveness of this approach can be shown in a recent study which identified herbs with potential beneficial effects on cognitive functioning [10]. Another line of research is to study structure–activity relationships of families of TCM-derived compounds, such as phenylethanoid glycosides, to gain insights of potential leads for drug design [11]. Recent advances in chemoinformatics have added to the tools available for TCM-based drug discovery. Techniques such as database construction, molecular similarity searching, virtual screening and inverse docking, provide hints and guidance on selecting diverse compounds for screening and optimization [12]. Several studies dealing with the virtual screening of Chinese herbs have led to the identification of compounds that are inhibitors of specific enzymes or receptors [13]. The combination of these techniques would enable a more unified and knowledge-based approach to assemble TCM-derived chemical libraries for efficient screening. One note of caution is that the holistic concepts of drug administration in TCM have not yet been fully incorporated into the current approaches for identification of candidate drugs from the traditional medicine. Further research into experimental and clinical sciences would be needed to explain how TCM can exert effects in multi-factorial and complex conditions like Alzheimer's and vascular dementia [14].

TCM products selected with different approaches are placed through different protocols to prepare for experimental screening. A typical procedure in use includes general solvent extraction and further fractionation with reverse-phase high performance liquid chromatography (HPLC) or normal phase-HPLC. The raw material is first extracted by a series of solvents with different polarities at distillation temperature. Extracts thus obtained are subjected to reverse-phase or normal phase fractionation according to the solvent properties. Conditions for chromatographic separation are optimized to ensure complete elution of various components. The eluted samples are collected either directly into 96-well plates or test tubes for redistribution into multiple 96-well plates. The initial microplate sets are often termed the "mother plates" in which the samples can be stored dried. The plates are stored at –20 °C and replicated later into multiple "daughter" microplate sets, each replicate set being designated for a particular HTS platform. Corresponding electronic data files can be created for each microplate prepared, thereby allowing the samples to be tracked through the subsequent processes.

26.3 Target Selection

A drug target refers to a key molecule that has a proven function in the pathophysiology of a disease, and the activity can be modulated by a drug. A consensus number of 324 drug targets had been proposed for all classes of approved therapeutic drugs in 2006 [15]. The central challenge of selecting good drug targets for TCM is posed by the lack of mechanistic insights into the synergistic effects of the mixture of constituents. We have employed the strategy to characterize the actions of herbal medicines at disease-specific molecular targets followed by subsequent identification of chemical compounds underlying these molecular effects.Following this line of reasoning, a research program was initiated to characterize *in vitro* molecular actions of TCM drugs that are often used for the treatment of stroke. The results indicate that these drugs possess activity at disparate molecular targets in the signaling pathways involved in *N*-methyl-D-aspartate (NMDA) receptor-mediated neuronal injury and death [16, 17]. Over the past few years, the systems biology approach has enabled researchers to investigate the actions of TCM in a dynamic biological system. Applications of various omic analytical techniques have been used to study the biological effects of Chinese herbal medicine in a wide range of conditions, such as cancer, fibrosis, gastrointestinal and cardiovascular diseases, metabolic and immune disorders, neurological and psychiatric conditions [18]. The recent use of metabonomics has yielded some successes in identifying early biochemical changes in Chinese medicine syndromes [19]. Further research endeavors in this direction would help to unravel the multi-target therapeutic concept of TCM and facilitate selection of ideal targets for TCM-based drug discovery.

While disease relevance is the main driver when choosing a therapeutic target, the process is also influenced by a complex balance of scientific, medical and strategic considerations [20]. The known drug targets could be broadly divided into four major classes. They are: (1) G Protein-Coupled Receptors (GPCRs), (2) nuclear (hormone) receptors, (3) ion channels, and (4) enzymes including the different protein kinases. Pharmacologically active compounds that act via GPCRs constitute the largest category of clinical drugs in current use. New estimates show that the global market for GPCRs targeting drugs will reach US$120.5 billion by 2017 [21]. There are three different G protein subtypes; each corresponds to the activation of a different intracellular second messenger system. Traditionally, this has meant that measuring the ability of a series of ligands to activate different GPCRs requires the establishment of three different assay platforms. We have developed a proprietary universal G protein adapter technology (US Patent 6,462,178) that allows GPCRs with different G protein-coupling specifications to be screened using a single assay format. Based on these considerations, our initial efforts were directed to screen for TCM extracts or compounds that act on GPCRs.

26.4 HTS Assay Development

The first step in the HTS process is the development of a testing method or screen. Screen development involves the miniaturization of prototype assays and the optimization of HTS operations. Often a biological assay in its earliest form is highly cumbersome and suitable only for validating biological activity. The assay readout may be only semi-quantitative and impractical for HTS. Assay methodologies have to be reduced to a simple microplate-based design, producing a quantitative readout which can be measured in a microplate-compatible instrument. For speed, ease of use and maximum reproducibility, the screening protocol should incorporate as few steps as possible. Moreover, *in vitro* assays are performed in artificial environments in which the biological system studied can be unstable or exhibiting an activity below its potential. We have miniaturized a number of assays to the microplate format, and subjected them to a rigorous validation process before they are used for the routine assay of TCM libraries. Hits derived from the assays described in the following paragraphs are confirmed by standard biochemical assays such as those that measure intracellular second messengers. Most HTS assays employed in the pharmaceutical industry are amenable to TCM-based drug discovery.

26.4.1 Cell-Based Assays

Most available drug targets for selected therapeutic areas can be incorporated into cell-based assays for screen development. This entails cloning of specific drug receptors and enzymes, expression of recombinant proteins, and establishment of recombinant cell lines. The cDNAs of the molecular targets are verified by restriction digests and the functional expressions of receptors are determined by transient transfections (Fig. 26.1). Where available, the human molecular targets are introduced into cultured cell lines for cell-based drug screening. In the absence of human constructs, rat or mouse molecular targets are preferred since more pharmacological data are available for these species. Particular efforts have been devoted to generate cell lines each stably expressing a $G\alpha_{16/z}$ chimera for high-throughput assays [22]. GPCRs represent the largest and most lucrative class of therapeutic targets as they control an enormous array of diverse biochemical, physiological and disease processes. The understanding of how GPCRs detect extracellular signals and transduce them to intracellular G proteins and second messenger pathways has led to the development and exploitation of chimeric and promiscuous G protein α subunits. These adaptors, such as qi5, $G\alpha_{16}$ and 16z44, allow signals from diverse GPCRs to be directed to the generation of an easily detectable product. This has enabled almost any GPCR to be incorporated into screening platforms such as fluorometric imaging plate reader (FLIPR), aequorin assays, yeast autocrine selection and microarray assays. The use of cell lines stably coexpressing chimeric $G\alpha$ protein [22–24] and GPCRs simplifies the drug screening process and aids in the deorphanization of new receptors.

26.4.2 Reporter Gene Assays

Amongst the different types of assays developed for the detection of bioactivity in extracts, fractions, as well as purified compounds of herbal origin, reporter gene assays are highly versatile and reliable. The reporter gene unit consists of a promoter and a reporter gene. The promoter region binds transcription factors that can be activated either by a broad range of biochemical pathways or by the selective activation of individual targets [25]. A variety of reporter genes allow users a choice of signal that can be tailored to the required sensitivity, the available detection apparatus, the cellular system employed, and the required compatibility with multiplexed assays. The reporter genes that are of particular relevance to natural product assay include: chloramphenicol acetyltransferase (CAT), β-galactosidase, aequorin, green fluorescent protein and its derivatives, luciferase, secreted placental alkaline phosphatase (SPAP) and β-lactamase. The latter two are called extracellular reporter genes for their ability to encode proteins that are secreted into the culture medium. The reader can refer to other reviews for more details on their basic methodology, capacities and limitations. We have successfully adapted a number of luciferase and aequorin assays into 96-well format for the screening of ligands for GPCRs and other targets [26]. In an early investigation, we have used a reporter gene assay to detect the presence of phytoestrogens in a TCM formula, Bupleurum and Peony Formula, which possesses a clinical history for the treatment of menopausal syndrome and menstrual irregularity. The assay employs two different cDNA components to detect responses of HeLa cells to oestrogenic substances. The first component, the Gal4-HEGO chimeric receptor, consists of the ligand binding domain of the oestrogen receptor linked to the DNA binding domain of the yeast transcription factor, Gal4. In addition, a luciferase reporter gene was used that consists of the firefly luciferase cDNA regulated by the rabbit β-globin basal promoter and five tandem consensus Gal4 response elements. Induction of reporter gene expression is dependent upon ligand activation of the chimeric receptor [27]. Using a 96-well format in a microplate luminometer, this assay proved to be sensitive to concentrations of the formula as low as 0.02 % (w/v), suggesting that this system is suitable for use as a quick, sensitive and quantifiable bioassay for detecting the presence of phytoestrogens in complex TCM formulas.

26.4.3 Cell Growth and Viability

Despite the numerous attempts to meet the huge needs, drug development for central nervous system (CNS) disorders remains challenging. A variety of compounds isolated from herbal preparations exhibited pro-survival effects in various neurotoxic models [28]. Although the underlying mechanisms have not been delineated for most compounds, TCM is undoubtedly a valuable source of neuroprotective agents. Neuronal loss occurs as an essential feature in acute or chronic neurodegenerative disorders. It has long been recognized that many factors can trigger secondary auto-destructive reactions within the CNS, including excitatory amino acids, eicosanoids, lipid degradation products, tissue cations, inflammation, and immune responses. These secondary injury factors are released or activated over a period from seconds to days after the primary insult and may act either sequentially or in parallel to cause delayed or expanding cell death. We have developed a number of assays in 96-well format for early recognition of drug candidates, including:

TrkA Fibroblast Survival Assay: Nerve growth factor (NGF) is one of the neurotrophins that supports the survival and differentiation of a variety of neurons. Any TCM-derived small molecules that mimic or induce NGF activity may represent potential treatment for neurodegenerative diseases. An effective way to screen for the NGF-like activities of TCM compounds is to evaluate their survival effects on cell lines expressing TrkA, the high-affinity receptor for NGF. Upon serum deprivation, the survival of these cells is dependent on the presence of NGF-like growth factors in TCM preparations. Quantification of cell survival is based on the tetrazolium salt (MTT) assay that specifically detects the metabolic activity.

Prevention of Neuronal Cell Apoptosis: Caspase-3, a member of the cysteine aspartase family, is activated early in apoptosis and involved in a cascade of proteolysis of several important molecules that facilitate the process of cell death. A colorimetric assay using neuronal cell lines can be used to evaluate the inhibitory effects of TCM preparations on the enzyme.

26.4.4 Enzymatic Assay

Another important treatment approach for CNS diseases involves reversible inhibition of acetycholinesterase (AChE), thereby combating the loss of acetylcholine caused by the death of cholinergic neurons [29]. We have established a high throughput assay to screen for potential inhibitors of AChE activity. The traditional colorimetric method involves a series of biochemical reactions. The presence of the AChE inhibitors in TCM preparations inhibits the breakdown of the substrate acetylcholine-iodide (ACh-I) to acetic acid and choline, a reaction catalyzed by the AChE; and hence preventing the reaction of acetic acid and the chromogen 5,5′-dithio-bis(2-nitrobenzoic) acid (DTNB) to form yellow anions of 5-thio-2-nitro-benzoic acid. An AChE inhibitory curve and the IC_{50} can be determined using a series of drug concentrations. We have optimized a new and rapid protocol that can monitor the inhibitory effect of the drugs using fluorescence multi-well plate scanner. The activity of AChE is indirectly monitored using a sensitive fluorogenic probe 10-acetyl-3,7-dihydroxyphenoxazine which detects H_2O_2, the secondary breakdown product of choline by choline oxidase. The reaction is performed in multi-well plates and the fluorescence emitted correlated with the potency of the drug. One of the advantages of using this one-step protocol is the significant reduction in the experimental time involved. Moreover, potential interference due to the inherent color of the samples, a problem often encountered with the colorimetric method, is minimized.

26.5 Functional Assays

Many TCM preparations are associated with a broad range of immunomodulatory activities. The activation of immune-responsive cells via a variety of membrane receptors is frequently coupled to the activation of various sequential kinase cascades [30]. In general, receptor activations often lead to the stimulation of mitogen-activated protein kinases (MAPKs). Different subtypes of MAPKs including extracellular signal-regulated kinase (ERK), c-Jun N-terminal kinase (JNK) and p38 MAPK regulate the stimulatory phosphorylation of transcription factors, such as Elk, c-Jun and ATF-2 [31]. Increased activities of these intracellular signaling events often result in the activation of immune-responsive cells, and the subsequent changes in cytokine production which serve as modulators for different phases of the immune response. Hence, detection of the signaling intermediates in their stimulatory phosphorylated forms and the production of cytokines can serve as useful indicators of whether administration of a pharmacological agent is linked to immune-modulatory effects. To prove the practicality of this approach, we have conducted different studies to demonstrate how TCM formula, herbal extract, and constituents regulate intracellular signaling pathways of immuno-responsive cells such as Jurkat T-lymphocytes and THP-1 monocytes [32–35]. These cell models can be utilized to screen for the active constituents in TCM preparations capable of modulating immuno-responses.

Other approaches may also prove useful in characterizing TCM products with anti-inflammatory activity. In the study of the popular TCM nutraceutical 'Kwei Ling Ko' (KLK; Tortoise shell-Rhizome jelly), we investigated the effects of KLK extract on peroxisome proliferation-activated receptor (PPAR) pathway. PPARs are members of the nuclear hormone receptor/transcription factor superfamily with multiple roles in adipocyte differentiation, glucose homeostasis, immunomodulation and anti-inflammatory regulation. As PPAR is required for adipocyte induction, we used adipogenesis as a possible screen for the activation of the PPAR pathway, and showed that KLK extract was able to induce the adipocyte differentiation of fibroblast cell lines. The activity of nuclear factor-κB (NFκB), a transcription factor responsible for the regulation of proinflammatory genes, was down-regulated in response to the extract. Luciferase reporter gene assays further demonstrated that KLK inhibited both basal and tumor necrosis factor-α-stimulated NFκB activation. These results give insights on possible mechanism through which some TCM preparations manifest their anti-inflammatory effects [36].

26.6 Cytotoxicity Assay

In view of the high attrition rate in drug discovery, there is a strong need to detect drug toxicity at early stages. *In vitro* cytotoxicity testing has become an integral aspect of drug discovery, and a variety of different cell lines have been used to predict organ specific toxicity [37, 38]. The human HepG2 hepatocellular carcinoma cell line is of particular interest as the liver is the primary site for drug metabolism. A wide array of assays has been developed to measure different indicators of cytotoxicity on the cell line [39]. Furthermore, the cells can provide useful mechanistic information regarding the chemotherapeutic potential for controlling malignant hepatocyte growth of liver cancer [40, 41]. We have employed the model to study the JNK and p38 subgroups of the MAPK family, which are closely associated with cellular death signals from membrane receptors and extracellular stress [42, 43]. An early study using this approach revealed that the TCM compound, protocatechuic acid, is capable of inducing JNK-dependent hepatocellular carcinoma cell death [44].

26.7 Pharmacokinetics

A large number of drug candidates exist that are both very potent and efficacious in eliciting highly selective responses at molecular targets *in vitro*, but are devoid of measurable effects when given *in vivo* and hence are of little or no utility as therapeutic agents. The possible causes are either that the compounds are not readily absorbed from the site of administration or are rapidly metabolized or excreted. Defining absorption, distribution, metabolism and excretion (ADME) of a drug is important not only to determine its ultimate utility in treating disease but also to predict ancillary effects such as drug/drug interaction. Implementation of high-throughput ADME screening strategies at early stage would facilitate early discrimination of leads and failures. Furthermore, new pharmacokinetics tools are available to provide reliable means for screening and identifying the potentially bioactive components contributing to the pharmacological effects of TCM [45]. We have incorporated the following *in vitro* models in our drug discovery program [46–48]. The results from these studies give important parameters for the further development of drug candidates.

Caco-2 model: The use of intestinal epithelial cell lines for oral absorption screening has become increasingly important at the early stage of drug discovery. Amongst various *in vitro* systems, the Caco-2 model has become the most popular choice due to its high throughput potential and reliability. The Caco-2 model is suitable for screening purposes such as the identification of drugs with potential absorption problems and to select drugs candidates with adequate absorption characteristics from a series of pharmacologically active molecules. It can provide comparative information on the intestinal absorption, metabolism and the mucosal toxicity of drug molecules with high throughput.

Rat liver microsomal model: Metabolic transformation of a drug determines the time it circulates in the body (half-life), which in turns determines the effectiveness and safety of the drug. For this reason, candidate drug compounds are often screened early in the discovery process for metabolic stability. Biotransformation occurs in many tissues, such as the kidneys and intestine, but by far the most important organ is the liver. A common system for measuring hepatic metabolism uses liver microsomes, a subcellular fraction containing major drug-metabolising enzymes, including the cytochrome P450 family and flavin monooxygenase. The liver microsomal system offers high throughput at a low cost, and is one of the best-characterized *in vitro* systems for drug biotransformation research that is simple to use.

Apart from the general characterization of the pharmacokinetics of TCM-based drug candidates, it would be advantageous to evaluate the ability of the drug candidate to penetrate the blood–brain barrier (BBB). The BBB consists of a continuous layer of endothelial cells joined by tight junctions at the cerebral vasculature. It represents a physical and enzymatic barrier to restrict and regulate the penetration of compounds into and out of the brain and maintain the homeostasis of the brain microenvironment. Brain penetration is essential for compounds where the site of action is within the CNS, whereas BBB penetration needs to be minimized for compounds that target peripheral sites to reduce potential CNS-related side effects. Therefore, it is critical during the drug discovery phase to design and select compounds having appropriate brain penetration properties for drug targets that reside within and outside the CNS. A number of *in vitro* BBB models are available for drug screening [49].

Monoculture of Primary Endothelial Cells: Primary cultures of brain vascular endothelial cells have been reported to be derived from various sources, namely, bovine, mouse, rat, porcine, non-human primate and human. Brain microvessels are isolated from cortex and digested to produce endothelial cells for growth in cell culture. The endothelial monolayer can be grown on porous membranes, which can be placed in side-by-side diffusion chambers for measurement of drug transport across the monolayer *in vitro*. The resulting cultures are usually of high purity (>95 %) with BBB properties well conserved.

Monoculture of Immortalized Brain Endothelial Cells: A number of immortalized brain endothelial cell lines are available. The advantages of using genetically engineered brain endothelial cell lines are that they are easy to culture and can be used in HTS assays. Furthermore, they provide a homogenous and phenotypically stable population of cells.

Co-cultures: The *in vitro* models that best mimic the anatomical conditions of BBB are co-cultures in which astrocytes and/or pericytes are included [50]. Several co-cultures have been established with primary astrocytes, glial cells, pericytes, neurons and blood cells, in all possible combinations with a variety of sources of BBB endothelial cells. These models represent more predictive tools to determine permeability of drug candidates across BBB.

26.8 Examples of Drug Leads Isolated from TCM

We have taken TCM extracts and compounds through the different HTS platforms, identified hits, verified lead candidates in secondary screens, determined the selectivity profiles of the lead candidates, and further characterized the lead candidates to a stage appropriate for the filing of patent applications. Several drug candidates have been advanced into scale-up manufacturing and pre-clinical studies. Some examples are listed below.

26.8.1 Melatoninergic Ligands

Melatonin (N-acetyl-5-methoxytryptamine), the pineal gland hormone, is widely distributed in mammalian tissues and exerts its action via two melatonin receptor sub-types, MT_1 and MT_2. Melatonin is known to play functional roles in regulating circadian rhythms and seasonal reproduction. In recent years, growing evidence has also linked melatonin to a variety of other body systems and disease states, thus highlighting its significance as a therapeutic agent. However, due to its properties, melatonin is ineffective in clinical use, thus prompting the development of melatoninergic ligands that mimic the actions of melatonin but in a manner that is more potent and specific for melatonin receptors. While there are over seventy patents on melatoninergic ligands, success in developing therapeutically effective melatoninergic ligands is limited [51].

We have successfully identified novel isoquinolinone compounds that act as agonists at the melatonin MT_1 and MT_2 receptors, but not at the *Xenopus* melatonin receptor. The compounds are derived from an herb from Northeastern China and Shanxi province that has long been used for the treatment of epilepsy and cerebral ischemia. Recent studies suggest that melatonin agonists are useful in the treatment of pain and in the prevention or treatment of convulsions, including epilepsy. The identification of isoquinolinone compounds with melatonin agonistic effects is thus in accordance with the clinical application of the herb. The target-selectivity of the compounds was subsequently confirmed by re-screening against multiple GPCRs in the FLIPR platform. The compounds only served as agonists at the melatonin MT_1 and MT_2 receptors amongst 30 different GPCRs, including representative candidates from aminergic, dopaminergic, serotonergic, muscarinic, GABAergic, and peptidergic receptors. Further characterizations were conducted using recombinant cell lines stably expressing either the melatonin MT_1 or MT_2 receptors. These confirmed their efficacy at these two receptor subtypes as well as indicating their potencies in comparison with current melatonin agonists. Several families of derivatives with distinctive structural features have been synthesized based on the natural single compounds [52]. Structure–activity relationship analyses of selected series of compounds have been conducted to obtain valuable information for further development of more selective and metabolically stable compounds, which could potentially be used for treating insomnia, cancer, diabetes and neurodegenerative diseases. A schematic presentation of the process is given in Fig. 26.2, which also represents the general workflow involved in many TCM-based HTS.

26.8.2 Telomerase Activators

Our capabilities and achievements in TCM-based drug discovery have led to a growing number of collaborative opportunities with major pharmaceutical companies. A notable example is the long-time strategic partnership with the San Francisco-based Geron Corporation. We have employed a systematic approach to screen TCM products with Geron's proprietary telomerase technology, and successfully identified small molecule compounds with strong telomerase modulation activity. Amongst these are astragaloside IV, the major active constituent of *Astragalus membranaceus*, and its alycone derivative cycloastragenol. Telomerase is a specific ribonucleoprotein reverse transcriptase capable of synthesizing or maintaining the length of a telomere through adding hexameric DNA repeats to telomeric ends [53]. Controlled telomerase activation has emerged as an attractive approach to restore the regenerative or functional capacity of cells afflicted with degenerative conditions or chronic stress [54–59]. In contrast to most normal somatic cells, which show little or no telomerase activity, immune cells

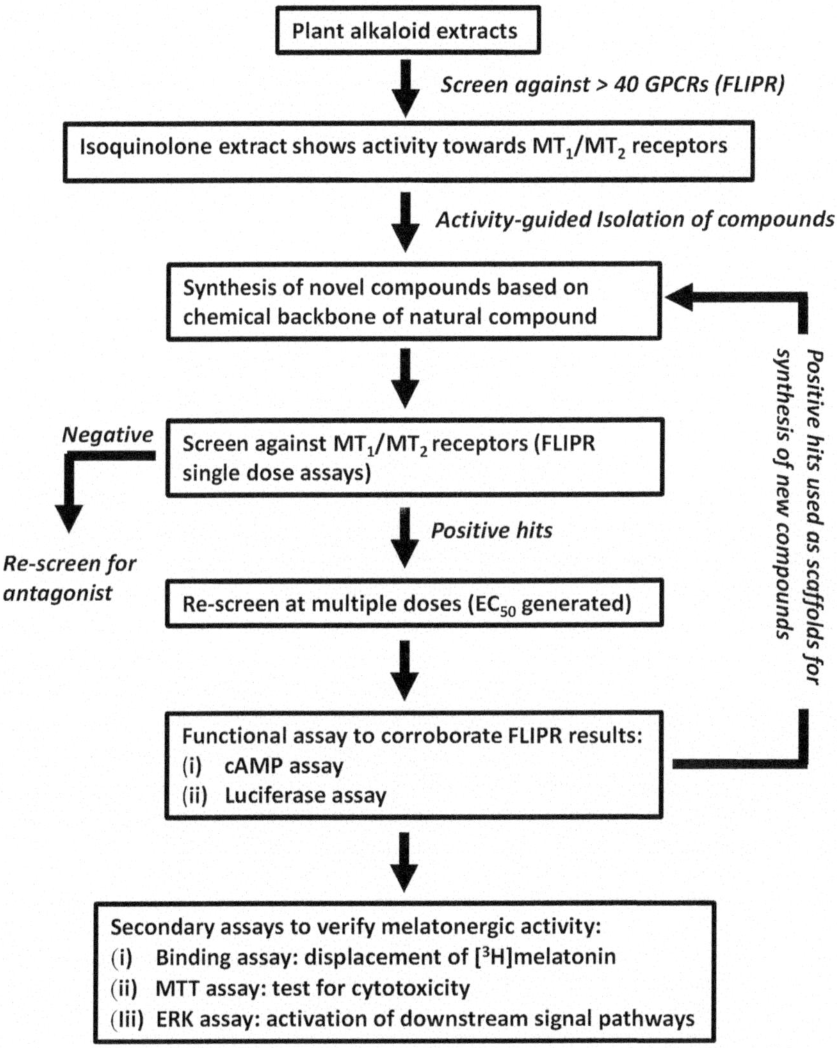

Fig. 26.2 Steps involved in the identification of novel melatoninergic compounds from Chinese medicine

up-regulate telomerase in concert with activation. During aging and chronic HIV-1 infection, there are high proportions of dysfunctional $CD8^+$ cytotoxic T lymphocytes with short telomeres. *In vitro* studies showed that cycloastragenol can transiently activate telomerase, retard telomere shortening, increase proliferative potential, and thereby enhance cytokine/chemokine production and antiviral response of $CD8^+$ T lymphocytes from HIV-infected patients [60]. Our study further revealed that both astragaloside IV and cycloastragenol can stimulate ERK phosphorylation in multiple cell types through activation of the Src/

MEK/ERK pathway [61]. Another study indicated that cycloastragenol can be efficiently absorbed through intestinal epithelium; but it undergoes extensive first-pass metabolism in rat and human liver microsomes, thereby reducing its oral bioavailability [46]. Nonetheless, these findings provide a new direction for screening potential drug leads and investigating mechanisms underlying the beneficial effects of TCM.

26.9 Future Perspectives

The pharmaceutical industry is currently facing challenges of epic proportions. The cost of developing a new drug has increased, as have total R&D expenditures, while the rate of introduction of new molecular entities has at best remained constant and attrition rates have risen sharply [62]. It indicates that the pool of easily discoverable new drugs has been temporarily depleted. Another probable reason is that there is a shift in demand toward drugs for chronic and degenerative illnesses. Lengthier and more extensive clinical trials are required to demonstrate the effects of these types of drugs. The major pharmaceutical companies are facing serious shortages of quality drug candidates in their product pipelines. TCM offers tremendous advantage by allowing a "knowledge-based" drug discovery program through which the screening of TCM preparations is guided by the wealth of empirical experience. The TCM-based HTS program outlined above allows thousands of compounds to be screened against a test assay, and achieving a hit rate much higher than random library screening. It is envisioned that the combination of TCM and new technologies will provide novel treatment regimens to combat the debilitating consequences of diseases and disorders.

Acknowledgments Support for the work came from the Innovation and Technology Fund (AF/178/97, AF/166/99, ITS/119/00, ITS/113/03), the University Grants Committee (AoE/B-15/01, AoE/B-15/01-II, T13-607/12R and HKUST 660108), the Hong Kong Jockey Club Charities Trust, and Geron Corporation.

References

1. Kitano H (2002) Computational systems biology. Nature 420:206–210
2. Wang M, Lamers RJ, Korthout HA, van Nesselrooij JH, Witkamp RF, van der Heijden R, Voshol PJ, Havekes LM, Verpoorte R, van der Greef J (2005) Metabolomics in the context of systems biology: bridging traditional Chinese medicine and molecular pharmacology. Phytother res 19:173–182
3. Chen X, Zhou H, Liu YB, Wang JF, Li H, Ung CY, Han LY, Cao ZW, Chen YZ (2006) Database of traditional Chinese medicine and its application to studies of mechanism and to prescription validation. Br J Pharmacol 149:1092–1103
4. Wang Y, Fan X, Qu H, Gao X, Cheng Y (2012) Strategies and techniques for multi-component drug design from medicinal herbs and traditional chinese medicine. Curr Top Med Chem 12:1356–1362

5. Li WF, Jiang JG, Chen J (2008) Chinese medicine and its modernization demands. Arch Med Res 39:246–251
6. Tu Y (2011) The discovery of artemisinin (qinghaosu) and gifts from Chinese medicine. Nat Med 17:1217–1220
7. Miller LH, Su X (2011) Artemisinin: discovery from the Chinese herbal garden. Cell 146:855–858
8. Graziose R, Lila MA, Raskin I (2010) Merging traditional Chinese medicine with modern drug discovery technologies to find novel drugs and functional foods. Curr Drug Discov Technol 7:2–12
9. Tian S, Wang J, Li Y, Xu X, Hou T (2012) Drug-likeness analysis of traditional Chinese medicines: 3. Prediction of drug-likeness using machine learning approaches. Mol Pharm PMID 22738405
10. May BH, Lu C, Bennett L, Hügel HM, Xue CC (2012) Evaluating the traditional Chinese literature for herbal formulae and individual herbs used for age-related dementia and memory impairment. Biogerontology 13:299–312
11. Fu GM, Pang H, Wong YH (2008) Naturally occurring phenylethanoid glycosides: potential leads for new therapeutics. Curr Med Chem 15:2592–2613
12. Li XJ, Kong DX, Zhang HY (2010) Chemoinformatics approaches for traditional Chinese medicine research and case application in anticancer drug discovery. Curr Drug Discov Technol 7:22–31
13. Barlow DJ, Buriani A, Ehrman T, Bosisio E, Eberini I, Hylands PJ (2012) In-silico studies in Chinese herbal medicines' research: evaluation of in silico methodologies and phytochemical data sources, and a review of research to date. J Ethnopharmacol 140:526–534
14. Ho YS, So KF, Chang RC (2011) Drug discovery from Chinese medicine against neurodegeneration in Alzheimer's and vascular dementia. Chin Med 6:15
15. Gashaw I, Ellinghaus P, Sommer A, Asadullah K (2011) What makes a good drug target? Drug Discov Today 16:1037–1043
16. Sun X, Chan LN, Gong X, Sucher NJ (2003) N-methyl-D-aspartate receptor antagonist activity in traditional Chinese stroke medicines. Neurosignals 12:31–38
17. Sucher NJ (2006) Insights from molecular investigations of traditional Chinese herbal stroke medicines: implications for neuroprotective epilepsy therapy. Epilepsy Behav 8:350–362
18. Buriani A, Garcia-Bermejo ML, Bosisio E, Xu Q, Li H, Dong X, Simmonds MS, Carrara M, Tejedor N, Lucio-Cazana J, Hylands PJ (2012) Omic techniques in systems biology approaches to traditional Chinese medicine research: present and future. J Ethnopharmacol 140:535–544
19. Sun H, Zhang A, Wang X (2012) Potential role of metabolomic approaches for chinese medicine syndromes and herbal medicine. Phytother Res 26:1466–1471
20. Knowles J, Gromo G (2003) A guide to drug discovery: Target selection in drug discovery. Nat Rev Drug Discov 2:63–69
21. Global Industry Analysts Inc (2012) World market for G-protein-coupled receptors (GPCRs) targeting drugs to reach US$120.5 billion by 2017. http://www.strategyr.com/press MCP-6269.asp. Accessed 20 Apr 2012
22. Liu AM, Ho MK, Wong CS, Chan JH, Pau AH, Wong YH (2003) $G\alpha_{16/z}$ chimeras efficiently link a wide range of G protein-coupled receptors to calcium mobilization. J Biomol Screen 8:39–49
23. New DC, Wong YH (2004) Characterization of CHO cells stably expressing a $G\alpha_{16/z}$ chimera for high-throughput screening of GPCRs. Assay Drug Dev Technol 2:269–280
24. New DC, Wong YH (2005) Chimeric and promiscuous G proteins in drug discovery and the deorphanization of GPCRs. Drug Design Rev Online 2:66–79
25. New DC, Miller-Martini DM, Wong YH (2003) Reporter gene assays and their applications to bioassays of natural products. Phytother Res 17:439–448
26. Liu AM, New DC, Lo RK, Wong YH (2009) Reporter gene assays. Methods Mol Biol 486:109–123

27. Miller-Martini DM, Chan RY, Ip NY, Sheu SJ, Wong YH (2001) A reporter gene assay for the detection of phytoestrogens in traditional Chinese medicine. Phytother Res 15:487–492
28. Song JX, Sze SC, Ng TB, Lee CK, Leung GP, Shaw PC, Tong Y, Zhang YB (2012) Anti-Parkinsonian drug discovery from herbal medicines: what have we got from neurotoxic models? J Ethnopharmacol 139:698–711
29. Stahl SM (2000) The new cholinesterase inhibitors for Alzheimer's disease, Part 2: illustrating their mechanisms of action. J Clin Psychiatry 61:813–814
30. Correll PH, Morrison AC, Lutz MA (2004) Receptor tyrosine kinases and the regulation of macrophage activation. J Leukoc Biol 75:731–737
31. Yang SH, Whitmarsh AJ, Davis RJ, Sharrocks AD (1998) Differential targeting of MAP kinases to the ETS-domain transcription factor Elk-1. EMBO J 17:1740–1749
32. Chan AS, Yip EC, Yung LY, Pang H, Luk SC, Pang SF, Wong YH (2005) Immuo-regulatory effects of CKBM on the activities of mitogen-activated protein kinases and the release of cytokines in THP-1 monocytic cells. Biol Pharm Bull 28:1645–1650
33. Chan AS, Yip EC, Yung LY, Pang H, Luk SC, Pang SF, Wong YH (2006) CKBM stimulates MAPKs but inhibits LPS-induced INF-γ production in lymphocytes. Phytother Res 20:725–731
34. Chan ASL, Pang H, Yip ECH, Tam YK, Wong YH (2006) The aqueous extract of Radix Glycyrrhizae stimulates mitogen-activated protein kinases and nuclear factor-κB in Jurkat T-cells and THP-1 monocytic cells. Am J Chin Med 34:263–278
35. Chan AS, Pang H, Yip EC, Tam YK, Wong YH (2005) Carvacrol and eugenol differentially stimulate intracellular Ca^{2+} mobilization and mitogen-activated protein kinases in Jurkat T-cells and monocytic THP-1 cells. Planta Med 71:634–639
36. Yip EC, Liu AM, Wong JT, Wong YH (2005) An aqueous extract of the popular Chinese nutraceutic Kwek Ling Ko (Tortoise shell-Rhzoma Jelly) activates the PPARγ pathway and down-regulates the NFκB activity. Phytomedicine 12:748–759
37. Schoonen WG, Westerink WM, de Roos JA, Débiton E (2005) Cytotoxic effects of 100 reference compounds on Hep G2 and HeLa cells and of 60 compounds on ECC-1 and CHO cells. I mechanistic assays on ROS, glutathione depletion and calcein uptake. Toxicol In Vitro 19:505–516
38. Schoonen WG, de Roos JA, Westerink WM, Débiton E (2005) Cytotoxic effects of 110 reference compounds on HepG2 cells and for 60 compounds on HeLa, ECC-1 and CHO cells. II mechanistic assays on NAD(P)H, ATP and DNA contents. Toxicol In Vitro 19:491–503
39. Miret S, De Groene EM, Klaffke W (2006) Comparison of in vitro assays of cellular toxicity in the human hepatic cell line HepG2. J Biomol Screen 11:184–193
40. Ansorena E, García-Trevijano ER, Martínez-Chantar ML, Huang ZZ, Chen L, Mato JM, Iraburu M, Lu SC, Avila MA (2002) S-Adenosylmethionine and methylthioadenosine are antiapoptotic in cultured rat hepatocytes but proapoptotic in human hepatoma cells. Hepatology 35:274–280
41. Sun D, Gong Y, Kojima H, Wang G, Ravinsky E, Zhang M, Minuk GY (2003) Increasing cell membrane potential and GABAergic activity inhibits malignant hepatocyte growth. Am J Physiol Gastrointest Liver Physiol 285:G12–G19
42. Ichijo H (1999) From receptors to stress-activated MAP kinases. Oncogene 18:6087–6093
43. Harper SJ, LoGrasso P (2001) Signalling for survival and death in neurones: the role of stress-activated kinases, JNK and p38. Cell Signal 13:299–310
44. Yip EC, Chan AS, Pang H, Tam YK, Wong YH (2006) Protocatechuic acid induces cell death in HepG2 hepatocellular carcinoma cells through a c-Jun N-terminal kinase-dependent mechanism. Cell Biol Toxicol 22:293–302
45. Wang X, Sun H, Zhang A, Jiao G, Sun W, Yuan Y (2011) Pharmacokinetics screening for multi-components absorbed in the rat plasma after oral administration traditional Chinese medicine formula Yin-Chen-Hao-Tang by ultra performance liquid chromatography-electrospray ionization/quadrupole-time-of-flight mass spectrometry combined with pattern recognition methods. Analyst 136:5068–5076
46. Zhu J, Lee S, Ho MK, Hu Y, Pang H, Ip FC, Chin AC, Harley CB, Ip NY, Wong YH (2010) In vitro intestinal absorption and first-pass intestinal and hepatic metabolism of cycloastragenol, a potent small molecule telomerase activator. Drug Metab Pharmacokinet 25:477–486

47. Zhu J, Hu Y, Ho MK, Wong YH (2011) 3-methoxylphenylpropyl amides as novel receptor subtype-selective melatoninergic ligands: characterization of physicochemical and pharmacokinetic properties. Xenobiotica 41:35–45
48. Zhu J, Hu Y, Ho MK, Wong YH (2012) Pharmacokinetics, oral bioavailability and metabolism of a novel isoquinolinone-based melatonin receptor agonist in rats. Xenobiotica 42:1138–1150
49. Prieto P, Blaauboer BJ, de Boer AG, Boveri M, Cecchelli R, Clemedson C, Coecke S, Forsby A, Galla HJ, Garberg P, Greenwood J, Price A, Tähti H (2004) European centre for the validation of alternative methods blood-brain barrier in vitro models and their application in toxicology. The report and recommendations of ECVAM Workshop 49. Altern Lab Anim 32:37–50
50. Wilhelm I, Fazakas C, Krizbai IA (2011) In vitro models of the blood-brain barrier. Acta Neurobiol Exp (Wars) 71:113–128
51. Mody SM, Hu Y, Ho MK, Wong YH (2007) In search of novel and therapeutically significant melatoninergic ligands. Recent Pat CNS Drug Discov 2:241–245
52. Hu Y, Ho MK, Chan KH, New DC, Wong YH (2010) Synthesis of substituted N-[3-(3-methoxyphenyl)propyl] amides as highly potent MT_2-selective melatonin ligands. Bioorg Med Chem Lett 20:2582–2585
53. Granger MP, Wright WE, Shay JW (2002) Telomerase in cancer and aging. Crit Rev Oncol Hematol 41:29–40
54. Funk WD, Wang CK, Shelton DN, Harley CB, Pagon GD, Hoeffler WK (2000) Telomerase expression restores dermal integrity to in vitro-aged fibroblasts in a reconstituted skin model. Exp Cell Res 258:270–278
55. Rudolph KL, Chang S, Millard M, Schreiber-Agus N, DePinho RA (2000) Inhibition of experimental liver cirrhosis in mice by telomerase gene delivery. Science 287:1253–1258
56. Dagarag M, Ng H, Lubong R, Effros RB, Yang OO (2003) Differential impairment of lytic and cytokine functions in senescent human immunodeficiency virus type 1-specific cytotoxic T lymphocytes. J Virol 77:3077–3083
57. Gu X, Zhang J, Brann DW, Yu FS (2003) Brain and retinal vascular endothelial cells with extended life span established by ectopic expression of telomerase. Invest Ophthalmol Vis Sci 44:3219–3225
58. Roy NS, Nakano T (2004) Telomerase immortalization of neuronally restricted progenitor cells derived from the human fetal spinal cord. Nat Biotechnol 22:297–305
59. Verra NC, Jorritsma A, Weijer K, Ruizendaal JJ, Voordouw A, Weder P, Hooijberg E, Schumacher TN, Haanen JB, Spits H, Luiten RM (2004) Human telomerase reverse transcriptase-transduced human cytotoxic T cells suppress the growth of human melanoma in immunodeficient mice. Cancer Res 64:2153–2161
60. Fauce SR, Jamieson BD, Chin AC, Mitsuyasu RT, Parish ST, Ng HL, Kitchen CM, Yang OO, Harley CB, Effros RB (2008) Telomerase-based pharmacologic enhancement of antiviral function of human $CD8^+$ T lymphocytes. J Immunol 181:7400–7406
61. Yung LY, Lam WS, Ho MK, Hu Y, Ip FC, Pang H, Chin AC, Harley CB, Ip NY, Wong YH (2012) Astragaloside IV and cycloastragenol stimulate the phosphorylation of extracellular signal-regulated protein kinase in multiple cell types. Planta Med 78:115–121
62. Pammolli F, Magazzini L, Riccaboni M (2011) The productivity in pharmaceutical R&D. Nat Rev Drug Discov 10:428–438

Author Biography

Professor Yung Hou WONG got his Ph.D at University of Cambridge and is currently a Chair Professor of the Division of Life Science at the Hong Kong University of Science and Technology, Hong Kong SAR. He is also the Director of Biotechnology Research Institute and Molecular Neuroscience Center. Professor Wong's research interests are in the structure and function of G protein-coupled receptors. By utilizing a combination of biochemical, immunological, and molecular biology approaches, he has made significant contributions to the understanding of the mechanisms of drug actions, especially those involved in the development of opiate tolerance and dependence. He has also created several universal G protein adapters that serve as powerful molecular tools for drug discovery.

Catherine Chi-Kit WONG is a Research Projects Coordinator at the Biotechnology Research Institute (BRI), Hong Kong University of Science and Technology. She received her Master of Arts in Biochemistry from the University of Texas at Austin, and started her professional career in biotechnology as a Quality Control Manager at EY Laboratories in 1989. Ms Wong joined BRI in 1997 after more than 3 years of project management experience in Chinese mainland. She has provided liaison and administrative support to the institute, and contributed effectively to writing grant proposals and disseminating research findings. Ms Wong currently also provides editorial assistance to the editors of Journal of Neuroscience, Journal of Molecular Signaling and Neurosignals.

Part VIII
Questions and Future Perspectives

Chapter 27
Achievements, Questions Arising and Future Outlook on the Path to Discover New Medicinal Compounds

Evandro Fei Fang and Tzi Bun Ng

Abstract Nature nourishes an inexhaustible drug bank which can be exploited to treat human diseases. On the way to find different natural drugs, significant strides have been made, including the discovery of antitumor drugs (such as paclitaxel/taxol), drugs against chronic diseases (such as ginsenosides, genipin), and drugs against infectious diseases (artemisinin) or even drugs that promote longevity (resveratrol, rapamycin, and spermidine). In this chapter we will give a brief description of the previous accomplishments and current problems on drug screening from natural sources. Further perspectives in this field will also be discussed.

27.1 Introduction

People are lucky to enjoy their lives in this new millennium which is characterized by a high living standard, convenient traffic to various parts of the world, state-of-the-art techniques which provide us with modern medical services and greatly prolong people's lifespan. Unfortunately, due to the tight working schedule, unhealthy diet and living style, as well as an increased exposure to a variety of environmental pollutants and carcinogenic agents, we are facing a severe chronic non-communicable disease burden (such as diabetes, cardiovascular diseases, and neurodegenerative diseases) while at the same time encountering attacks of infectious diseases (exemplified by AIDS, SARS, bird flu, swine influenza) [1, 2]. Furthermore, not only patients but also normal people would like to extend their life expectancy [3]. There has been a long

E. F. Fang (✉)
Laboratory of Molecular Gerontology, National Institute on Aging, National Institutes of Health, Baltimore, MD 21224, USA
e-mail: fange@mail.nih.gov; evandro.fang@nih.gov; fangfei1030@yahoo.com.cn

T. B. Ng
School of Biomedical Sciences, Faculty of Medicine, The Chinese University of Hong Kong, Hong Kong, People's Republic of China
e-mail: b021770@mailserv.cuhk.edu.hk

E. F. Fang and T. B. Ng (eds.), *Antitumor Potential and Other Emerging Medicinal Properties of Natural Compounds*, DOI: 10.1007/978-94-007-6214-5_27,

history in medicinal researchers building our ongoing knowledge base in the aforementioned diseases. Natural compounds provide a promising approach.

27.2 Achievements on Drug Discovery From Nature

In view of the finding that more than half of all the drugs worldwide originate from natural products and their synthetic derivatives, it is gratifying that we can locate drug candidates from nature for currently un-curable diseases. Scientists have found many natural compounds with promising health benefits from plants, bacteria, animals, and even trace and rare metals and elements.

27.2.1 Antitumor Compounds

Because of the daunting challenge that curing cancer poses on human health, there are a plethora of studies on antitumor drugs with successful examples that have been established. Firstly, green tea (the leaves of *Camellia sinensis)* is the first used beverage in Asia, and the fourth in USA. Epigallocatechin-3-gallate (Fig. 27.1a) is the major polyphenol in green tea and shows potent antitumor effects. The underlying mechanism of its chemopreventive effects is ascribed to cell cycle arrest and the induction of apoptosis in tumor cells [4]. Secondly, a pentahydroxy glucosyl flavone, gossypin (Fig. 27.1b) purified from *Hibiscus vitifolius,* has shown antitumor activity in a variety of tumor cells in vitro and in vivo. In addition to the inhibition of topoisomerase I and II, the antitumor effects of gossypin are also attributed to its ability to inhibit the NF-κB activation pathway [5]. Thirdly, as one of the internationally best-selling cancer drugs, paclitaxel (commerciallynamed taxol, Fig. 27.1c) is a natural compound acquired from the needles of the endangered Pacific yew (*Taxus brevifolia*) [6]. It was approved by the FDA in 1992 for treatment of ovarian cancer, and since then the use of this drug has been extended to breast, lung, head and neck cancer. One more example that deserves mention is daidzein (Fig. 27.1d), an isoflavone phytoestrogen from soybean (*Glycine max*). The biologically active forms of daidzein are equol and *O*-desmethylangolensin (*O*-DMA) produced by intestinal bacteria. Daidzein has been recorded with the ability to reduce the risk of breast and prostate cancers [7].

Some anti-leukaemia medicines have been discovered from nature, such as Indirubin (Fig. 27.1e), homoharringtonine (Fig. 27.1f) and arsenic trioxide. Indirubin is a red-coloured compound, minor but active ingredient of the herbal drug Danggui Longhui Wan used to treat chronic myelocytic leukaemia (CML). Hoessel et al. unveiled the molecular mechanisms associated with the inhibition of cyclin-dependent kinases, and the inhibition of proliferation in tumor cells due to arrest in the G2/M phase [8]. On the other hand, homoharringtonine is a natural compound from the evergreen tree *Cephalotaxus harringtonia*, which shows

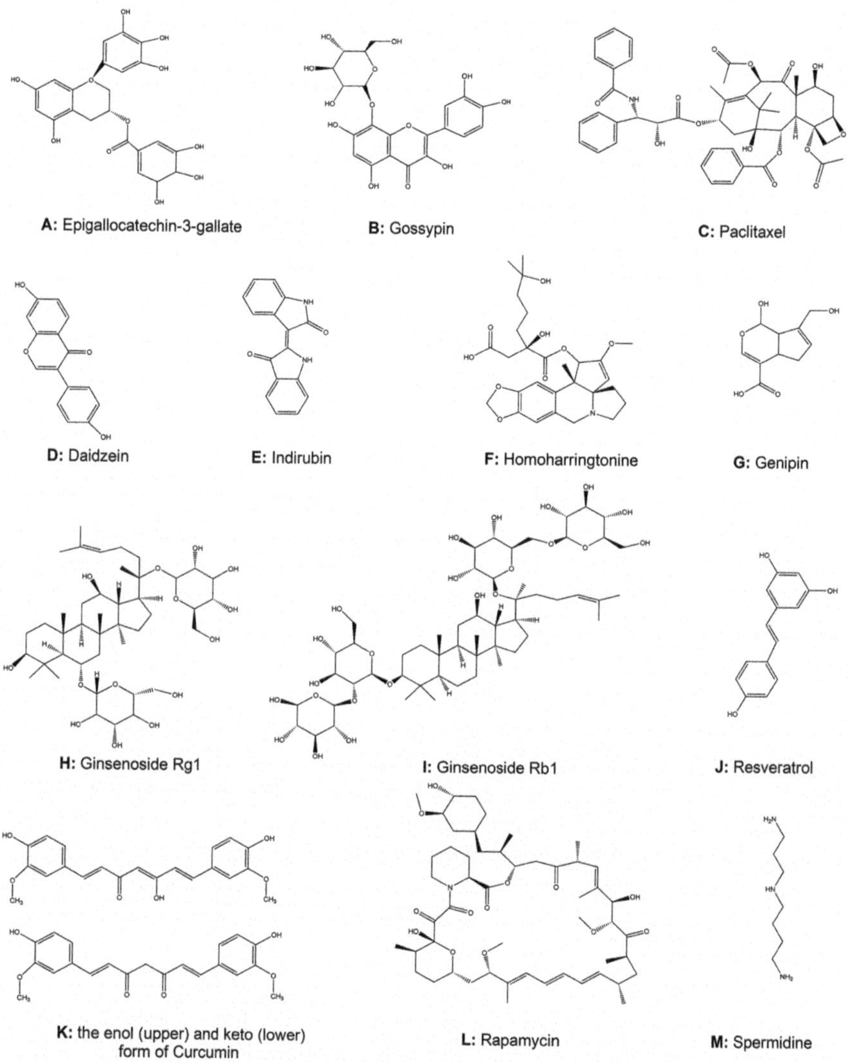

Fig. 27.1 Structures of different natural compounds with medicinal benefits. For details of the medicinal activities of each compound, please refer to the manuscript for details

promise for treating chronic lymphocytic leukemia (CLL). Homoharringtonine can reduce expression of the intrinsically short-lived antiapoptotic protein myeloid cell leukemia-1 (Mcl-1) as well as induce apoptosis in CLL [9]. In addition, the Chinese medicinal formula Realgar-*Indigo naturalis* is effective in treating human acute promyelocytic leukaemia (APL) with a complete remission rate of over 95 %. The major active ingredients are tetraarsenic tetrasulfide, indirubin and tanshinone IIA, and their leukemia killer function is mainly caused by intensified ubiquitination/degradation of myelocytic leukemia-retinoic acid receptor alpha (PML-RARα)

and G1/G0 phase arrest [10]. However, the most famous and universally accepted anti-APL therapy is all-trans retinoic acid (ATRA) with arsenic trioxide (As_2O_3). This combination therapy leads to lasting remission of APL. As_2O_3 functions by binding to the RBCC domain of PML-RARα, followed by PML oligomerization, finally resulting in enhanced SUMOylation and degradation [11].

27.2.2 Compounds Targeting on Chronic Diseases

Liver diseases constitute a substantial percentage of human chronic diseases. Genipin (Fig. 27.1g) has been regarded as a potent therapeutic agent for different cholestatic liver diseases. It is an aglycone metabolite of geniposide produced as a result of the action of intestinal bacteria. Geniposide serves as an effective constituent of *Gardenia jasminoides* used as a component of the choleretic and hepatoprotective drug Inchin-ko-to in China and Japan [12]. The medicinal effect of genipin in the liver is attributed to its ability to enhance multidrug resistance-associated protein 2 (Mrp2)-mediated bile formation and organic anion transport [12]. Another example is ginsenosides from Ginseng (genus *Panax*). As an important component of herbal medicine as well as one of the most extensively used botanical plants in the west, ginseng has been used to promote a protective immune response in patients and the regulation of angiogenesis. Dr. Fan and colleagues found that there were two main ginsenosides, Rg1 and Rb1 (Fig. 27.1h, i): the first one exhibited angiogenic activity whereas the latter inhibited the earliest step in angiogenesis [13]. The promotion of functional neovascularization, endothelial cell proliferation and tubulogenesis has been validated in both in vitro and in vivo studies [13]. An additional example is curcumin (Fig. 27.1k) which is a constituent in the daily used spice turmeric (*Curcuma longa rhizomes*). Curcumin has been studied extensively. It demonstrates anti-oxidant, anti-inflammatory, anti-angiogenic, and anti-tumor activities [14].

27.2.3 Autophagy-Inducing agents and Their Health Benefits

Autophagy encompasses the different routes that cells employ to deliver damaged organelles and unused long-lived proteins to lysosomes for degradation, which in turn provides nutrition for the cells. It has been reported that the defect of different autophagies (including macroautophagy, microautophagy, and chaperone-mediated autophagy) or even a specific type of macroautophagy (such as mitophagy) constitutes the etiology of different accelerated-aging diseases, tumors, and neurological diseases [15, 16]. Observational and intervention studies in cells, animals, and some conducted on humans have suggested the anti-ageing, antitumor, and anti-neurodegenerative activities of some autophagy-inducing agents such as resveratrol, rapamycin and spermidine. Their structures are shown in Fig. 27.1j, l, and m, respectively.

The compound which has formed the subject of the most comprehensive investigations is resveratrol, a natural phenol produced by different plants, especially grapes. Resveratrol manifests anti-diabetic activity, extends the lifespan of different species by the increase of insulin sensitivity, SIRT1, AMPK and PGC-1α activity, mitochondrial number/function and the improvement of motor function [17, 18]. The second autophagy-inducing natural compound rapamycin is a compound extracted from the bacterium *Streptomyces hygroscopicus* present in a soil sample from Easter island (Rapa Nui).It exhibits therapeutic potential against a pre-aging disease Cockayne syndrome by up-regulation of mitophagy [3, 16]. The third example is spermidine which is a polyamine involved in cellular metabolism. It is rich in human semen and also in grapefruits, and peas (*Pisum sativum*) [19, 20]. An interesting report by Eisenberg and coworkers disclosed that administration of spermidine extended the lifespan in various ageing models by the regulation of epigenetic modifications, induction of autophagy and concurrently inhibition of necrosis [21].

27.2.4 Natural Compounds Against Infectious Diseases

Malaria, AIDS and tuberculosis are the three most devastating infectious diseases worldwide, for they cause high morbidity and mortality [22]. Malaria is the most common life-threatening infection accounting for around 1 million deaths every year. Artemisinin, the world's most important anti-malaria medication, is a natural compound harvested from Qinghao (*Artemisia annua*) by Chinese scientists in the 1970s [23]. Artemisinin could be located in mitochondria of *Plasmodium falciparum*, followed by the damage of mitochondrial function, production of reactive oxygen species, and finally induced death of *Plasmodium falciparum* [24]. Some herbal plants also show anti-HIV activity such as Tian Hua-fen (*Trichosanthes kirilowii*) [25] and bitter melon (*Momordica charantia*) [26]. Their active constituents are trichosanthin and MAP30 (for details please refer to the chapter by Prof. Sylvia Lee-Huang). In addition, crude extracts of some herbal plants (*Polyalthia sp., Haplophragma adenophyllum, Annona reticulate L., Aegle marmelos* L. Correa.) showed a tuberculosis inhibition rate exceeding 90 %. However, the bioactive components as well as the molecular mechanisms have not been fully elucidated and deserve further studies [27].

27.3 Current Questions on Drug Discovery

Scientific studies over the centuries suggest that nature nourishes an inexhaustible drug bank for combatting human diseases [28]. On the way to discovering different natural drugs, though significant achievements have been made, some questions remain to be answered and new issues should be addressed.

27.3.1 Establish Good in vitro, ex vivo, and in vivo Screening Models to Mimic Gut Environment

The current in vitro drug screening methods are effective for some but not all of the bioactive natural compounds. For instance, humans cannot absorb the original forms of ginsenoside Rb1, geniposide [12] and daidzein [7] directly. Only with the help from the digestive action of different gut bacteria, their respective active forms (20-*O*-β-D-glucopyranosyl-20(S)-protopanaxadiol; genipin; equol and *O*-DMA) are then taken up by the gut to exhibit their medicinal activities. In view of this, it is suspected that some promising medicinal compounds may be missed when the common in vitro drug screening system is employed. Besides exerting effects on drug bioavailability, adsorption, and toxicity, recent studies also provide evidence that the gut microbiota play a role in the onset and progression of some diseases, such as diabetes, fatty liver phenotype, and inflammatory bowel disease [29].The human gut microbiota consist of trillions of microbes with a collective genome (named microbiome) that contains 100 fold more genes than the host genome [29]. Future work to unveil the mystery of gut microbiota and their pros and cons to human health will undoubtedly assist us to establish a more comprehensive drug testing system.

27.3.2 Synergistic Effects of Some Natural Compounds

Results from recent investigations on drug discovery indicate that there is synergism in the actions of some compounds, which have no medicinal effects respectively, but the synergistic effects emerge if they are applied together. Examples can be found in Realgar-*Indigo caturalis* formula and Sho-saiko-to. Realgar-*Indigo caturalis* formula has been confirmed to be a good treatment for APL. There are different functions of the three active ingredients: tetraarsenic tetrasulfide, indirubin and tanshinone IIA. Tetraarsenic tetrasulfide works as the 'sovereign' as it takes up the role to bind on and damage PML-RARα. On the other hand, indirubin assumes the role as the 'assistant' to exhibit an anti-proliferative effect on leukemia cells. The 'minister' goes to tanshinone IIA which inhibits the migration of tumor cells [10]. In addition, the anti-hepatocellular carcinoma efficacy of Sho-saiko-to is attributed to different medicinal components whereas no single component can contribute to the effect [30]. It deserves mention here that the 'isolation of one active compound'-dominated screening methodology may suffer from the drawback of missing the point when applied to medicinal formulas [31]. A new drug screening system targeting on this point would help to optimize screening efficiency.

27.3.3 Harmony with Nature on the Modernization of Ethnomedicine

As a complex multi-disciplinary system, ethnomedicine refers to the investigation of traditional medical practice with cultural interpretations, and addresses the healthcare-seeking process as well as healing practices [32]. Representatives of ethnomedicine encompass, among others, Traditional Chinese Medicine (TCM), Siddha, and Ayurveda, which offer rich pickings for modern drug development. Though ethnomedicine has some literature records and shows long-purported health benefits with little side effects, the chemical composition, dosages and toxicity of the plants/components in the ethnomedcal formulas are not well-known/clearly defined, and are sometimes illegal. For example, *Aristolochia* spp. had been used as herbal remedies in mainland China and Taiwan for a long time, but were banned by the Chinese government in 2003 as this type of herbs contains the carcinogenic compound aristolochic acid (AA). Chen and colleagues carried out a molecular epidemiologic study in Taiwan, and lifted the veil on the mechanism which happened in the upper urinary tract. The active form of AA aristolactam bound to DNA, followed by mutation in the *TP53* tumor-suppressor gene and two oncogenes *FGFR3* and *HRAS*, and finally caused urothelial cancer [33]. Current and future toxicological studies on herbal medicines will help to ensure the safety of the treatment.

It is no surprise to us that some precious components in TCM formulas originate from endangered animals, such as tigers, bears, rhinoceroses, and some turtle and tortoise species. Though China banned domestic trade of tiger tissues in 1993, the allegedly large health benefits of tiger parts as well as high-return black market are the incentives for the continual sale of tiger products [34]. Another example is bear bile which is believed to mitigate fever, protect the liver, improve eyesight, down gallstones, and others. Unfortunately, the way to collect bile from a reared bear is painful, distressing, and may cause disease in the animal [35]. Furthermore, many turtle and tortoise species are at risk since they are over-captured for their shells which are a main component in some jelly formulas believed to reinforce human body immunity and restrain tumorigenesis [36]. Well-designed animal experiments (and/or clinical trials) should be conducted to investigate the alleged medicinal benefits of these samples. If the samples are devoid of the alleged activities, such therapies should be banned, and public education should be launched. Even if the applications pan out well, due to the scarcity of natural supplies we should promote the use of substitutes generated by modern technology. It is noteworthy that the antimalarial drug artemisinin [37], and the wide-spectrum antitumor drug paclitaxel/Taxol [38] can be synthesized in the laboratory. These improvements could accelerate production and lead to a significant reduction in the price.

27.3.4 Mechanistic Studies: The Crosstalk Between Apoptosis and Autophagy

Apoptosis and autophagy are two types of cell death both linked to the antitumor and other chemotherapeutic mechanisms of natural compounds. For instance, the antitumor mechanisms of trichosanthin [39], MAP30 [40] and RNase MC2 [41, 42] are attributed to the induction of both caspase-8 and caspase-9 regulated apoptotic cascades. Regarding autophagy, besides the above mentioned anti-ageing and anti-neurodegenerative activities, it is also implicated in the antitumor action of some medicinal components. However, in contrast to the explicitly defined role of apoptosis in the antitumor effects of some drugs, the functions of autophagy in human health are controversial. It has been noticed that (a) in some circumstances autophagy is the only way to elicit the antitumor activity [43]; (b) sometimes the synergistic effects of apoptosis and autophagy induce tumor cell death [44]; (c) whereas there are also reports demonstrating that autophagy diminishes the antitumor effect [45]. In view of the findings that (a) ASPP1 and ASPP2 regulate the tumor-suppressive function of p53 and (b) iASPP (an inhibitory member of the ASPP family) is a key inhibitor of p53 [46, 47], and (c) p53 is associated with both cell death types [48], the functions of ASPP1, ASPP2, iASPP, and p53 in the crosstalk merit investigations. We may be able to depict a panorama of the crosstalk between apoptosis and autophagy soon.

27.4 Future Perspectives

As a melting pot of eastern and western cultures, Hong Kong also plays a significant role in the discovery of new drugs of natural origins [49]. There are some major institutes such as Institute of Chinese Medicine at The Chinese University of Hong Kong, School of Chinese Medicine at The University of Hong Kong, and the Biotechnology Research Institute at Hong Kong University of Science and Technology. Out of China, GP-TCM (Good Practice in Traditional Chinese Medicine Research in the Post-genomic Era), the first EU-funded Coordination Action consortium dedicated to TCM research has been established. In addition to TCM, there are some other sources for the purification of natural medicinal compounds and the work is being undertaken as exemplified by the dedicated efforts of researchers at different universities and institutions referred to in this book. In future drug development, one should extend the simple purification and testing procedure to a wider scope: give consideration to synergistic effects, human gut microbiota, high-qualify clinical trials, patent applications, and even environment/animal protection to drug substitution.

On the road to drug discovery, some new ideas/concepts are generated such as medical systems biology [50] and a new health care system centered on the 4Ps (Predictive, Preventive, Personalized, and participatory). Such new ideas will assist the direction in drug screening. The repertoire of natural medicinal compounds is a treasure. But they represent merely a small portion discovered from nature and a great many await to be discovered.

27.5 Disclaimer

This article was written in a personal capacity (E.F.F.) and does not represent the opinions of the US Food and Drug Administration, the US Department of Health and Human Services, or the US Federal Government.

Acknowledgments We thank Prof. Vilhelm A. Bohr and Dr. Peter Sykora at NIH for invaluable feedback and critically reading of the manuscript, and Dr. Chuan-hao Li (Department of Chemistry of The Chinese University of Hong Kong, and Department of Chemistry of Anhui University) for drawing Fig. 27.1. This research was supported in part by the Intramural Research Progarm of the NIH, National Institute on Aging, USA.

References

1. Vijaykrishna D, Smith GJ, Pybus OG, Zhu H, Bhatt S, Poon LL et al (2011) Long-term evolution and transmission dynamics of swine influenza A virus. Nature 473:519–522
2. Howlader N, Noone AM, Krapcho M, Neyman N, Aminou R, Altekruse SF et al (2012) SEER Cancer Statistics Review, 1975–2009 (Vintage 2009 Populations). National Cancer Institute. Bethesda, MD.based on November 2011 SEER data submission, posted to the SEER web site, April 2012
3. Fang EF, Scheibye-Knudsen M, Bohr VA, Ng TB (2012) The anti-aging efficacy of natural compounds. Med Aromat Plants 1:1–2
4. Ahmad N, Feyes DK, Nieminen AL, Agarwal R, Mukhtar H (1997) Green tea constituent epigallocatechin-3-gallate and induction of apoptosis and cell cycle arrest in human carcinoma cells. J Natl Cancer Inst 89:1881–1886
5. Kunnumakkara AB, Nair AS, Ahn KS, Pandey MK, Yi Z, Liu M et al (2007) Gossypin, a pentahydroxy glucosyl flavone, inhibits the transforming growth factor beta-activated kinase-1-mediated NF-kappaB activation pathway, leading to potentiation of apoptosis, suppression of invasion, and abrogation of osteoclastogenesis. Blood 109:5112–5121
6. Service RF (2000) Hazel trees offer new source of cancer drug. Science 288:27–28
7. Atkinson C, Frankenfeld CL, Lampe JW (2005) Gut bacterial metabolism of the soy isoflavone daidzein: exploring the relevance to human health. Exp Biol Med 230:155–170
8. Hoessel R, Leclerc S, Endicott JA, Nobel ME, Lawrie A, Tunnah P et al (1999) Indirubin, the active constituent of a Chinese antileukaemia medicine, inhibits cyclin-dependent kinases. Nat Cell Biol 1:60–67
9. Chen R, Guo L, Chen Y, Jiang Y, Wierda WG, Plunkett W (2011) Homoharringtonine reduced Mcl-1 expression and induced apoptosis in chronic lymphocytic leukemia. Blood 117:156–164
10. Wang L, Zhou GB, Liu P, Song JH, Liang Y, Yan XJ et al (2008) Dissection of mechanisms of Chinese medicinal formula realgar-indigo naturalis as an effective treatment for promyelocytic leukemia. Proc Natl Acad Sci USA 105:4826–4831
11. Zhang XW, Yan XJ, Zhou ZR, Yang FF, Wu ZY, Sun HB et al (2010) Arsenic trioxide controls the fate of the PML-RARalpha oncoprotein by directly binding PML. Science 328:240–243
12. Shoda J, Miura T, Utsunomiya H, Oda K, Yamamoto M, Kano M et al (2004) Genipin enhances Mrp2 (Abcc2)-mediated bile formation and organic anion transport in rat liver. Hepatology 39:167–178
13. Sengupta S, Toh SA, Sellers LA, Skepper JN, Koolwijk P, Leung HW et al (2004) Modulating angiogenesis: the yin and the yang in ginseng. Circulation 110:1219–1225
14. Maheshwari RK, Singh AK, Gaddipati J, Srimal RC (2006) Multiple biological activities of curcumin: a short review. Life Sci 78:2081–2087
15. Rubinsztein DC, Marino G, Kroemer G (2011) Autophagy and aging. Cell 146:682–695
16. Scheibye-Knudsen M, Ramamoorthy M, Sykora P, Maynard S, Lin PC, Minor RK et al (2012) Cockayne syndrome group B protein prevents the accumulation of damaged mitochondria by promoting mitochondrial autophagy. J Exp Med 209:855–869
17. Baur JA, Pearson KJ, Price NL, Jamieson HA, Lerin C, Kalra A et al (2006) Resveratrol improves health and survival of mice on a high-calorie diet. Nature 444:337–342
18. Lagouge M, Argmann C, Gerhart-Hines Z, Meziane H, Lerin C, Daussin F et al (2006) Resveratrol improves mitochondrial function and protects against metabolic disease by activating SIRT1 and PGC-1 alpha. Cell 127:1109–1122

19. Klein H, Priebe A, Jager HJ (1979) Putrescine and spermidine in peas: effects of nitrogen source and potassium supply. Physiol Plant 45:497–499
20. Suzuki O, Matsumoto T, Katsumata Y, Oya M (1982) A new enzymatic method for quantitation of spermine in human semen Zeitschrift fur Rechtsmedizin. J Leg Med 88:67–73
21. Eisenberg T, Knauer H, Schauer A, Buttner S, Ruckenstuhl C, Carmona-Gutierrez D et al (2009) Induction of autophagy by spermidine promotes longevity. Nat Cell Biol 11:1305–1314
22. Hotez PJ, Molyneux DH, Fenwick A, Ottesen E, Ehrlich SS, Sachs JD (2006) Incorporating a rapid-impact package for neglected tropical diseases with programs for HIV/AIDS, tuberculosis, and malaria. PLoS Med 3:e102
23. Tu Y (2011) The discovery of artemisinin (qinghaosu) and gifts from Chinese medicine. Nat Med 17:1217–1220
24. Wang J, Huang L, Li J, Fan Q, Long Y, Li Y et al (2010) Artemisinin directly targets malarial mitochondria through its specific mitochondrial activation. PLoS ONE 5:e9582
25. Fang EF, Ng TB, Shaw PC, Wong RN (2011) Recent progress in medicinal investigations on trichosanthin and other ribosome inactivating proteins from the plant genus Trichosanthes. Curr Med Chem 18:4410–4417
26. Fang EF, Ng TB (2011) Bitter gourd (*Momordica charantia*) is a cornucopia of health: a review of its credited antidiabetic, anti-HIV, and antitumor properties. Curr Mol Med 11:417–436
27. Elkington BG, Southavong B, Sydara K, Souliya O, Vanthanouvong M, Nettavong K et al (2009) Biological evaluation of plants of Laos used in the treatment of tuberculosis in Lao traditional medicine. Pharm Biol 47:26–33
28. Orlikova B, Diederich M (2012) Power from the garden: plant compounds as inhibitors of the hallmarks of cancer. Curr Med Chem 19:2061–2087
29. Jia W, Li H, Zhao L, Nicholson JK (2008) Gut microbiota: a potential new territory for drug targeting. Nat Rev Drug Discov 7:123–129
30. Oka H, Yamamoto S, Kuroki T, Harihara S, Marumo T, Kim SR et al (1995) Prospective study of chemoprevention of hepatocellular carcinoma with Sho-saiko-to (TJ-9). Cancer 76:743–749
31. Stone R (2008) Biochemistry. Lifting the veil on traditional Chinese medicine. Science 319:709–710
32. Williams LA (2006) Ethnomedicine. West Indian Med J 55:215–216
33. Chen CH, Dickman KG, Moriya M, Zavadil J, Sidorenko VS, Edwards KL et al (2012) Aristolochic acid-associated urothelial cancer in Taiwan. Proc Natl Acad Sci USA 109:8241–8246
34. Guo J (2007) Tigers in trouble: year of the tiger. Nature 449:16–18
35. Sheng X, Zhang H, Weng Q (2012) Traditional Chinese medicine: China's bear farms prompt public outcry. Nature 484:455
36. Murphy MB (2007) Traditional remedies put animal species at risk. Nature 448:644
37. Levesque F, Seeberger PH (2012) Continuous-flow synthesis of the anti-malaria drug artemisinin. Angew Chem 51:1706–1709
38. Mukaiyama T, Shiina I, Iwadare H, Saitoh M, Nishimura T, Ohkawa N et al (1999) Asymmetric total synthesis of taxol. Chem-A Eur J 5:121–161
39. Fang EF, Zhang CZ, Zhang L, Wong JH, Chan YS, Pan WL et al. Trichosanthin inhibits breast cancer cell proliferation in both cell lines and nude mice by promotion of apoptosis. PloS One (In press)
40. Fang EF, Zhang CZ, Wong JH, Shen JY, Li CH, Ng TB (2012) The MAP30 protein from bitter gourd (*Momordica charantia*) seeds promotes apoptosis in liver cancer cells in vitro and in vivo. Cancer Lett 324(1):66–74
41. Fang EF, Zhang CZ, Fong WP, Ng TB (2012) RNase MC2: a new *Momordica charantia* ribonuclease that induces apoptosis in breast cancer cells associated with activation of MAPKs and induction of caspase pathways. Apoptosis: Int J Program Cell Death 17:377–387
42. Fang EF, Zhang CZ, Zhang L, Fong WP, Ng TB (2012) In vitro and in vivo anticarcinogenic effects of RNase MC2, a ribonuclease isolated from dietary bitter gourd, toward human liver cancer cells. Int J Biochem Cell Biol 44:1351–1360

43. Xie CM, Chan WY, Yu S, Zhao J, Cheng CH (2011) Bufalin induces autophagy-mediated cell death in human colon cancer cells through reactive oxygen species generation and JNK activation. Free Radic Biol Med 51:1365–1375
44. Saiki S, Sasazawa Y, Imamichi Y, Kawajiri S, Fujimaki T, Tanida I et al (2011) Caffeine induces apoptosis by enhancement of autophagy via PI3 K/Akt/mTOR/p70S6 K inhibition. Autophagy 7:176–187
45. Carew JS, Nawrocki ST, Kahue CN, Zhang H, Yang C, Chung L et al (2007) Targeting autophagy augments the anticancer activity of the histone deacetylase inhibitor SAHA to overcome Bcr-Abl-mediated drug resistance. Blood 110:313–322
46. Bergamaschi D, Samuels Y, O'Neil NJ, Trigiante G, Crook T, Hsieh JK et al (2003) iASPP oncoprotein is a key inhibitor of p53 conserved from worm to human. Nat Genet 33:162–167
47. Samuels-Lev Y, O'Connor DJ, Bergamaschi D, Trigiante G, Hsieh JK, Zhong S et al (2001) ASPP proteins specifically stimulate the apoptotic function of p53. Mol Cell 8:781–794
48. Maiuri MC, Zalckvar E, Kimchi A, Kroemer G (2007) Self-eating and self-killing: crosstalk between autophagy and apoptosis. Nat Rev Mol Cell Biol 8:741–752
49. Cyranoski D (2001) Hong Kong seeks secrets of Chinese medicine. Nature 412:7
50. van der Greef J, Stroobant P, van der Heijden R (2004) The role of analytical sciences in medical systems biology. Curr Opin Chem Biol 8:559–565

Index

E. F. Fang and T. B. Ng (eds.), *Antitumor Potential and Other Emerging Medicinal Properties of Natural Compounds*, DOI: 10.1007/978-94-007-6214-5,

The manufacturer's authorised representative in the EU is Springer Nature Customer Service Centre GmbH, Europaplatz 3, 69115 Heidelberg, Germany. If you have any concerns regarding our products, please contact ProductSafety@springernature.com

Printed and bound by CPI Group (UK) Ltd, Croydon, CR0 4YY
25/09/2026
02222847-0002